Selected Works of Wen-Tsun Wu

Selected Works *of* *Wen-Tsun Wu*

Wen-Tsun Wu

Academia Sinica, China

World Scientific

NEW JERSEY • LONDON • SINGAPORE • BEIJING • SHANGHAI • HONG KONG • TAIPEI • CHENNAI

Published by

World Scientific Publishing Co. Pte. Ltd.
5 Toh Tuck Link, Singapore 596224
USA office: 27 Warren Street, Suite 401-402, Hackensack, NJ 07601
UK office: 57 Shelton Street, Covent Garden, London WC2H 9HE

British Library Cataloguing-in-Publication Data
A catalogue record for this book is available from the British Library.

SELECTED WORKS OF WEN-TSUN WU

ISBN-13 978-981-279-107-8
ISBN-10 981-279-107-8

Printed in Singapore by B & JO Enterprise

Foreword

The present "Selected Papers" may be considered as a brief survey of my scientific career in mathematical sciences.

My researches in mathematical sciences are consisting of two stages. The researches in the first stage, started in 1947, are in pure mathematics, mainly in algebraic topology, occasionally also in algebraic geometry. This ended actually in 1965, the beginning of cultural revolution. See Nos. 1–5 of "Selected Papers". During the cultural revolution there were however some sporadic research works in pure mathematics, with papers published a little later. See Nos. 6, 7, 14, 15, 18 of "Selected Papers". Such researches stopped completely at the end of cultural revolution, *viz.* the year 1976.

The second stage of my mathematical researches took place during the cultural revolution. It took place owing to my learning of the history of our proper mathematics in ancient times. See No. 17 of "Selected Papers".

During the cultural revolution I was sent to some computer-manufacture company to learn and work with laborers. Being striken by the powerfulness of computers I began to consider of applying computers to the study of mathematics. It results in a method of proving geometry theorems by means of computers. Extending further the method it gave rise to the subject what I called the Mathematics Mechanization which had an immense varieties of applications in science and technology, besides the mathematics itself. See Nos. 16, 19–30 of the "Selected Papers".

For some general description of my scientific career one may refer to the book "The Road of WU Wen-tsun", written in Chinese by Professors HU and SHI, published by Shanghai Science-Technology Press, year 2002.

Wen-tsun Wu
Dec. 27, 2007

Contents

Foreword v

1. On the product of sphere bundles and the duality theorem modulo two 1
2. Classes caractéristiques et i-carrès d'une variétè 15
3. Les i-carrés dans une variétè grassmanniénne 19
4. On the realization of complexes in Euclidean spaces I 23
5. On the realization of complexes in Euclidean spaces III 71
6. On universal invariant forms 85
7. Theory of I^*-functor in algebraic topology — Effective calculation and axiomatization of I^*-functor on complexes 99
8. On the decision problem and the mechanization of theorem-proving in elementary geometry 117
9. Toward mechanization of geometry — Some comments on Hilbert's "Grundlagen der Geometrie" 139
10. The out-in complementary principle 153
11. A constructive theory of differential algebraic geometry based on works of J. F. Ritt with particular applications to mechanical theorem-proving in differential geometries 177
12. Basic principles of mechanical theorem-proving in elementary geometries 195
13. On zeros of algebraic equations — An application of Ritt principle 225
14. On the planar imbedding of linear graphs I 231

15. On the planar imbedding of linear graphs II 245

16. A mechanization method of geometry and its applications I 259

17. Recent studies of the history of Chinese mathematics 273

18. On Chern numbers of algebraic varieties with arbitrary singularities 285

19. Mechanical derivation of Newton's gravitational laws from Kepler's laws 295

20. A mechanization method of geometry and its applications II 303

21. A mechanization method of geometry and its applications III 307

22. On the foundation of algebraic differential geometry 325

23. On the genetic zero and Chow basis of an irreducible ascending set 345

24. Mechanical theorem proving of differential geometries and some of its applications in mechanics 367

25. On a finiteness theorem about optimization problems 387

26. On surface-fitting problem in CAGD 401

27. Central configurations in planet motions and vortex motions 411

28. On algebraico-differential equations-solving 425

29. On the construction of Groebner basis of a polynomial ideal based on Riquier–Janet theory 437

30. On "good" bases of algebraico-differential ideals 455

ANNALS OF MATHEMATICS
Vol. 49, No. 3, July, 1948

ON THE PRODUCT OF SPHERE BUNDLES AND THE DUALITY THEOREM MODULO TWO

BY WU WEN-TSUN

(Received August 15, 1947)

Introduction*

Given two sphere bundles $\mathfrak{S}_1$ and $\mathfrak{S}_2$ over the base complexes K_1 and K_2 respectively, it is possible to define in a natural way a "product bundle" over the product complex $K_1 \times K_2$. When $K_1 = K_2 = K$ (say), the part of the product bundle over the diagonal of the product complex $K \times K$ is the product bundle in the sense of Whitney.[1] We shall prove in the present paper that a certain duality theorem holds for the product bundle over $K_1 \times K_2$ and that Whitney's duality theorem for sphere bundles follows from this more general duality theorem as a consequence. (Throughout the paper coefficients mod 2 will be used.) The idea of this proof seems to be quite different from Whitney's original one, of which only a brief sketch is known.[2]

The paper is divided into three sections. In §1 some preliminary considerations and theorems on vector fields are given. A duality theorem for the product bundle over $K_1 \times K_2$ is then proved in §2. §3 is devoted to a proof of Whitney's duality theorem.

§1

1. We recall in this paragraph the definition of a bundle of linear spaces or more simply, a vector bundle.

A complex K with cells $\sigma_1, \sigma_2, \cdots$ and a ν-dimensional vector space V are given. To each point p of K a ν-dimensional vector space $V(p)$ is associated so that $V(p)$ and $V(q)$ are disjoint if p and q are distinct points of K. Suppose there are non-degenerate linear mappings $\xi_{\sigma_i p}$ of V on $V(p)$ for every σ_i of K and every point p of σ_i with the following condition satisfied: For p common to σ_i and σ_j, $\xi^{-1}_{\sigma_i p}\xi_{\sigma_j p}$ gives a continuous map of $\sigma_i \cap \sigma_j$ into the group of non-degenerate linear mappings of V on itself. Then we can make the union of all the spaces $V(p)$ into a single topological space $\mathfrak{V}$ in a natural way so that for every cell $\sigma_i \in K$ the topological product $V \times \sigma_i$ is homeomorphic to the union of all $V(p)$ for which $p \in \sigma_i$. This homeomorphism is in fact given by $\xi_{\sigma_i}(p, x) = \xi_{\sigma_i p}(x)$, where $x \in V$.

We shall introduce the following terminologies:

$\mathfrak{V}$, the vector bundle;

K, the base complex;

---

* The problems in this paper were suggested to me by Professor S. S. Chern, with whom I have many helpful discussions. To him are expressed here my thanks.

[1] WHITNEY, Lectures in Topology. Harvard Univ., 1941, p. 131.

[2] WHITNEY, Proc., Nat. Acad. Sci., 26 (1940), pp. 143–148.

V, the director space;

$V(p)$, the vector space over p;

ν, the bundle-dimension of $\mathfrak{B}$;

$\xi_{\sigma p}$, the coordinate system in σ.

2. Let $\mathfrak{B}$ be the vector bundle defined in paragraph 1. By a continuous mapping of K or a subcomplex of K into $\mathfrak{B}$ we shall always mean one that maps the points p of K or a subcomplex of K into the respective $V(p)$ over p. A set of continuous mappings $\varphi_1, \cdots \varphi_m$ of a subcomplex L of K into $\mathfrak{B}$ is said to form an m-field $\Phi = \{\varphi_1, \cdots \varphi_m\}$ over L. We say that Φ is continuous (or discontinuous) at a point $p \in L$ if $\varphi_1(p), \cdots \varphi_m(p)$ are linearly independent (or linearly dependent) in $V(p)$ and that Φ is a continuous m-field over L if it is continuous at every point $p \in L$.

Let K^r be the r-dimensional skeleton of K. As is well known,[3] for $m \leqq \nu$ continuous m-field $\{\varphi_1, \cdots \varphi_m\}$ always exists over $K^{\nu-m}$. Now orient V and the cells of the complex arbitrarily and consider any $(\nu - m + 1)$-dimensional oriented cell σ_i. For points p in $\partial\sigma_i$, $\xi_{\sigma_i p}^{-1}\varphi_1(p), \cdots \xi_{\sigma_i p}^{-1}\varphi_m(p)$ together give a map φ_{σ_i} of $\partial\sigma_i$ into the Stiefel manifold[4] $V_{\nu,m}$ of all ordered sets of m linearly independent vectors in V. The characteristic[4] $d(\sigma_i)$ of this mapping is either an integer or is defined only mod 2. In any way, the chain

$$w = \sum_i d(\sigma_i)\sigma_i$$

when reduced mod 2 if necessary, is a $(\nu - m + 1)$-dimensional cocycle mod 2 of K the class of which is independent of the particular choice of the continuous m-field and the orientations of the cells of K. The classes thus obtained will be called the characteristic classes and their cocycles characteristic cocycles. We denote them by W^r, $r = 1, 2, \cdots \nu$.

For convenience we shall define W^0 to be the class containing the cocycle I, which is the sum of all vertices of K. We also put all $W^r = 0$, for $r > \nu$.

3. We shall put $[a - b] = a - b$ for $a \geqq b$ and $[a - b] = 0$ for $b > a$. We shall prove that in a vector bundle $\mathfrak{B}$ it is possible to construct on K an m-field $\Phi = \{\varphi_1, \cdots \varphi_m\}$, ($m$ not necessarily $\leqq \nu$) which satisfies the following conditions (C_r), $r = 0, 1, 2, \cdots$:

(C_r). Let p be any point in K^r. The conditions are:

CASE 1. If $m + r \leqq \nu$, then $\varphi_1, \cdots \varphi_m$ are linearly independent at p;

CASE 2. If $m + r > \nu$, then there is an integer $0 \leqq i \leqq m - [\nu - r]$ such that $\varphi_1(p), \cdots \varphi_{[\nu-r]+i}(p)$ are linearly independent while

$$\varphi_{[\nu-r]+i+1}(p) = \cdots = \varphi_m(p) = 0.$$

Such a field will be called a canonical field. We shall construct it successively over $K^0, K^1, \cdots$ as follows:

1^0. Construction over $K^0, \cdots K^{[\nu-m]}$.

[3] See for example STIEFEL, Comm. Math. Helv. 8 (1936), 331.

[4] STIEFEL, loc. cit., 310–323. The characteristic will be denoted sometimes by Char $_{\sigma_i}\{\varphi_1 \cdots \varphi_m\}$, Char $_{\varphi_{\sigma_i}}\sigma_i$, etc.

If $m < \nu$, we take $\varphi_1(p), \cdots \varphi_m(p)$ at a vertex p to be any m linearly independent vectors in $V(p)$ and then extend successively to $K^{\nu-m}$. If $m \geqq \nu$ and p a vertex, we take $\varphi_1(p), \cdots \varphi_\nu(p)$ to be ν linearly independent vectors in $V(p)$, while we set $\varphi_{\nu+1}(p) = \cdots = \varphi_m(p) = 0$. In this way $(C_0), \cdots (C_{[\nu-m]})$ evidently hold.

2^0. Construction over K^r, $r > [\nu - m]$, assuming that $\varphi_1, \cdots \varphi_m$ have been constructed over K^{r-1} with (C_{r-1}) satisfied.

Consider any r-dimensional cell σ^r. We can denote its points by $tp, 0 \leqq t \leqq 1$, where p is a point on $\partial\sigma^r$ and $0 \cdot p = 0$ is a fixed interior point of σ.

$\{\varphi_1, \cdots \varphi_{[\nu-r]}\}$ being defined and continuous on ∂^r we can extend it continuously into the interior of σ^r. Then $\bar{\varphi}_1^*(tp) = \xi_{\sigma p}^{-1}\varphi_1(tp), \cdots \bar{\varphi}_{[\nu-r]}^*(tp) = \xi_{\sigma p}^{-1}\varphi_{[\nu-r]}(tp)$ are $[\nu - r]$ continuous mappings of σ^r into V so that for each tp these are $[\nu - r]$ linearly independent vectors. We can find $\nu - [\nu - r]$ further mappings $\bar{\varphi}_{[\nu-r]+1}^*(tp), \cdots \bar{\varphi}_\nu^*(tp)$ of σ^r into V so that for every tp in σ^r, $\bar{\varphi}_1^*(tp), \cdots \bar{\varphi}_\nu^*(tp)$ are linearly independent and form a positive system in V, assuming that V has been oriented. Put $\xi_{\sigma p}\bar{\varphi}_i^*(tp) = \bar{\varphi}_i(tp)$, $i = 1, 2, \cdots \nu$, we get a ν-field $\{\bar{\varphi}_1, \cdots \bar{\varphi}_\nu\}$ continuous over σ^r. We have moreover

$$\bar{\varphi}_i(tp) = \varphi_i(tp), \qquad \text{for } i = 1, \cdots [\nu - r].$$

Let

$$\varphi_i(p) = \sum_{j=1}^{\nu} a_j^{(i)}(p)\bar{\varphi}_j(p), \qquad i = [\nu - r] + 1, \cdots m$$

where $a_j^{(i)}(p)$ are real numbers. We define

$$\varphi_i(tp) = \sum_{j=1}^{\nu} ta_j^{(i)}(p)\bar{\varphi}_j(tp), \qquad i = [\nu - r] + 1, \cdots m.$$

The field $\{\varphi_1, \cdots \varphi_m\}$ is then extended over σ^r. Doing this for all σ^r, we get an m-field over K^r.

The only places in σ^r where discontinuity occurs are either 1) 0, or 2) $tp, t \neq 0$ for which $\{\varphi_1, \cdots \varphi_m\}$ is discontinuous at p.

In case 1), $\{\varphi_1, \cdots \varphi_{[\nu-r]}\}$ is continuous, while $\varphi_{[\nu-r]+1}(0) = \cdots = \varphi_m(0) = 0$. Condition (C_r) is thus satisfied with $i = 0$.

In case 2), there exists by induction an integer i so that $\{\varphi_1, \cdots \varphi_{[\nu-r+1]+i}\}$ is continuous at p, while

$$\varphi_{[\nu-r+1]+i+1}(p) = \cdots = \varphi_m(p) = 0.$$

This is true when p is replaced by $tp, t \neq 0$. As

$$[\nu - r + 1] = \begin{cases} [\nu - r] + 1, & \text{for } r < \nu + 1 \\ [\nu - r], & \text{for } r \geqq \nu + 1 \end{cases}$$

we see that condition (C_r) is satisfied.

We remark in passing that in case $r \leqq \nu$, not all of

$$a_{\nu-r+1}^{(\nu-r+1)}(p), \cdots a_\nu^{(\nu-r+1)}(p), \qquad p \in \partial\sigma^r$$

are zero. For we have

$$\varphi_i(p) = \bar{\varphi}_i(p), \qquad i = 1, 2, \cdots \nu - r$$

$$\varphi_{\nu-r+1}(p) = \sum_{j=1}^{\nu} a_j^{(\nu-r+1)}(p)\cdot\bar{\varphi}_j(p).$$

As $\varphi_{\nu-r+1}(p)$ is linearly independent of $\varphi_i(p)$, $i = 1, 2, \cdots \nu - r$, for $p \in \partial\sigma^r$, the matrix of their components with respect to the field $\{\bar{\varphi}_1, \cdots \bar{\varphi}_\nu\}$ over σ^r, namely,

$$\begin{pmatrix} 1 & 0 & \cdots & 0 & 0 & \cdots & 0 \\ 0 & 1 & \cdots & 0 & 0 & \cdots & 0 \\ \cdots & \cdots & \cdots & \cdots & \cdots & \cdots & \cdots \\ 0 & 0 & \cdots & 1 & 0 & \cdots & 0 \\ a_1^{(\nu-r+1)} & a_2^{(\nu-r+1)} & \cdots & a_{\nu-r}^{(\nu-r+1)} & a_{\nu-r+1}^{(\nu-r+1)} & \cdots & a_\nu^{(\nu-r+1)} \end{pmatrix}$$

must be of rank $\nu - r + 1$.

4. Given two vector bundles $\mathfrak{B}_1$, $\mathfrak{B}_2$, of which the base complexes and the other symbols are distinguished by the subscripts 1 and 2, we shall define a third bundle $\mathfrak{B}$ according to the following table:

Base complex: $K = K_1 \times K_2$;

Director space: $V = V_1 \oplus V_2$;[5]

Vector space over $p_1 \times p_2$: $V(p_1 \times p_2) = V_1(p_1) \oplus V_2(p_2)$;[5]

Bundle dimension: $\nu = \nu_1 + \nu_2$;

Coordinate system in $\sigma_1 \times \sigma_2$:

$$\xi_{\sigma_1\times\sigma_2, p_1\times p_2}(\mathfrak{x}_1 + \mathfrak{x}_2) = \xi_{\sigma_1 p_1, 1}(\mathfrak{x}_1) + \xi_{\sigma_2 p_2, 2}(\mathfrak{x}_2), \text{ where } \mathfrak{x}_1 \in V_1,\ \mathfrak{x}_2 \in V_2.$$

This bundle $\mathfrak{B}$, as a topological space, is a topological product of $\mathfrak{B}_1$ and $\mathfrak{B}_2$. In fact, by means of the coordinate system $\xi_{\sigma_1\times\sigma_2, p\times p_2}(\mathfrak{x}_1 + \mathfrak{x}_2)$ in $\sigma_1 \times \sigma_2$, we map the point $\xi_{\sigma_1\times\sigma_2, p_1\times p_2}(\mathfrak{x}_1 + \mathfrak{x}_2)$ into the point $(\xi_{\sigma_1 p_1, 1}(\mathfrak{x}_1), \xi_{\sigma_2 p_2, 2}(\mathfrak{x}_2))$ of $\mathfrak{B}_1 \times \mathfrak{B}_2$. This mapping is clearly topological. We can therefore write $\mathfrak{B}_1 \times \mathfrak{B}_2$ for $\mathfrak{B}$ without confusion and shall call $\mathfrak{B}$ the product bundle of $\mathfrak{B}_1$ and $\mathfrak{B}_2$.

Let the characteristic classes of $\mathfrak{B}$, $\mathfrak{B}_1$ and $\mathfrak{B}_2$ be respectively denoted by W, W_1, W_2, with the convention made at the end of paragraph 2. Then our main theorem is the following

THEOREM I. *The characteristic classes of the product bundle $\mathfrak{B} = \mathfrak{B}_1 \times \mathfrak{B}_2$ are expressible in terms of those of $\mathfrak{B}_1$ and $\mathfrak{B}_2$. More precisely, we have the formula*

$$W^r = \sum_{i=0}^{r} W_1^i \times W_2^{r-i}, \qquad r = 1, 2 \cdots \nu. \tag{4.1}$$

The multiplication of cohomology classes occurred in this formula may be explained as follows:

Let

$$C_1 = \sum_i a_{i1}\sigma_{i1}, \qquad C_2 = \sum_j a_{j2}\sigma_{j2}$$

[5] $V_1 \oplus V_2$ means the join of the two vector spaces. We assume that $V(p_1 \times p_2)$ and $V(q_1 \times q_2)$ are disjoint for $p_1 \times p_2 \neq q_1 \times q_2$ even if $p_1(p_2)$ may coincide with $q_1(q_2)$.

be any two chains of K_1 respectively K_2, where a_{i1}, a_{j2} are elements of a coefficient ring, (in the case considered above is the ring of residue classes mod 2). We define a product chain $C = C_1 \times C_2$ of $K_1 \times K_2$ by

$$C = \sum_{i,j} a_{i1} a_{j2} (\sigma_{i1} \times \sigma_{j2}).$$

Since

$$\delta(\sigma_{i1} \times \sigma_{j2}) = \delta\sigma_{i1} \times \sigma_{j2} + (-1)^p \sigma_{i1} \times \delta\sigma_{j2}, \qquad p = \dim. \sigma_{i1}$$

we have

$$\delta C = \delta C_1 \times C_2 + (-1)^p C_1 \times \delta C_2,$$

if C_1 is a p-dimensional chain.

It follows readily that we can define the product class of two cohomology classes in a unique way. It is this product that we denote by $W_1 \times W_2$ in (4.1).

5. If two vector bundles $\mathfrak{B}_1$, $\mathfrak{B}_2$ of which the symbols are again distinguished by subscripts 1 and 2 are defined on the same base complex K, a third bundle $\breve{\mathfrak{B}}$ over the same K can be defined according to the following table:

Base complex: K;
Director space: $\breve{V} = V_1 \oplus V_2$;[6]
Vector space over p: $\breve{V}(p) = V_1(p) \oplus V_2(p)$;[6]
Bundle-dimension: $\nu = \nu_1 + \nu_2$;
Coordinate system in σ:

$$\breve{\xi}_{\sigma p}(\mathfrak{x}_1 + \mathfrak{x}_2) = \xi_{\sigma p,1}(\mathfrak{x}_1) + \xi_{\sigma p,2}(\mathfrak{x}_2), \text{ where } \mathfrak{x}_1 \in V_1 \cdot \mathfrak{x}_2 \in V_2.$$

$\breve{\mathfrak{B}}$ will be called the span bundle of $\mathfrak{B}_1$ and $\mathfrak{B}_2$ and we shall write $\breve{\mathfrak{B}} = \mathfrak{B}_1 \smile \mathfrak{B}_2$. This notation is suggested by the following "duality theorem" of Whitney:

THEOREM II. *For the span bundle $\breve{\mathfrak{B}} = \mathfrak{B}_1 \smile \mathfrak{B}_2$ we have*

$$\breve{W}^r = \sum_{i=0}^{r} W_1^i \smile W_2^{r-i}, \qquad r = 1, 2 \cdots \nu \tag{5.1}$$

where $\breve{W}$, W_1, W_2 are the respective characteristic classes of $\breve{\mathfrak{B}}$, $\mathfrak{B}_1$, $\mathfrak{B}_2$ with the convention of paragraph 3.

§2

6. Throughout §2 the notations of paragraph 4 will be used. Let $\Phi_1 = \{\varphi_{1,1}, \cdots \varphi_{m,1}\}$ and $\Phi = \{\varphi_{1,2}, \cdots \varphi_{m,2}\}$ be canonical m-fields on K_1, K_2 as defined in paragraph 3, $m \leqq \nu$. We now construct an m-field $\Phi = \{\varphi_1, \cdots \varphi_m\}$ on $K = K_1 \times K_2$ by setting

$$\varphi_i(p_1 \times p_2) = \varphi_{i,1}(p_1) + \varphi_{m-i+1,2}(p_2), \qquad i = 1, 2, \cdots m.$$

Φ is continuous on the skeleton $K^{\nu-m}$ of K.

[6] As in 5) the spaces $V(p)$ are assumed to be disjoint from each other.

PROOF. First consider the points $p_1 \times p_2$ in the cell $\sigma_1^{\nu_1 - m_1} \times \sigma_2^{\nu_2 - m_2}$, where $0 \leqq m_1 \leqq \nu_1$, $0 \leqq m_2 \leqq \nu_2$ and $m_1 + m_2 = m$.

Suppose that Φ is discontinuous at $p_1 \times p_2$. Then there would exist real numbers $a_1, \cdots a_m$ not all zero such that

$$a_1\varphi_1(p_1 \times p_2) + \cdots + a_m\varphi_m(p_1 \times p_2) = 0 \tag{6.1}$$

that is, (6.2)

$$a_1\varphi_{1,1}(p_1) + \cdots + a_m\varphi_{m,1}(p_1) = 0$$

and

$$a_1\varphi_{m,2}(p_2) + \cdots + a_m\varphi_{1,2}(p_2) = 0.$$

Since the fields Φ_1 and Φ_2 are canonical, there are integers i_1 and i_2 such that

1) $\varphi_{1,1}(p_1), \cdots \varphi_{m_1+i_1,1}(p_1)$ are linearly independent and $\varphi_{1,2}(p_2), \cdots \varphi_{m_2+i_2,2}(p_2)$ are linearly independent.

2)

$$\varphi_{m_1+i_1+1,1}(p_1) = \cdots = \varphi_{m,1}(p_1) = 0$$

$$\varphi_{m_2+i_2+1,2}(p_2) = \cdots = \varphi_{m,2}(p_2) = 0.$$

By 2), (6.2) becomes

$$a_1\varphi_{1,1}(p_1) + \cdots + a_{m_1+i_1}\varphi_{m_1+i_1,1}(p_1) = 0$$

$$a_m\varphi_{1,2}(p_2) + \cdots + a_{m-m_2-i_2+1}\varphi_{m_2+i_2,2}(p_2) = 0.$$

By 1),

$$a_1 = \cdots = a_{m_1+i_1} = 0$$

$$a_m = \cdots = a_{m-m_2-i_2+1} = 0.$$

As $m - m_2 = m_1$, it follows that all the a's are zero, which is a contradiction.

Consider next the points $p_1 \times p_2$ in a cell $\sigma_1^{\nu_1+m_1} \times \sigma_2^{\nu_2 - m_2}$, where $0 \leqq m_1$, $0 \leqq m_2 \leqq \nu_2$ and $m_2 - m_1 = m$.

We then have $m \leqq m_2$ and Φ_2 is continuous on $\sigma_2^{\nu_2 - m_2}$. Hence from the second equation of (6.2) we would have again $a_1 = \cdots = a_m = 0$, which is also a contradiction.

The other case is similar, and thus our assertion is proved.

7. Before evaluating the characteristics from the field constructed in the above paragraph, we shall prove in this section a lemma on the degree of mapping.

Let S_i', S_i be spheres of dimensions n_i which bound the cells V_i', V_i, $i = 1, 2$. Denote by S', S the joins of the pairs of spheres S_1', S_2' and S_1, S_2 respectively, i.e.,

$$\begin{aligned} S' &= S_1' \times V_2' + S_2' \times V_1' \\ S &= S_1 \times V_2 + S_2 \times V_1. \end{aligned} \tag{7.1}$$

The points of $S'(S)$ can be conveniently denoted by

$$t_1x_1' \times t_2x_2' \qquad (t_1x_1 \times t_2x_2)^{7}$$

[7] We can equally well denote these points by $t_1x_1' + t_2x_2'$ $(t_1x_1 + t_2x_2)$.

where $x'_i \in S'_i$ $\quad (x_i \in S_i)$,

$$t_1 = 1 \text{ and } 0 \leqq t_2 \leqq 1 \text{ or } t_2 = 1 \text{ and } 0 \leqq t_1 \leqq 1.$$

Or, $0'_i(0_i)$ being the centers of $V'_i(V_i)$, $S'(S)$ is composed of the sets $x'_1 \times 0'_2x'_2$ and $0'_1x'_1 \times x'_2$ ($x_1 \times 0_2x_2$ and $0_1x_1 \times x_2$).

Let S'_i (S_i) be oriented. Orient V'_i (V_i) in coherence with S'_i (S_i), then (7.1) defines an orientation of S' (S).

Suppose we are given continuous mappings f_i of S'_i into S_i with degree d_i. Define a mapping f of S' into S by

$$f(t_1x'_1 \times t_2x'_2) = t_1f_1(x'_1) \times t_2f_2(x'_2). \tag{7.2}$$

Then we have the following

LEMMA. *The degree of the mapping f is given by*

$$d = d_1d_2. \tag{7.3}$$

PROOF. Subdivide S'_i, S_i ($i = 1, 2$) into sufficiently fine simplexes and deform f_1, f_2 into simplicial approximations f^*_i. Next we subdivide S', S into cells so that the ground-cells of S, say, are of the form

$$(0_1\sigma_1) \times \sigma_2, \qquad \sigma_1 \times (0_2\sigma_2)$$

where σ_i are ground-simplexes of the subdivisions of $S_i (i = 1, 2)$.

Deform f to f^* so that during the deformation relations analogous to (7.2) always hold, with the final result

$$f^*(t_1x'_1 \times t_2x'_2) = t_1f^*_1(x'_1) \times t_2f^*_2(x'_2).$$

We shall determine the degree of mapping of f^*.

For this purpose consider any oriented ground-cell of S, say

$$(0_1\sigma_1) \times \sigma_2.$$

For ground-simplexes τ'_{1i} of S'_1 and τ'_{2j} of S'_2, we have

$$f^*[(0'_1\tau'_{1i}) \times \tau'_{2j}] = \pm (0_1\sigma_1) \times \sigma_2 \tag{7.4}$$

if and only if

$$f^*_1(\tau'_{1i}) = \pm \sigma_1, \qquad f^*_2(\tau'_{2j}) = \pm \sigma_2. \tag{7.5}$$

Let the number of simplexes τ_{1i}, τ_{2j} for which (7.5) hold with positive (negative) sign be P_1, P_2, (N_1, N_2) so that

$$P_1 - N_1 = d_1, \qquad P_2 - N_2 = d_2.$$

Then the number of cells $(0'_1\tau'_{1i}) \times \tau'_{2j}$ for which (7.4) holds with positive (negative) sign is $P = P_1P_2 + N_1N_2$ ($N = P_1N_2 + P_2N_1$). Hence

$$\begin{aligned} d &= P - N = (P_1 - N_1)(P_2 - N_2) \\ &= d_1d_2. \end{aligned}$$

The lemma is thus proved.

8. We now come to the determination of characteristics for the field defined in paragraph 6. For this purpose let us consider a $(\nu - m + 1)$-cell $\sigma = \sigma_1^{r_1} \times \sigma_2^{r_2}$ of K, where $r_1 + r_2 = \nu - m + 1$. We suppose first that $0 < r_1 \leqq \nu_1$, $0 < r_2 \leqq \nu_2$.

As shown in paragraph 3, there are continuous mappings $\bar{\varphi}_{i,1}$, $i = 1, 2, \cdots \nu_1$ of $\sigma_1^{r_1}$ in $\mathfrak{B}_1$ and continuous mappings $\bar{\varphi}_{j,2}$, $j = 1, 2, \cdots \nu_2$ of $\sigma_2^{r_2}$ in $\mathfrak{B}_2$ satisfying the following conditions:

1^0. $\bar{\varphi}_{1,1}, \cdots \bar{\varphi}_{\nu_1,1}$ are linearly independent at every point $tp_1 \in \sigma_1$ and $\bar{\varphi}_{1,2}, \cdots \bar{\varphi}_{\nu_2,2}$ are linearly independent at every point $tp_2 \in \sigma_2$, where we denote as usual by p_1 a point in $\partial\sigma_1$, p_2 a point in $\partial\sigma_2$ and $0 \leqq t \leqq 1$.

2^0.
$$\varphi_{i,1}(tp_1) = \bar{\varphi}_{i,1}(tp_1) \text{ for } i = 1, 2, \cdots \nu_1 - r_1$$
$$\varphi_{j,2}(tp_2) = \bar{\varphi}_{j,2}(tp_2) \text{ for } j = 1, 2, \cdots \nu_2 - r_2$$

3^0. If
$$\varphi_{i,1}(p_1) = a_{1,1}^{(i)}(p_1)\bar{\varphi}_{1,1}(p_1) + \cdots + a_{\nu_1,1}^{(i)}(p_1)\bar{\varphi}_{\nu_1,1}(p_1),$$
$i = \nu_1 - r_1 + 1, \cdots \nu_1$
$$\varphi_{j,2}(p_2) = a_{1,2}^{(j)}(p_2)\bar{\varphi}_{1,2}(p_\nu) + \cdots + a_{\nu_2,2}^{(j)}(p_\nu)\bar{\varphi}_{\nu_2,2}(p_\nu),$$
$j = \nu_2 - r_2 + 1, \cdots \nu_2$

then
$$\varphi_{i,1}(tp_1) = ta_{1,1}^{(i)}(p_1)\bar{\varphi}_{1,1}(tp_1) + \cdots + ta_{\nu_1,1}^{(i)}(p_1)\bar{\varphi}_{\nu_1,1}(tp_1),$$
$i = \nu_1 - r_1 + 1, \cdots \nu_1$
$$\varphi_{j,2}(tp_2) = ta_{1,2}^{(j)}(p_\nu)\bar{\varphi}_{1,2}(tp_\nu) + \cdots + ta_{\nu_2,2}^{(j)}(p_\nu)\bar{\varphi}_{\nu_2,2}(tp_\nu),$$
$j = \nu_2 - r_2 + 1, \cdots \nu_2$

4^0. If V_1 and V_2 are definitely oriented, $\bar{\varphi}^*_{1,1}(tp_1), \cdots \bar{\varphi}^*_{\nu_2,1}(tp_1)$ form a positive system in V_1 and $\bar{\varphi}^*_{1,2}(tp_2), \cdots \bar{\varphi}^*_{\nu_2,2}(tp_2)$ form a positive system in V_2, where
$$\bar{\varphi}^*_{i,k}(tp_k) = \xi^{-1}_{\sigma_k, tp_k, k}\bar{\varphi}_{i,k}(tp_k),\ i = 1, 2, \cdots \nu_k\ ;\ k = 1, 2.$$

Whence the vectors
$$\bar{\varphi}_{i,1}(t_1p_1), \bar{\varphi}_{j,2}(t_2p_2),\ i = 1, 2, \cdots \nu_1\ ;\ j = 1, 2, \cdots \nu_2,$$
form a basis in $V(t_1p_1 \times t_2p_2)$.

Write for simplicity
$$a_{i,k}^{(j)}(p_k) = a_{i,k}^{(j)},\ i = 1, \cdots \nu_k\ ;\ j = \nu_k - r_k + 1, \cdots \nu_k\ ;\ k = 1, 2.$$

Then we have for every point $t_1p_1 \times t_2p_2$ in $\partial\sigma$, where
$$t_1 = 1 \text{ and } 0 \leqq t_2 \leqq 1 \text{ or } t_2 = 1 \text{ and } 0 \leqq t_1 \leqq 1$$

$$(\Phi)\begin{cases}\varphi_i(t_1p_1\times t_2p_2)=\bar\varphi_{i.1}(t_1p_1)+t_2a_{1.2}^{(m+1-i)}\bar\varphi_{1.2}(t_2p_2)+\cdots+t_2a_{\nu_2.2}^{(m+1-i)}\bar\varphi_{\nu_2.2}(t_2p_2), \\ \qquad\qquad 1\leqq i\leqq \nu_1-r, \\ \varphi_i(t_1p_1\times t_2p_2)=t_1a_{1.1}^{(i)}\bar\varphi_{1.1}(t_1p_1)+\cdots+t_1a_{\nu_1.1}^{(i)}\bar\varphi_{\nu_1.1}(t_1p_1)+\bar\varphi_{m+1-i.2}(t_2p_2), \\ \qquad\qquad \nu_1-r_1+2\leqq i\leqq m_1 \\ \varphi_{\nu_1-r_1+1}(t_1p_1\times t_2p_2)=t_1a_{1.1}^{(\nu_1-r_1+1)}\bar\varphi_{1.1}(t_1p_1)+\cdots+t_1a_{\nu_1.1}^{(\nu_1-r_1+1)}\bar\varphi_{\nu_1.1}(t_1p_1) \\ \qquad\qquad +t_2a_{1.2}^{(\nu_2-r_2+1)}\bar\varphi_{1.2}(t_2p_2)+\cdots+t_2a_{\nu_2.2}^{(\nu_2-r_2+1)}\bar\varphi_{\nu_2.2}(t_2p_2).\end{cases}$$

Now take arbitrary positive basis in V_1 and V_2, say $\mathfrak{x}_{1.1},\cdots\mathfrak{x}_{\nu_1.1}$ and $\mathfrak{x}_{1.2},\cdots\mathfrak{x}_{\nu_2.2}$. As $\bar\varphi^*_{i.1}$, $\bar\varphi^*_{j.2}$ are defined both on the boundary and in the interior of σ_1 and σ_2 respectively, we can deform continuously the set of mappings $\{\bar\varphi^*_{1.1},\cdots\bar\varphi^*_{\nu_1.1},\bar\varphi^*_{1.2},\cdots\bar\varphi^*_{\nu_2.2}\}$ of $\partial\sigma$ in V into a set of constant mappings which orders to any point in $\partial\sigma$ the system of constant vectors $\mathfrak{x}_{1.1},\cdots\mathfrak{x}_{\nu_1.1},\mathfrak{x}_{1.2},\cdots\mathfrak{x}_{\nu_2.2}$. This deformation then induces a continuous deformation of the m-field $\Phi=\{\varphi_1,\cdots\varphi_m\}$ over $\partial\sigma$ into a continuous m-field $\Theta=\{\theta_1,\cdots\theta_m\}$ over $\partial\sigma$ given by the following set of equations:

$$(\Theta)\begin{cases}\theta_i(t_1p_1\times t_2p_2)=\xi_{\sigma,\,t_1p_1\times t_2p_2}\theta_i^*(t_1p_1\times t_2p_2), \qquad i=1,2,\cdots m \\ \theta_i^*(t_1p_1\times t_2p_2)=\mathfrak{x}_{i.1}+t_2a_{1.2}^{(m+1-i)}\mathfrak{x}_{1.2}+\cdots+t_2a_{\nu_2.2}^{(m+1-i)}\mathfrak{x}_{\nu_2.2}, \\ \qquad\qquad 1\leqq i\leqq\nu_1-r_1 \\ \theta_i^*(t_1p_1\times t_2p_2)=t_1a_{1.1}^{(i)}\mathfrak{x}_{1.1}+\cdots+t_1a_{\nu_1.1}^{(i)}\mathfrak{x}_{\nu_1.1}+\mathfrak{x}_{m+1-i.2}, \\ \qquad\qquad \nu_1-r_1+2\leqq i\leqq m \\ \theta^*_{\nu_1-r_1+1}(t_1p_1\times t_2p_2)=t_1a_{1.1}^{(\nu_1-r_1+1)}\mathfrak{x}_{1.1}+\cdots+t_1a_{\nu_1.1}^{(\nu_1-r_1+1)}\mathfrak{x}_{\nu_1.1} \\ \qquad\qquad +t_2a_{1.2}^{(\nu_2-r_2+1)}\mathfrak{x}_{1.2}+\cdots+t_2a_{\nu_2.2}^{(\nu_2-r_2+1)}\mathfrak{x}_{\nu_2.2}.\end{cases}$$

It follows from the remark at the end of paragraph 3, that for points $t_1p_1\times t_2p_2\in\partial\sigma$, at least one of

$$t_1a_{\nu_1-r_1+1.1}^{(\nu_1-r_1+1)},\cdots t_1a_{\nu_1.1}^{(\nu_1-r_1+1)},\qquad t_2a_{\nu_2-r_2+1.2}^{(\nu_2-r_2+1)},\cdots t_2a_{\nu_2.2}^{(\nu_2-r_2+1)}$$

is not 0. Hence we can deform continuously the m-field Θ into a continuous m-field $\Psi=\{\psi_1,\cdots\psi_m\}$ given by

$$(\Psi)\begin{cases}\psi_i(t_1p_1\times t_2p_2)=\xi_{\sigma,\,t_1p_1\times t_2p_2}\psi_i^*(t_1p_1\times t_2p_2), \qquad i=1,2,\cdots m \\ \psi_i^*(t_1p_1\times t_2p_2)=\mathfrak{x}_{i.1}, \qquad i=1,2,\cdots\nu_1-r_1 \\ \psi_i^*(t_1p_1\times t_2p_2)=\mathfrak{x}_{m+1-i.2}, \qquad i=\nu_1-r_1+2,\cdots m \\ \psi^*_{\nu_1-r_1+1}(t_1p_1\times t_2p_2)=t_1a_{\nu_1-r_1+1.1}^{(\nu_1-r_1+1)}\mathfrak{x}_{\nu_1-r_1+1.1}+\cdots+t_1a_{\nu_1.1}^{(\nu_1-r_1+1)}\mathfrak{x}_{\nu_1.1} \\ \qquad\qquad +t_2a_{\nu_2-r_2+1.2}^{(\nu_2-r_2+1)}\mathfrak{x}_{\nu_2-r_2+1.2}+\cdots+t_2a_{\nu_2.2}^{(\nu_2-r_2+1)}\mathfrak{x}_{\nu_2.2}.\end{cases}$$

Moreover, there is no loss of generality in assuming that

$$[a_{\nu_1-r_1+1.1}^{(\nu_1-r_1+1)}]^2+\cdots+[a_{\nu_1.1}^{(\nu_1-r_1+1)}]^2=1$$
$$[a_{\nu_2-r_2+1.2}^{(\nu_2-r_2+1)}]^2+\cdots+[a_{\nu_2.2}^{(\nu_2-r_2+1)}]^2=1$$

for otherwise we can bring about this by a further deformation. (See the remark at the end of paragraph 3.)

For $p_1 \in \partial\sigma_1$, $p_2 \in \partial\sigma_2$, let us put

$$\psi^*_{i,1}(p_1) = \mathfrak{x}_{i,1}, \qquad i = 1, 2, \cdots \nu_1 - r_1$$

$$\psi^*_{\nu_1-r_1+1,1}(p_1) = a^{(\nu_1-r_1+1)}_{\nu_1-r_1+1,1}\,\mathfrak{x}_{\nu_1-r_1+1,1} + \cdots + a^{(\nu_1-r_1+1)}_{\nu_1,1}\,\mathfrak{x}_{\nu_1,1}$$

$$\psi^*_{j,2}(p_2) = \mathfrak{x}_{j,2}, \qquad j = 1, 2, \cdots \nu_2 - r_2$$

$$\psi^*_{\nu_2-r_2+1,2}(p_2) = a^{(\nu_2-r_2+1)}_{\nu_2-r_2+1,2}\,\mathfrak{x}_{\nu_2-r_2+1,2} + \cdots + a^{(\nu_2-r_2+1)}_{\nu_2,2}\,\mathfrak{x}_{\nu_2,2}$$

$$\xi_{\sigma_1 p_1}\psi^*_{i,1}(p_1) = \psi_{i,1}(p_1), \qquad i = 1, 2, \cdots \nu_1 - r_1 + 1$$

$$\xi_{\sigma_2 p_2}\psi^*_{j,2}(p_2) = \psi_{j,2}(p_2), \qquad j = 1, 2, \cdots \nu_2 - r_2 + 1.$$

Then we can prove in a similar manner that $\Psi_1 = \{\psi_{1,1}, \cdots \psi_{\nu_1-r_1+1,1}\}$ and $\Psi_2 = \{\psi_{1,2}, \cdots \psi_{\nu_2-r_2+1,2}\}$ form continuous $(\nu_1 - r_1 + 1)$- and $(\nu_2 - r_2 + 1)$-fields on $\partial\sigma_1$ and $\partial\sigma_2$ and are respectively continuously deformable from the fields $\Phi_1 = \{\varphi_{1,1}, \cdots \varphi_{\nu_1-r_1+1,1}\}$ and $\Phi_2 = \{\varphi_{1,2}, \cdots \varphi_{\nu_2-r_2+1,2}\}$.

Put[8] $$\text{Char.}_{\partial\sigma_k}\Psi_k = d(\sigma_k), \qquad k = 1, 2$$

and $$\text{Char.}_{\partial\sigma}\Psi = d(\sigma),$$

then, from what has proved above, we have

$$\text{Char.}_{\partial\sigma_k}\Phi_k = d(\sigma_k), \qquad k = 1, 2$$

$$\text{Char.}_{\partial\sigma}\Phi = d(\sigma).$$

The vectors $\psi^*_{\nu_1-r_1+1,1}(p_1)$, $\psi^*_{\nu_2-r_2+1,2}(p_2)$ and $\psi^*_{\nu_1-r_1+1}$ now define respectively a map f_1 of $\partial\sigma_1$ in an $(r_1 - 1)$-dimensional sphere S_1, a map f_2 of $\partial\sigma_2$ in an $(r_2 - 1)$-dimensional sphere S_2, and a map f of $\partial\sigma$ in the join of S_1 and S_2, of which the degrees are respectively say d_1, d_2 and d. As these maps are connected by the relation

$$f(t_1 p_1 \times t_2 p_2) = t_1 f_1(p_1) \times t_2 f_2(p_2)$$

it follows from paragraph 7,

$$d = d_1 d_2.$$

Since ψ^*_i, $\psi^*_{i,1}$, $\psi^*_{j,2}$ are constant vectors for $i \neq \nu_1 - r_1 + 1$, $j \neq \nu_2 - r_2 + 1$, d, d_1 and d_2 are respectively equal, or congruent mod 2, to the characteristics $d(\sigma)$, $d(\sigma_1)$ and $d(\sigma_2)$. Hence

$$d(\sigma) \equiv d(\sigma_1)\cdot d(\sigma_2) \mod 2. \tag{8.1}$$

Next consider a $(\nu - m + 1)$-cell $\sigma = \sigma_1^{r_1} \times \sigma_2^{r_2}$ of K where $r_2 > \nu_2$, $r_1 > 0$. We must then have $\nu_1 - r_1 + 1 > m$. The field $\{\varphi_{1,1}, \cdots \varphi_{m,1}\}$ is thus contin-

[8] See note 4).

uous at every point of $\sigma_1^{r_1}$. Using the same reasoning as before, we see that the given field $\{\varphi_1, \cdots \varphi_m\}$ is deformable to a second one $\{\psi_1, \cdots \psi_m\}$ so that

$$\xi_{\sigma, t_1 p_1 \times t_2 p_2}^{-1} \psi_i(t_1 p_1 \times t_2 p_2) = \mathfrak{x}_{i,1}, \qquad i = 1, 2, \cdots m.$$

Hence in this case

$$d(\sigma) = 0. \tag{8.2}$$

The case $r_1 > \nu_1$ is similar.

For the last case where $r_1 = 0$, $r_2 \leqq \nu_2$ or $r_2 = 0$, $r_1 \leqq \nu_1$ we can prove in the same way that

$$d(\sigma) \equiv d(\sigma_2) \bmod 2 \tag{8.3}$$

respectively $\qquad d(\sigma) \equiv d(\sigma_1) \bmod 2$

where $d(\sigma_i)$ is the characteristic of the field $\{\varphi_{1,i}, \cdots \varphi_{\nu_i - r_i + 1, i}\}$ on $\partial\sigma_i$, and $d(\sigma)$ is that of the field $\{\varphi_1, \cdots \varphi_m\}$ on $\partial\sigma$.

9. The proof of Theorem I is now immediate. The canonical fields on K_1, K_2 constructed in the preceding sections give chains

$$\begin{aligned} w_1^{r_1} &= \sum_k d(\sigma_{k,1}^{r_1}) \cdot \sigma_{k,1}^{r_1}, \qquad r_1 = 0, 1, \cdots \nu_1 \\ w_2^{r_2} &= \sum_k d(\sigma_{k,2}^{r_2}) \cdot \sigma_{k,2}^{r_2}, \qquad r_2 = 0, 1, \cdots \nu_2 \end{aligned} \tag{9.1}$$

in which $d(\sigma_{k,i}^{r_i})$ is the characteristic of the field $\{\varphi_{1,i}, \cdots \varphi_{\nu_i - r_i + 1, i}\}$ on $\partial\sigma_{k,i}^{r_i}$, $\sigma_{k,i}^{r_i} \in K_i$. Also the m-field $\{\varphi_1, \cdots \varphi_m\}$ defines a chain

$$w^r = \sum_k d(\sigma_k^r) \cdot \sigma_k^r, \qquad r = \nu - m + 1. \tag{9.2}$$

When reduced mod 2 if necessary, these chains give the characteristic cocyles of the respective bundles.

According to (8.1), (8.2) and (8.3), the coefficients of these chains are connected by the following relations:

$$\begin{aligned} d(\sigma_{j,1}^{r_1} \times \sigma_{k,2}^{r_2}) &\equiv d(\sigma_{j,1}^{r_1}) \cdot d(\sigma_{k,2}^{r_2}) \bmod 2, \qquad \text{for } r_1 \leqq \nu_1 \text{ and } r_2 \leqq \nu_2 \\ d(\sigma_{j,1}^{r_1} \times \sigma_{k,2}^{r_2}) &= 0, \qquad \text{for } r_1 > \nu_1, r_2 > 0 \quad \text{or} \quad r_2 > \nu_2, r_1 > 0 \\ d(\sigma_{j,1}^{0} \times \sigma_{k,2}^{i_2}) &\equiv d(\sigma_{k,2}^{r_2}) \bmod 2, \qquad \text{for } r_2 \leqq \nu_2 \\ d(\sigma_{j,1}^{r_1} \times \sigma_{k,2}^{0}) &\equiv d(\sigma_{j,2}^{r_1}) \bmod 2, \qquad \text{for } r_1 \leqq \nu_1. \end{aligned} \tag{9.3}$$

Let us put

$$w_i^{r_i} = \sum_k d(\sigma_{k,i}^{r_i}) \cdot \sigma_{k,i}^{r_i} = 0 \quad \text{for} \quad r_i > \nu_i, i = 1, 2$$

and

$$w_i^0 = \sum_k d(\sigma_{k,i}^0) \cdot \sigma_{k,i}^0 = \sum_k \sigma_{k,i}^0, \qquad i = 1, 2.$$

This is in agreement with the convention made in the end of paragraph 2. Then the equations (9.3) can be mingled into a single one:

$$d(\sigma_{j,1}^{r_1} \times \sigma_{k,2}^{r_2}) \equiv d(\sigma_{j,1}^{r_1}) \cdot d(\sigma_{k,2}^{r_2}) \bmod 2.$$

It follows therefore from (9.1) and (9.2) that

$$w^r = \sum_{1=0}^{r} w_1^i \times w_2^{r-i} \qquad \bmod 2$$

i.e.,

$$W^r = \sum_{1=0}^{r} W_1^i \times W_2^{r-i}.$$

§3.

10. This section will be devoted to a proof of Whitney's duality theorem mod 2. In preparing for the proof we shall make a few remarks.

Our first remark is concerned with the definition of the cup product in a complex K by means of our product defined in paragraph 4. This method of introducing the cup product is due to Lefschetz,[9] but we shall summarize for our purpose the main result in a simplified form.

Let β, γ be two cohomology classes of K. Then $\beta \times \gamma$ is a cohomology class of $K \times K$. The diagonal mapping

$$d: K \rightarrow K \times K,$$

defined by

$$d \cdot x \rightarrow x \times x, \quad x \in |K|,$$

induces a chain transformation of K into $K \times K$ and hence a homomorphism d^* of the cohomology groups of $K \times K$ into those of K. The theorem of Lefschetz asserts that $d^*(\beta \times \gamma) = \beta \cup \gamma$, where the latter is the cup product.

11. Our second remark is related to the notion of an induced sphere or vector bundle. Let a complex L be mapped simplicially into K by f. To a point $q \in L$ we associate the vector space $q \times V(f(q))$. The union of all these vector spaces can be made in a natural way to a vector bundle over L, the coordinate systems $\eta_{\tau_i q}$, for $q \in \tau_i \in L$, being defined by $\eta_{\tau_i q} = \xi_{f(\tau_i)f(q)}$. We shall call this vector bundle the bundle over L *induced* by the mapping f. It follows immediately from the definition of the induced vector bundle that $f^* W^r$, $r = 0, 1, \cdots$, are the characteristic cohomology classes of L, where W^r are the same of K and f^* the homomorphism of the cohomology groups of K into those of L induced by f.

12. With all the above preparations the proof of Theorem II follows immediately:

Let $\mathfrak{B}_1$, $\mathfrak{B}_2$ be bundles over K, and let $\mathfrak{B}_1 \times \mathfrak{B}_2$ be the product bundle over $K \times K$. Let $d: K \rightarrow K \times K$ be the diagonal map, and $\tilde{\mathfrak{B}}$ the bundle over K

[9] Lefschetz, Algebraic Topology. Amer. Math. Soc. Colloquium publ., 1942, pp. 173–181.

induced by d. It is clear that $\breve{\mathfrak{B}}$ is the span bundle of $\mathfrak{B}_1$ and $\mathfrak{B}_2$. From paragraphs 10 and 11 it follows respectively that

$$d^*(W_1^i \times W_2^{r-i}) = W_1^i \cup W_2^{r-i}$$

and that

$$d^* W^r = \breve{W}^r.$$

Applying the homomorphism d^* to the formula (4.1) we are therefore led to (5.1). This proves Theorem II.

Institute of Mathematics
Academia Sinica
Shanghai, China

Reprinted with permission from *C. R. Acad. Sci. Paris*, 230 (1950) 508–511.

TOPOLOGIE ALGÉBRIQUE. — *Classes caractéristiques et i-carrés d'une variété.*

Note (*) de M. **Wu Wen-tsun**, présentée par M. Élie Cartan.

1. Soit M un espace topologique vérifiant les conditions suivantes :

a. Le groupe $H^n(M)$ (1) est de rang 1 dont la base est X_1^n;

b. On a $H^p(M) \approx \mathrm{Hom}[H^{n-p}(M), Z_2]$ dont l'isomorphisme est établi par le cup produit $X^p(Y^{n-p})X_1^n = X^p \cup Y^{n-p}$, $X^p \in H^p(M)$, $Y^{n-p} \in H^{n-p}(M)$.

Pa exemple, une variété compacte de dimension n est un tel espace. Dans un tel espace on peut définir un système de classes $U^p \in H^p(M)$, $0 \leqq 2p \leqq n$, par les équations suivantes :

$$U^p \cup Y^{n-p} = Sq^p Y^{n-p} \quad (2) \quad [\text{pour } Y^{n-p} \text{ quelconque de } H^{n-p}(M)]. \tag{1}$$

Nous les appellerons les *classes canoniques* ou les U-*classes* de M. Les classes W^i, $0 \leqq i \leqq n$, définies par

$$W^i = \sum_p Sq^{i-p} U^p \tag{2}$$

seront alors appelées les *classes caractéristiques* ou les W-*classes* de M. On a, par exemple, $W^0 = U^0 = 1$, $W^1 = U^1$, $W^2 = U^2 + U^1 \cup U^1$, etc.

Le nom des classes caractéristiques est justifié par le théorème suivant :

Théorème. — *Pour une variété compacte* M *les* W-*classes ainsi définies s'identifient aux classes caractéristiques de Stiefel-Whitney de cette variété.*

2. La démonstration de ce théorème s'appuie sur un théorème de Thom (3), et le lemme suivant, démontré par H. Cartan (3) :

Lemme. — *Dans un espace-produit* $M \times M'$ *on a*

$$Sq^i(X \otimes Y) = \sum_j Sq^j X \otimes Sq^{i-j} Y, \qquad X \in H^*(M), \qquad Y \in H^*(M').$$

(*) Séance du 30 janvier 1950.

(1) $H^*(M)$ [$H^p(M)$] = le groupe de cohomologie (de dimension p) de l'espace M. Le groupe des coefficients sera exclusivement le groupe Z_2 des entiers mod 2 sauf, mention du contraire. La classe unité de $H^0(M)$ est désignée par 1.

(2) Nous adoptons ici la nouvelle notation de Steenrod pour les *i-carrés* : $Sq^p X^q = Sq_{q-p} X^q$. *Cf.* Steenrod, *Annals of Math.*, **48**, 1947, p. 290-319.

(3) *Voir* la Note précédente de Thom sur les *variétés plongées* et *i-carrés* (même numéro des *Comptes rendus*) et la Note précédente de H. Cartan sur *une théorie axiomatique des i-carrés* (*Comptes rendus*, **230**, 1950, p. 425).

On en déduit que, dans un même espace M, *on a*

$$Sq^i(X \cup Y) = \sum_j Sq^j X \cup Sq^{i-j} Y, \qquad X, Y \in H^*(M).$$

Prenons maintenant une base $\{X^p_\alpha\}$ de $H^*(M)$ dans la variété M, supposée de dimension n, telle que $X^p_\alpha \cup X^{n-p}_\beta = \delta_{\alpha\beta} X^n_1$. La classe $\Delta^n \in H^n(M \times M)$ correspondant à la diagonale de l'espace-produit $M \times M$ s'exprime alors par $\Delta^n = \sum_{\alpha,p} (X^p_\alpha \otimes X^{n-p}_\alpha)$. D'après le lemme précédent, on a donc

$$Sq^i \Delta^n = \sum_{\alpha,p,j} Sq^{i-j} X^p_\alpha \otimes Sq^j X^{n-p}_\alpha.$$

D'autre part, soit $\sum_\alpha a^i_\alpha X^i_\alpha$ la classe caractéristique de Stiefel-Whitney de dimension i de la variété M qui est aussi la classe de Stiefel-Whitney de la structure normale de M par rapport à $M \times M$. On a, d'après la formule (6) de Thom ([3]), $Sq^i \Delta^n = \sum_\alpha \psi^* a^i_\alpha X^i_\alpha$, où ψ^* applique $H^i(M)$ dans $H^{n+i}(M \times M)$. On en déduit

$$Sq^i \Delta^n = \sum_{\alpha,\mu,q} a^i_\alpha X^q_\mu \otimes (X^i_\alpha \cup X^{n-q}_\mu).$$

En considérant les termes de la forme $X^i_\mu \otimes X^n_1$ dans les deux expressions de $Sq^i \Delta^n$, on trouve que

$$\sum_\mu a^i_\mu X^i_\mu = \sum_p Sq^{i-p} U^p,$$

c'est-à-dire la classe de Stiefel-Whitney de dimension i coïncide avec la classe W^i, définie par (1) et (2). C. Q. F. D.

3. Le théorème précédent montre que les classes caractéristiques de Stiefel-Whitney d'une variété compacte de dimension n sont complètement déterminées par les classes canoniques U^p, $0 \leq 2p \leq n$, et par conséquent par la structure des cup produits et les i-carrés de cette variété. On peut en déduire d'autres propriétés concernant les classes de Stiefel-Whitney, ainsi qu'il suit ([4]):

a. Les classes W^i *pour* $2i > n$ *sont complètement déterminées par les classes* W^i *pour* $0 \leq 2i \leq n$, *et par les opérations de carrés.*

b. $W^n = 0$ pour n impair; $W^n = Sq^k U^k = U^k \cup U^k$ pour $n = 2k$ pair; $W^2 = W^1 \cup W^1$ pour $n = 3$; $W^1 \cup W^1 \cup W^2 = 0$ pour $n = 4$ (on peut même démontrer $W^1 \cup W^2 = 0$ pour $n \leq 5$).

c. Pour M *orientable* et $n = 2k$ pair, U^k est une classe de première espèce, c'est-à-dire, U^k est déduite d'une classe aux coefficients entiers par réduction

---

([4]) *Cf.* H. Whitney, *Michigan Lectures*, 1941, p. 101-141.

mod 2. Pour $n = 4$ la classe $W^2 = U^2 + U^1 \cup U^1 = U^2$ est alors de première espèce et par conséquent la troisième classe de Stiefel-Whitney (aux coefficients entiers) est nulle; ce n'est pas le cas en général pour $n > 4$, comme le montre la variété orientable de dimension 5 construite de la façon suivante : M^5 est le produit topologique d'un plan projectif complexe P et d'un segment $I = [0, 1]$ avec l'identification $(x, y, z) \times (0) = (\bar{x}, \bar{y}, \bar{z}) \times (1)$, où x, y, z sont des nombres complexes, coordonnées homogènes de P, et $\bar{x}, \bar{y}, \bar{z}$ leurs complexes conjugués.

d. Définissons un autre système de classe U^p (ici $0 \leqq p \leqq n$) par récurrence par les équations $\bar{U}^0 = U^0 = 1$ et $\sum_i \bar{U}^i \cup U^{p-i} = 0$, pour $p > 0$. Les classes $\overline{W}$ définies par $\overline{W}^i = \sum_p Sq^{i-p}\, \bar{U}^p$ $(0 \leqq i \leqq n)$ satisfont alors aux équations $\overline{W}^0 = 1$ et $\sum_i \overline{W}^i \cup W^{p-i} = 0$ pour $p > 0$. Cela veut dire que les classes $\overline{W}^i$ ne sont autres que les classes caractéristiques duales de M introduites par Whitney. On a $\overline{W}^n = \sum_p Sq^{n-p}\, \bar{U}^p = \sum_p U^{n-p} \cup \bar{U}^p = 0$, d'après (1).

e. D'après H. Cartan, $U^p = 0$ pour p impair et M orientable. On en déduit que $W^{n-1} = 0$ pour M orientable et $n = 4k + 2$, ce qui est aussi une conséquence de *c*.

(Extrait des *Comptes rendus des séances de l'Académie des Sciences*, t. 230, p. 508-511, séance du 6 février 1950.)

GAUTHIER-VILLARS, IMPRIMEUR-LIBRAIRE DES COMPTES RENDUS DES SÉANCES DE L'ACADÉMIE DES SCIENCES
135356-50 Paris — Quai des Grands-Augustins, 55.

Reprinted with permission from *C. R. Acad. Sci. Paris*, 230 (1950) 918–920.

TOPOLOGIE ALGÉBRIQUE. — *Les i-carrés dans une variété grassmannienne.*

Note de M. **Wu Wen-tsün**, présentée par M. Élie Cartan.

1. L'anneau de coefficients de l'anneau de cohomologie $H^*(M)$ d'un espace M sera dans ce qui suit exclusivement l'anneau des entiers mod 2.

Soient W^i, $i \geq 0$ quelconque, les W-classes (classes caractéristiques de Stiefel-Whitney) d'une s. f. s. (structure fibrée sphérique) avec la convention $W^0 = 1$ (classe unité de la base), et $W^i = 0$ si $i > m$, $m - 1$ étant la dimension de la fibre sphère. Nous allons démontrer la formule suivante :

$$Sq^r W^s = \sum_t C^t_{s-r+t-1} W^{r-t} W^{s+t} \qquad (s \geq r > 0), \tag{1}$$

où C^q_p = coefficient binomial pour $p \geq q > 0$, $= 0$ pour $p < q > 0$, et $= 1$ pour $p = -1$ et $q = 0$ (tous sont réduits mod 2).

Signalons d'abord quelques conséquences de cette formule : définissons, dans la base, un système de classes U^p ($p \geq 0$ quelconque) par les équations suivantes :

$$W^i = \sum_p Sq^{i-p} U^p, \qquad i \geq 0 \text{ quelconque}; \tag{2}$$

nous les appellerons les *classes canoniques* de la structure considérée. Si la s. f. s. est en particulier la structure tangente associée à une variété différentiable M de dimension m, on voit, en comparant avec les équations (1) et (2) d'une Note précédente [1], que le nom de classes canoniques est justifié; de plus, parmi toutes les s. f. s. (aux fibres S^{m-1}) sur la variété M comme base, la structure tangente de M possède la propriété remarquable suivante :

$$U^p = 0 \qquad \text{pour} \quad 2p > m, \tag{3}$$

De (1) et (3) on déduit :

a. Pour une structure orientable on a $U^{2k+1} = 0$, k quelconque, ce qui généralise un théorème de H. Cartan [1],

b. Pour la structure tangente d'une variété différentiable de dimension m, on a $W^1 W^{m-2} = 0$ si $m = 4k$; $W^1 W^{m-3} = 0$; $W^1 W^{m-1} = 0$ si $m = 4k+1$; $W^m = W^1 W^{m-1}$ si $m = 4k+2$; $W^1 W^{m-1} = 0$, $W^{m-1} = W^1 W^{m-2}$ si $m = 4k+3$.

---

(1) *Comptes rendus*, **230**, 1950, p. 508-511.

2. Soit $G_{n,m}$ la variété grassmannienne des m-éléments linéaires dans un espace euclidien R^{n+m} de dimension $n+m$ passant par l'origine de R^{n+m}. On sait (2) que l'*anneau* $H^*(G_{n,m})$ est engendré par les classes W^i de la s. f. s. $\mathcal{G}_{n,m}$ (fibres S^{m-1}) de base $G_{n,m}$ canoniquement associée à $G_{n,m}$. De plus, comme m'a fait remarquer H. Cartan :

LEMME 1. — *Soit* $\varphi_p(W^i)$ *un polynôme non identiquement nul en* $W^1, \ldots W^m$ *tel que pour chaque terme* $W^{i_1} \ldots W^{i_k}$ *de ce polynôme on ait* $i_1 + \ldots + i_k = p \leqq n$. *Alors* $\varphi_p(W^i)$ *est un élément non nul de* $H^*(G_{n,m})$.

Supposons alors que R^{n+m} soit le produit de deux espaces euclidiens $R_j^{n_j+m}$ de dimension $n_j + m_j$ ($j = 1, 2$). Soient G_{n_j,m_j} ($j = 1, 2$) les variétés grassmanniennes définies respectivement dans $R_j^{n_j+m_j}$. Pour $X_j \in G_{n_j,m_j}$ soit $X \in G_{n,m}$ le joint de X_1 et X_2, on a alors une application canonique

$$f : \quad G_{n_1,m_1} \times G_{n_2,m_2} \to G_{n,m}$$

définie par $f(X_1 \times X_2) = X$. En désignant par W_j^i ($j = 1, 2$) respectivement les W-classes des structures $\mathcal{G}_{n_j,m_j}$, on a :

LEMME 2. — *Le type d'homologie mod 2 de* f *est déterminé par* (3) :

$$f^* W^i = \sum_j W_1^j \otimes W_2^{i-j} \qquad (i \geqq 0 \text{ quelconque})$$

Comme conséquence des lemmes 1 et 2, en conservant les notations, on a :

LEMME 3. — *Pour* $p \leqq n_1$ *et* n_2, $\varphi_p(W^i)$ *est un élément non nul de* $H^*(G_{n,m})$ *si et seulement si* $f^* \varphi_p(W^i)$ *est un élément non nul de* $H^*(G_{n_1,m_1} \times G_{n_2,m_2})$.

3. *Démonstration de* (1). — Nous poserons

$$\varphi_{r,s}(W^i) = Sq^r W^s + \sum_t C^t_{s-r+t-1} W^{r-t} W^{s+t}.$$

La formule (1), ou, ce qui revient au même, la formule $\varphi_{r,s}(W_j^i) = 0$, étant évidente pour $m = 1$, nous supposerons par induction qu'elle est exacte pour les structures dont les fibres sphères ont une dimension $< m-1$, où $m > 1$. Soient maintenant W^i, W_j^i respectivement les W-classes des structures $\mathcal{G}_{n,m}$ et $\mathcal{G}_{n_j,m_j}$ ($j = 1, 2$) où $n = n_1 + n_2$, $n_j \geqq r + s$, $m_1 = m - 1$, $m_2 = 1$. De la formule $f^* Sq^i = Sq^i f^*$, d'un théorème de H. Cartan (4), et du lemme 2 du

---

(2) S. CHERN, *Annals of Math.*, **49**, 1948, p. 362-372.

(3) Nous remarquons que le théorème de Whitney sur le produit de deux structures fibrées sphériques est une conséquence de ce lemme dont la démonstration est donnée dans ma Thèse, Strasbourg, 1949.

(4) *Comptes rendus*, **230**, 1950, p. 425-427.

paragraphe 2, on déduit

$$f^\star \varphi_{r,s}(W^i) = \varphi_{r,s}(W_1^2) \otimes 1 + \varphi_{r,s-1}(W_1^i) \otimes W_2^1 + \varphi_{r-1,s-1}(W_1^i) \otimes (W_2^1)^2.$$

D'après l'hypothèse d'induction on a donc $f^\star \varphi_{r,s}(W^i) = 0$ et par conséquent $\varphi_{r,s}(W^i) = 0$ d'après le lemme 3. La structure $\mathcal{G}_{n,m}$ étant universelle pour n assez grand, on a $\varphi_{r,s}(W^i) = 0$ pour une s. f. s. quelconque. La formule (1) est ainsi démontrée par induction.

Soient en particulier W^i les W-classes de la structure $\mathcal{G}_{n,m}$ sur la base $G_{n,m}$. L'anneau $H^\star(G_{n,m})$ étant engendré par les classes W^i, on voit que la formule (1) détermine complètement les i-carrés dans $G_{n,m}$ en les exprimant comme des polynômes en W^i.

(Extrait des *Comptes rendus des séances de l'Académie des Sciences*, t. **230**, p. 918-920, séance du 6 mars 1950.)

GAUTHIER-VILLARS, IMPRIMEUR-LIBRAIRE DES COMPTES RENDUS DES SÉANCES DE L'ACADÉMIE DES SCIENCES.
135687-50 Paris. — Quai des Grands-Augustins, 55.

SCIENTIA SINICA

Vol. VII, No. 3, 1958

MATHEMATICS

ON THE REALIZATION OF COMPLEXES IN EUCLIDEAN SPACES I*

WU WEN-TSÜN (吳文俊)

(Institute of Mathematics, Academia Sinica)

ABSTRACT

It was early known that any n-dimensional abstract complex may be realized in a $(2n+1)$-dimensional euclidean space R^{2n+1}. From this theorem, whose proof is quite simple, it follows that the $(2n+1)$-dimensional euclidean space contains in reality all imaginable n-dimensional complexes. However, the complete recognization of all n-dimensional complexes in an euclidean space of a given dimension m where $m<2n+1$, is a problem much more difficult which cannot, it seems, be solved completely in the near future. Among the miscellaneous results so far obtained along this line the most remarkable one is no doubt that of Van Kampen[3,4] and Flores[5], who first proved the existence of n-dimensional complexes which, even under further subdivisions, cannot be realized in an R^{2n}.

The invariant by means of which Van Kampen was able to conclude the non-realizability of a (finite simplicial) n-dimensional complex in an R^{2n} may be described as follows. Denote the k-dimensional simplexes of the given n-dimensional complex K by S_i^k. Any two simplexes of K with no vertices in common will be said to be disjoint. Let A be the set of all unordered index pairs (i, j), corresponding to pairs of disjoint n-dimensional simplexes S_i^n and S_j^n. Construct a vector space $\mathfrak{L}$ on the ring of integers with dimension equal to the number of elements in A. Any vector of $\mathfrak{L}$ may then be represented by a system of integers (α_{ij}) where $(i, j) \in A$. To each pair of disjoint simplexes S_a^{n-1} and S_l^n in K a certain vector $V_{la}=(\alpha_{ij})$ of $\mathfrak{L}$ may be determined in the following manner. If both $i, j \neq l$ or one of them, say $j=l$, but S_a^{n-1} is not a face of S_i^n, then we put $\alpha_{ij}=0$. Otherwise we put $\alpha_{il}=\pm 1$ (with sign conveniently chosen). Two vectors P, P' of $\mathfrak{L}$ will then be said to be equivalent, if $P-P'$ is a certain linear combination with integral coefficients of

---

*First published in Chinese in *Acta Mathematica Sinica*, Vol. V, No. 4, pp. 505—552, 1955.

vectors of form V_{la} above defined. The vectors of $\mathfrak{L}$ are thus distributed in such equivalence classes.

Take now an arbitrary simplicial subdivision K_1 of K and try to realize K_1 in R^{2n} as far as possible. We shall obtain then some "almost true" realization such that parts $'S_i^k$ and $'S_j^l$, corresponding to disjoint S_i^k and S_j^l of K will be disjoint in R^{2n} when $k+l<2n$, while they intersect only in isolated points when $k=l=n$. With respect to an orientation arbitrarily chosen of R^{2n}, $'S_i^n$ and $'S_j^n$ determine then a definite intersection number $\pm\alpha_{ij}$ (with sign conveniently chosen). These numbers determine in turn a vector $P=(\alpha_{ij})$ of $\mathfrak{L}$. Van Kampen's work shows that, whatever be the subdivision K_1 of K and the "almost true" realization of K_1 in R^{2n}, the corresponding vectors P always belong to one and the same equivalence class in $\mathfrak{L}$. It follows that this equivalence class is an invariant of the complex K. It is evident that the belonging of the zero vector to this invariant equivalence class is a necessary condition for the existence of "true" realization of K in R^{2n}. It is this invariant which has enabled Van Kampen to assure the existence of n-dimensional complexes non-realizable in R^{2n}. On the other hand, Van Kampen failed to ascertain whether the above necessary condition is also sufficient and the problem of characterizing n-dimensional complexes in R^{2n} remains unsettled up to the present. Moreover the method of Van Kampen-Flores cannot be seen to be readily generalizable to the realizability in R^m, m being arbitrary. We remark also that whether Van Kampen's invariant is a "topological" invariant of the space of K, or even whether it is a combinatorial invariant of K_1, cannot be decided from his works.

At the time of Van Kampen and Flores the cohomology theory has not yet been created. To get a deeper insight of their results we will reformulate them in the modern terminology of cohomology. Their statements will then become clear and natural as follows. From the given simplicial complex (of any dimension) let us construct a sub-complex $\tilde{K}^*$ of $K\times K$, consisting of all cells $\sigma\times\tau$ such that σ, τ are disjoint in K. Identify each pair $\sigma\times\tau$ and $\tau\times\sigma$ of $\tilde{K}^*$ to the same cell $\sigma*\tau=\tau*\sigma$, we get a cell complex K^*. Suppose that the cells $\sigma\in K$ are oriented and let us orient the cells $\sigma*\tau$ of K^* as $\sigma\times\tau$ in the product complex $K\times K$, such that

$$\sigma*\tau=(-1)^{\dim\sigma\dim\tau}\,\tau*\sigma. \tag{1}$$

Then for dim $K=n$, any vector P in $\mathfrak{L}$ may be regarded as an integral $2n$-dimensional cocycle of K^*, and the equivalence of vectors in $\mathfrak{L}$ is the same as the cohomologousness of their corresponding cocycles. It follows that Van Kampen's invariant is essentially an integral cohomology class in K^*.

From this reformulation we may naturally extend Van Kampen's method to the realization problem of complexes of arbitrary dimension in euclidean space of arbitrary dimension m. For this let us take an arbitrary simplicial subdivision K_1 of K and try to realize K_1 as much as possible in R^m, such that any two simplexes of K_1 are in general position whenever possible. Let the chain in R^m thus obtained, corresponding to any $\sigma \in K$, be denoted by σ'. Then, with respect to a fixed orientation of R^m, to any two disjoint simplexes σ, τ in K with sum of dimensions just equal to m, there corresponds a definite intersection number $\emptyset(\sigma', \tau')$. Let $I_{(m)}$ be either the additive group of integers I or the group of integers mod 2 I_2, depending on m, and $\rho_{(m)}$ the corresponding identity or reduction mod 2. Then an m-dimensional cochain $\varphi \in C^m(K^*, I_{(m)})$ may be defined by

$$\varphi(\sigma * \tau) = \varepsilon_r \rho_{(m)} \emptyset(\sigma', \tau'), \text{ dim } \sigma = r, \text{ dim } \tau = m - r, \tag{2}$$

where $\varepsilon_r = +1$ or -1, depending on r. To make φ a cocycle on coefficient group $I_{(m)}$ and to make the definition of φ consistent with (1) we should take ε_r such that

$$\rho_{(m)} \varepsilon_r + \rho_{(m)} \varepsilon_{r+1} = 0,$$

$$\rho_{(m)} \varepsilon_r = \rho_{(m)} \varepsilon_{m-r}.$$

If we make the further restriction that $\varepsilon_0 = +1$, then to make the above equations consistent, we may take

$$I_{(m)} = \begin{cases} I, & \text{when } m = \text{even}, \\ I_2, & \text{when } m = \text{odd} \end{cases}$$

and to choose ε_r to be, say $(-1)^r$. We thus obtain an integral cocycle φ in the case. that m be even, while only a mod 2 cocycle φ in the contrary case Just as in the special case considered by Van Kampen, it turns out that these cocycles, whatever the subdivision K_1 and the "almost true'' realization may be, always belong to one and the same cohomology class $\Phi^m \in H^m(K^*, I_{(m)})$. Moreover, it may be shown that so far as $m>1$, any cocycle in Φ^m may be realized as one arisen from some subdivision K_1 of K and some almost true realization of K_1 in R^m. However, this is not true for $m=1$, as seen from very simple examples.

The series of classes $\Phi^m \in H^m(K^*, I_{(m)})$ will be called in the present work the *imbedding classes* of K. The vanishing of Φ^m is evidently a necessary condition for the realizability of K in R^m. We have $2\Phi^m=0$, when m is even; but in general Φ^m are nontrivial and thus they serve as effective tools in the study of realization problems.

We remark that we may define, just as in (2), with respect to any simplicial subdivision K_1 of K and any "almost true" realization of K_1 in R^m, a certain integral m-dimensional cochain $\tilde{\varphi}$ in $\tilde{K}^*$ by

$$\tilde{\varphi}(\sigma \times \tau) = (-1)^{\dim \sigma} \varnothing(\sigma', \tau'), \ \sigma \times \tau \in \tilde{K}^*, \ \dim \sigma + \dim \tau = m.$$

It is true that $\tilde{\varphi}$ is always an *integral* cocycle and its class $\tilde{\Phi}^m$ is uniquely determined. However, it turns out that $\tilde{\Phi}^m$ is always 0 (what is not easily seen from the definition itself) and therefore the complex $\tilde{K}^*$ is not so useful as K^*, so far as the realization problem is concerned.

Let $R^1 \subset R^2 \subset \cdots \subset R^m$ be a sequence of linear subspaces of increasing dimensionality in R^m. By trying to realize the complex K in a certain canonical manner such that $K^0 \subset R^1$, $K^1 \subset R^3$, etc., representative cocycles in Φ^m may be explicitly constructed. This not only gives the means to compute effectively these classes in every concrete case, but also makes it possible to derive a series of properties of Φ^m which are not easy to foresee, e.g., $\frac{1}{2}\delta\Phi^{2m-1}=\Phi^{2m}$, $\Phi^i \cup \Phi^j = \Phi^{i+j}$ mod 2. This also enables us to determine, for some particular complexes generalized from those of Van Kampen, exactly the lowest dimension of R^m in which they may be realized. It seems that this cannot be done with any other known methods.

At last we should point out that the realization problem is in reality "topological", but not "homotopic" in character. For example, a segment and a triangle have the same homotopy type, but the former may be realized in R^1 while the latter cannot. It follows that the problem cannot be completely solved without the aid of topological invariants which are in general not invariants of homotopy types. In a previous paper[6] the author has described a general method of constructing such invariants. The above-mentioned groups $H^m(K^*, G)$ (and $H^m(\tilde{K}^m, G)$) are particular cases of these invariants and we may thus legitimitely write $H^{m,2}(K, G)$ or $H^{m,2}(P, G)$ instead of $H^m(K^*, G)$ where $P=\bar{K}$ is the space of K. Similarly we write $H^m(\tilde{K}^*, G) = \tilde{H}^{m,2}(K, G) = \tilde{H}^{m,2}(P, G)$. Based on [6] we may prove that $\Phi^m \in H^{m,2}(K, I_{(m)}) \approx H^{m,2}(P, I_{(m)})$ are not only combinatorial invariants of K but also topological invariants of P, an important point completely disregarded by Van Kampen in the special case studied by him. On the other hand, Φ^m are not invariants of homotopy type of P. It seems that this is the very reason for the successfulness of methods, originated from Van Kampen and developed here.

We restrict ourselves in the present paper to give a basis of the whole theory and leave to later considerations the study of relations of the imbedding classes with Steenrod squares and also with Stiefel-Whitney classes in the case of a manifold. We leave also to a later occasion the proof of the sufficiency of our condition for the realizability in certain extreme cases.

§1. Linear Realization of Complexes in Euclidean Spaces

In what follows, K will be a finite euclidean simplicial complex,[1] and R^m a euclidean space of dimension m.

Suppose given in R^m a euclidean simplicial complex K', which is isomorphic to K under the correspondence $T : K \to K'$, then we shall say that $K' = TK$ is a *linear realization* of K in R^m. Denote the topological map induced by T of the spaces $\bar{K}$, $\bar{K}'$ of K, K' by $\bar{T} : \bar{K} \equiv \bar{K}'$, then $\bar{T}$ or T will be called a *linear imbedding* of K in R^m.

It is known that any abstract simplicial complex of dimension r may be realized as the associated abstract complex of a euclidean simplicial complex in R^{2r+1} of dimension $2r+1$, but not necessarily so in euclidean spaces of lower dimension[3–5]. From this we may draw two conclusions. First, the problem of existence of euclidean complex in an R^m associated with a given abstract simplicial complex, is equivalent to the problem of linear realizability of euclidean simplicial complexes in R^m. For that reason, whenever we speak of complexes, we mean euclidean complexes in a euclidean space of sufficiently high dimension, and a subdivision will always mean a euclidean subdivision. Secondly, a euclidean complex K in general has no linear realization in R^m if $m < 2 \dim K + 1$. To study this problem, we shall recapitulate and introduce some concepts as follows.

Let σ^r, τ^s be two euclidean simplexes in R^m, of dimensions r and s respectively. If for any r'-dimensional face σ' of σ^r and any s'-dimensional face τ' of τ, the linear subspace determined by σ' and τ' has a dimension $\min(r'+s'+1, m)$, or in other words, if any $r'+1$ vertices of σ^r and $s'+1$ vertices of τ^s are linearly independent so far as $r'+s'+1 \leqslant m$, then σ^r, σ^s are said to be *in general position.*

Suppose given in R^m a set of points $v_1', \cdots, v_n'$ and a set of geometric simplexes[2] $\sigma_1', \cdots, \sigma_s'$ spanned by these points of which the totality K' satisfies the following conditions: 1°. If σ_i' is spanned by $v_{i_0}', \cdots, v_{i_k}'$ and $k \leqq m$, then $v_{i_0}', \cdots, v_{i_k}'$ are linearly independent so that σ_i' may be considered as a euclidean simplex. 2°. If $\sigma_i' \in K'$, then any face of σ_i' is in K' too. 3°. If σ_i', $\sigma_j' \in K'$ are both euclidean simplexes and have no vertices in common, then σ_i', σ_j' are in general position. In such case we shall say that K' is an *almost euclidean simplicial complex* in R^m, and $\bar{K}' = \sum_i \bar{\sigma}_i'$ is defined as the *space* of K'.

---

1) We consider only finite complexes in this work, so that the modifier "finite" will be omitted throughout.

2) For the definition of geometric simplex cf. [2], pp. 607. The geometric simplex spanned by $a_0, \cdots, a_r$ of R^m will be denoted by $(a_0, \cdots, a_r)$.

Suppose that to a euclidean simplicial K we have in R^m an almost euclidean simplicial complex K' isomorphic to K, i.e., K, K' have same number of vertices v_i, v_i', $i = 1, \cdots, n$ and the 1–1–correspondence $v_i \leftrightarrow v_i'$ between these vertices is such that $(v_{i_0} \cdots v_{i_r}) \in K$ is equivalent to $(v'_{i_0}, \cdots, v'_{i_r}) \in K'$. Let the induced correspondence be $T : K \to K'$, then we will define $K' = TK$ as an *almost linear realization* of K in R^m. The continuous map $\bar{T} : \bar{K} \to \bar{K}'$ induced by T and also T itself will then be called an *almost linear imbedding* of K in R^m.

Evidently a linear realization (or linear imbedding) of K in R^m is also an almost linear realization (or almost linear imbedding) of K in R^m, but the converse is not true. It is easy to see that K has almost linear realizations in R^m of arbitrary dimension m, though it has linear realizations only in R^m of dimension m sufficiently high. We shall introduce in what follows some invariants of K through its almost linear realizations in R^m with the aim to study the linear realizability of K in R^m.

Since a complex is equivalent to its subdivisions from the point of view of combinatorial topology, we shall introduce the following concepts.[1)]

Suppose given a simplicial subdivision K_1 of K and a linear (or almost linear) realization $K_1' = TK_1$ of K_1 in R^m, then we shall say that K_1' is a *semi-linear* (or *almost semi-linear*) *realization* of K in R^m *through its subdivision* K_1, and T or the induced topological map $\bar{T} : \bar{K} \equiv \bar{K}_1'$ (or continuous map $\bar{T} : \bar{K} \to \bar{K}_1'$) will be defined as a *semi-linear* (or *almost semi-linear*) *imbedding* of K in R^m *through its subdivision* K_1.

Let K be an almost euclidean simplicial complex in R^m and o be a point of R^m. The $(r + 1)$-dimensional geometric simplex $o\sigma$ spanned by o and any r-dimensional geometric simplex σ of K will be called the central projection of σ from o. The totality of all such simplexes $o\sigma$ and the simplexes of K form a simplicial complex, called the *central projection* of K from o and denoted by oK. In general, oK is not an almost euclidean complex even if K be so. However, we have the following

(A) **Lemma.** If K is an almost euclidean simplicial complex in R^m and L a subcomplex of K, then there exist points o in R^m such that $oL + K$ is an almost euclidean complex. Moreover, such points o may be chosen in any neighbourhood of any point o' of R^m.

Proof. Consider any pair of simplexes σ^r, τ^s in K having no vertices in common, for which the sum of dimensions $r+s$ is $\leqslant m-2$.

---

1) From some examples given by Cairns and Van Kampen[7, 8], it may be seen that the problem of realization would be topologically meaningless by considering only the original complex without introducing further subdivisions.

Since K is almost euclidean, the linear subspace $P(\sigma, \tau)$ determined by σ, τ has a dimension $r+s+1 \leqslant m-1$. Again the linear subspace $P(\sigma)$ determined by any simplex σ of dimension $r \leqslant m-1$ in K has a dimension $r \leqslant m-1$. Hence in any neighbourhood of o' there exist points not belonging to any of such linear subspaces $P(\sigma, \tau)$ or $P(\sigma)$. Evidently any such point may be chosen to be a point o as required in the Lemma.

(B) **Lemma.** Let K be a euclidean simplicial complex, L a subcomplex of K, and L_* the subcomplex formed of all simplexes of L which has no vertex in common with any simplex of $K-L$. Let $f: \bar{K} \rightarrow R^m$ be a continuous map such that $f/\bar{L}$ is an almost linear imbedding of L in R^m, i.e., $f(\bar{L})$ is the space of an almost euclidean complex L' in R^m : $\bar{L}' = f(\bar{L})$ and L' is isomorphic to L under the map f. Then for any $\varepsilon > 0$, there is an ε-approximation $\bar{\bar{T}} : \bar{K} \rightarrow R^m$ of f such that $\bar{\bar{T}}$ is an almost semi-linear imbedding of K in R^m through a subdivision K_1, $\bar{\bar{T}}/\bar{L} \equiv f$, and K_1 has a subdivision L_1 of L as a subcomplex which coincides with L_* on $\bar{L}_*$.

Proof. Since K is finite, we have $\delta > 0$ such that for any two points $x, y \in \bar{K}$, $\rho(x, y) < \delta$ would imply $\rho(f(x), f(y)) < \varepsilon/5$. Take now a simplicial subdivision K_0 of K such that L_* is a subcomplex of K_0 and any simplex of K_0 on $\overline{K_0 - L}$ has a diameter $< \delta$. The part of K_0 on $\bar{L}$ is a subdivision of L which will be denoted by L_0. Let K_1 be the subdivision of K_0 obtained by constructing central subdivisions[1] of simplexes of $K_0 - L_*$, the centre of $\sigma \in K_0$ being o_σ. The part of K_1 on $\bar{L}$ will be denoted by L_1. Under f, L_0 and L_1 will correspond respectively to a simplicial subdivision L_0' of L' and its central subdivision L_1'. By convenient choice of K_0 and centres o_σ, we may make L_0' and L_1' the almost euclidean complexes which will be supposed to be so. Arrange now the simplexes in K_0 but not in L_0 in an order $\sigma_1 < \cdots < \sigma_n$, such that those of lower dimension will precede those of higher dimension, but otherwise arbitrary. By (A), we may take successively points $o_1', \cdots, o_n'$ to satisfy the following conditions:

1°. o_i' is in the $\varepsilon/5$ neighbourhood of $f(o_{\sigma_i})$.

2°. If we define $T(o_{\sigma_i}) = o_i'$, $i = 1, \cdots, n$, and $T(\sigma) = f(\sigma)$ for $\sigma \in L_1$, then T determines an almost semi-linear imbedding of K_0 in R^m through K_1.

Evidently $\bar{T}/\bar{L} \equiv f$. Let $x \in \bar{\tau} = (v_0 \cdots v_r) \in K_1 - L_1$, v_i being vertices of K_1, then

1) By a central subdivision we mean a subdivision analogous to the construction in barycentric subdivision. The only difference is that here any interior point of corresponding simplex, not necessarily the barycenter, may be used as the centre of projection in the construction.

$$\rho(\overline{T}(v_i), \overline{T}(v_k)) \leqslant \rho(\overline{T}(v_i), f(v_i)) + \rho(f(v_i), f(v_k)) + \rho(f(v_k), \overline{T}(v_k))$$
$$< \frac{\varepsilon}{5} + \frac{\varepsilon}{5} + \frac{\varepsilon}{5} = 3\frac{\varepsilon}{5},$$
$$\rho(\overline{T}(x), \overline{T}(v_0)) \leqslant \text{Diam } \overline{T}\,\bar{\tau} = \max_{i,k} \rho(\overline{T}(v_i), \overline{T}(v_k)) < 3\frac{\varepsilon}{5},$$
$$\rho(\overline{T}(x), f(x)) \leqslant \rho(\overline{T}(x), \overline{T}(v_0)) + \rho(\overline{T}(v_0), f(v_0)) + \rho(f(v_0), f(x))$$
$$< 3\frac{\varepsilon}{5} + \frac{\varepsilon}{5} + \frac{\varepsilon}{5} = \varepsilon.$$

Hence $\overline{T}$ is an ε-approximation of f and (B) is proved.

§2. The Imbedding Cochain of an Almost Semi-linear Realization

Let R^m be a euclidean space of dimension m with fixed orientation, K, L be two euclidean simplicial complexes in R^m and $x = \sum a_i \sigma_i$, $y = \sum b_j \tau_j$ be two chains on integer coefficients in K, L of dimension r, s respectively with $r + s = m$. The subcomplexes of K, L determined by those σ_i, τ_j for which a_i, b_j are $\neq 0$ will be denoted by $|x|$ and $|y|$ respectively. Suppose that $|x|$, $|y|$ are in general position, i.e., any simplex σ of $|x|$ and any simplex τ of $|y|$ are in general position, then with respect to the oriented R^m, the chains x, y have an intersection number ([2] Chap. 11)

$$\emptyset(x, y) = \sum a_i b_j \emptyset(\sigma_i, \tau_j),$$

which is bilinear and possesses the following three properties (dim $x=r$, dim $y=s$ and all intersection numbers are supposed to be defined):

$$\emptyset(x, y) = (-1)^{rs} \emptyset(y, x), \qquad r + s = m, \tag{1}$$

$$\emptyset(x, \partial y) = (-1)^r \emptyset(\partial x, y), \qquad r + s = m + 1, \tag{2}$$

and finally, change the orientation of R^m and denote the intersection number with respect to this otherwise oriented R^m by $\emptyset'$, then

$$\emptyset(x, y) = -\emptyset'(x, y), \qquad r + s = m. \tag{3}$$

We may also extend the definition of intersection number $\emptyset(x, y)$ of chains x, y with sum of dimensions $r + s = m$ for which only the conditions $\overline{\partial x} \cap \bar{y} = \bar{x} \cap \overline{\partial y} = \emptyset$ are satisfied by considering them as singular chains. The properties (1), (2) and (3) hold still for such intersection numbers.

In what follows $\emptyset_m$ will be used to denote the intersection number or the intersection number reduced mod 2, according as m is even or odd.

For an arbitrary euclidean simplicial complex K, define now two abstract complexes $\tilde{K}^*$ and K^* as follows. First, we define $\tilde{K}^*$ as the subcomplex of the product complex $K \times K$ consisting of all cells $\sigma \times \tau$ for which $\sigma, \tau \in K$ have no vertices in common. Orient each cell $\sigma \times \tau$ of $\tilde{K}^*$ in the usual manner we would have

$$\partial(\sigma \times \tau) = \partial\sigma \times \tau + (-1)^r \sigma \times \partial\tau, \qquad r = \dim \sigma \tag{4}$$

(σ, τ are oriented cells of K). In $\tilde{K}^*$, $\sigma \times \tau \to \tau \times \sigma$ defines a cell map t of period 2 and having no fixed cells, which induces a chain map given by

$$t_{\#}(\sigma \times \tau) = (-1)^{rs} (\tau \times \sigma), \qquad r = \dim \sigma, \qquad s = \dim \tau. \tag{5}$$

With respect to t, $\tilde{K}^*$ has a modular complex $\tilde{K}^*/t = K^*$, which is obtained by identifying each pair of cells $\sigma \times \tau, \tau \times \sigma$ in $\tilde{K}^*$ with a single cell $\sigma * \tau$ (or what is the same, $\tau * \sigma$). Orient $\sigma * \tau$ now as $\sigma \times \tau$, then by (4), (5) we have ($r = \dim \sigma$, $s = \dim \tau$):

$$\partial(\sigma * \tau) = \partial\sigma * \tau + (-1)^r \sigma * \partial\tau, \tag{6}$$

$$\sigma * \tau = (-1)^{rs} \tau * \sigma. \tag{7}$$

We may regard $\tilde{K}^*$ and K^* as euclidean complexes. Then by [6], the homotopy type of $\tilde{K}^*$ and K^*, in particular the homology and cohomology groups of $\tilde{K}^*$ and K^*, are all topological invariants of $\bar{K} = P$, *a fortiori* combinatorial invariants of K. For this reason we shall adopt the following notations: $H^r(\tilde{K}^*, G) = \tilde{H}^{r,2}(K, G) = \tilde{H}^{r,2}(P, G)$, $H^r(K^*, G) = H^{r,2}(K, G) = H^{r,2}(P, G)$.

Consider now any almost semi-linear realization $K_1' = TK_1$ of a simplicial complex K in R^m through a simplicial subdivision K_1 of K. Let the chain map induced by the subdivision $K \to K_1$ be Sd, and the chain map from K_1 to K_1' induced by T be $T_{\#}$. Write for simplicity $T_{\#}\,Sd$ by T. Similar notations will be used throughout this work. For any oriented cell $\sigma \times \tau$ of dimension m in $\tilde{K}^*$, $T\sigma$ and $T\tau$ are in general position since σ, τ have no common vertex in K and T is almost semi-linear. It follows that we may define an integer by

$$\tilde{\varphi}_T(\sigma \times \tau) = (-1)^{\dim \sigma}\, \emptyset(T\sigma, T\tau). \tag{8}$$

Then $\tilde{\varphi}_T$ is a cochain in $\tilde{K}^*$ on integer coefficients.

Let

$$I_{(m)} = \begin{cases} I, & m \text{ even} \\ I_2, & m \text{ odd} \end{cases}$$

Let $\rho_{(m)} : I \to I_{(m)}$ be the identity or the reduction mod 2. Further let $\emptyset_{(m)} = \rho_{(m)} \emptyset$ so that $(-1)^r \emptyset_m = (-1)^s \emptyset_m$ for $r + s = m$. Then we have always

$$(-1)^s \varnothing_m(T\tau, T\sigma) = (-1)^s \cdot (-1)^{rs} \varnothing_m(T\sigma, T\tau) = (-1)^{rs}(-1)^r \varnothing_m(T\sigma, T\tau)$$

for any $\sigma * \tau \in K^*$ with $\dim \sigma = r$, $\dim \tau = s$ and $r+s=m$. Comparing with (7) we see that ($\sigma * \tau \in K^*$, $\dim \sigma + \dim \tau = m$)

$$\varphi_T(\sigma * \tau) = (-1)^{\dim \sigma} \varnothing_m(T\sigma, T\tau) \tag{9}$$

defines unambiguously in K^* a cochain $\varphi_T \in C^m(K^*, I_{(m)})$.

Let $\xi \times \eta$ be any cell of dimension $m+1$ in $\tilde{K}^*$. Then

$$\begin{aligned}\delta\tilde{\varphi}_T(\xi \times \eta) &= \tilde{\varphi}_T \partial(\xi \times \eta) = \tilde{\varphi}_T(\partial\xi \times \eta) + (-1)^{\dim \xi} \tilde{\varphi}_T(\xi \times \partial\eta) = \\ &= (-1)^{\dim \xi - 1} \varnothing(T\,\partial\xi, T\eta) + \varnothing(T\xi, T\,\partial\eta) = \\ &= (-1)^{\dim \xi - 1} \cdot (-1)^{\dim \xi} \varnothing(T\xi, \partial T\eta) + \varnothing(T\xi, T\partial\eta) = 0\,.\end{aligned}$$

Similarly we have $\delta\varphi_T(\xi * \eta) = 0 \mod I_{(m)}$. Hence $\tilde{\varphi}_T$, φ_T are all m-dimensional cocycles of $\tilde{K}^*$ and K^* on coefficient groups I and $I_{(m)}$ respectively. If T is a semi-linear imbedding, then $\tilde{\varphi}_T$ and φ_T are evidently 0. Hence from the definition of $\tilde{\varphi}_T$ and φ_T we see that they may serve as a measure of T to the deviation from a true semi-linear imbedding. We shall accordingly define $\tilde{\varphi}_T$ and φ_T as the *imbedding cochains* of the almost semi-linear imbedding (or realization) T. The above results may then be written as follows:

Theorem 1. With respect to R^m with a fixed orientation, the imbedding cochains $\tilde{\varphi}_T \in C^m(\tilde{K}^*)$ and $\varphi_T \in C^m(K^*, I_{(m)})$ of an almost semi-linear realization T of a euclidean simplicial complex K in R^m are all cocycles (and may be thus called the *imbedding cocycles* of T).

The definition of imbedding cocycles depends on the orientation of R^m. By (3) we have

Theorem 2. With respect to R^m with the two opposite orientations, the two imbedding cocycles $\tilde{\varphi}_T$ and $\tilde{\varphi}'_T$ (or φ_T and φ'_T) of an almost semi-linear imbedding T of K in R^m differ at most by a sign:

$$\tilde{\varphi}_T = -\tilde{\varphi}'_T\,, \qquad \varphi_T = -\varphi'_T\,. \tag{10}$$

§3. Definition of Imbedding Classes

Let K, R^m be the same as in the preceding section, K_1, K_2 be two simplicial subdivisions of K, and $T_1K_1 = K'_1$, $T_2K_2 = K'_2$ be two almost semi-linear realizations of K in R^m through K_1 and K_2 respectively. The aim of the present section is to prove that the imbedding cocycles of T_1 and T_2 are cohomologous to each other. We shall suppose in what follows that $\bar{\bar{T}}_1\bar{K}_1$ and $\bar{\bar{T}}_2\bar{K}_2$ are disjoint. As this may be achieved by at most a parallel translation of $\bar{\bar{T}}_2\bar{K}_2$ and as the imbedding cocycles $\tilde{\varphi}_{T_2}=\tilde{\varphi}_2$, $\varphi_{T_2}=\varphi_2$ remain unchanged after the parallel translation, there will be no loss of generality in making this supposition.

Arrange now the simplexes of K in an order $\sigma_1 < \sigma_2 < \cdots$, such that simplexes of lower dimension will precede those of higher dimension, but otherwise arbitrary. Let $[1, 2]$ be the closed interval $1 \leqslant t \leqslant 2$, and $K \times [1, 2]$ the complex with usual cell decomposition which will be considered as a euclidean complex. Let J_0 be the set of all indices i for which $\dim \sigma_i = 0$, $J_r = J_{r-1} + (r)$, $L_0 = K \times (1) + K \times (2) + \sum_{i \in J_0} \sigma_i' \times [1, 2]$ and $L_r = L_{r-1} + \sigma_r \times [1, 2]$, $r = 1, 2, \cdots$. We shall construct now for each $r = 0, 1, 2, \cdots$ a simplicial subdivision $L_{r,0}$ of L_r and an almost semi-linear realization $H_r L_{r,0} = L_{r,0}'$ of L_r in R^m through the subdivision $L_{r,0}$ such that the following conditions are satisfied:

1°. $L_{0,0} = K_1 \times (1) + K_2 \times (2) + \sum_{i \in J_0} \sigma_i \times [1, 2]$.

2°. $H_0(\tau_j \times (j)) = T_j(\tau_j)$, $\tau_j \in K_j$, $j = 1, 2$ and for $i \in J_0$, $\bar{H}(\bar{\sigma}_i \times [1, 2])$ is a simple broken line l_i.

3°. $L_{r-1,0}$ is a subcomplex of $L_{r,0}$ and $H_r / L_{r-1,0} \equiv H_{r-1}$.

4°. If $i, j \in J_r$, $\dim \sigma_i + \dim \sigma_j = m - 2$, and σ_i, σ_j have no vertex of K in common, then $\bar{H}_r(\bar{\sigma}_i \times [1, 2] \cap \bar{H}_r(\bar{\sigma}_j \times [1, 2]) = \emptyset$.

For the construction let us first draw in R^m for each $i \in J_0$ a simple broken line l_i joining $T_j(\bar{\sigma}_i \times (j))$, $j = 1, 2$, such that these l_i together with $T_1 K_1 + T_2 K_2$ form an almost euclidean complex. In the case $m \geqslant 2$, we shall choose l_i to be disjoint from each other. Define now $L_{0,0}$ and H_0 according to 1°, 2°, then 1°—4° are all satisfied for them. Suppose now $L_{i,0}$ and H_i have been constructed for $i \leqslant r - 1$ which satisfy 1°—4° and let us define $L_{r,0}$ and H_r as follows. If $r \in J_0$, then $L_{r,0} = L_{0,0}$, $H_r \equiv H_0$ having been defined. Furthermore, as the case $\dim \sigma_r > m - 2$ is trivial, we shall suppose in what follows $\dim \sigma_r = e \leqslant m - 2$ and > 0.

Put $d = m - 2 - e \leqslant m - 3$. Any two simplexes ξ', η' in $L_{r-1,0}'$ not belonging to $H_{r-1}(K_1 \times (1) + K_2 \times (2))$ and having a dimension $\leqslant d + 1$ have an intersection which determines a linear subspace $P(\xi', \eta')$ with a dimension $\leqslant \max(-1, 2(d+1) - m)$. Let ζ' be any simplex in $L_{r-1,0}'$ lying on $\bar{H}_{r-1}(\bar{\sigma}_r \times (1) + \bar{\sigma}_r \times (2) + \bar{\sigma}_r' \times [1, 2])$, then ζ' and $P(\xi', \eta')$ will determine a linear subspace $P(\xi', \eta', \zeta')$ with a dimension $\leqslant \max(-1, 2(d+1) - m) + e + 1 = \max(d + 1, e) \leqslant m - 2$. Take now a point o_r in the interior of $\bar{\sigma}_r \times [1, 2]$ and form the central projection of the boundary of $\bar{\sigma}_r \times [1, 2]$ with centre o_r, thus obtaining a simplicial subdivision $L_{r,1}$ of $L_{r-1,0} + \sigma_r \times [1, 2]$ which contains $L_{r-1,0}$ as a subcomplex. By § 1 (A) we may choose a point o_r' in R^m not lying on any linear subspaces $P(\xi', \eta', \zeta')$ such that $H_r' / L_{r-1,0} \equiv H_{r-1}$, $H_r'(o_r) = o_r'$ will induce an almost semi-linear realization $H_r' L_{r,1} = L_{r,1}'$ of L_r in R^m through $L_{r,1}$. Then $L_{r,1}$ and H_r' satisfy 1°, 2°, 3° and for $i, j \in J_{r-1}$ also satisfy 4°. Suppose now $i_1, i_2, \cdots \in J_{r-1}$ be the totality of indices for which $\dim \sigma_{i_\mu} = d$ and σ_{i_μ}, σ_r have no common vertex in K. Then for each μ, $\bar{H}'_r(\bar{\sigma}_r \times [1, 2]) \cap \bar{H}'_r(\bar{\sigma}_{i_\mu} \times [1, 2])$ consists of at most finite number of points $p'_{1\mu}, \cdots, p'_{s_\mu,\mu}$. Owing to the choice of o'_r, $\bar{H}'^{-1}_r(p'_{h\mu})$ consists of only one point in $\sum_\lambda \bar{\sigma}_{i_\mu} \times [1, 2]$, say $p_{h\mu}$, where $h = 1, \cdots, s_\mu$. Consider now any fixed index $i = i_\lambda$. In $L_{r-1,0}$, let us take any d-dimensional simplex η lying on $\bar{\sigma}_i \times (2)$. Take also point x interior to η with $\bar{H}'_r(x) = x'$. If $d=0$, we shall denote by p_1 that point among $p_{h\lambda} = p_h$ ($h=1, \cdots, s_\lambda = s$ on $\sigma_i \times [1,2]$) which is nearest to $x = \sigma_i \times (2)$ from $\sigma_i \times (1)$ to x, and denote by B the part of $\sigma_i \times [1, 2]$ from p_1 to x. Then $\bar{H}'_r(B) = B' \subset l_i$ is a simple broken line. If $d>0$, then as $2(d+1)-m \leqslant (d+1)-2$, we may still join p_1 to x by a simple broken line B lying wholly on $\bar{\sigma}_i \times [1, 2]$ such that for any point $y \in \bar{\sigma}_{i_\mu} \times [1, 2]$, we shall have $\bar{H}'_r(y) \notin B' = \bar{H}'_r(B)$ so far as $y \notin B$. It follows that, whatever the case may be, B has always a neighborhood N in $\bar{\sigma}_i \times [1, 2]$ such that $\bar{H}'_r(N) = N'$ is disjoint from $\bar{H}'_r(\sum_\mu \bar{\sigma}_{i_\mu} \times [1, 2] - N)$; $\bar{H}'_r/N$ is one to one, and B does not pass through the points $p_2, \cdots, p_s$. Let the vertices of the broken line B be successively $x, x_1, \cdots, x_k = p_1$ and suppose that the segment xx_1 is in the $(d+1)$-dimensional simplex ξ_1, the segment $x_{j-1}\, x_j$ is in the $(d+1)$-dimensional simplex ξ_j of $L_{r-1,0}$ lying on $\bar{\sigma}_i \times [1, 2]$ ($j=2, \cdots, k$), and x_j is an interior point of the d-dimensional simplex η_j which is the face common to ξ_j and ξ_{j+1} ($j=1, 2, \cdots, k-1$). Put $\bar{H}'_r(x_j) = x'_j$, $\bar{H}'_r(\xi_j) = \xi'_j$, $\bar{H}'_r(\eta_j) = \eta'_j$, $\bar{H}'_r(\eta) = \eta'$. Prolong $x'_1 x'$ to x'_0 such that $x' x'_0$ meets $\bar{H}'_r(\sum_\mu \bar{\sigma}_{i_\mu} \times [1, 2])$ only in x'. Denote by $z_1, \cdots, z_t$ the totality of points in $\bar{H}'^{-1}_r(p'_1) \cap \bar{\sigma}_r \times [1, 2]$ and denote by ζ_i the $(e+1)$-dimensional simplex of $L_{r,1}$ lying on $\bar{\sigma}_r \times [1, 2]$ which contains z_i in its interior. The integer t will be called the multiplicity index of p'_1. Consider any $\zeta_1 = \zeta$, and denote by P'_k the $(e+1)$-dimensional linear subspace determined by $\zeta' = \bar{H}'_r(\zeta)$. Through each x'_j ($0 \leqslant j < k$) draw now an $(m-1)$-dimensional linear subspace P'_j such that for $0 < j < k$, P'_{j1} contains η'_j with $\bar{\xi}'_j$, $\bar{\xi}'_{j+1}$ on opposite sides of P'_j, and P'_0 meets $x'_0\, x'_1$ only in x'_0. Take an $(e+1)$-dimensional simplex τ containing $z = z_1$ in its interior and contained in $\bar{\zeta}$, with diameter less than a sufficiently small $\varepsilon>0$. Put $\bar{\tau}' = \bar{H}'_r(\bar{\tau})$. For any $y \in \bar{\tau}$ we shall draw a broken line $B'_y = y'_k \cdots y'_1\, y'_0\, x'_0$ such that $y'_j \in P'_j$, $y'_k = \bar{H}'_r(y)$ and $y'_j\, y'_{j-1}$ is parallel to $x'_j\, x'_{j-1}$ ($j = 1, \cdots, k$). Evidently for ε sufficiently small we may make $B'_y \cap \sum_\mu \bar{H}'_r(\bar{\sigma}_{i_\mu} \times [1, 2]) = \emptyset$ for $y \in \bar{\tau}$. After the choice of such an ε we may define a continuous map $\bar{H}''_r : \bar{L}_{r,1} \rightarrow R^m$ by $\bar{H}''_r/\bar{L}_{r,1} - \bar{\tau} \equiv \bar{H}'_r$, while for any $y \in \bar{\tau}$, $\bar{H}''_r$ maps linearly yz_1 to the segment B'_y. By § 1 (B), we may construct an arbitrarily small approximation $\bar{H}'''_r$ of $\bar{H}''_r$ ($\bar{H}'''_r/\bar{L}_{r,1} - \bar{\tau} \equiv \bar{H}''_r$) such that $\bar{H}'''_r\, \bar{L}_{r,1}$ is the space of an almost euclidean complex, and $L_{r,1}$ has a simplicial subdivision $L_{r,2}$ having $L_{r-1,0}$ as a subcomplex, while $\bar{H}'''_r$ is the continuous map associated with the simplicial map $H'''_r\,(L_{r,2}) = L'_{r,2}$. By taking the approximation sufficiently small, we may make $\bar{H}'''_r(\bar{\sigma}_r \times$

$\times [1,2]) \cap \bar{H}_r'''(\bar{\sigma}_i \times [1,2])$ consist of only $k-1$ points $p_2', \cdots, p_k'$, or though of the same number of points $p_1', \cdots, p_k'$ as before, the multiplicity index of p_1' is decreased by 1, while the number of intersecting points and the multiplicity indices of $\bar{H}_r'''(\bar{\sigma}_r \times [1,2])$ and $\bar{H}_r'''(\bar{\sigma}_{i_\mu} \times [1,2])$ (besides at p_1') are all unchanged.

Proceeding successively with the same process, we shall make the number of intersections of the images in R^m of $\bar{\sigma}_r \times [1,2]$ and $\bar{\sigma}_i \times [1,2]$ reduce to 0. Using the same procedure to each σ_{i_μ} we get finally a complex $L_{r,0}$ and a realization $H_r L_{r,0} = L'_{r,0}$ which satisfies the conditions 1°—4°.

By induction on r, we get finally a simplicial subdivision L of $K \times [1,2]$ and an almost semi-linear realization $HL = L'$ of $K \times [1,2]$ in R^m through L. By 4°, this H will satisfy the following condition: If $\dim \sigma_i + \dim \sigma_j = m-2$ and σ_i, σ_j have no common vertex in K, then

$$\bar{H}(\bar{\sigma}_i \times [1,2]) \cap \bar{H}(\bar{\sigma}_j \times [1,2]) = \emptyset . \tag{1}$$

Now let the chain map induced by the subdivision of $K \times [1,2]$ into L be denoted by Sd, and that induced by H be denoted by $H_\#$. Orient $[1,2]$ by the direction from 1 to 2 and put for simplicity

$$h\,\sigma_i = (-1)^{\dim \sigma_i} H_\# Sd(\sigma_i \times [1,2]) ,$$

then

$$\partial h\,\sigma_i = T_2\sigma_i - T_1\sigma_i - h\,\partial\sigma_i , \qquad \sigma_i \in K . \tag{2}$$

By (1), we get further: If $\dim \sigma_i + \dim \sigma_j = m-2$ and σ_i, σ_j have no common vertex in K, then

$$\emptyset(h\sigma_i, h\sigma_j) = 0 . \tag{3}$$

Define now a cochain $\tilde{\psi} \in C^{m-1}(\tilde{K}^*)$ as follows:

$$\tilde{\psi}(\sigma_i \times \sigma_j) = \emptyset(\partial h\,\sigma_i, h\sigma_j) , \ \dim \sigma_i + \dim \sigma_j = m-1 , \ \sigma_i \times \sigma_j \in \tilde{K}^* . \tag{4}$$

Since

$$\begin{aligned} \emptyset_m(\partial h\sigma_i, h\sigma_j) &= (-1)^{\dim \sigma_j + 1} \emptyset_m(h\sigma_i, \partial h\sigma_j) = \\ &= (-1)^{\dim \sigma_j + 1} \cdot (-1)^{(\dim \sigma_j + 1)\dim \sigma_i} \emptyset_m(\partial h\sigma_j, h\sigma_i) = \\ &= (-1)^{\dim \sigma_i \dim \sigma_j} \emptyset_m(\partial h\sigma_j, h\sigma_i) , \end{aligned}$$

we know, by comparing with (7) of § 2, that we may define unambiguously a cochain $\psi \in C^{m-1}(K^*, I_{(m)})$ by

$$\psi(\sigma_i * \sigma_j) = \emptyset_m(\partial h\sigma_i, h\sigma_j), \ \dim \sigma_i + \dim \sigma_j = m-1, \ \sigma_i * \sigma_j \in K^* . \tag{5}$$

For any $\sigma_k \times \sigma_l \in \tilde{K}^*$ with $\dim \sigma_k + \dim \sigma_l = m$ we have now

$$\delta\psi(\sigma_k\times\sigma_l)=\tilde{\psi}\,\partial(\sigma_k\times\sigma_l)=\tilde{\psi}(\partial\sigma_k\times\sigma_l)+(-1)^{\dim\sigma_k}\tilde{\psi}(\sigma_k\times\partial\sigma_l)=$$
$$=\emptyset(\partial h\,\partial\sigma_k,h\sigma_l)+(-1)^{\dim\sigma_k}\emptyset(\partial h\sigma_k,h\partial\sigma_l)\,,$$

or

$$\delta\tilde{\psi}(\sigma_k\times\sigma_l)=(-1)^{\dim\sigma_k}[\emptyset(h\partial\sigma_k,\partial h\sigma_l)+\emptyset(\partial h\sigma_k,h\partial\sigma_l)]\,. \tag{6}$$

On the other hand, since $\bar{T}_1\bar{K}_1$ and $\bar{T}_2\bar{K}_2$ are disjoint,

$$\emptyset(T_1\sigma_k,\ T_2\sigma_l)=\emptyset(T_2\sigma_k,\ T_1\sigma_l)=0\,. \tag{7}$$

By (2), (3), (4), (7) and the definitions of $\tilde{\varphi}_{T_i}=\tilde{\varphi}_i$ ($i=1$, 2) we get

$$\begin{aligned}(-1)^{\dim\sigma_k}[\tilde{\varphi}_1(\sigma_k\times\sigma_l)+\tilde{\varphi}_2(\sigma_k\times\sigma_l)]&=\emptyset(T_1\sigma_k,\ T_1\sigma_l)+\emptyset(T_2\sigma_k,\ T_2\sigma_l)=\\&=\emptyset(T_2\sigma_k-T_1\sigma_k-h\partial\sigma_k,\ T_2\sigma_l-T_1\sigma_l-h\partial\sigma_l)-\emptyset(h\partial\sigma_k,h\partial\sigma_l)+\\&\quad+\emptyset(T_2\sigma_k-T_1\sigma_k,\ h\partial\sigma_l)+\emptyset(h\partial\sigma_k,\ T_2\sigma_l-T_1\sigma_l)=\\&=\emptyset(\partial h\sigma_k,\partial h\sigma_l)-\emptyset(h\partial\sigma_k,h\partial\sigma_l)+\emptyset(\partial h\sigma_k+h\partial\sigma_k,h\partial\sigma_l)+\\&\quad+\emptyset(h\partial\sigma_k,\partial h\sigma_l+h\partial\sigma_l)=\emptyset(\partial h\sigma_k,h\partial\sigma_l)+\emptyset(h\partial\sigma_k,\partial h\sigma_l)\,.\end{aligned}$$

Comparing with (6) we get

$$\tilde{\varphi}_1(\sigma_k\times\sigma_l)+\tilde{\varphi}_2(\sigma_k\times\sigma_l)=\delta\tilde{\psi}(\sigma_k\times\sigma_l)\,.$$

Since $\sigma_k\times\sigma_l$ is an arbitrary m-cell of $\tilde{K}^*$, we have

$$\tilde{\varphi}_1+\tilde{\varphi}_2\sim 0\,. \tag{8}$$

In particular, if T_2 is obtained from T_1 by parallel translations in R^m so that $\tilde{\varphi}_1=\tilde{\varphi}_2$, (8) becomes

$$2\,\tilde{\varphi}_1\sim 0\,. \tag{9}$$

From (8), (9) we get therefore

$$\tilde{\varphi}_1\sim\tilde{\varphi}_2\,. \tag{10}$$

Similarly, we have also

$$2\,\varphi_1\sim 0\,,\qquad \mathrm{mod}\ \mathrm{I}_{(m)}\,, \tag{11}$$

$$\varphi_1\sim\varphi_2\,,\qquad \mathrm{mod}\ \mathrm{I}_{(m)}\,. \tag{12}$$

The above imbedding cocycles $\tilde{\varphi}_T$, φ_T are defined with respect to R^m with a fixed orientation. If we reverse the orientation of R^m and denote the corresponding imbedding cocycles of T by $\tilde{\varphi}'_T$, φ'_T, then we have by Theorem 2 of § 2, $\tilde{\varphi}'_T=-\tilde{\varphi}_T$, $\varphi'_T=-\varphi_T$. Hence by (9), (11) we have

$$\tilde{\varphi}'_T\sim\tilde{\varphi}_T\,, \tag{13}$$

$$\varphi'_T\sim\varphi_T\,,\qquad \mathrm{mod}\ \mathrm{I}_{(m)}\,. \tag{14}$$

From (9)–(14) we get therefore the following two theorems:

Theorem 3. The imbedding cocycles $\tilde{\varphi}_T$ and φ_T of an almost semi-linear realization T of a euclidean simplicial complex K in an oriented R^m each belong to a fixed cohomology class $\tilde{\Phi}^m \in \tilde{H}^{m,2}(K)$ and $\Phi^m \in H^{m,2}(K, I_{(m)})$. Moreover, these classes are independent of the chosen orientation of R^m and the chosen realization T.

Definition. The cohomology class Φ^m in the above theorem will be called the m-dimensional *imbedding class* of K, $m > 0$.

Theorem 4. All imbedding classes on integer coefficients have order 2:

$$2\,\Phi^m = 0 \qquad (m \text{ even} > 0)\,. \tag{15}$$

Remark. We have also $2\tilde{\Phi}^m = 0$, $m > 0$. However, we shall prove later that we have always $\tilde{\Phi}^m = 0$ (cf. Theorem 9 of § 5 and Theorem 16 of § 8). Hence in reality $\tilde{\Phi}^m$ are of no significance at all.

From the definition of imbedding cocycles and imbedding classes we have

Theorem 5. A necessary condition that a euclidean simplicial complex K may have a linear (or semi-linear) realization in R^m, is that

$$\Phi^m = 0\,. \tag{16}$$

We shall see later that in certain cases, this condition is also sufficient.

Evidently $\tilde{K}^*$ is a two-sheeted covering complex of K^*. Denote the projection by π:

$$\pi(\sigma \times \tau) = \sigma * \tau \qquad (\sigma \times \tau \in \tilde{K}^*)\,.$$

Then by the definition of $\tilde{\varphi}_T$, φ_T we have evidently $\pi^{\#}\varphi_T = \rho_{(m)}\tilde{\varphi}_T$. Hence we have

Theorem 6. Let π be the covering projection of $\tilde{K}^*$ on K^*, then

$$\pi^*\,\Phi^m = \tilde{\Phi}^m\,, \qquad (m \text{ even} > 0) \tag{17}$$

$$\pi^*\,\Phi^m = \rho_2\,\tilde{\Phi}^m\,, \qquad (m \text{ odd}) \tag{18}$$

in which ρ_2 denotes reduction mod 2. Suppose for the moment $\tilde{\Phi}^m = 0$ (cf. the remark above), then (17) and (18) may be reduced simply to

$$\pi^*\,\Phi^m = 0\,, \qquad m > 0\,. \tag{19}$$

§ 4. The Realizability of Any Cocycle in the Imbedding Classes

We have proved that the m-dimensional imbedding cochains $\tilde{\varphi}_T$ (or φ_T) of K are all cocycles and belong to one and the same cohomology class $\tilde{\Phi}^m \in \tilde{H}^{m,2}(K)$ (or $\Phi^m \in H^{m,2}(K, I_{(m)})$. Conversely, by the definition of $\tilde{\varphi}_T$ in § 2, any "imbedding cocycle" $\tilde{\varphi}_T$ in $\tilde{\Phi}^m$ must satisfy the conditions ($\sigma_i \times \sigma_j \in \tilde{K}^*$, $\dim \sigma_i + \dim \sigma_j = m$)

$$\tilde{\varphi}_T(\sigma_i \times \sigma_j) = (-1)^{m+\dim \sigma_i \dim \sigma_j} \tilde{\varphi}_T(\sigma_j \times \sigma_i) .$$

Hence if we change $\tilde{\varphi}_T$ into an arbitrary coboundary $\delta \tilde{\psi}$, $\tilde{\psi} \in C^{m-1}(\tilde{K}^*)$, the cocycle $\tilde{\varphi}_T + \delta \tilde{\psi}$ of $\tilde{\Phi}^m$ thus obtained is in general no more an imbedding cocycle, and is not necessarily realized as one of an almost semi-linear realization of K in R^m. On the contrary, for the class Φ^m we have the following

Theorem 7. If $m > 1$, then any cocycle in Φ^m may be realized as an imbedding cocycle. In other words, there must exist an almost semi-linear realization of K in the oriented R^m, with any given cocycle in Φ^m as its imbedding cocycle.

Remark. This theorem is not true for $m = 1$. For example, let K be a one-dimensional complex consisting of three vertices a, b, c and two segments ab, ac. Since K may be realized in R^1, we have $\Phi^1 = 0$. Hence defining a mod 2 cochain $\psi \in C^\circ(K^*, I_2)$ by $\psi(b * c) = 1$, $\psi(a * b) = \psi(a * c) = 0$, we would have $\varphi = \delta \psi \in \Phi^1$. Suppose that there exists an almost semi-linear realization T of K in R^1 with imbedding cocycle $\varphi_T = \varphi$. Let $T(a) = a'$, $T(b) = b'$ and $T(c) = c'$. Then since $\varphi_T(b * (ac)) = \delta \psi (b * (ac)) = \psi (b * c) \neq 0$, we have $\rho_2 \emptyset(b', T(ac)) \neq 0$ so that b' must lie between a' and c'. Similarly, since $\varphi_T(c * (ab)) \neq 0$, c' must lie between a' and b'. But this is impossible. Consequently $\varphi \in \Phi^1$ cannot be realized as an imbedding cocycle.

Proof of Theorem 7. Consider any almost semi-linear realization $T_0 K_0 = K_0'$ of K in oriented R^m through a subdivision K_0 of K with corresponding imbedding cocycle $\varphi_{T_0} = \varphi_0 \in \Phi^m$. Denote the simplexes of K by $\sigma_1, \sigma_2 \cdots$ with $\dim \sigma_i = d_i$. Consider an arbitrary but fixed $(m-1)$-cell $\sigma_i * \sigma_j$ in $K^* : d_i + d_j = m - 1$, and define a cochain $\chi_{i,j} \in C^{m-1}(K^*, I_{(m)})$ by

$$\chi_{i,j}(\sigma_k * \sigma_l) = \begin{cases} 0, & \sigma_k * \sigma_l \neq \sigma_i * \sigma_j \text{ or } \sigma_j * \sigma_i, \\ 1, & \sigma_k * \sigma_l = \sigma_i * \sigma_j . \end{cases} \tag{1}$$

Our object is to modify T_0 to an almost semi-linear realization of K in the oriented R^m through a subdivision K_1 of K such that the imbedding cocycle of T_1 is

$$\varphi_{T_1} = \varphi_1 = \varphi_0 + c\,\delta\,\chi_{i,j}, \tag{2}$$

where $c = \pm 1$ is arbitrarily but previously assigned. Since $\sigma_i * \sigma_j$ is an arbitrary $(m-1)$-cell of K^*, we may start from φ_0 and arrive at any cocycle in Φ^m, and our theorem would be proved in this manner.

For this purpose let us remark first that, as $m>1$ by hypothesis, we may suppose $d_i > 0$. Consider any simplexes of K_0 of dimension d_i, d_j with $\bar{\tau}_i \subset \bar{\sigma}_i$, $\bar{\tau}_j \subset \bar{\sigma}_j$ respectively. Let $\tau_i' = T_0\tau_i$, $\tau_j' = T_0\tau_j$. In each of τ_i, τ_j take an interior point x_0, x such that $x_0' = \overline{T}_0(x_0) \notin \overline{T}_0(\overline{K_0^{m-1} - \bar{\tau}_i})$ $x' = \overline{T}_0(x) \notin \overline{T}_0(\overline{K_0^{m-1} - \bar{\tau}_j})$. Construct a simple broken line B, with successive vertices $x_0', x_1', \cdots, x_n'$, such that the following conditions are satisfied:

1°. $x' \in x_{n-1}' x_n'$, $x_{n-1}' x_n'$ is orthogonal to the linear subspace R determined by τ_j' and $B \cap \bar{\tau}_j' = (x')$.

2°. B is disjoint to the space of $|T_0 K_0 - T_0 St_0 \tau_i - T_0 St_0 \tau_j|^{m-2}$ in which St_0 denotes the star in K_0. Finally,

3°. $x_0' x_1'$ is orthogonal to the linear subspace Q_0 determined by τ_i' and meets $\overline{T}_0 \overline{K_0^{m-1}}$ only in the point x_0'.

Let Q be the d_i-dimensional linear subspace in R^m passing through x' and completely orthogonal to R and $x_{n-1}' x_n'$. Let T be an orthogonal transformation of R^m, transforming Q_0 to Q, x_0' to x', and the $x_0' x_1'$ direction to $x' x_n'$ direction. For any point $y_0 \in St_0 \tau_i$ and any $\varepsilon \geqslant 0$ and $\leqslant 1$, let $y_0(\varepsilon)$ be the point on $x_0 y_0$ with $x_0 y_0(\varepsilon) : x_0 y_0 = \varepsilon$. Let B_{y_0} be the broken line with successive vertices $y_0' = \overline{T}_0(y_0)$, x_0', x_1', $\cdots$, x_{n-1}', y_{n-1}', $y' = T(y_0')$, y_n' and x_n', where $y_{n-1}' y'$, $y' y_n'$ are parallel and equal to $x_{n-1}' x'$, $x' x_n'$ respectively. From 1°—3° and the above construction we see that

4°. If ε is sufficiently small, then for any $y_0 \in \overline{St_0 \tau_i}$, $B_{y_0(\varepsilon)}$ is disjoint from the space of $|T_0 K_0 - T_0 St_0 \tau_i - T_0 St_0 \tau_j|^{m-2}$.

5°. If $\varepsilon > 0$ is sufficiently small, then for any $y_0 \in \bar{\tau}_i$, $B_{y_0(\varepsilon)}$ is disjoint from $\bar{\tau}_j'$.

Now for any $\tau_k \in Cl\,St_0 \tau_i$, let $\tau_{k,\varepsilon}$ be the contraction of τ_k with centre x_0 and ratio of contraction $\varepsilon : 1$ $(0 \leqslant \varepsilon \leqslant 1)$. In particular, $\tau_{k,1} = \tau_k$, $\bar{\tau}_{k,0} = (x_0)$. Let L_ε be the complex formed of all $\tau_{k,\varepsilon}$ for which $\tau_k \in Cl\,St_0 \tau_i$. Construct a simplicial subdivision K_{00}, with both $K_0 - St_0 \tau_i$ and L_ε as its subcomplexes. In $\overline{K}_{00} \times [0, 1]$ identify each segment $(z) \times [0, 1]$ with a single point, where $z \in \overline{K_{00} - St_{00} \tau_{i,\varepsilon}}$ (St_{00} being star in K_{00}), obtaining thus a space $\overline{M}_{\varepsilon 0}$ and a natural map $f_\varepsilon : \overline{K}_{00} \times [0, 1] \to \overline{M}_{\varepsilon 0}$. Under f_ε, $K_{00} \times [0, 1]$ will induce on $M_{\varepsilon 0}$ a cell decomposition such that f_ε is a cell-map. Denote this cell decomposition of $\overline{M}_{\varepsilon 0}$ by $M_{\varepsilon 0}$. Then the parts of $M_{\varepsilon 0}$ on $f_\varepsilon(\overline{K}_{00} \times (0))$

and $f_\varepsilon(\bar{K}_{00}\times(1))$ are isomorphic to $K_{00}\times(0)$ and $K_{00}\times(1)$ respectively under the map f_ε.

Define now a map $\bar{H}^0:\bar{M}_{\varepsilon 0}\to R^m$ as follows. First, for any $y_0\in\overline{St_0\tau_i}$, let α_{y_0} be a linear map of $[0, 1]$ to B_{y_0} such that $\alpha_{y_0}(i/2\,(n+2))$, $i=0, 1, \cdots, n+2$ are successively $y_0', x_0', \cdots, x_{n-1}', y_{n-1}', y_n'$ while $\alpha_{y_0}(1)=x_n'$ (for the symbols cf. above). If $z\in f_\varepsilon(\overline{K_{00}-St_{00}\,\tau_{i,\varepsilon}})$, then define $\bar{H}^0(z)=\bar{T}_0(z)$. Finally, let $\tau_k\in St_0\,\tau_i$, y_0 be any point $\in\bar{\tau}_{k,\varepsilon}$, but not interior to $\bar{\tau}_{i,\varepsilon}$, then for any $t\in[0, 1]$, $\bar{H}^0$ linearly map $f_\varepsilon(y_0x_0, t)$ to the broken line

$$B_{y_0,t}=\alpha_{y_0}\left[0,\ \frac{t}{2}\right]+\alpha_{y_0}\left(\frac{t}{2}\right)\alpha_{x_0}\left(\frac{t}{2}\right),$$

such that

$$\bar{H}^0 f_\varepsilon(y_0(\eta), t)=\alpha_{y_0}(1-\eta), \qquad 1-\frac{t}{2}\leqslant\eta\leqslant 1,$$

$$\bar{H}^0 f_\varepsilon(y_0(\eta), t)=\alpha_{y_0(2\eta/(2-t))}\left(\frac{t}{2}\right), \qquad 0\leqslant\eta\leqslant 1-\frac{t}{2}.$$

The map $\bar{H}^0$ thus defined is evidently continuous. Let $g_t:\bar{K}\to\bar{K}\times[0,1]$ be the map $g_t(z)=(z, t)$, $z\in\bar{K}$, and $\bar{H}_t^0:\bar{K}\to R^m$ be the map $\bar{H}_t^0=\bar{H}^0 f_\varepsilon g_t$. Then $\bar{H}_0^0\equiv\bar{T}_0$, and $\bar{H}_1^0$ maps y_0x_0 to B_{y_0}, where $y_0\in\bar{\tau}_{k,\varepsilon}-\operatorname{Int}\bar{\tau}_{i,\varepsilon}$, $\tau_k\in St_0\,\tau_i$. By 4°, 5° we have then

4_0°. If ε is sufficiently small, then for any $\tau_k\in St_0\,\tau_i$ $\tau_l\in|K_0-St_0\,\tau_i-St_0\,\tau_j|^{m-2}$, we have

$$\bar{H}_t^0(\bar{\tau}_{k,\varepsilon})\cap\bar{T}_0\bar{\tau}_l=\emptyset, \qquad t\in[0, 1].$$

5_0°. If $\varepsilon>0$ is sufficiently small, then $\bar{H}_1^0(\bar{\tau}_{i,\varepsilon})$ is disjoint from $\bar{T}_0\bar{\tau}_j$.

Choose now $\varepsilon>0$ sufficiently small such that 4_0° and 5_0° are both satisfied. By § 1 (B), there exist a simplicial subdivision M_ε of $M_{\varepsilon 0}$ and an arbitrarily small approximation $\bar{H}:\bar{M}_{\varepsilon 0}\to R^m$ of $\bar{H}^0$ such that $\bar{H}\equiv\bar{H}^0/f_\varepsilon(\bar{K}_{00}\times(0))$, and $\bar{H}$ be an almost semi-linear imbedding of $M_{\varepsilon 0}$ in R^m through M_ε. This approximation $\bar{H}$ of $\bar{H}^0$ may be chosen so small that the following conditions corresponding to 4_0° and 5_0° will be supposed to be satisfied.

6°. For any $\tau_k\in St_0\,\tau_i$ and $\tau_l\in|K_0-St_0\,\tau_i-St_0\,\tau_j|^{m-2}$ we have $(\bar{H}_t=\bar{H}f_\varepsilon g_t)$

$$\bar{H}_t(\bar{\tau}_{k,\varepsilon})\cap\bar{T}_0\bar{\tau}_l=\emptyset, \qquad t\in[0, 1].$$

7°. $\bar{H}_1(\bar{\tau}_{i,\varepsilon})\cap\bar{T}_0\bar{\tau}_j=\emptyset$.

Define now a map $\bar{T}_1:\bar{K}\to R^m$ by $\bar{T}_1=\bar{H}_1 f_\varepsilon g_1$, then $\bar{T}_1$ determines an almost semi-linear realization T_1 of K in R^m through a certain subdivision K_1. Let Sd and Sd' be the chain maps induced by the subdivisions K or $K_0 \to K_{00}$ and $M_{\varepsilon 0} \to M_\varepsilon$ respectively. For any chain c of K or K_0, $H\, Sd' f_\varepsilon(Sd\, c \times [0, 1])$ is a chain in R^m, which will be denoted for simplicity by $(-1)^{\dim c}\, h(c)$. Then we have

$$\partial h(c) = T_1(c) - T_0(c) - h\partial(c)\,, \qquad c \in C^*(K_0) \text{ or } C^*(K)\,. \tag{3}$$

$$h(\tau) = 0\,, \qquad \tau \in K_0 \text{ but } \notin St_0\, \tau_i\,. \tag{4}$$

By 5°, 7° and the construction we see easily $\emptyset_m(h\tau_i, T_0\sigma_j) = \pm 1$. By conveniently choosing the orthogonal transformation T of R^m to be orientation-preserving or orientation-reversing, we may always make[1] ($c = \pm 1$ as in (1))

$$\emptyset(h\tau_i\,, T_0\,\sigma_j) = (-1)^{d_i+1}\, c\,. \tag{5}$$

Let $\tau_k \in St_0\, \tau_i$, $\tau_l \in K_0 - St_0\, \tau_i - St_0\, \tau_j$, where $\dim \tau_k + \dim \tau_l = m - 1$. Then since $\dim \tau_k \geqslant \dim \tau_i > 0$ we have $\dim \tau_l \leqslant m-2$ and $\tau_l \in |K_0 - St_0\, \tau_i - St_0\, \tau_j|^{m-2}$. Hence, by 6°,

$$\emptyset(h\tau_k, T_0\, \tau_l) = 0\,. \tag{6}$$

Let the imbedding cocycle of T_1 be $\varphi_{T_1} = \varphi_1$. We shall prove that for any cell $\sigma_k * \sigma_l \in K^*$ with $d_k + d_l = m$, we have

$$\varphi_1(\sigma_k * \sigma_l) - \varphi_0(\sigma_k * \sigma_l) = c\, \delta\, \chi_{i,j}(\sigma_k * \sigma_l)\,. \tag{7}$$

Case I. σ_k, $\sigma_l \notin St\, \sigma_i$.

In that case $\overline{T}_1 / \bar{\sigma}_k \equiv \overline{T}_0$, $\overline{T}_1 / \bar{\sigma}_l \equiv \overline{T}_0$. Hence $\varphi_1(\sigma_k * \sigma_l) = \varphi_0(\sigma_k * \sigma_l)$. As $\delta\chi_{i,j}(\sigma_k * \sigma_l) = 0$, we have (7).

Case II. $\sigma_k \in St\, \sigma_i$, $\sigma_l = \sigma_j$.

We have $d_k = d_i + 1$, and $\overline{T}_1 / \bar{\sigma}_l \equiv \overline{T}_1 / \bar{\sigma}_j \equiv \overline{T}_0 / \bar{\sigma}_j$. Let

$$\partial \sigma_k = a\sigma_i + \zeta\,, \qquad a = \pm 1\,, \tag{8}$$

(ζ contains no term involving σ_i), then

$$\delta\, \chi_{i,j}(\sigma_k * \sigma_l) = \chi_{i,j}(\partial\sigma_k * \sigma_l) = a\, \chi_{i,j}(\sigma_i * \sigma_j) = a\,. \tag{9}$$

Let $\tau_k \in St_0\, \tau_i$ be the d_k–dimensional simplex of K_0 with $\bar{\tau}_k \subset \bar{\sigma}_k$, and let τ_i, τ_k be oriented concordantly with σ_i, σ_k respectively, then by (8) we get

$$\partial\, \tau_k = a\, \tau_i + \zeta'\,, \qquad a = \pm 1\,, \tag{10}$$

(ζ' contains no term involving τ_i). By (3)–(6) and (10), we get

---

1) This is not necessarily possible when $m=1$ (cf. the remark before the proof).

$$(-1)^{d_k}[\varphi_1(\sigma_k*\sigma_l)-\varphi_0(\sigma_k*\sigma_l)]=\emptyset_m(T_1\sigma_k,T_1\sigma_l)-\emptyset_m(T_0\sigma_k,T_0\sigma_l)=$$
$$=\emptyset_m(T_1\sigma_k-T_0\sigma_k,T_0\sigma_j)=\emptyset_m(T_1\tau_k-T_0\tau_k,T_0\tau_j)=$$
$$=\emptyset_m(\partial h\tau_k,T_0\sigma_j)+\emptyset_m(h\partial\tau_k,T_0\sigma_j)=\pm\emptyset_m(h\tau_k,\partial T_0\sigma_j)+$$
$$+a\,\emptyset_m(h\tau_i,T_0\sigma_l)+\emptyset_m(h\zeta',T_0\sigma_i)=(-1)^{d_j+1}ac=(-1)^{d_k}ac.$$

Comparing with (9), we get (7).

Case III. $\sigma_l\in St\,\sigma_i$, $\sigma_k=\sigma_j$.

According to Case II, we have

$$\varphi_1(\sigma_k*\sigma_l)-\varphi_0(\sigma_k*\sigma_l)=(-1)^{d_k d_l}[\varphi_1(\sigma_l*\sigma_k)-\varphi_0(\sigma_l*\sigma_k)]=$$
$$=(-1)^{d_k d_l}c\,\delta\chi_{i,j}(\sigma_l*\sigma_k)=c\,\delta\chi_{i,j}(\sigma_k*\sigma_l),$$

i.e., the equation (7).

Case IV. $\sigma_k=\sigma_i$, $\sigma_l\in St\,\sigma_j$.

We have then $d_l=d_j+1$. Let (ζ contains no term involving σ_j)

$$\partial\sigma_l=a\sigma_j+\zeta,\qquad a=\pm1.$$

Then

$$\delta\chi_{i,j}(\sigma_k*\sigma_l)=\delta\chi_{i,j}(\sigma_i*\sigma_l)=(-1)^{d_i}\chi_{i,j}(\sigma_i*\partial\sigma_l)=$$
$$=(-1)^{d_i}a\,\chi_{i,j}(\sigma_i*\sigma_j)=(-1)^{d_i}a.$$

Since $\overleftrightarrow{T}_1/\bar\sigma_l\equiv\overleftrightarrow{T}_0/\bar\sigma_l$, we have as before

$$(-1)^{d_i}[\varphi_1(\sigma_k*\sigma_l)-\varphi_0(\sigma_k*\sigma_l)]=\emptyset_m(T_1\sigma_i-T_0\sigma_i,T_0\sigma_l)=$$
$$=\emptyset_m(\partial h\sigma_i,T_0\sigma_l)+\emptyset_m(h\partial\sigma_i,T_0\sigma_l)=(-1)^{d_i+1}\emptyset_m(h\sigma_i,T_0\partial\sigma_l)=$$
$$=(-1)^{d_i+1}[a\,\emptyset_m(h\sigma_i,T_0\sigma_j)+\emptyset_m(h\sigma_i,T_0\zeta)]=ac.$$

By comparing, we get again (7).

Case V. $\sigma_l=\sigma_i$, $\sigma_k\in St\,\sigma_j$.

By Case IV we have

$$\varphi_1(\sigma_k*\sigma_l)-\varphi_0(\sigma_k*\sigma_l)=(-1)^{d_k d_l}[\varphi_1(\sigma_l*\sigma_k)-\varphi_0(\sigma_l*\sigma_k)]=$$
$$=(-1)^{d_k d_l}c\,\chi_{i,j}(\sigma_l*\sigma_k)=c\,\chi_{i,j}(\sigma_k*\sigma_l),$$

i.e., the equation (7).

Case VI. The other cases.

We have then necessarily $\sigma_k\in St\,\sigma_i$, $\sigma_k\neq\sigma_i$, $\sigma_l\neq\sigma_j$, or $\sigma_k=\sigma_i$, $\sigma_l\notin St\,\sigma_j$, or the cases with the role of k, l interchanged. For the two preceding cases we have always $\sigma_l\notin St\,\sigma_j$. Hence, let $\tau_k\in St_0\,\tau_i$ be the d_k-dimensional simplex of K_0 with $\bar\tau_k\subset\bar\sigma_k$ and let τ_i, τ_k, be oriented as σ_i, σ_k, we would have

$$\emptyset_m(\partial h\tau_k,T_0\sigma_l)=\emptyset_m(h\partial\tau_k,T_0\sigma_l)=0.$$

As $T_1\sigma_l = T_0\sigma_l$, we have $\varphi_1(\sigma_k * \sigma_l) - \varphi_0(\sigma_k * \sigma_l) = 0$. On the other hand, that $\delta\chi_{i,j}(\sigma_k * \sigma_l) = 0$ is evident. Hence (7) is true. The same holds on interchanging the role of k and l.

From the all possible cases considered above, we see that (7) is always true for any $\sigma_k * \sigma_l \in K^*$ with $d_k + d_l = m$. Hence (2) is established and our theorem is completely proved.

§5. Relations between Φ^{2m-1} and Φ^{2m} : $\frac{1}{2}\delta\Phi^{2m-1} = \Phi^{2m}$

In this section the vertices of a euclidean simplicial complex K will be arranged in a definite order $a_1 < a_2 < \cdots < a_N$, and any simplex $\sigma \in K$ will be written in the normal form $\sigma = (a_{i_0} \cdots a_{i_r})$ with $i_0 < < \cdots < i_r$ and oriented accordingly. The dimension of σ will be denoted by $d(\sigma)$, the barycentre of σ by o_σ, and the barycentric subdivision of K by K_1. The simplexes of K will also be arranged in an order $<$ such that $\sigma = (a_{i_0} \cdots a_{i_r}) < \tau = (a_{j_0} \cdots a_{j_s})$ if and only if either $d(\sigma) = r < d(\tau) = s$, or $r = s$ and t exists with $i_0 = j_0, \cdots, i_{t-1} = j_{t-1}$ but $i_t < j_t$.

Theorem 8. Between the imbedding classes $\Phi^{2m-1} \in H^{2m-1,2}(K, I_2)$ and $\Phi^{2m} \in H^{2m,2}(K)$ of K we have the following relation:

$$\frac{1}{2}\delta\,\Phi^{2m-1} = \Phi^{2m}. \tag{1}$$

Proof. Let R^{2m} be a euclidean space of dimension $2m$, with a rectangular system of coordinates $(x_1, \cdots, x_{2m})$. Let R^s be the linear subspace of dimension s defined by $x_{s+1} = \cdots = x_{2m} = 0$ which is separated by R^{s-1} into two parts $R^s_+ : x_s > 0$ and $R^s_- : x_s < 0$, and will be oriented according as the ordered sequence of coordinates $x_1, \cdots, x_s$. Let l_s $(1 \leqslant s \leqslant m-1)$ be the line $x_1 = \cdots = x_{2s} = 1$, $x_{2s+2} = \cdots = x_{2m} = 0$ so that $l_s \subset R^{2s+1}$ and $l_s \cap R^{2s-1} = \emptyset$. Define an almost semi-linear realization $TK_1 = K_1'$ of K in R^{2m} through K_1 as follows. Let us take on R^1 a set of mutually different points $A_1, A_2, \cdots, A_N$ and on l_s a set of mutually different points $A_{i_0 \cdots i_s}$ $(1 \leqslant i_0 < \cdots < i_s \leqslant N)$, $1 \leqslant s \leqslant m-1$. Then

$$T(a_i) = A_i,$$
$$T(o_\sigma) = A_{i_0 \cdots i_s} = A_\sigma (\sigma = (a_{i_0} \cdots a_{i_s}) \in K)$$

defines uniquely a semi-linear realization T of K^{m-1} in $R^{2m-1} \subset R^{2m}$ through its barycentric subdivision. For any $\sigma = (a_{i_0} \cdots a_{i_s}) \in K$ with $d(\sigma) = s \geqslant m$ we may choose by §1 (A) a point $A_\sigma = A_{i_0 \cdots i_s}$ in R^{2m}_+ so that on defining

$$T(o_\sigma) = A_\sigma, \qquad \sigma \in K, \qquad d(\sigma) \geqslant m,$$

we may extend the above defined semi-linear realization to an *almost* semi-linear realization $TK_1 = K_1'$ of K in R^{2m} through K_1. Similarly, for any $\sigma \in K$ with $d(\sigma) \geqslant m$, we may also choose a point A_σ' in R_+^{2m-1} such that

$$\begin{cases} T'(a_i) = T(a_i) = A_i\,, & \\ T'(o_\sigma) = T(o_\sigma) = A_\sigma\,, & d(\sigma) \leqslant m-1\,, \quad \sigma \in K\,, \\ T'(o_\sigma) = A_\sigma'\,, & d(\sigma) \geqslant m\,, \quad \sigma \in K\,, \end{cases}$$

will define an *almost* semi-linear realization $T'K_1 = K_1''$ of K in R^{2m-1} through K_1.

With respect to R^{2m} and R^{2m-1} already oriented, T defines an imbedding cocycle $\varphi_T = \varphi^{2m} \in \Phi^{2m}$ and T' defines an imbedding cocycle $\varphi_{T'} = \varphi^{2m-1} \in \Phi^{2m-1}$. Let $\emptyset^s$ and Lk^s denote the intersection number and the linking number in the oriented R^s respectively. If $\sigma * \tau \in K^*$, $d(\sigma) + d(\tau) = 2m-1$ and $d(\sigma) < m-1$, $d(\tau) > m$, then by construction $T'\bar{\sigma} = T\bar{\sigma} \subset R^{2m-3}$, $T'(Int\bar{\tau}) \cap R^{2m-3} = \emptyset$, so that $\overline{T'\sigma} \cap \overline{T'\tau} = \emptyset$ and we have

$$\emptyset^{2m-1}(T'\sigma, T'\tau) = 0\,, \qquad d(\sigma) < m-1\,, \qquad d(\sigma) + d(\tau) = 2m-1\,. \quad (2)$$

Define now $\varphi_0^{2m-1} \in C^{2m-1}(K^*)$ by ($\sigma * \tau \in K^*$, $d(\sigma) + d(\tau) = 2m-1$, $d(\sigma) < d(\tau)$):

$$\varphi_0^{2m-1}(\sigma * \tau) = (-1)^m\, \emptyset^{2m-1}(T'\sigma, T'\tau)\,. \tag{3}$$

Then it is easy to see that (3) remains true for $d(\sigma) > d(\tau)$. By (2) we have then ($\sigma * \tau \in K^*$, $d(\sigma) + d(\tau) = 2m-1$):

$$\varphi_0^{2m-1}(\sigma * \tau) = \varphi_0^{2m-1}(\tau * \sigma) = 0\,, \qquad d(\sigma) < m-1\,, \qquad d(\tau) > m\,, \quad (3)'$$

and

$$\rho_2\, \varphi_0^{2m-1} = \varphi^{2m-1}\,, \tag{4}$$

where ρ_2 denotes reduction mod 2. Our object is to prove that

$$\delta \varphi_0^{2m-1} = 2\, \varphi^{2m}\,. \quad (4')$$

For this purpose let $\xi * \eta \in K^*$, $d(\xi) + d(\eta) = 2m$, and consider the various possible cases as follows:

Case I. $d(\xi) < d(\eta)$ so that $d(\xi) < m$, $d(\eta) > m$.

We have the

$$\delta\, \varphi_0^{2m-1}(\xi * \eta) = \varphi_0^{2m-1}\, \partial(\xi * \eta) =$$

$$= \varphi_0^{2m-1}(\partial \xi * \eta) + (-1)^{d(\xi)}\, \varphi_0^{2m-1}(\xi * \partial\eta) =$$

$$= (-1)^m \mathfrak{L}^{2m-1}(T'\partial\xi, T'\eta) + (-1)^m \cdot (-1)^{d(\xi)} \mathfrak{L}^{2m-1}(T'\xi, T'\partial\eta) =$$
$$= (-1)^m \mathfrak{L}^{2m-1}(T'\partial\xi, T'\eta) + (-1)^m \mathfrak{L}^{2m-1}(\partial T'\xi, T'\eta) =$$
$$= 2(-1)^m \mathfrak{L}^{2m-1}(T'\partial\xi, T'\eta).$$

As $d(\xi) < m$, any face in $\partial\xi$ has a dimension $< m-1$, so that by (2) we have

$$\delta\varphi_0^{2m-1}(\xi * \eta) = 0. \tag{5}$$

On the other hand, as $d(\xi) < m$, we have by construction $\overline{T}\overline{\xi} \subset R^{2m-1}$, $\overline{T}(Int\overline{\eta}) \cap R^{2m-1} = \emptyset$, so that $\overline{T}\overline{\xi} \cap \overline{T}\overline{\eta} = \emptyset$ and we have

$$\varphi^{2m}(\xi * \eta) = (-1)^{d(\xi)} \mathfrak{L}^{2m}(T\xi, T\eta) = 0. \tag{6}$$

Comparing (5) and (6), we get

$$\delta\varphi_0^{2m-1}(\xi * \eta) = 2\varphi^{2m}(\xi * \eta) = 0. \tag{7}$$

Case II. $d(\xi) > d(\eta)$ so that $d(\xi) > m$, $d(\eta) < m$.

We have then

$$\delta\varphi_0^{2m-1}(\xi * \eta) = (-1)^{d(\xi)\, d(\eta)} \delta\varphi_0^{2m-1}(\eta * \xi),$$
$$\varphi^{2m}(\xi * \eta) = (-1)^{d(\xi)\, d(\eta)} \varphi^{2m}(\eta * \xi).$$

Hence by (5) and (6), we again get (7).

Case III. $d(\xi) = d(\eta) = m$.

By (3) we have then

$$\delta\varphi_0^{2m-1}(\xi * \eta) = \varphi_0^{2m-1}(\partial\xi * \eta) + (-1)^m \varphi_0^{2m-1}(\xi * \partial\eta) =$$
$$= (-1)^m \mathfrak{L}^{2m-1}(T'\partial\xi, T'\eta) + \mathfrak{L}^{2m-1}(T'\xi, T'\partial\eta).$$

or

$$\delta\varphi_0^{2m-1}(\xi * \eta) = 2\mathfrak{L}^{2m-1}(T'\xi, T\partial\eta). \tag{8}$$

On the other hand, let B_η be the reflection of A_η with respect to R^{2m-1}, then $A_\eta T\partial\eta - B_\eta T\partial\eta$ is a cycle on integer coefficients, $B_\eta\overline{T}\overline{\eta}'$ is disjoint from $\overline{T}\overline{\xi}$, and $T\partial\xi = T'\partial\xi$. Hence

$$\varphi^{2m}(\xi * \eta) = (-1)^m \mathfrak{L}^{2m}(T\xi, T\eta) = (-1)^m \mathfrak{L}^{2m}(T\xi, A_\eta T\partial\eta) =$$
$$= (-1)^m \mathfrak{L}^{2m}(T\xi, A_\eta T\partial\eta - B_\eta T\partial\eta) =$$
$$= (-1)^m Lk^{2m}(\partial T\xi, A_\eta T\partial\eta - B_\eta T\partial\eta) =$$
$$= (-1)^m Lk^{2m}(\partial T'\xi, A_\eta T\partial\eta - B_\eta T\partial\eta) =$$
$$= (-1)^m \mathfrak{L}^{2m}(T'\xi, A_\eta T\partial\eta - B_\eta T\partial\eta) =$$
$$= (-1)^m \mathfrak{L}^{2m}(T'\xi, A_\eta T\partial\eta).$$

It may be seen that the last expression is the same as $\phi^{2m-1}(T'\xi, T\partial\eta)$, so that we have

$$\varphi^{2m}(\xi * \eta) = \emptyset^{2m-1}(T'\xi, T\partial\eta). \tag{9}$$

Comparing (8) and (9) we get

$$\delta\, \varphi_0^{2m-1}(\xi * \eta) = 2\varphi^{2m}(\xi * \eta). \tag{10}$$

From (7) and (10), we see that for any $\xi * \eta \in K^*$ with $d(\xi) + d(\eta) = 2m$, we have always (10). Hence we get (4)′. From (4) and (4)′ we get (1) and the theorem is proved.

Theorem 9. $\tilde{\Phi}^{2m} = 0$.

Proof. Let $\pi : \tilde{K}^* \to K^*$ be the covering projection. Then by Theorem 6,

$$\pi^* \Phi^{2m-1} = \rho_2 \tilde{\Phi}^{2m-1}, \qquad \pi^* \Phi^{2m} = \tilde{\Phi}^{2m}.$$

As $\pi^* \cdot \frac{1}{2}\delta = \frac{1}{2}\delta \cdot \pi^*$, we get by Theorem 8

$$\frac{1}{2}\delta(\rho_2 \tilde{\Phi}^{2m-1}) = \tilde{\Phi}^{2m}.$$

As $\tilde{\Phi}^{2m-1}$ is a cohomology class on integer coefficients, we have $\tilde{\Phi}^{2m} = 0$.

§6. Explicit Expressions of Certain Representative Cocycles in Φ^{2m-1}

We will make the same assumptions and use the same notations about K as in the preceding section. For any $\sigma * \tau \in K^*$, let $\{\sigma * \tau\}$ denote the cochain on integral coefficients of K^* which takes the value 1 on the cell $\sigma * \tau$ and the value 0 on all other cells of K^*. The purpose of this section is to prove the following

Theorem 10. The $(2m-1)$–dimensional imbedding class Φ^{2m-1} of K has a representative cocycle

$$\varphi^{2m-1} = \rho_2 \sum \{(a_{i_0} \cdots a_{i_{m-1}}) * (a_{j_0} \cdots a_{j_m})\}, \tag{1}$$

in which $\sum$ is extended over all possible sets of indices (i, j) for which $j_0 < i_0 < j_1 < \cdots < i_{m-1} < j_m$. Similarly, $\tilde{\Phi}^{2m-1}$ has also a representative cocycle $\tilde{\varphi}^{2m-1}$ given by

$$\tilde{\varphi}^{2m-1} = \sum [\{(a_{i_0} \cdots a_{i_{m-1}}) \times (a_{j_0} \cdots a_{j_m})\} - \{(a_{j_0} \cdots a_{j_m}) \times (a_{i_0} \cdots a_{i_{m-1}})\}], \tag{1'}$$

in which $\sum$ has the same meaning as in (1).

The proof of this theorem will be divided into several steps as follows.

1°. **Lemma.** Let $R^{2h-3} \subset R^{2h-1}$ and σ, τ be euclidean simplexes of dimension r, s respectively, with $r + s = 2h - 2$. Suppose $\bar{\sigma}^{\cdot}$ and $\bar{\tau}^{\cdot}$ to be disjoint, and T a semi-linear imbedding of $\sigma^{\cdot} + \tau^{\cdot}$ in R^{2h-3}. Let l be a line in R^{2h-1} not meeting and also not parallel to R^{2h-3}, and A_0, A_1, A_2 be three mutually different points on l. Orient R^{2h-3} as an oriented simplex ξ in R^{2h-3} and orient R^{2h-1} as the oriented simplex $A_1 A_2 \xi$. Then

$$\text{Ø}^{2h-1}(A_1 A_2 T\partial\sigma, A_0 T\partial\tau) =$$

$$= \begin{cases} (-1)^r Lk^{2h-3}(T\partial\sigma, T\partial\tau), & \text{if } A_0 \text{ lies between } A_1, A_2, \\ 0, & \text{if otherwise;} \end{cases} \tag{2}$$

or what is the same,

$$\text{Ø}^{2h-1}(A_0 T\partial\tau, A_1 A_2 T\partial\sigma) =$$

$$= \begin{cases} Lk^{2h-3}(T\partial\tau, T\partial\sigma), & \text{if } A_0 \text{ lies between } A_1, A_2, \\ 0, & \text{if otherwise,} \end{cases} \tag{2$'$}$$

in which Ø^k and Lk^k denote respectively the intersection number and the linking number in the oriented R^k $(k = 2h-1, 2h-3)$.

Proof. Suppose first A_0 does not lie between A_1 and A_2. If $A_1 A_2 T\bar{\sigma}^{\cdot}$ and $A_0 T\bar{\tau}^{\cdot}$ have an intersecting point, then there must exist points x, y in $T\bar{\sigma}^{\cdot}$, $T\bar{\tau}^{\cdot}$ respectively and a point z on l between A_1, A_2, such that the segments zx and $A_0 y$ will meet in the above point. As $A_0 \neq z$ and $x \neq y$ by hypothesis, the two lines $A_0 z = l$ and $xy \subset R^{2h-3}$ would have intersecting points, contrary to supposition. Hence $A_1 A_2 T\bar{\sigma}^{\cdot}$ and $A_0 T\bar{\tau}^{\cdot}$ are disjoint and we get the lower half of (2) or (2)$'$.

Next suppose A_0 lies between A_1, A_2. We may then take in R^{2h-3} a point $A \notin T(\bar{\sigma}^{\cdot} + \bar{\tau}^{\cdot})$. Prolong $A_0 A$ to A_0' and join $A_1 A_0'$, $A_0' A_2$. Let $C = A_0' A_1 + A_1 A_2 + A_2 A_0'$. Take also in R^{2h-3} a point O' such that $O'(T\sigma^{\cdot} + T\tau^{\cdot})$ is an almost euclidean complex. Define a semi-linear imbedding T' of σ in R^{2h-3} through its barycentric subdivision by $T'/\sigma^{\cdot} \equiv T$ and $T'(O_\sigma) = O'$. Then

$$Lk^{2h-3}(T\partial\sigma, T\partial\tau) = Lk^{2h-3}(T'\partial\sigma, T\partial\tau) =$$
$$= Lk^{2h-3}(\partial T'\sigma, T\partial\tau) = \text{Ø}^{2h-3}(T'\sigma, T\partial\tau).$$

Since the 2-dimensional simplex $A_1 A_2 A_0'$ and R^{2h-3} meet in the single point A, we see according to the chosen orientations of R^{2h-3}, R^{2h-1} that

$$\text{Ø}^{2h-3}(T'\sigma, T\partial\tau) = \text{Ø}^{2h-1}(CT'\sigma, T\partial\tau) = \text{Ø}^{2h-1}(CT'\sigma, \partial(A_0 T\partial\tau)) =$$
$$= (-1)^r \text{Ø}^{2h-1}(\partial(CT'\sigma), A_0 T\partial\tau) = (-1)^r \text{Ø}^{2h-1}(C\partial T'\sigma, A_0 T\partial\tau) =$$
$$= (-1)^r \text{Ø}^{2h-1}(CT\partial\sigma, A_0 T\partial\tau).$$

Hence

$$Lk^{2h-3}(T\partial\sigma, T\partial\tau) = (-1)^r\, \emptyset^{2h-1}(CT\partial\sigma, A_0\, T\partial\tau) =$$
$$= (-1)^r\, [\emptyset^{2h-1}(A_1\, A_2\, T\partial\sigma, A_0\, T\partial\tau) + \emptyset^{2h-1}(A_0'\, A_1\, T\partial\sigma, A_0\, T\partial\tau) +$$
$$+ \emptyset^{2h-1}(A_2\, A_0'\, T\partial\sigma, A_0\, T\partial\tau)]\,. \tag{3}$$

Denote by R^{2h-2} the linear subspace determined by A_0A_0' and R^{2h-3}. Consider any point z of $A_0'A_1 + A_0'A_2$, any point x of $\overline{T\sigma}^{\cdot}$, and any point y of $\overline{T\tau}^{\cdot}$. In case $z \neq A_0'$, zx will meet R^{2h-2} only in $x \in R^{2h-3}$ while $A_0y \subset R^{2h-2}$, $A_0y \cap R^{2h-3} = (y) \neq (x)$. In case $z = A_0'$, $A_0'x$ and A_0y will lie wholly in R^{2h-2} but on opposite sides of R^{2h-3}. Hence whatever $z \in A_0'A_1 + A_1'A_2$, $x \in \overline{T\sigma}^{\cdot}$ and $y \in \overline{T\tau}^{\cdot}$ may be, zx and A_0y are always non-intersecting. Consequently $A_0'A_1\overline{T\sigma}^{\cdot} + A_0'A_2\overline{T\sigma}^{\cdot}$ and $A_0\overline{T\tau}'$ are disjoint and we have

$$\emptyset^{2h-1}(A_0'\, A_1\, T\partial\sigma, A_0\, T\partial\tau) = \emptyset^{2h-1}(A_2\, A_0'\, T\partial\sigma, A_0\, T\partial\tau) = 0\,.$$

It follows that (3) becomes the upper half of (2) and our Lemma is proved.

2°. Let R^{2m-1} be a euclidean space of dimension $2m-1$, having a rectangular system of coordinates $(x_1, \cdots, x_{2m-1})$. Let $R^s\,(1 \leqslant s \leqslant 2m-1)$ be the s–dimensional linear subspace of R^{2m-1} defined by $x_{s+1} = \cdots = x_{2m-1} = 0$, l_s $(1 \leqslant s \leqslant m-1)$ the line $x_1 = \cdots = x_{2s} = 1$, $x_{2s+2} = \cdots = x_{2m-1} = 0$, such that $l_s \subset R^{2s+1}$ and l_s meet R^{2s} in the single point $O_s = (\underbrace{1, \cdots, 1}_{2s}, \underbrace{0, \cdots, 0}_{2m-2s-1})$. Corresponding to each $s \geqslant 1$ and $\leqslant m-1$ and each $\sigma = (a_{i_0} \cdots a_{i_s}) \in K$, we shall take a point $A_{i_0 \cdots i_s} = A_\sigma$ with $x_{2s+1}(A_\sigma) > 0$ such that $\sigma < \tau$ (τ is another s-dimensional simplex of K) would imply $x_{2s+1}(A_\sigma) < x_{2s+1}(A_\tau)$, or in simpler form, $A_\sigma < A_\tau$. Take also points $A_i = (i, \underbrace{0, \cdots, 0}_{2m-2})$ on R^1. Define now an almost semi-linear imbedding T of K in R^{2m-1} through its barycentric subdivision K_1 by

$$T(a_i) = A_i\,,$$
$$T(o_\sigma) = A_\sigma\,, \quad \sigma \in K\,, \quad 0 < \dim\sigma \leqslant m-1,$$

and for $\sigma \in K$, let $\dim\sigma \geqslant m$, $T(o_\sigma) = A_\sigma$ be a point conveniently chosen in $R_+^{2m-1}(x_{2m-1} > 0)$. We have then

$$T(a_{i_0} \cdots a_{i_h}) = A_{i_0 \cdots i_h} T\partial(a_{i_0} \cdots a_{i_h})\,, \tag{4}$$

and

$$T\partial(a_{i_0} \cdots a_{i_h}) = \partial \sum_{r,s} (-1)^{r+s} A_{i_0 \cdots \hat{i}_r \cdots i_h} A_{i_0 \cdots \hat{i}_s \cdots i_h} T(a_{i_0} \cdots \hat{a}_{i_r} \cdots \hat{a}_{i_s} \cdots a_{i_h}), \tag{5}$$

in which $h \leqslant m$ and $\hat{a}_i(\hat{i}_t)$ means that $a_i(i_t)$ does not appear in the corresponding sequence.

Proof. (4) is quite evident. The right hand side of (5) is

$$=-\sum_{r,s}(-1)^{r+s}A_{i_0\cdots\hat{i}_r\cdots i_h}T(a_{i_0}\cdots\hat{a}_{i_r}\cdots\hat{a}_{i_s}\cdots a_{i_h})+$$

$$+\sum_{r,s}(-1)^{r+s}A_{i_0\cdots\hat{i}_s\cdots i_h}T(a_{i_0}\cdots\hat{a}_{i_r}\cdots\hat{a}_{i_s}\cdots a_{i_h})+$$

$$+\sum_{r,s}(-1)^{r+s}A_{i_0\cdots\hat{i}_r\cdots i_h}A_{i_0\cdots\hat{i}_s\cdots i_h}\partial T(a_{i_0}\cdots\hat{a}_{i_r}\cdots\hat{a}_{i_s}\cdots a_{i_h})\,.$$

Denote the three terms successively by $\sum_1$, $\sum_2$ and $\sum_3$, then

$$\sum\nolimits_3=\sum_{r,s,t}(-1)^{r+s+t}(A_{i_0\cdots\hat{i}_r\cdots i_h}A_{i_0\cdots\hat{i}_s\cdots i_h}-A_{i_0\cdots\hat{i}_r\cdots i_h}A_{i_0\cdots\hat{i}_t\cdots i_h}+$$

$$+A_{i_0\cdots\hat{i}_s\cdots i_h}A_{i_0\cdots\hat{i}_t\cdots i_h})T(a_{i_0}\cdots\hat{a}_{i_r}\cdots\hat{a}_{i_s}\cdots\hat{a}_{i_t}\cdots a_{i_h})\,.$$

As $A_{i_0\cdots\hat{i}_\lambda\cdots i_h}(\lambda=r,s,t)$ are all on the line l_{h-1}, we have $(r<s<t)$

$$A_{i_0\cdots\hat{i}_r\cdots i_h}A_{i_0\cdots\hat{i}_s\cdots i_h}-A_{i_0\cdots\hat{i}_r\cdots i_h}A_{i_0\cdots\hat{i}_t\cdots i_h}+A_{i_0\cdots\hat{i}_s\cdots i_h}A_{i_0\cdots\hat{i}_t\cdots i_h}=0\,.$$

Hence

$$\Sigma_3=0\,. \tag{6}$$

Next we have

$$\sum\nolimits_1+\sum\nolimits_2=\sum_{r,s}(-1)^{r+r}A_{i_0\cdots\hat{i}_r\cdots i_h}[-T(a_{i_0}\cdots\hat{a}_{i_r}\cdots\hat{a}_{i_s}\cdots a_{i_h})+T(a_{i_0}\cdots\hat{a}_{i_s}\cdots\hat{a}_{i_r}\cdots a_{i_h})]=$$

$$=\sum(-1)^r A_{i_0\cdots\hat{i}_r\cdots i_h}T\partial(a_{i_0}\cdots\hat{a}_{i_r}\cdots a_{i_h})=$$

$$=T\sum_r(-1)^r(a_{i_0}\cdots\hat{a}_{i_r}\cdots a_{i_h})\,,$$

or

$$\sum\nolimits_1+\sum\nolimits_2=T\partial(a_{i0}\cdots a_{i_h})\,. \tag{6$'$}$$

From (6) and (6)′ we get (5).

3°. Let us orient R^s, $s=1,\cdots,2m-1$, as its coordinate sequence $x_1,\cdots,x_s$ and orient l_s by the increasing values of x_s. For any $0\leqslant h\leqslant m-1$ and any two simplexes $\sigma=(a_{i_0}\cdots a_{i_{m-1}})$, $\tau=(a_{j_0}\cdots a_{j_m})$ of K having no vertices in common, let

$$\mathcal{O}^{2m-2h-1}(T(a_{i_h}\cdots a_{i_{m-1}}),\sum_{r,s}(-1)^{r+s}A_{j_h\cdots\hat{j}_r\cdots j_m}A_{j_h\cdots\hat{j}_s\cdots j_m}T(a_{j_h}\cdots\hat{a}_{j_r}\cdots\hat{a}_{j_s}\cdots a_{j_m}))=I_h\,. \tag{7$_h$}$$

Then

$$I_h=\begin{cases}(-1)^{(m-h)(m-h-1)/2}, & j_h<i_h<j_{h+1}<\cdots<i_{m-1}<j_m\,,\\ 0\,, & \text{otherwise}.\end{cases} \tag{8$_h$}$$

Proof. For $h=m-1$, we have

$$I_{m-1} = \emptyset^1(T(a_{i_{m-1}}), -A_{j_m}A_{i_{m-1}}) = \emptyset^1(A_{i_{m-1}}, A_{i_{m-1}}A_{j_m}) .$$

Since $A_{i_{m-1}}$, $A_{j_{m-1}}$, A_{j_m} are all on the line R^1 and $A_{i_{m-1}}$ lies between $A_{j_{m-1}}$ and A_{j_m} if and only if $j_{m-1} < i_{m-1} < j_m$, we have

$$I_{m-1} = \begin{cases} +1, & j_{m-1} < i_{m-1} < j_m , \\ 0, & i_{m-1} < j_{m-1} \text{ or } i_{m-1} > j_m . \end{cases}$$

Hence $(8)_{m-1}$ is true. Suppose now $(8)_{h+1}, \cdots, (8)_{m-1}$ have been proved and let use prove $(8)_h$ as follows.

Case I. $i_h < j_h$ or $i_h > j_{h+1}$.

By construction $A_{i_h \cdots i_{m-1}}$, $A_{j_h \cdots \hat{j}_r \cdots j_m}$ and $A_{j_h \cdots \hat{j}_s \cdots j_m}$ are all on the line $l_t(t = m - h - 1)$ and for $i_h < j_h$, we have always $A_{i_h \cdots i_{m-1}} < A_{j_h \cdots \hat{j}_r \cdots j_m}$, $r = h, \cdots m$, while for $i_h > j_{h+1}$, we have always $A_{i_h \cdots i_{m-1}} > A_{j_h \cdots \hat{j}_r \cdots j_m}$, $r = h, \cdots, m$. Hence whatever be k and $r \neq s (h \leqslant k \leqslant m - 1, h \leqslant r < s \leqslant m)$, we have

$$A_{i_h \cdots i_{m-1}} \overline{T}(a_{i_h} \cdots \hat{a}_{i_k} \cdots a_{i_{m-1}}) \cap A_{j_h \cdots \hat{j}_r \cdots j_m} A_{j_h \cdots \hat{j}_s \cdots j_m} \overline{T}(a_{j_h} \cdots \hat{a}_{j_r} \cdots \hat{a}_{j_s} \cdots a_{j_m}) = \emptyset .$$

Hence by (4) we get

$$I_h = \emptyset^{2m-2h-1}(A_{i_h \cdots i_{m-1}} T \sum_k (-1)^{k-h} (a_{i_h} \cdots \hat{a}_{i_k} \cdots a_{i_{m-1}}) ,$$
$$\sum_{r,s} (-1)^{r+s} A_{j_h \cdots \hat{j}_r \cdots j_m} A_{j_h \cdots \hat{j}_s \cdots j_m} T(a_{j_h} \cdots \hat{a}_{j_r} \cdots \hat{a}_{j_s} \cdots a_{j_m})) = 0 .$$

This is the lower half of $(8)_h$.

Case II. $j_h < i_h < j_{h+1}$.

Again by (4) we get

$$I_h = \emptyset^{2m-2h-1}[A_{i_h \cdots i_{m-1}} \partial T(a_{i_h} \cdots a_{i_{m-1}}) ,$$
$$\sum_{r,s} (-1)^{r+s} A_{j_h \cdots \hat{j}_r \cdots j_m} A_{j_h \cdots \hat{j}_s \cdots j_m} T(a_{j_h} \cdots \hat{a}_{j_r} \cdots \hat{a}_{j_s} \cdots a_{j_m})] .$$

Since $A_{i_h \cdots i_{m-1}} < A_{j_h \cdots \hat{j}_r \cdots j_m}$ for $r > h$, we have for $r, s > h$,

$$\emptyset^{2m-2h-1}[A_{i_h \cdots i_{m-1}} \partial T(a_{i_h} \cdots a_{i_{m-1}}), A_{j_h \cdots \hat{j}_r \cdots j_m} A_{j_h \cdots \hat{j}_s \cdots j_m} T(a_{j_h} \cdots \hat{a}_{j_r} \cdots \hat{a}_{j_s} \cdots a_{j_m})] = 0.$$

As

$$A_{j_{h+1} \cdots j_m} A_{j_h \cdots \hat{j}_s \cdots j_m} = A_{j_{h+1} \cdots j_m} A_{j_h j_{h+2} \cdots j_m} + A_{j_h j_{h+2} \cdots j_m} A_{j_h \cdots \hat{j}_s \cdots j_m} ,$$

the expression for I_h may be further simplified as

$$I_h = \emptyset^{2m-2h-1} \Big[A_{i_h \cdots i_{m-1}} \partial T(a_{i_h} \cdots a_{i_{m-1}}) ,$$
$$(-1)^h \sum_s (-1)^s A_{j_{h+1} \cdots j_m} A_{j_h \cdots \hat{j}_s \cdots j_m} T(a_{j_{h+1}} \cdots \hat{a}_{j_s} \cdots a_{j_m}) \Big] =$$

$$= (-1)^h \mathscr{L}^{2m-2h-1}\Big[A_{i_h\cdots i_{m-1}}\partial T(a_{i_h}\cdots a_{i_{m-1}}),$$

$$\sum_s (-1)^s A_{j_{h+1}\cdots j_m} A_{j_h j_{h+2}\cdots j_m} T(a_{j_{h+1}}\cdots \hat{a}_{j_s}\cdots a_{j_m})\Big] =$$

$$= -\mathscr{L}^{2m-2h-1}[A_{i_h\cdots i_{m-1}}\partial T(a_{i_h}\cdots a_{i_{m-1}}),\ A_{j_{h+1}\cdots j_m} A_{j_h j_{h+1}\cdots j_m}\partial T(a_{j_{h+1}}\cdots a_{j_m})].$$

Applying the Lemma of 1°, and (4), (5) we get

$$I_h = Lk^{2m-2h-3}[\partial T(a_{i_h}\cdots a_{i_{m-1}}),\ \partial T(a_{j_{h+1}}\cdots a_{j_m})] =$$

$$= Lk^{2m-2h-3}\Big[\sum_k (-1)^{k-h} T(a_{i_h}\cdots \hat{a}_{i_k}\cdots a_{i_{m-1}}),$$

$$\partial \sum_{r,s} (-1)^{r+s} A_{j_{h+1}\cdots \hat{j}_r\cdots j_m} A_{j_{h+1}\cdots \hat{j}_s\cdots j_m} T(a_{j_{h+1}}\cdots \hat{a}_{j_r}\cdots \hat{a}_{j_s}\cdots a_{j_m})\Big] =$$

$$= (-1)^{m+1}\mathscr{L}^{2m-2h-3}\Big[\sum_k (-1)^k A_{i_h\cdots \hat{i}_k\cdots i_{m-1}} T\partial(a_{i_h}\cdots \hat{a}_{i_k}\cdots a_{i_{m-1}}),$$

$$\sum_{r,s} (-1)^{r+s} A_{j_{h+1}\cdots \hat{j}_r\cdots j_m} A_{j_{h+1}\cdots \hat{j}_s\cdots j_m} T(a_{j_{h+1}}\cdots \hat{a}_{j_r}\cdots \hat{a}_{j_s}\cdots a_{j_m})\Big].$$

For $k > h$, we have $i_h < j_{h+1}$ and $A_{i_h\cdots \hat{i}_k\cdots i_{m-1}} < A_{j_{h+1}\cdots \hat{j}_r\cdots j_m}$ $r = h+1, \cdots, m$, hence $(k > h)$

$$\mathscr{L}^{2m-2h-3}[A_{i_h\cdots \hat{i}_k\cdots i_{m-1}} T\partial(a_{i_h}\cdots \hat{a}_{i_k}\cdots a_{i_{m-1}}),$$

$$A_{j_{h+1}\cdots \hat{j}_r\cdots j_m} A_{j_{h+1}\cdots \hat{j}_s\cdots j_m} T(a_{j_{h+1}}\cdots \hat{a}_{j_r}\cdots \hat{a}_{j_s}\cdots a_{j_m})] = 0,$$

and I_h may be simplified as

$$I_h = (-1)^{m+h+1}\mathscr{L}^{2m-2h-3}\Big[A_{i_{h+1}\cdots i_{m-1}} T\partial(a_{i_{h+1}}\cdots a_{i_{m-1}}),$$

$$\sum_{r,s} (-1)^{r+s} A_{j_{h+1}\cdots \hat{j}_r\cdots j_m} A_{j_{h+1}\cdots \hat{j}_s\cdots j_m} T(a_{j_{h+1}}\cdots \hat{a}_{j_r}\cdots \hat{a}_{j_s}\cdots a_{j_m})\Big] =$$

$$= (-1)^{m+h+1}\mathscr{L}^{2m-2h-3}\Big[T(a_{i_{h+1}}\cdots a_{i_{m-1}}),$$

$$\sum_{r,s} (-1)^{r+s} A_{j_{h+1}\cdots \hat{j}_r\cdots j_m} A_{j_{h+1}\cdots \hat{j}_s\cdots j_m} T(a_{j_{h+1}}\cdots \hat{a}_{j_r}\cdots \hat{a}_{j_s}\cdots a_{j_m})\Big] =$$

$$= (-1)^{m+h+1} I_{h+1}.$$

By induction hypothesis $(8)_{h+1}$, we get $(8)_h$.

4°. We now prove Theorem 1° as follows.

Take the point $A'_{i_0\cdots i_{m-1}} : x_{2m-1}(A'_{i_0\cdots i_{m-1}}) < 0$ symmetric to $A_{i_0\cdots i_{m-1}}$ with respect to $o_{m-1} = l_{m-1} \cap R^{2m-2}$ and define a semi-linear realization T'_σ of $\sigma = (a_{i_0}\cdots a_{i_{m-1}}) \in K$ in R^{2m-1} through the barycentric subdivision of σ by

$$T'_\sigma(a_i) = A_i\,; \quad i = i_0, \cdots, i_{m-1}\,; \quad T'_\sigma|\sigma^\cdot \equiv T\,; \quad T'_\sigma(0_\sigma) = A'_\sigma .$$

Then $T\sigma - T'_\sigma\sigma$ is a cycle. For any $\tau \in K$ having no vertex in common with σ where $d(\tau) + d(\sigma) \leqslant 2m - 1$, we have $\bar{T}'_\sigma\bar{\sigma} \cap \bar{T}\bar{\tau} = \emptyset$, hence $\emptyset^{2m-1}(T'_\sigma\sigma, T\tau) = 0$.

Let $\sigma = (a_{i_0} \cdots a_{i_{m-1}})$, $\tau = (a_{j_0} \cdots a_{j_m})$, $\sigma \times \tau \in \tilde{K}^*$, then

$$(-1)^{m-1}\tilde{\varphi}_T^{2m-1}(\sigma \times \tau) = \emptyset^{2m-1}(T\sigma\,, T\tau) = \emptyset^{2m-1}(T\sigma - T'_\sigma\sigma, T\tau) =$$

$$= \emptyset^{2m-1}\Big[T\sigma - T'_\sigma\sigma\,,\ T\tau - \sum_{r,s}(-1)^{r+s}A_{j_0\cdots\hat{j}_r\cdots j_m}A_{j_0\cdots\hat{j}_s\cdots j_m}T(a_{j_0}\cdots\hat{a}_{j_r}\cdots\hat{a}_{j_s}\cdots a_{j_m})\Big] +$$

$$+ \emptyset^{2m-1}\Big[T\sigma - T'_\sigma\sigma\,, \sum_{r,s}(-1)^{r+s}A_{j_0\cdots\hat{j}_r\cdots j_m}A_{j_0\cdots\hat{j}_s\cdots j_m}T(a_{j_0}\cdots\hat{a}_{j_r}\cdots\hat{a}_{j_s}\cdots a_{j_m})\Big].$$

By (5), $T\sigma - T'_\sigma\sigma$ and $T\tau - \sum_{r,s}(-1)^{r+s}A_{j_0\cdots\hat{j}_r\cdots j_m}A_{j_0\cdots\hat{j}_s\cdots j_m}T(a_{j_0}\cdots\hat{a}_{j_r}\cdots\hat{a}_{j_s}\cdots a_{j_m})$ are both cycles, so that the first term of the right hand side in the above expression vanishes. As $\bar{T}'_\sigma\bar{\sigma}$ is disjoint from $A_{j_0\cdots\hat{j}_r\cdots j_m}A_{j_0\cdots\hat{j}_s\cdots j_m}\bar{T}(a_{j_0}\cdots\hat{a}_{j_r}\cdots\hat{a}_{j_s}\cdots a_{j_m})$, the above expression may be simplified as

$$(-1)^{m-1}\tilde{\varphi}_T^{2m-1}(\sigma \times \tau) = \emptyset^{2m-1}\Big[T\sigma\,,$$
$$\sum_{r,s}(-1)^{r+s}A_{j_0\cdots\hat{j}_r\cdots j_m}A_{j_0\cdots\hat{j}_s\cdots j_m}T(a_{j_0}\cdots\hat{a}_{j_r}\cdots\hat{a}_{j_s}\cdots a_{j_m})\Big].$$

By (7), (8) of 3°, we get

$$\tilde{\varphi}_T^{2m-1}(\sigma \times \tau) = \begin{cases}(-1)^{(m-1)(m-2)/2}, & j_0 < i_0 < j_1 < \cdots < i_{m-1} < j_m\,,\\ 0\,, & \text{otherwise.}\end{cases} \tag{9}$$

From this we get further

$$\tilde{\varphi}_T^{2m-1}(\tau \times \sigma) = -\tilde{\varphi}_T^{2m-1}(\sigma \times \tau) =$$

$$= \begin{cases} -(-1)^{(m-1)(m-2)/2}\,, & j_0 < i_0 < j_1 < \cdots < i_{m-1} < j_m\,,\\ 0\,, & \text{otherwise.}\end{cases} \tag{9$'$}$$

Next let $\sigma \times \tau \in \tilde{K}^*$, $d(\sigma) + d(\tau) = 2m - 1$, while $d(\sigma) < m - 1$, $d(\tau) > m$. Then $\bar{T}\bar{\sigma} \subset R^{2m-3}$ and $\bar{T}(\mathrm{Int}\,\bar{\tau}) \cap \bar{T}\bar{\sigma} = \emptyset$. Hence

$$\tilde{\varphi}_T^{2m-1}(\sigma \times \tau) = 0\,, \quad d(\sigma) < m - 1\,, \quad d(\tau) > m\,. \tag{10}$$

Similarly,

$$\tilde{\varphi}_T^{2m-1}(\tau \times \sigma) = 0\,, \quad d(\sigma) < m - 1\,, \quad d(\tau) > m\,. \tag{10$'$}$$

From (9), (9)′, (10) and (10)′ we get

$$\tilde{\varphi}_T^{2m-1} = \sum (-1)^{(m-1)(m-2)/2}[\{(a_{i_0}\cdots a_{i_{m-1}}) \times (a_{j_0}\cdots a_{j_m})\} - $$
$$- \{(a_{j_0}\cdots a_{j_m}) \times (a_{i_0}\cdots a_{i_{m-1}})\}],$$

in which $\sum$ is extended over all possible sets of indices (i, j) for which $j_0 < i_0 < j_1 < \cdots < i_{m-1} < j_m$. Now by Theorem 4 and the remark below $\tilde{\varphi}_T$ and $-\tilde{\varphi}_T$ are both cocycles in $\tilde{\Phi}^{2m-1}$. Hence $\tilde{\Phi}^{2m-1}$ has a representative cocycle as given by (1)′.

Similarly we have a representative cocycle in Φ^{2m-1} as given in (1). Our theorem is thus completely proved.

§7. Explicit Expressions for Certain Representative Cocycles in Φ^{2m}

The notations will be the same as in the preceding section.

Theorem 11. In Φ^{2m} there is a representative cocycle

$$\varphi^{2m} = \sum \{(a_{i_0}\cdots a_{i_m})*(a_{j_0}\cdots a_{j_m})\}, \tag{1}$$

in which $\sum$ is extended over all possible sets of indices (i, j) such that $i_0 < j_0 < \cdots < i_m < j_m$.

Proof. By Theorem 10 of §6, Φ^{2m-1} has a representative cocycle

$$\varphi^{2m-1} = \rho_2 \sum \{(a_{i_0}\cdots a_{i_{m-1}})*(a_{j_0}\cdots a_{j_m})\},$$

in which $\sum$ is extended over all possible sets of indices (i, j) such that $j_0 < i_0 < j_1 < \cdots < i_{m-1} < j_m$. Define now $\varphi_0^{2m-1} \in C^{2m-1}(K^*)$ by

$$\varphi_0^{2m-1} = \sum \{(a_{i_0}\cdots a_{i_{m-1}})*(a_{j_0}\cdots a_{j_m})\}, \tag{2}$$

in which $\sum$ is as before, then

$$\rho_2\varphi_0^{2m-1} = \varphi^{2m-1}. \tag{3}$$

By Theorem 7 of §5, $\Phi^{2m} = \frac{1}{2}\delta\Phi^{2m-1}$, hence Φ^{2m} has a representative cocycle φ_0^{2m} such that

$$\delta\varphi_0^{2m-1} = 2\varphi_0^{2m}. \tag{4}$$

We prove now φ_0^{2m} is the same as φ^{2m} in (1) as follows.

Let $\sigma * \tau = (a_{i_0}\cdots a_{i_p}) * (a_{j_0}\cdots a_{j_q}) \in K^*$, $i_0 < j_0$, $p + q = 2m$, then

$$\delta\varphi_0^{2m-1}(\sigma*\tau) = \varphi_0^{2m-1}(\partial(a_{i_0}\cdots a_{i_p})*(a_{j_0}\cdots a_{j_q})) +$$
$$+ (-1)^p\varphi_0^{2m-1}((a_{i_0}\cdots a_{i_p})*\partial(a_{j_0}\cdots a_{j_q}))\,, \qquad (5)$$

or

$$\delta\varphi_0^{2m-1}(\sigma*\tau) = \sum\nolimits_1 + (-1)^p \sum\nolimits_2 \,,$$

where

$$\left.\begin{aligned}\sum\nolimits_1 &= \sum_r (-1)^r\varphi_0^{2m-1}((a_{i_0}\cdots \hat{a}_{i_r}\cdots a_{i_p})*(a_{j_0}\cdots a_{j_q}))\,,\\ \sum\nolimits_2 &= \sum_r (-1)^r\varphi_0^{2m-1}((a_{j_0}\cdots \hat{a}_{j_r}\cdots a_{j_q})*(a_{i_0}\cdots a_{i_p}))\,.\end{aligned}\right\} \qquad (6)$$

Consider now various possible cases as follows (i_0 always $< j_0$):

Case I. $p = q = m$, $i_0 < j_0 < \cdots < i_m < j_m$.

By (2) we have

$$\varphi_0^{2m-1}((a_{i_0}\cdots \hat{a}_{i_r}\cdots a_{i_m})*(a_{j_0}\cdots a_{j_m})) = \begin{cases}1\,, & r=0\,,\\ 0\,, & r>0\,;\end{cases}$$

$$\varphi_0^{2m-1}((a_{j_0}\cdots \hat{a}_{j_r}\cdots a_{j_m})*(a_{i_0}\cdots a_{i_m})) = \begin{cases}1\,, & r=m\,,\\ 0\,, & r<m\,.\end{cases}$$

Hence $\sum_1 = 1$, $\sum_2 = (-1)^m$, and (5) becomes

$$\delta\varphi_0^{2m-1}(\sigma*\tau) = 2\,.$$

Comparing with (1) we get

$$\delta\varphi_0^{2m-1}(\sigma*\tau) = 2\varphi^{2m}(\sigma*\tau)\,. \qquad (7)$$

Case II. $p = q = m$, and $i_0 < j_0 < \cdots < i_m < j_m$ does not hold.

In that case there exists either an index s with no j_k satisfying $i_s < j_k < i_{s+1}$ so that

$$\varphi_0^{2m-1}(a_{j_0}\cdots \hat{a}_{j_r}\cdots a_{j_m})*(a_{i_0}\cdots a_{i_m})) = 0\,, \qquad (8)$$

or an index s with no i_k satisfying $j_s < i_k < j_{s+1}$. In the second alternative we have still (8) for $r \neq s$, $s+1$, while for $r = s$, $s+1$, $\varphi_0^{2m-1}((a_{j_0}\cdots \hat{a}_{j_s}\cdots a_{j_m}) * (a_{i_0}\cdots a_{i_m}))$ and $\varphi_0^{2m-1}((a_{j_0}\cdots \hat{a}_{j_{s+1}}\cdots a_{j_m}) * (a_{i_0}\cdots a_{i_m}))$ have the same value 0 or 1 so that $\sum_2 = (-1)^s + (-1)^{s+1} = 0$. Hence we have always $\sum_2 = 0$.

As $i_0 < j_0$, we have $\varphi_0^{2m-1}((a_{i_0}\cdots \hat{a}_{i_r}\cdots a_{i_m}) * (a_{j_0}\cdots a_{j_m})) = 0$ for

$r > 0$. It is also $= 0$ for $r = 0$ since $j_0 < i_1 < j_1 < \cdots < i_m < j_m$ is not true. Hence we have always $\sum_1 = 0$.

It follows that $\delta\varphi_0^{2m-1}(\sigma * \tau) = 0$. As $\varphi^{2m}(\sigma * \tau) = 0$ by (1), we get (7).

Case III. $p < m - 1$ or $q < m - 1$.

In that case each term of $\sum_1$, $\sum_2$ is 0 and we get still (7) by comparing with (1).

Case IV. $p = m - 1$, $q = m + 1$.

As $i_0 < j_0 < j_1$, we have $\sum_2 = 0$. Comparing with (1), we get (7).

Case V. $p = m + 1$, $q = m - 1$.

In that case $\sum_2 = 0$, while

$$\sum_1 = \sum_r (-1)^r \varphi_0^{2m-1}((a_{j_0} \cdots a_{j_{m-1}}) * (a_{i_0} \cdots \hat{a}_{i_r} \cdots a_{i_{m+1}})) .$$

Now we have necessarily an index s with no j_k satisfying $i_s < j_k < i_{s+1}$ $(0 \leqslant s \leqslant m)$. Then $\varphi_0^{2m-1}((a_{j_0} \cdots a_{j_{m-1}}) * (a_{i_0} \cdots \hat{a}_{i_r} \cdots a_{i_{m+1}}))$ are all 0 for $r \neq s$, $s + 1$ while for $r = s$, $s + 1$ they are both of same value 0 or 1. We have thus always $\sum_1 = 0$. Comparing with (1) we get again (7).

Thus whatever the case may be, we get always (7) for any $\sigma * \tau \in K^*(d(\sigma) + d(\tau) = 2m)$. Hence φ_0^{2m} in (4) coincides with φ^{2m} of (1) and the theorem is proved.

§ 8. Relations between Φ^i and Their Topological Invariance

As before let the vertices of K be arranged in a fixed order $a_1 < a_2 < \cdots$, and all simplexes of K be written in normal form $(a_{i_0} \cdots a_{i_k})$, with $i_0 < i_1 < \cdots < i_k$. Since $\tilde{K}^*$ is a two-sheeted covering complex of K^*, we may define as in [9] §1 chain transformations

$$t\colon\ C(\tilde{K}^*, G) \to C(\tilde{K}^*, G) ,$$
$$\pi\colon\ C(\tilde{K}^*, G) \to C(K^*, G) ,$$

and

$$\bar{\pi}\colon\ C(K^*, G) \to C(\tilde{K}^*, G) ,$$

such that for $(a_{i_0} \cdots a_{i_p})$, $(a_{j_0} \cdots a_{j_q}) \in K$ with no vertices in common, we have

$$t((a_{i_0} \cdots a_{i_p}) \times (a_{j_0} \cdots a_{j_q})) = (-1)^{pq}((a_{j_0} \cdots a_{j_q}) \times (a_{i_0} \cdots a_{i_p})) , \tag{1}$$

$$\pi((a_{i_0} \cdots a_{i_p}) \times (a_{j_0} \cdots a_{j_q})) = (a_{i_0} \cdots a_{i_p}) * (a_{j_0} \cdots a_{j_q}) , \tag{2}$$

$$\pi((a_{i_0}\cdots a_{i_p})*(a_{i_0}\cdots a_{i_q})) = (a_{i_0}\cdots a_{i_p}) \times (a_{j_0}\cdots a_{j_q}) + \\ + (-1)^{pq}(a_{j_0}\cdots a_{j_q}) \times (a_{i_0}\cdots a_{i_p}) . \quad (3)$$

Put $s = 1 + t$, $d = 1 - t$, and let π', $\bar{\pi}'$, t', s', d' be the dual cochain transformations of π, $\cdots$. For any classes $X \in H^r(K^*, I_2)$ and $Y \in H^r(K^*)$, let us take $\tilde{x}_0$, $\tilde{x}_1 \in C^*(\tilde{K}^*, I_2)$ and $\tilde{y}_0$, $\tilde{y}_1$, $\tilde{y}_2 \in C^*(\tilde{K}^*)$ such that

$$\bar{\pi}'\tilde{x}_0 \in X , \quad \delta\tilde{x}_0 = d'\tilde{x}_1 , \tag{4}$$

$$\bar{\pi}'\tilde{y}_0 \in Y , \quad \delta\tilde{y}_0 = d'\tilde{y}_1 , \quad \delta\tilde{y}_1 = s'\tilde{y}_2 . \tag{4$'$}$$

Then $\bar{\pi}'\tilde{x}_1$, and $\bar{\pi}'\tilde{y}_2$ are cocycles mod 2 and integral respectively whose classes are independent of the choice of x, y and may be denoted by $\mu^* X \in H^{r+1}(K^*, I_2)$ and $\nu^* Y \in H^{r+2}(K^*)$. By [9] §1 (or [10] §1), we have (1 denotes the unit class mod 2 or integral in K^*)

$$\mu^* X = \mu^* 1 \cup X . \tag{5}$$

Similarly we have (cf. [10] §1):

$$\nu^* Y = \nu^* 1 \cup Y , \tag{6}$$

and

$$\rho_2\nu^* = (\mu^*)^2\rho_2 . \tag{7}$$

We shall now first establish the following

Theorem 12. Let Φ^0 be the integral unit class of K^*, then

$$\Phi^{2i+1} = \mu^*\rho_2\Phi^{2i} , \qquad i \geqslant 0 , \tag{8}$$

$$\Phi^{2i+2} = \nu^*\Phi^{2i} , \qquad i \geqslant 0 , \tag{9}$$

$$\rho_2\Phi^{2i+2} = \mu^*\Phi^{2i+1} , \qquad i \geqslant 0 . \tag{10}$$

Proof. Define $\tilde{\varphi}^i \in C^i(\tilde{K}^*)$, $i \geqslant 0$, as follows:

$$\tilde{\varphi}^{2m-1}((a_{i_0}\cdots a_{i_{m-1}}) \times (a_{j_0}\cdots a_{j_m})) = \begin{cases} 1 , & j_0 < i_0 < \cdots < j_m , \\ 0 , & \text{otherwise} ; \end{cases} \tag{11}$$

$$\tilde{\varphi}^{2m-1}((a_{i_0}\cdots a_{i_p}) \times (a_{j_0}\cdots a_{j_q})) = 0 , \quad p + q = 2m - 1 , \quad (p, q) \neq (m-1, m), \tag{11$'$}$$

$$\tilde{\varphi}^{2m}((a_{i_0}\cdots a_{i_m}) \times (a_{j_0}\cdots a_{j_m})) = \begin{cases} 1 , & i_0 < j_0 < \cdots < j_m , \\ 0 , & \text{otherwise} ; \end{cases} \tag{12}$$

$$\tilde{\varphi}^{2m}((a_{j_0}\cdots a_{j_p}) \times (a_{i_0}\cdots a_{i_q})) = 0 , \quad p + q = 2m , \quad (p, q) \neq (m, m) . \tag{12$'$}$$

Then we shall show that

$$\delta\tilde{\varphi}^{2m-1} = s'\tilde{\varphi}^{2m} , \tag{I}$$

and

$$\delta\tilde{\varphi}^{2m} = (-1)^m d'\tilde{\varphi}^{2m+1}. \tag{II}$$

Let us first prove (I). Consider any $(a_{i_0} \cdots a_{i_p}) \times (a_{j_0} \cdots a_{j_q}) \in \tilde{K}^*$ $p + q = 2m$, then we have

$$\delta\tilde{\varphi}^{2m-1}((a_{i_0} \cdots a_{i_p}) \times (a_{j_0} \cdots a_{j_q})) = \sum\nolimits_1 + (-1)^p \sum\nolimits_2 , \tag{13}$$

where

$$\sum\nolimits_1 = \sum_r (-1)^r \tilde{\varphi}^{2m-1}((a_{i_0} \cdots \hat{a}_{i_r} \cdots a_{i_p}) \times (a_{j_0} \cdots a_{j_q})) , \tag{13$_1$}$$

$$\sum\nolimits_2 = \sum_r (-1)^r \tilde{\varphi}^{2m-1}((a_{i_0} \cdots a_{i_p}) \times (a_{j_0} \cdots \hat{a}_{j_r} \cdots a_{j_q})) . \tag{13$_2$}$$

For the calculation of (13) let us consider various possible cases as follows.

Case I. $(p, q) \neq (m, m)$.

In that case $(p-1, q) \neq (m-1, m)$, hence $\sum_1 = 0$ by (11)'. If $(p, q) \neq (m-1, m+1)$, then by (11)' $\sum_2 = 0$ too. Suppose now $p = m-1$, $q = m+1$, then there exists an index s with no i_k satisfying $j_s < i_k < j_{s+1}$. For $r \neq s, s+1$ we have then $\tilde{\varphi}^{2m-1}((a_{i_0} \cdots a_{i_{m+1}}) \times (a_{j_0} \cdots \hat{a}_{j_r} \cdots a_{j_{m+1}})) = 0$, while for $r = s, s+1$, $\tilde{\varphi}^{2m-1}(a_{i_0} \cdots a_{i_{m-1}}) \times (a_{j_0} \cdots \hat{a}_{j_s} \cdots a_{j_m}))$ and $\tilde{\varphi}^{2m-1}((a_{i_0} \cdots a_{i_{m-1}}) \times (a_{j_0} \cdots \hat{a}_{j_{s+1}} \cdots a_{j_m}))$ are either both 0 or both 1. Hence $\sum_2 = 0$ always and we have

$$\delta\tilde{\varphi}^{2m-1}((a_{i_0} \cdots a_{i_p}) \times (a_{j_0} \cdots a_{j_q})) = 0 , \quad (p, q) \neq (m, m), \quad p + q = 2m .$$

Case II. $(p, q) = (m, m)$.

By (11)', $\sum_2$ is evidently 0.

If there is an s with no i_k satisfying $j_s < i_k < j_{s+1}$, then $\sum_1 = 0$ by (11).

If there is an s with no j_k satisfying $i_s < j_k < i_{s+1}$, then by (11), $\tilde{\varphi}^{2m-1}((a_{i_0} \cdots \hat{a}_{i_r} \cdots a_{i_m}) \times (a_{j_0} \cdots a_{j_m})) = 0$ for $r \neq s$, $s+1$, while for $r = s$, $s+1$, they are either both 0 or both 1. Hence $\sum_1 = 0$ always.

If there is no such s, then either $i_0 < j_0 < \cdots < i_m < j_m$ or $j_0 < i_0 < \cdots < j_m < i_m$. By (10), we have then

$$\tilde{\varphi}^{2m-1}((a_{i_0} \cdots \hat{a}_{i_r} \cdots a_{i_m}) \times (a_{j_0} \cdots a_{j_m})) = \begin{cases} 1, & r = 0 \\ 0, & r > 0 \end{cases} \quad (i_0 < j_0 < \cdots < i_m < j_m) ,$$

$$= \begin{cases} 1, & r = m \\ 0, & r < m \end{cases} \quad (j_0 < i_0 < \cdots < j_m < i_m) .$$

Hence

$$\sum\nolimits_1 = \begin{cases} 1, & i_0 < j_0 < \cdots < i_m < j_m, \\ (-1)^m, & j_0 < i_0 < \cdots < j_m < i_m, \\ 0, & \text{otherwise}. \end{cases}$$

Combining the above cases together, we have

$$\delta\tilde{\varphi}^{2m-1}((a_{i_0}\cdots a_{i_p}) \times (a_{j_0}\cdots a_{j_q})) = \begin{Bmatrix} 1, & p=q=m, & i_0<j_0<\cdots<i_m<j_m, \\ (-1)^m, & p=q=m, & j_0<i_0<\cdots<j_m<i_m, \\ 0, & & \text{otherwise}. \end{Bmatrix} \quad (14)$$

Next, by (1) and (12), (12)′, we have

$$s'\tilde{\varphi}^{2m}((a_{i_0}\cdots a_{i_m}) \times (a_{j_0}\cdots a_{j_m})) =$$
$$= \tilde{\varphi}^{2m}((a_{i_0}\cdots a_{i_m}) \times (a_{j_0}\cdots a_{j_m})) + (-1)^m\tilde{\varphi}^{2m}((a_{j_0}\cdots a_{j_m}) \times (a_{i_0}\cdots a_{i_m})) =$$
$$= \begin{cases} 1, & i_0 < j_0 < \cdots < i_m < j_m, \\ (-1)^m, & j_0 < i_0 < \cdots < j_m < i_m, \\ 0 & \text{otherwise}. \end{cases}$$
$$s'\tilde{\varphi}^{2m}((a_{i_0}\cdots a_{i_p}) \times (a_{j_0}\cdots a_{j_q})) = 0, \qquad (pq) \neq (m, m).$$

Comparing with (14), we get (I).

Next let us prove (II) as follows. Consider any cell $(a_{i_0}\cdots a_{i_p}) \times (a_{j_0}\cdots a_{j_q}) \in \tilde{K}^*$, $p + q = 2m + 1$, we have

$$\delta\tilde{\varphi}^{2m}((a_{i_0}\cdots a_{i_p}) \times (a_{j_0}\cdots a_{j_q})) = \sum\nolimits_1 + (-1)^p \sum\nolimits_2, \qquad (15)$$

where

$$\sum\nolimits_1 = \sum (-1)^r\tilde{\varphi}^{2m}((a_{i_0}\cdots \hat{a}_{i_r}\cdots a_{i_p}) \times (a_{j_0}\cdots a_{j_q})), \qquad (16)_1$$

$$\sum\nolimits_2 = \sum (-1)^r\tilde{\varphi}^{2m}((a_{i_0}\cdots a_{i_p}) \times (a_{j_0}\cdots \hat{a}_{j_r}\cdots a_{j_q})). \qquad (16)_2$$

Case I. $(p, q) \neq (m, m+1)$ or $(m+1, m)$.

We have evidently $\sum_1 = \sum_2 = 0$ by (12)′.

Case II. $(p, q) = (m, m+1)$.

By (12)′ we have $\sum_1 = 0$, while by (12) we have by the same method as above

$$\sum\nolimits_2 = \begin{cases} 1, & j_0 < i_0 < \cdots < i_m < j_{m+1}, \\ 0 & \text{otherwise}. \end{cases}$$

Case III. $(p, q) = (m+1, m)$.

In that case we have $\sum_2 = 0$, while

$$\sum_1 = \begin{cases} (-1)^{m+1}, & i_0 < j_0 < \cdots < j_m < i_{m+1}, \\ 0, & \text{otherwise.} \end{cases}$$

Combining together all the preceding cases, we get

$$\delta\tilde{\varphi}^{2m}((a_{i_0}\cdots a_{i_p}) \times (a_{j_0}\cdots a_{j_q})) =$$
$$= \left\{ \begin{array}{lll} (-1)^m, & (p, q) = (m, m+1), & j_0 < i_0 < \cdots < i_m < j_{m+1}, \\ (-1)^{m+1}, & (p, q) = (m+1, m), & i_0 < j_0 < \cdots < j_m < i_{m+1}, \\ 0, & & \text{otherwise.} \end{array} \right\} \tag{17}$$

Next by (1), (11) and (11)′, we have

$$d'\tilde{\varphi}^{2m+1}((a_{i_0}\cdots a_{i_p}) \times (a_{j_0}\cdots a_{j_q})) =$$
$$= \tilde{\varphi}^{2m+1}((a_{i_0}\cdots a_{i_p}) \times (a_{j_0}\cdots a_{j_q})) - (-1)^{pq}\tilde{\varphi}^{2m+1}((a_{j_0}\cdots a_{j_q}) \times (a_{i_0}\cdots a_{i_p})) =$$
$$= \begin{cases} 1, & (p, q) = (m, m+1), \quad j_0 < i_0 < \cdots < i_m < j_{m+1}, \\ -1, & (p, q) = (m+1, m), \quad i_0 < j_0 < \cdots < j_m < i_{m+1}, \\ 0, & \text{otherwise.} \end{cases}$$

Combining with (17), we get (II).

Define now $\varphi^i \in C^i(K^*)$ by

$$\varphi^i = \pi'\tilde{\varphi}^i, \qquad i \geqslant 0. \tag{18}$$

Then by (3), (11), (11)′, (12) and (12)′, we have

$$\varphi^{2m-1}((a_{i_0}\cdots a_{i_p})*(a_{j_0}\cdots a_{j_q})) =$$
$$= \begin{cases} 1, & p = m-1, \quad q = m, \quad j_0 < i_0 < j_1 < \cdots < i_{m-1} < j_m, \\ 0, & \text{otherwise,} \end{cases} \qquad (i_0 > j_0); \tag{19}$$

$$\varphi^{2m}((a_{i_0}\cdots a_{i_p})*(a_{j_0}\cdots a_{j_q})) =$$
$$= \begin{cases} 1, & p = q = m, \quad i_0 < j_0 < \cdots < i_m < j_m, \\ 0, & \text{otherwise,} \end{cases} \qquad (i_0 < j_0). \tag{20}$$

By Theorem 10 of §6 and Theorem 11 of §7,

$$\rho_2\varphi^{2m-1} \in \Phi^{2m-1}, \qquad m > 0, \tag{21}$$

and

$$\varphi^{2m} \in \Phi^{2m}, \qquad m > 0. \tag{22}$$

Moreover φ^0 is the integral unit cocycle on K^*. Hence from (I), (II), (18), (21), (22) and the definitions of μ^* and ν^*, we get (8), (10) and also $\nu^*\Phi^{2m} = (-1)^m \Phi^{2m+2}$. By Theorem 4 of §3, $2\Phi^{2m+2} = 0$. Hence the last equation is the same as (9) and our theorem is proved.

From the above theorem and (5), (6), (7) we get also the following theorems:

Theorem 13.

$$\Phi^{2i} \cup \Phi^{2j} = \Phi^{2i+2j}, \tag{23}$$

$$\rho_2 \Phi^{2i} \cup \Phi^{2j+1} = \Phi^{2i+2j+1}, \tag{24}$$

$$\Phi^{2i+1} \cup \Phi^{2j+1} = \rho_2 \Phi^{2i+2j+2}. \tag{25}$$

Theorem 14. Denote by $(\ )^i$ the i–fold powers by cup products, then

$$\Phi^{2i} = (\Phi^2)^i, \qquad i > 0, \tag{26}$$

$$\Phi^{2i+1} = (\Phi^1)^{2i+1}, \qquad i \geqslant 0, \tag{27}$$

$$\rho_2 \Phi^{2i} = (\Phi^1)^{2i}, \qquad i > 0. \tag{28}$$

Theorem 15. If $\Phi^m = 0$, then for any $i > 0$, we have $\Phi^{m+i} = 0$.

Theorem 16. $\tilde{\Phi}^{2m-1} = 0$, for $m > 0$.

Proof. Define $\tilde{\varphi}^{2m-1}$ by (11) and (11)′, we have by Theorem 10 of § 6 ($\sum$ being extended over all possible sets of indices (i, j) with $j_0 < i_0 < j_1 < \cdots < i_{m-1} < j_m$):

$$d'\tilde{\varphi}^{2m-1} = \sum \{(a_{i_0} \cdots a_{i_{m-1}}) \times (a_{j_0} \cdots a_{j_m})\} - \sum \{(a_{j_0} \cdots a_{j_m}) \times (a_{i_0} \cdots a_{i_{m-1}})\} \in \tilde{\Phi}^{2m-1}.$$

But by (II) we have $d'\tilde{\varphi}^{2m-1} \sim 0$. Hence $\tilde{\Phi}^{2m-1} = 0$.

Combining this theorem with Theorem 9 of § 5, we see that $\tilde{\Phi}^m$ are always $= 0$ $(m > 0)$. Hence $\tilde{\Phi}^m$ are practically useless. On the other hand, as we shall see in the two following sections, Φ^m are generally $\neq 0$ and play an important role in the study of realization of complexes.

Theorem 17. All the imbedding classes $\Phi^m \in H^{m,2}(K, I_{(m)}) = H^{m,2}(P, I_{(m)})$, $m > 0$, of a complex K are topological invariants of the polyhedron $\bar{K} = P$.

Proof. Let L be another simplicial subdivision of P. As the construction of K^* and $\tilde{K}^*$ from K, let L^*, $\tilde{L}^*$ be the corresponding complexes constructed from L and let $\omega : \tilde{L}^* \to L^*$ be the covering projection. By [6], we know that the spaces $\bar{\tilde{L}}^*$, $\bar{\tilde{K}}^*$ of the complexes $\tilde{L}^*$, $\tilde{K}^*$ have same homotopy type, and the same is true for the spaces $\bar{L}^*$, $\bar{K}^*$ of L^*, K^*. Moreover, the identity of these homotopy types may be realized by continuous maps $\tilde{f} : \bar{\tilde{L}}^* \to \bar{\tilde{K}}^*$ and $f : \bar{L}^* \to \bar{K}^*$ such that $\pi\tilde{f} = = f\omega$. It follows that μ^*, ν^* are commutative with f^* (cf. [9] § 1 or [10] § 1) and by Theorem 12 we have (Φ^0 denotes the integral unit class):

$$f^* \Phi^{2i}(K) = f^*(\nu^*)^i \Phi^\circ(K) = (\nu^*)^i f^* \Phi^\circ(K) = (\nu^*)^i \Phi^\circ(L) = \Phi^{2i}(L)\,,$$

$$f^* \Phi^{2i+1}(K) = f^*(\mu^*)^{2i+1} \rho_2 \Phi^\circ(K) = (\mu^*)^{2i+1} f^* \rho_2 \Phi^\circ(K) =$$

$$= (\mu^*)^{2i+1} \rho_2 \Phi^\circ(L) = \Phi^{2i+1}(L)\,.$$

Since $f^* : H^{m,2}(K, I_{(m)}) \approx H^{m,2}(L, I_{(m)})$ may be considered as the identity homomorphism of $H^{m,2}(P, I_{(m)})$, the last two equations show that $\Phi^m(K)$ and $\Phi^m(L)$, $m > 0$, are identical elements in $H^{m,2}(P, I_{(m)})$. In other words, $\Phi^m(K) \in H^{m,2}(P, I_{(m)})$, $m > 0$, are independent of the subdivision K of P and are therefore topological invariants of P.

The above theorem may also be slightly extended as follows.

Let $P \subset Q$ be a regular pair of finite polyhedrons so that P, Q have simplicial subdivisions L, K respectively for which L is a regular subcomplex of K (cf. [6]). Construct complexes L^* and K^* as before, then L^* is a subcomplex of K^* and the inclusion map i will induce homomorphisms

$$i^* : H^m(K^*, G) \to H^m(L^*, G)\,,$$

or

$$i^* : H^{m,2}(K, G) \to H^{m,2}(L, G)\,. \tag{29}$$

As in [6], these homomorphisms are really independent of the choice of the subdivisions K, L and may thus be written as

$$i^* : H^{m,2}(Q, G) \to H^{m,2}(P, C)\,. \tag{30}$$

As in the preceding theorem we may then prove the following

Theorem 18. Let $P \subset Q$ (or $L \subset K$) be a regular pair of finite polyhedrons (or a regular pair of finite simplicial complexes). Define i^* as the homomorphisms in (29), (30) induced by the inclusion map $i : P \subset Q$ (or $i : L \subset K$), then

$$i^* \Phi^m(Q) = \Phi^m(P)\,,$$

or

$$i^* \Phi^m(K) = \Phi^m(L)\,.$$

§ 9. Complexes Realizable in R^{m+1} but not in R^m

Given integers $n > 0$ and $N \geqslant n$. Let us take $N+1$ linearly independent points $a_0, \cdots, a_N$ in R^N which span an N-dimensional simplex Δ_N. The n-dimensional skeleton of Δ_N is an n-dimenional complex $K_{N,n}$. Using the notations $K \subset R^m$ and $K \not\subset R^m$ to denote that K can or cannot be semi-linearly realized in R^m, we have the following

Theorem 19.

$$K_{N,n} \subset R^{2n+1}, \qquad N \geqslant 2n+2. \qquad (1) \qquad\qquad K_{N,n} \not\subset R^{2n}, \qquad N \geqslant 2n+2. \qquad (1)'_n$$

$$K_{m+2,n} \subset R^{m+1}, \quad 2n \geqslant m \geqslant n. \qquad (2) \qquad\qquad K_{m+2,n} \not\subset R^{m}, \quad 2n \geqslant m \geqslant n. \quad (2)'_{m,n}$$

$$K_{n+1,n} \subset R^{n+1}. \qquad (3) \qquad\qquad K_{n+1,n} \not\subset R^{n}. \qquad (3)'$$

$$K_{n,n} \subset R^{n}. \qquad (4) \qquad\qquad K_{n,n} \not\subset R^{n-1}. \qquad (4)'$$

In the proof below, a_i will be arranged in the order $a_0 < \cdots < a_N$ and φ^m will denote the representative cocycle in Φ^m as asserted in Theorem 10 of § 6 and Theorem 11 of § 7. All simplexes of $K_{N,n}$ will also be supposed to be written in normal forms $(a_{i_0} \cdots a_{i_k}) : 0 \leqslant i_0 < \cdots < i_k \leqslant N$.

Proof of (1).

This is a classical result, of which the proof is quite simple (cf. [1] Chap. 1 and [2] Chap. 3 § 2).

Proof of (2).

In R^{m+1} let us take $m+2$ linearly independent points $a'_0, \cdots, a'_{m+1}$ which span a simplex Δ'_{m+1}, and a point a'_{m+2} in the interior of Δ'_{m+1}. Let $K'_{m+2,n}$ be the complex formed of all k-dimensional simplexes $(0 \leqslant k \leqslant n)$ with vertices taken from a'_i $(i = 0, 1, \cdots, m+2)$. Then $T(a_i) = a'_i$, $i = 0, 1, \cdots, m+2$, define a linear realization $K'_{m+2,n} = TK_{m+2,n}$ of $K_{m+2,n}$ in R^{m+1}.

(3) and (4) are evident.

Before proceeding to the proof of (1)′—(4)′, let us first remark that (1)′ is the well-known result of Van Kampen and Flores[3-5] (3)′ states that an n-sphere is not imbeddable in R^n and (4)′ states that an n-simplex is not imbeddable in R^{n-1}. Both (3)′ and (4)′ are classical results, of which the proof of the former depends on Alexander's duality theorem, and that of the latter is a consequence of a theorem of Brouwer connected with theory of dimension. In what follows we shall give (1)′—(4)′ a unified proof which makes it plausible that in all these cases the non-imbeddability is owing to the same fact, namely, the corresponding imbedding class Φ^m is $\neq 0$ (cf. Theorem 5 of § 3).

Proof of (1)′.

Evidently it is sufficient to prove that $K_{2n+2,n} \not\subset R^n$.

For this let us consider in $K^*_{2n+2,n}$ the following integral chain

$$z = \sum_{i_0 < j_0} (a_{i_0} \cdots a_{i_n}) * (a_{j_0} \cdots a_{j_n}), \tag{5}$$

or what is the same,

$$z = \frac{1}{2} \sum \varepsilon_{i_0 j_0} (a_{i_0} \cdots a_{i_n}) * (a_{j_0} \cdots a_{j_n}), \tag{5$'$}$$

in which the preceding $\sum$ is extended over all possible cells with $i_0 < j_0$, the second $\sum$ is extended over all possible cells, and $\varepsilon_{i_0 j_0} = +1$ or $(-1)^n$ according as $i_0 < j_0$ or $i_0 > j_0$.

Let us now calculate ∂z. Consider any cell $(a_{k_0} \cdots a_{k_{n-1}}) * (a_{l_0} \cdots a_{l_n}) \in K^*_{2n+2,n}$. Let r, s be the two remaining numbers after removing $k_0, \cdots, k_{n-1}, l_0, \cdots, l_n$ from $0, 1, \cdots, 2n+2$. Suppose $r < s$ and $k_0 < \cdots < k_{\alpha-1} < r < k_\alpha < \cdots < k_{\beta-1} < s < k_\beta < \cdots < k_{n-1}$. Then the term $(a_{k_0} \cdots a_{k_{n-1}}) * (a_{l_0} \cdots a_{l_n})$ in ∂z is produced from the following terms in (5)′:

$$\partial\,(a_{k_0} \cdots a_{k_{\alpha-1}} a_r a_{k_\alpha} \cdots a_{k_{n-1}}) * (a_{l_0} \cdots a_{l_n}),$$
$$\partial\,(a_{k_0} \cdots a_{k_{\beta-1}} a_s a_{k_\beta} \cdots a_{k_{n-1}}) * (a_{l_0} \cdots a_{l_n}),$$
$$(a_{l_0} \cdots a_{l_n}) * \partial\,(a_{k_0} \cdots a_{k_{\alpha-1}} a_r a_{k_\alpha} \cdots a_{k_{n-1}}),$$
$$(a_{l_0} \cdots a_{l_n}) * \partial\,(a_{k_0} \cdots a_{k_{\beta-1}} a_s a_{k_\beta} \cdots a_{k_{n-1}}).$$

Hence the coefficient λ of the term $(a_{k_0} \cdots a_{k_{n-1}}) * (a_{l_0} \cdots a_{l_n})$ in ∂z is given by the following:

for $k_0 < r$,

$$\lambda = \frac{1}{2}\,[(-1)^\alpha \varepsilon_{k_0 l_0} + (-1)^\beta \varepsilon_{k_0 l_0} + (-1)^n \cdot (-1)^\alpha \cdot (-1)^{n(n-1)} \varepsilon_{l_0 k_0} + $$
$$+ (-1)^n \cdot (-1)^\beta \cdot (-1)^{n(n-1)} \varepsilon_{l_0 k_0}] = [(-1)^\alpha + (-1)^\beta]\, \varepsilon_{k_0 l_0},$$

for $r < k_0 < s$,

$$\lambda = \frac{1}{2}\,[\varepsilon_{r l_0} + (-1)^\beta \varepsilon_{k_0 l_0} + (-1)^n \cdot (-1)^{n(n-1)} \varepsilon_{l_0 r} + $$
$$+ (-1)^n \cdot (-1)^\beta \cdot (-1)^{n(n-1)} \varepsilon_{l_0 k_0}] = \varepsilon_{r l_0} + (-1)^\beta \varepsilon_{k_0 l_0},$$

and for $r < s < k_0$,

$$\lambda = \frac{1}{2}\,[\varepsilon_{r l_0} + \varepsilon_{s l_0} + (-1)^n \cdot (-1)^{n(n-1)} \varepsilon_{l_0 r} + $$
$$+ (-1)^n \cdot (-1)^{n(n-1)} \varepsilon_{l_0 s}] = \varepsilon_{r l_0} + \varepsilon_{s l_0}.$$

Consequently we have always $\lambda \equiv 0 \bmod 2$, and $\rho_2 z$ is a mod 2 cycle.

For any $r \geqslant 0$ and $\leqslant 2n+2$, define now

$$\alpha_r(s) = \begin{cases} s, & 0 \leqslant s \leqslant r-1, \\ s+1, & r \leqslant s \leqslant 2n+1. \end{cases}$$

By Theorem 11 of § 7, the $2n$–dimensional imbedding class Φ^{2n} of $K_{2n+2,n}$ has a representative cocycle

$$\varphi^{2n} = \sum_{r=0}^{2n+2} \{(a_{\alpha_r(0)} a_{\alpha_r(2)} \cdots a_{\alpha_r(2n)}) * (a_{\alpha_r(1)} a_{\alpha_r(3)} \cdots a_{\alpha_r(2n+1)})\}.$$

Hence $\varphi^{2n}(z) = 2n+3$ and $\rho_2 \varphi^{2n}(\rho_2 z) \neq 0$. It follows that $\rho_2 \Phi^{2n} \neq 0$ and we have $K_{2n+2,n} \not\subset R^{2n}$. i.e., (1)′, by Theorem 5 of § 3.

Proof of $(2)'_{2n-1,n}$, *i. e.*, $K_{2n+1,n} \not\subset R^{2n-1}$.

This may be derived from $(1)'$. Suppose we have a realization $T : K_{2n+1,n} \subset R^{2n-1}$. Consider R^{2n-1} as a linear subspace of R^{2n} and $K_{2n+1,n}$ as a subcomplex of $K_{2n+2,n}$. Take a point $a'_{2n+2} \in R^{2n}$ and $\notin R^{2n-1}$. Then by setting $T(a_{2n+2}) = a'_{2n+2}$, we may extend T to a realization $T : K_{2n+2,n} \subset R^{2n}$, contrary to $(1)'$. Hence $K_{2n+1,n} \not\subset R^{2n-1}$.

we may also give a direct proof as follows:

By Theorem 10 of § 6, Φ^{2n-1} has a representative cocycle

$$\varphi^{2n-1} = \rho_2 \sum \{(a_{i_0} \cdots a_{i_{n-1}}) * (a_{j_0} \cdots a_{j_n})\} =$$

$$= \rho_2 \sum_{r=0}^{2n+1} \{(a_{\alpha_r(1)}\, a_{\alpha_r(3)} \cdots a_{\alpha_r(2n-1)}) * (a_{\alpha_r(0)}\, a_{\alpha_r(2)} \cdots a_{\alpha_r(2n)})\},$$

in which the first $\sum$ is extended over all possible sets of indices (i, j) with $0 \leqslant j_0 < i_0 < j_1 < \cdots < i_{n-1} < j_n \leqslant 2n + 1$, and α_r is defined by

$$\alpha_r(s) = \begin{cases} s, & 0 \leqslant s \leqslant r+1, \\ s+1, & r \leqslant s \leqslant 2n. \end{cases}$$

Consider now in K^* a $(2n-1)$-dimensional integral chain

$$z = \sum_{j_0>0} (a_{i_0} \cdots a_{i_{n-1}}) * (a_{j_0} \cdots a_{j_n}),$$

in which $\sum$ is extended over all possible sets of indices (i, j) with $j_0 > 0$. It is easy to see that $\rho_2 z$ is a mod 2 cycle and $\varphi^{2n-1}(\rho_2 z) = 1$ mod 2. It follows that $\varphi^{2n-1} \not\sim 0$ or $\Phi^{2n-1} \neq 0$ and $K_{2n+1,n} \not\subset R^{2n-1}$, as we require to prove.

Proof of $(1)'_{m,n}$.

When $n = 1$, we have $m = 1, 2$ and $(2)'$ becomes $K_{3,1} \not\subset R^1$ and $K_{4,1} \not\subset R^2$ which is $(2)'_{2n-1,n}$ and $(1)'_n$ in the case $n = 1$ and hence is already proved.

Suppose now $(2)'_{m,n-1}$ for $2(n-1) \geqslant m \geqslant n-1$ has already been proved. Prove now $(2)_{m,n}$ as follows.

When $m = 2n$, $(2)'_{m,n}$ is the same as $(1)'_n$ and has been proved. The case $m = 2n-1$ has also been proved. We may suppose therefore $m \leqslant 2n-2$. In that case $K_{m+2,n}$ has a subcomplex $K_{m+2,n-1}$. By induction hypothesis $K_{m+2,n-1} \not\subset R^m$, hence we have *à fortiori* $K_{m+2,n} \not\subset R^m$, what we require to prove.

We may also reason as follows. Consider the following assertion:

$$\Phi^m(K_{m+2,n}) \neq 0, \quad 2n \geqslant m \geqslant n \qquad (2)''_{m,n}.$$

When $m = 2n$ or $m = 2n - 1$ and in particular for $n = 1$, we know already that $(2)''_{m,n}$ is true. Suppose now $(2)''_{m,n-1}$ is true and $m \leqslant 2(n-1)$. Then by induction hypothesis $\Phi^m(K_{m+2,n-1}) \neq 0$. As $K_{m+2,n-1}$ is a subcomplex of $K_{m+2,n}$, we have by Theorem 18 of § 8, $\Phi^m(K_{m+2,n}) \neq 0$ and $(2)''_{m,n}$ is also true.

From $(2)''_{m,n}$ we get $(2)'_{m,n}$. From this reasoning it is seen that the truth of $(2)'_{m,n}$ is again due to $\Phi^m \neq 0$.

Proof of (3)'.

Suppose first $n = 2n'$. Then

$$\varphi^n = \varphi^{2n'} = \{(a_0\, a_2 \cdots a_{2n'}) * (a_1\, a_3 \cdots a_{2n'+1})\} \in \Phi^n,$$

and

$$z = \rho_2 \sum_{i_0 < j_0} (a_{i_0} \cdots a_{i_{n'}}) * (a_{j_0} \cdots a_{j_{n'}}) + \rho_2 \sum_{r=1}^{n'} \sum (a_{k_0} \cdots a_{k_{n'-r}}) * (a_{l_0} \cdots a_{l_{n'+r}}),$$

may be easily seen to be a mod 2 cycle, in which the first $\sum$ is extended over all possible sets of indices (i, j) with $i_0 < j_0$, and the second $\sum$ in the second term is extended over all possible sets of indices (k, l). As $\rho_2\varphi^{2n}(z) = 1$ mod 2, we have $\varphi^n \not\sim 0$ or $\Phi^n \neq 0$ and hence $K_{n+1,n} \not\subset R^n$.

Suppose next $n = 2n' - 1$, then

$$\varphi^n = \varphi^{2n'-1} = \rho_2 \{(a_1\, a_3 \cdots a_{2n'-1}) * (a_0\, a_2 \cdots a_{2n'})\} \in \Phi^n.$$

Moreover

$$z = \rho_2 \sum_{r=0}^{n'-1} \sum (a_{i_0} \cdots a_{i_{n'-r-1}}) * (a_{j_0} \cdots a_{j_{n'+r}})$$

is a mod 2 cycle, in which $\sum$ is extended over all possible sets of indices (i, j). Since $\varphi^n(z) = 1$ mod 2 we have $\varphi^n \not\sim 0$ or $\Phi^n \neq 0$ and again $K_{n+1,n} \not\subset R^n$.

Proof of (4)'.

First, (4) may be derived from (3)'. Consider R^{n-1} as a linear subspace of R^n and $K_{n,n}$ as a subcomplex of $K_{n+1,n}$. If there exists a realization $T : K_{n,n} \subset R^{n-1}$, then on taking a point $a'_{n+1} \in R^n$ but $\notin R^{n-1}$ and setting $T(a_{n+1}) = a'_{n+1}$, we get an extension of T to a realization $T : K_{n+1,n} \subset R^n$, contrary to (3)'. Hence $K_{n,n} \not\subset R^{n-1}$.

Given now a direct proof by the unified method as follows.

Suppose first n be even: $n = 2n'$. By Theorem 10 of § 6 we have

$$\varphi^{n-1} = \varphi^{2n'-1} = \rho_2 \{(a_1\, a_3 \cdots a_{2n'-1}) * (a_0\, a_1 \cdots a_{2n'})\} \in \Phi^{n-1}.$$

Put

$$z = \sum_{r=1}^{n'} \sum (a_{i_0} \cdots a_{i_{n'-r}}) * (a_{j_0} \cdots a_{j_{n'+r+1}}),$$

in which $\sum$ is extended over all possible sets of indices (i, j). Then $\rho_2 z$ is a mod 2 cycle and $\varphi^{n-1}(\rho_2 z) = 1 \bmod 2$. Hence $\Phi^{n-1} \neq 0$ and we have $K_{n,n} \not\subset R^{n-1}$

Next suppose n be odd: $n = 2n' + 1$. By Theorem 11 of §7 we have

$$\varphi^{n-1} = \varphi^{2n'} = \{(a_0\, a_2 \cdots a_{2n'}) * (a_1\, a_3 \cdots a_{2n'+1})\} \in \Phi^{2n}.$$

Put

$$z = \sum_{r=1}^{n'} \sum (a_{i_0} \cdots a_{i_{n'-r}}) * (a_{j_0} \cdots a_{j_{n'+r}}) + \sum_{i_0 < j_0} (a_{i_0} \cdots a_{i_{n'}}) * (a_{j_0} \cdots a_{j_{n'}}),$$

in which the second $\sum$ is extended over all possible sets of indices (i, j) and the last $\sum$ is extended over all possible sets of indices (i, j) with $i_0 < j_0$. Then $\rho_2 z$ is a mod 2 cycle and $\rho_2 \varphi^{n-1}(\rho_2 z) = 1$ mod 2. Hence $\Phi^{n-1} \neq 0$ and we have again $K_{n,n} \not\subset R^{n-1}$.

Our theorem is now completely proved. From the proof we have furthermore the following theorems:

Theorem 20. The necessary and sufficient condition for $K_{N,n} \subset R^m$ is $\Phi^m(K_{N,n}) = 0$

Theorem 21. To any $n > 0$ and $m \leqslant 2n$ and $\geqslant n-1$, there exist complexes $K(m, n) \subset R^{m+1}$. But $\not\subset R^m$. In other words, for $n - 1 \leqslant m \leqslant 2n$, R^{m+1} contains always more n–dimensional conplexes than does R^m.

The complexes $K(m, n)$ in this theorem may be taken to be $K_{m+2,n}$ for $2n \geqslant m \geqslant n$ and $K_{n,n}$ for $m = n - 1$.

10. Another Example of Van Kampen[3] and Its Generalization

In this section we shall apply the theory developed in the preceding sections to give an alternative proof of the non-imbeddability of another n–dimensional complex in R^{2n} also due to Van Kampen. We give also its generalizations.

The n–dimensional complex K_n in this second example of Van Kampen is constructed in the following manner. Consider $n + 1$ sets of triple of points $a_i^{(0)}, a_i^{(1)}, a_i^{(2)}$ $(i = 0, 1, \cdots, n)$. Take one point from each of these $n + 1$ sets to form an n–simplex, say $(a_0^{(i_0)} a_1^{(i_1)} \cdots a_n^{(i_n)})$ $(i = 0, 1$ or $2)$. Then K_n is formed by all these simplexes as well as all their faces. We have then

Theorem 22. (Van Kampen)[3] The Van Kampen complex K_n defined above is non-imbeddable in R^{2n}.

Proof. Arrange the vertices of K_n in an order such that $a_i^{(k)} < a_j^{(l)}$ if and only if either $i < j$ or $i = j$ and $k < l$ $(i, j = 0, 1, \cdots, n;\ k, l = 0, 1$ or $2)$. By Theorem 11 of § 7, the $2n$-dimensional imbedding class $\Phi^{2n} \in H^{2n,2}(K_n)$ of K_n has with respect to this ordering of vertices a representative cocycle

$$\varphi^{2n} = \sum \{(a_0^{(i_0)} a_1^{(i_1)} \cdots a_n^{(i_n)}) * (a_0^{(j_0)} a_1^{(j_1)} \cdots a_n^{(j_n)})\},$$

in which $\sum$ is extended over all possible sets of indices (i, j) with $i_0 < j_0, i_1 < j_1, \cdots, i_n < j_n$. Next construct a $2n$-dimensional chain

$$z = \sum (a_0^{(i_0)} a_1^{(i_1)} \cdots a_n^{(i_n)}) * (a_0^{(j_0)} a_1^{(j_1)} \cdots a_n^{(j_n)}),$$

in which $\sum$ is extended over all possible sets of indices (i, j) with $i_0 < j_0$. It is easy to see that $\rho_2 z$ is a mod 2 cycle and $\varphi^{2n}(z) = 3^{n+1} = 1$ mod 2. Hence $\Phi^{2n} \neq 0$ and $K_n \not\subset R^{2n}$, what we require to prove.

Let us now extend Van Kampen's example in the following manner. Suppose we are given $p+1$ sets $(p \geqslant -1)$ of triple of points $a_i^{(0)}, a_i^{(1)}, a_i^{(2)}$, $i = 0, 1, \cdots, p$; and q sets $(q \geqslant 0)$ of pairs of points $a_j^{(0)}$, $a_j^{(1)}$, $j = p+1, \cdots, p+q$. Take one point from each of these $p+q+1$ sets and form a $(p+q)$–dimensional simplex. The complex formed by all these simplexes as well as their faces will be denoted by $A_{p,q}$. In particular, $A_{p,0}$ is the above–defined Van Kampen's complex K_p, and $A_{-1,q}$ is a subdivision of the $(q-1)$–dimensional sphere.

Theorem 23. For $p \geqslant 0$, we have $A_{p,q} \subset R^{2p+q+1}$, but $\Phi^{2p+q}(A_{p,q}) \neq 0$, so that $A_{p,q} \not\subset R^{2p+q}$.

Theorem 24. If $n \leqslant m \leqslant 2n$, then the n–dimensional complex $\subset R^{m+1}$ but $\not\subset R^m$ as stated in Theorem 21 may also be taken as $A_{m-n,2n-m}$.

Theorem 25. The necessary and sufficient condition for $A_{p,q} \subset R^m$ is $\Phi^m = 0$.

The last two theorems are both simple consequences of Theorem 23. In order to prove Theorem 23, we shall prove first a general result as follows. Given a simplicial complex L and two points b_0, b_1, the set of all simplexes $b_0\sigma$, $b_1\sigma$ and τ $(\sigma, \tau \in L)$ forms a simplicial complex K, written as $K = b_0 L + b_1 L$. In fact, K is the join complex of L and the 0–sphere $\{b_0 b_1\}$. We have then

Theorem 26. Let K be the join complex of the simplicial complex L and the 0–sphere $S^0 = \{b_0, b_1\}$. If $\rho_2 \Phi^{2m}(L) \neq 0$, then $\Phi^{2m+1}(K) \neq 0$. If $\Phi^{2m-1}(L) \neq 0$, then $\rho_2 \Phi^{2m}(K) \neq 0$.

Proof. Let us arrange the vertices of L in an order $a_1 < a_2 < \cdots$ and the vertices of K in the order $b_0 < b_1 < a_1 < a_2 < \cdots$. With respect to such ordering of vertices, the imbedding classes $\Phi^m(L)$, $\Phi^m(K)$ of L, K would have respectively representative cocycles $\varphi^m(L)$ and $\varphi^m(K)$ as follows.

$$\varphi^{2m}(L) = \sum\nolimits_1 \{(a_{i_0} \cdots a_{i_m}) * (a_{j_0} \cdots a_{j_m})\}, \tag{1}$$

$$\varphi^{2m-1}(L) = \rho_2 \sum\nolimits_2 \{(a_{i_0} \cdots a_{i_{m-1}}) * (a_{j_0} \cdots a_{j_m})\}, \tag{2}$$

$$\begin{aligned}\varphi^{2m+1}(K) = {} & \rho_2 \sum_k \sum_1 \{(a_{i_0} \cdots a_{i_m}) * (b_k a_{j_0} \cdots a_{j_m})\} + \\ & + \rho_2 \sum\nolimits_2 \{(b_1 a_{i_0} \cdots a_{i_{m-1}}) * (b_0 a_{j_0} \cdots a_{j_m})\} + \\ & + \rho_2 \sum\nolimits_3 \{(a_{i_0} \cdots a_{i_m}) * (a_{j_0} \cdots a_{j_{m+1}})\},\end{aligned} \tag{3}$$

$$\begin{aligned}\varphi^{2m}(K) = {} & \sum_k \sum\nolimits_2 \{(b_k a_{i_0} \cdots a_{i_{m-1}}) * (a_{j_0} \cdots a_{j_m})\} + \\ & + \sum\nolimits_4 \{(b_0 a_{i_0} \cdots a_{i_{m-1}}) * (b_1 a_{j_0} \cdots a_{j_{m-1}})\} + \\ & + \sum\nolimits_1 \{(a_{i_0} \cdots a_{i_m}) * (a_{j_0} \cdots a_{j_m})\}.\end{aligned} \tag{4}$$

In these equations $\sum_1, \cdots, \sum_4$ are to be extended over all possible sets of indices (i, j) satisfying respectively:

$$\begin{aligned} & i_0 < j_0 < i_1 < \cdots < i_m < j_m, && (\textstyle\sum_1) \\ & j_0 < i_0 < j_1 < \cdots < i_{m-1} < j_m, && (\textstyle\sum_2) \\ & j_0 < i_0 < j_1 < \cdots < i_m < j_{m+1}, && (\textstyle\sum_3) \\ & i_0 < j_0 < i_1 < \cdots < i_{m-1} < j_{m-1}. && (\textstyle\sum_4) \end{aligned}$$

To any chain $z = \sum_i c_i(\sigma_i * \tau_i) \in C_r(L^*)$, define now a chain $B_0 z \in C_{r+1}(K^*)$ by

$$B_0 z = \sum_i c_i\,(\sigma_i * b_0 \tau_i + b_0 \sigma_i * \tau_i).$$

Then $B_0 : C_r(L^*) \to C_{r+1}(K^*)$ is a homomorphism satisfying

$$\partial B_0 z = B_0 \partial z \qquad \text{mod } 2. \tag{5}$$

For any $\sigma * \tau \in L^*$, $\dim \sigma + \dim \tau = r$, $c \in C_r(L^*)$, we have also

$$\{\sigma * b_0 \tau\}(B_0 c) = \{b_0 \sigma * \tau\}(B_0 c) = \{\sigma * \tau\}(c) \qquad \text{mod } 2, \tag{6}$$

and

$$\{\sigma * b_1 \tau\}(B_0 c) = \{b_1 \sigma * \tau\}(B_0 c) = 0. \tag{7}$$

Suppose now $\rho_2 \Phi^{2m}(L) \neq 0$. Then in L^* there must exist a mod 2 cycle $\rho_2 z$, $z \in C_{2m}(L^*)$, such that $\rho_2 \varphi^{2m}(L)(\rho_2 z) \neq 0$. By (1), (3), (6) and (7) we get

$$\varphi^{2m+1}(K)(\rho_2 B_0 z) = \rho_2 \varphi^{2m}(L)(\rho_2 z) \neq 0.$$

By (5), $\rho_2 B_0 z$ is a mod 2 cycle in K^*. Hence the last equation shows that $\varphi^{2m+1}(K) \not\sim 0$ or $\Phi^{2m+1}(K) \neq 0$. In the same manner we may prove that $\Phi^{2m-1}(L) \neq 0$ implies $\rho_2 \Phi^{2m}(K) \neq 0$.

Proof of Theorem 23.

When $q = 0$, $A_{p,0}$ is the same as the Van Kampen complex K_p in Theorem 22. By that theorem we have already $\rho_2 \Phi^{2p}(A_{p,0}) \neq 0$. As

$$A_{p,q} = a_q^{(0)} A_{p,q-1} + a_q^{(1)} A_{p,q-1},$$

by induction on applying successively Theorem 26 we obtain that $\Phi^{2p+q}(A_{p,q}) \neq 0$. Consequently $A_{p,q} \not\subset R^{2p+q}$.

On the other hand, let us take a rectangular system of coordinates $(x_1, \cdots, x_{2p+q+1})$ in R^{2p+q+1}. Let R^{2p+1} be the linear subspace of R^{2p+q+1} defined by $x_1 = \cdots = x_q = 0$, and R_i^1 $(i = 1, 2, \cdots, q)$ the line defined by $x_j = 0$ $j \neq i$. Let us take on each line R_i^1 two points $b_i^{(0)} = (\underbrace{0, \cdots, 0}_{i-1}, 1, 0, \cdots, 0)$ and $b_i^{(1)} = (\underbrace{0, \cdots, 0}_{i-1}, -1, 0, \cdots, 0)$. As $A_{p,0}$ is a complex of dimension p, there must exist a realization $T : A_{p,0} \subset R^{2p+1}$. On setting $T(a_{p+i}^{(0)}) = b_i^{(0)}$, $T(a_{p+i}^{(1)}) = b_i^{(1)}$, $i = 1, 2, \cdots, q$, we get then an extension of $T : A_{p,q} \subset R^{2p+q+1}$. The theorem is thus completely proved.

Remark. The theorem 19 of §9 concerning the realization problem of complex $K_{N,n}$ in R^m is in reality settled by known method (principally the method of Van Kampen-Flores). However, the realization problem of the complex $A_{p,q}$ in R^m, so far as the author knows, seems impossible to be settled by any known methods (Alexander, Van Kampen, Flores, Thom, etc.).

References

[1] Pontrjagin, L. 1947 *Combinatorial topology.*
[2] Alexandroff, P. and Hopf, H. 1935 *Topologie* I.
[3] Van Kampen, E. R. 1932 *Abh. Math. Sem. Hamburg*, **9**, 72–78.
[4] Van Kampen, Berichtigung dazu, *loc. cit.* 152–153.
[5] Flores, A. I. 1935 *Erg. math. Kolloqu.* **6**, 4–7.
[6] Wu Wen-tsün. 1953 *Acta Math. Sinica.* **3**, 261–290.
[7] Cairns, S. S. 1940 *Annals of Math.* **41**, 792–795.
[8] Van Kampen, E. R. 1941 *Lectures in topology*, 311–314.
[9] Wu Wen-tsün. 1955 *Acta Math. Sinica.* **5**, 37–63.
[10] Wu Wen-tsün. 1951 *Sur les puissances de Steenrod, Colloque de Topologie* (Strasbourg).

ACTA MATHEMATICA SINICA
Volume Eight (1958), No. 1, pp. 79–94

ON THE REALIZATION OF COMPLEXES IN EUCLIDEAN SPACES. III

WU WEN-JUN (WU WEN-CHÜN)
(Institute of Mathematics, Academia Sinica)

Let K be a finite simplicial complex. We can always view K as a Euclidean complex in a Euclidean space of sufficiently high dimension N. Let its underlying space be denoted by $\overline{K}$. In studying whether K can be imbedded in the Euclidean space R^m of some fixed dimension m, we have introduced the following definitions (in [1] the notation is slightly different).

Let $T: \overline{K} \to R^m$ be a topological mapping such that for every $\sigma \in K$, T/σ is a linear mapping, then T is called a linear imbedding of K. If $T: \overline{K} \to R^m$ is a linear imbedding of some simplicial subdivision K' of K, then T is called a semilinear imbedding of K in R^m through the subdivision K'. Again, if $T: \overline{K} \to R^m$ is a continuous mapping such that for any $\sigma \in K$, T/σ is a linear mapping (perhaps degenerate), T/σ is nondegenerate if $\dim \sigma \leq m$, and for any simplices σ and τ with no common vertices, and with $\dim \sigma + \dim \tau \leq m$, $T(\sigma)$ and $T(\tau)$ are in general position, then T is called a linear pseudoimbedding. If $T: \overline{K} \to R^m$ is a linear pseudoimbedding of some subdivision K', then call T a semilinear pseudoimbedding of K through K'.

In [1], for a semilinear pseudoimbedding of K in R^m, we have introduced a system of invariants $\Phi^m(K) \in H^m(K^*, I_{(m)})$, $m > 0$, called the imbedding index of K, where K^* is the reduced two-fold symmetric product of K, $I_{(m)}$ is the additive group of integers if m is even, integers mod 2 if m is odd. We have also proved that $\Phi^m(K) = 0$ is a necessary condition for K to be semilinearly imbeddable in R^m. The purpose of this paper is to prove that these conditions are also sufficient in certain extreme cases.

More explicitly, let K^1 be a 1-dimensional complex, then $\Phi^1(K^1) = 0$ is the necessary and sufficient condition for K^1 to be semilinearly imbeddable in R^1. This is obvious. Similarly, $\Phi^2(K^1) = 0$ is a necessary and sufficient condition for K^1 to be semilinearly imbeddable in R^2. As was pointed out in [2], this is another way of stating Kuratowski's Theorem. Otherwise, if the dimension n of K is greater than 2, then $\Phi^{2n}(K) = 0$ is a necessary and sufficient condition for K to be semilinearly imbeddable in R^{2n}. This is the main theorem of this paper (see §2, Theorem 1). This theorem was first studied in the work of Van Kampen (see [3]). The invariant introduced by Van Kampen in [3] is another way of expressing $\rho_2\Phi^{2n}(K)$, where ρ_2 is reduction mod 2. There are errors in both his statement and proof of the theorem (see [4]). But the method of our proof of the theorem still follows mainly the original proof of Van Kampen; where the mistake occurs, we use a construction of Whitney's to correct it.

In the following, a complex will always mean a finite Euclidean simplicial complex in Euclidean space.

§1. Several constructions

In order to prove the main theorem of this paper, namely Theorem 1 of §2, we need the

Received February 4, 1957.

following simple constructions:

A. Tube construction. Let C be an infinitely differentiable simple arc in R^m (i.e. the image in R^m of a line segment under an infinitely differentiable topological mapping), with endpoints a_0 and a_1. Let L_0^n and L_1^n be two n-dimensional linear subspaces orthogonal to C at a_0 and a_1, respectively, let S_0^{n-1} and S_1^{n-1} be $(n-1)$-spheres in L_0^n and L_1^n, with centers a_0 and a_1, and each with sufficiently small radius $\epsilon > 0$. Assume further that each S_i^{n-1} is oriented. We will prove that if $m \geq n+2$, there always exists an n-dimensional differentiable tube T in the ϵ-neighborhood of C, i.e. T is the image under a differentiable topological mapping of the topological product of an $(n-1)$-spheres with a line segment, such that the two ends of T are S_0^{n-1} and S_1^{n-1} respectively, and after suitably orienting T, $\partial T = S_1^{n-1} - S_0^{n-1}$.

To prove this, write the points of C as a_t, $0 \leq t \leq 1$. Let L_t^{m-1} be the $(m-1)$-dimensional linear subspace orthogonal to C at a_t. Let the Grassmann manifold of all oriented n-planes of L_t^{m-1} through a_t be denoted by $\widetilde{R}^{(t)}_{m-n-1,n}$. Then there exists a fiber space L on C in which the fibers are the image of the Grassmann manifold $\widetilde{R}^{(t)}_{m-n-1,n}$, whose projection $\pi: L \to C$ is defined by $\pi^{-1}(a_t) = \widetilde{R}^{(t)}_{m-n-1,n}$. Since C is a simple arc, this fiber bundle has a product structure. Also from the assumption that $m-n-1 \geq 1$, $\widetilde{R}^{(t)}_{m-n-1,n}$ is an arcwise connected space, so any cross section on a_0, a_1 may be extended to a cross section on C. In particular, take the cross section $f(a_0) = L_0^n$, $f(a_1) = L_1^n$ on a_0, a_1, where the orientation of L_0^n, L_1^n corresponds to that of S_0^{n-1}, S_1^{n-1}, it may be extended to a cross section f on C, $f(a_t) = L_t^n$, such that L_t^n is an oriented n-plane of L_t^{m-1} through a_t. From a theorem of Steenrod, we can make f into an infinitely differentiable cross section. Let S_t^{n-1} be an $(n-1)$-sphere of radius ϵ, center a_t in L_t^n, and T be the space formed by the S_t^{n-1}'s for $0 \leq t \leq 1$. Then for sufficiently small ϵ, T is a differentiable tube with ends S_0^{n-1} and S_1^{n-1} in the ϵ-neighborhood of C. Furthermore, with a suitable orientation, we have $\partial T = S_1^{n-1} = S_0^{n-1}$ as required.

B. Whitney construction. Let σ_1 and σ_2 be two n-simplices in the complex K with no common vertex and $n > 2$, and let T be a continuous mapping of $\overline{K}$ into R^{2n} satisfying the following conditions:

1°. The restriction of T to the $(n-1)$-skeleton of K is a linear imbedding.

2°. The restriction of T to the interior of any n-simplex of K is a differentiable topological mapping.

3°. T has only double points, but no triple points.

4°. $T(\sigma_1)$ and $T(\sigma_2)$ intersect in an even number of points $q_1, \cdots, q_r, q_1', \cdots, q_r'$. At each point q_i or q_i', the tangent planes of $T(\sigma_1)$ and $T(\sigma_2)$ intersect only in the point q_i or q_i'. Furthermore, with respect to a fixed orientation of R^{2n}, the intersection numbers of $T(\sigma_1)$ and $T(\sigma_2)$ (σ_1 and σ_2 are already oriented) are $+1$ at each q_i, and -1 at each q_i'.

Whitney's construction (see [5], §§10–12) permits us to alter T into another continuous mapping T', preserving properties 1°–3° and changing 4° to

$\overline{4}$°. $T'(\sigma_1)$ and $T'(\sigma_2)$ do not meet. Furthermore, the alteration only occurs on σ_1 and σ_2, and does not change the other double points. In other words, we have

$\overline{5}$°. T and T' agree on $\overline{K} - \operatorname{Int}\sigma_1 - \operatorname{Int}\sigma_2$. Furthermore, apart from eliminating the double points q_i and q_i', T and T' have the same remaining double points.

We briefly state the construction of T'.

We may construct a simple differentiable arc B_i ($i = 1, 2$), connecting $q_1 = q$ and $q_1' = q'$ in

$T(\sigma_i)$ such that B_i does not go through any other double points in $T(\sigma_i)$, and B_1 and B_2 have no common points other than the two endpoints q and q'. Let E^{2n} be a $2n$-dimensional Euclidean space with Cartesian coordinates $(x_1, \cdots, x_{2n})$, E^2 will be the (x_1, x_2) plane. Let A_1 be the line segment $0 \leq x_1 \leq 1$, $x_2 = x_3 = \cdots = x_{2n} = 0$, whose endpoints are $r = (0, \cdots, 0)$ and $r' = (1, 0, \cdots, 0)$, A_2 be a smooth curve: $x_2 = \lambda(x_1)$, $0 \leq x_1 \leq 1$, connecting r and r' in $E^2(x_2 \geq 0)$. Let $A = A_1 + A_2$, and τ be a 2-cell formed by a sufficiently small neighborhood of A and the interior of A. It follows from [5], §10, that for $n > 2$, we may construct a differentiable mapping ψ of τ, such that $\psi(r) = q$, $\psi(r') = q'$, $\psi(A_1) = B_1$, $\psi(A_2) = B_2$; the differentiable 2-cell $\psi(\tau) = \sigma$ intersects $T(\sigma_i)$ only at B_i for $i = 1, 2$, it does not intersect $T(\bar{K} - \operatorname{Int}\sigma_1 = \operatorname{Int}\sigma_2)$; and at any point of B_i, the tangent planes of $T(\sigma_i)$ and of σ only intersect at the tangent line of B_i at that point.

Since the intersection numbers of the two oriented cells $T(\sigma_1)$ and $T(\sigma_2)$ at q and q' are +1 and 1 respectively, and $n > 2$, it follows from [5], §11, that we may define a system of vector fields $W_1(q^*), \cdots, W_{2n}(q^*)$ on $\psi(\tau) = \sigma$, where all the $W_i(q^*)$ are linearly independent at each $q^* \in \sigma$, depending continuously and smoothly on q^*, and satisfying the following conditions:

(a) Let $e_i(r^*)$, $i = 1, \cdots, 2n$ be the unit vectors parallel to x_i at $r^* \in \tau$. Then $W_1(q^*)$ and $W_2(q^*)$ are the images of $e_1(r^*)$ and $e_2(r^*)$ respectively, under the vector mapping induced by ψ, where $q^* = \psi(r^*) \in \sigma$.

(b) When $q^* \in B_1$, $W_3(q^*), \cdots, W_{n+1}(q^*)$ are the tangent vectors of $T(\sigma_i)$ at q^*.

(c) When $q^* \in B_2$, $W_{n+2}(q^*), \cdots, W_{2n}(q^*)$ are the tangent vectors of $T(\sigma_2)$ at q^*.

Now for $r^* \in \tau$, define

$$\psi\left(r^* + \sum_{i=3}^{2n} a_i e_i(r^*)\right) = \psi(r^*) + \sum_{i=3}^{2n} a_i w_i(\psi(r^*)).$$

Then in a sufficiently small neighborhood U of $\tau \subset E^{2n}$, ψ is one-to-one. Now in each σ_i, take a sufficiently small neighborhood M_i of $C_i = T^{-1}(B_i)$; such that $T(M_i) \subset V = \psi(U)$, $i = 1, 2$. Let π be the projection: $(x_1, x_2, \cdots, x_{2n}) \to (x_1, 0, x_3, \cdots, x_{2n})$. We may assume M_i to be sufficiently small so that $\pi(N_1)$ and $\pi(N_2)$ only intersect on the x_1-axis, where $N_i = \psi^{-1}(T(M_i))$, $i = 1, 2$. This is always possible in view of conditions (a), (b) and (c).

Now take $\epsilon > 0$ sufficiently small, and construct a continuously differentiable real function $\nu(x)$, such that

$$|\nu(x)| \leqslant 1,\ \nu(0) = 1, \text{ and when } |x| \geq \epsilon^2,\ \nu(x) = 0$$

Construct another continuously differentiable function $x_2 = \mu(x_1)$ such that

$$\epsilon > \mu(x_1) - \lambda(x_1) > 0,\ 0 \leqslant x_1 \leqslant 1$$
$$\mu(x_1) = 0,\ x_1 < -\epsilon \text{ or } x_1 > 1 + \epsilon.$$

For any point $(x_1, \cdots, x_{2n}) = r^*$ in N_2, let

$$\theta(r^*) = r^* - \nu(x_3^2 + \cdots + x_{2n}^2)\mu(x_1)e_2 .$$

Then $\theta(N_2)$ does not meet N_1. Now alter $T\colon \bar{K} \to R^{2n}$ to $T_1\colon \bar{K} \to R^{2n}$, so that T_1 agrees with T on $\bar{K} - M_2$, and on M_2,

$$T_1/M_2 = \theta\psi^{-1}T.$$

Then, if ϵ is sufficiently small, T_1 is obviously continuous, and is a differentiable topological mapping on σ_1 and σ_2, does not have $q = q_1$ and $q' = q_1'$ as double points, but otherwise has the same double points as T.

Now use the same technique on $(q, q') = (q_2, q_2'), \cdots, (q_r, q_r')$ to finally obtain a mapping T', satisfying 1°–3°, $\bar{4}°$ and $\bar{5}°$ as required.

C. Van Kampen construction. Let σ_1 and σ_2 be two n-simplices of a $2n$-dimensional complex K, $n > 2$, having common vertices. Let T be a continuous mapping of $\bar{K}$ to R^{2n} satisfying the following conditions:

1°. T is a semilinear pseudoimbedding of K through some subdivision K'.

2°. T is a semilinear imbedding of the subcomplex K_1 (or K_2) defined by σ_1 (or σ_2) and its faces through the subdivision K_1' (or K_2') of the above K' restricted to σ_1 (or σ_2).

3°. T has only double points, but no triple points.

In this situation, Van Kampen's construction (see proof of Lemma 2 of [4]) permits us to change T to another continuous mapping T' satisfying the following conditions:

$\bar{1}°$. T' is a semilinear pseudoimbedding of K through some subdivision.

$\bar{2}°$. T' is a semilinear imbedding of the subcomplex L formed by σ_1, σ_2 and their faces through the subcomplex induced by the above subdivision of K restricted to L.

$\bar{3}°$. T' coincides with T on $\bar{K} - \operatorname{Int}\sigma_1 - \operatorname{Int}\sigma_2$. Moreover, T' has no triple points and has the same double points as T except that the original double points common to $T(\sigma_1)$ and $T(\sigma_2)$ have been removed.

Following the original construction of Van Kampen, we reconstruct the above T' as follows:

Let $x_i = T(x_{1i}) = T(x_{2i})$, $i = 1, \cdots, r$, where $x_{1i} \in \operatorname{Int}\sigma_1$ and $x_{2i} \in \operatorname{Int}\sigma_2$, be the double points of T formed by the intersection of $T(\sigma_1)$ and $T(\sigma_2)$. We will change T step by step to diminish the number of common double points in $T(\sigma_1)$ and $T(\sigma_2)$, and make the last mapping so obtained satisfy $\bar{1}°$–$\bar{3}°$.

For this let O be a common vertex of σ_1 and σ_2, and let τ_1 be an n simplex of σ_1 belonging to K_1' and having O as a vertex. In $T(\sigma_1)$ construct a broken line l_1 from x_1 to an interior point x_1' of $T(\tau_1)$ such that the broken line does not go through $x_2, \cdots, x_r$ or any other double points in $T(\sigma_1)$. Also construct a sufficiently small linear tube C_1 of l_1 with one end the boundary of a sufficiently small neighborhood V_1 of x_1 in $T(\sigma_2)$ and the other end the boundary of a sufficiently small n-dimensional convex neighborhood V_1' which intersects $T(\tau_1)$ only at x_1', and such that C_1 does not contain any double points of T. Now alter T to T_1 so that T_1 maps $\sigma_2 \cap T^{-1}(V_1)$ to $C_1 + V_1'$ and on the remaining $\bar{K} - \sigma_2 \cap \operatorname{Int} T^{-1}(V_1)$, T_1 is the same as T. Then this new mapping T_1 still has properties 1°–3°. Only its double points $x_1, \cdots, x_r$ have been changed to x_1', $x_2, \cdots, x_r$. The same technique can be applied to $x_2, \cdots, x_r$. Hence we may assume from the start that the double points $x_1, \cdots, x_r$ all lie in $T(\tau_1)$ and on any line segment $O'x_i$ has no double points except for x_i. Here $O' = T(O)$. Similarly, by slightly moving $T(\sigma_2)$ if necessary, we may assume that T has the following additional properties:

4°. For any $\tau' \in K_2'$ which does not have O as a vertex the linear subspace spanned by $T(\tau')$ does not pass through O'.

5°. For any $\tau \in K'$ and any $\tau' \in K_2'$ such that τ' does not have O as a vertex, and τ, τ' have no common vertex, then O' and the subspace spanned by $T(\tau)$ and that spanned by $T(\tau')$ are in general position.

Now let ξ_1 be the n-dimensional simplex in K_2' with x_{21} as an interior point. Obviously, ξ_1 does not have O as a vertex. Hence there exists a chain of n-simplices, $\xi_1, \cdots, \xi_s$ in K_2', such that every two consecutive simplices ξ_i, ξ_{i+1} $(i = 1, \cdots, s-1)$ have an $(n-1)$-simplex η_i in common, ξ_s has O as a vertex, but none of the other ξ_i's $(i < s)$ has O as a vertex. It follows from 5° that for any simplex $\tau \in K'$, $\tau \neq \xi_1$, when τ and ξ_1 have no vertex in common or only one, the linear subspace $L(O, \tau)$ spanned by O' and $T(\tau)$ intersect the linear subspace $L(\xi_1)$ spanned by $T(\xi_1)$ in a line segment $s(\tau)$ at most. When the common face of τ and ξ_1 has dim ≥ 1, $L(O, \tau)$ and $L(\xi_1)$ intersect in this common face only. Hence, because of the assumption that $n > 2$, we can construct a line segment in $T(\xi_1)$ through x_1, which intersects the interior of $T(\eta_1)$, but intersects all the line segments $s(\tau)$ only at x_1, and does not pass through any other double points in $T(\xi_1)$. Denote by y_1 the intersection of this line segment with $T(\eta_1)$, and by y_0 the intersection with another $(n-1)$-face $T(\eta_0)$ of $T(\xi_1)$. Again from 4° and 5°, for any simplex $\tau \in K'$, $\tau \neq \xi_2$, the linear subspace spanned by O' and $T(\tau)$ intersect $L(\xi_2)$ at most in a line segment or a face of $T(\xi_2)$. Since $n > 2$, in the interior of $T(\xi_2)$ we may construct a line through y_1 which does not go through any double points, which does not intersect any other such line segments, and which intersects $T(\eta_2)$ at y_2. Continuing this construction, we obtain a broken line $l = y_0 y_1 \cdots y_{s-1}$, where y_i is an interior point of $T(\eta_i)$. Now in $T(\eta_0)$, construct a sufficiently small $(n-1)$-simplex η_0' with y_0 as an interior point. Through each point y_0' on the boundary of η_0', construct a straight line parallel to $y_0 y_1$, which intersect $T(\eta_1)$ at y_1'; through y_1' construct a straight line parallel to $y_1 y_2$, which intersect $T(\eta_2)$ at y_2'; follow the same method, construct $y_3', \cdots, y_{s-1}'$. Then for each $i = 1, 2, \cdots, s-1$, all the y_i' form the boundary of an $(n-1,$-simplex η_i' in $T(\eta_i)$, all the line segments $y_{i-1}' y_i'$ form a tube ξ_i' in $T(\xi_i)$ with $y_{i-1} y_i$ as axis. Let $\xi' = \xi_1' + \cdots + \xi_{s-1}'$, and $C = \xi' + \eta_0'$. Then, because of the construction, 4°, 5° and the assumption that $O'x_1$ contains no double point other than x_1, we know that if we choose η_0' sufficiently small, C does not contain any double points. If we project C with O' as center, the resultant cone $\tilde{C}$ is an n-cell, which only intersects $T(\bar{K})$ on $C + O'\eta_{s-1}'$, and only intersects $T(\bar{K}_1)$ at O'.

Let $I(\xi')$ be the portion of $T(\xi_1) + \cdots + T(\xi_{s-1})$ enclosed by $\xi' + \eta_0' + \eta_{s-1}'$. Now alter T to T_1 so that T_1 coincides with the original T on $\bar{K} - \sigma_2 \cap \operatorname{Int} T^{-1}(I(\xi') + O'\eta_{s-1}')$, and T_1 maps $\sigma_2 \cap T^{-1}(I(\xi') + O'\eta_{s-1}')$ to $\tilde{C}$. Then the new mapping so obtained still satisfies 1°–3°, but x_1 is no longer a double point of T_1. T_1 has no new double points.

Now use this method on each of $x_2, \cdots, x_r$; we obtain a mapping T', which satisfies $\overline{1}°$–$\overline{3}°$ as required.

§2. Main theorem–the necessary and sufficient condition for $K^n \subset R^{2n}$ when $n > 2$

In [1], we have introduced a system of invariants, $\Phi^m(K) \in H^m(K^*, I_{(m)})$, $m > 0$, of a finite simplicial complex K, called the m-dimensional imbedding index of K. Here K^* is the reduced two-fold symmetric product of K, $I_{(m)}$ is the additive group of integers or the mod 2 integer group, according as m is even or odd. We have also proved that $\Phi^m(K) = 0$ is a necessary condition that K be semilinearly imbeddable in R^m. In [2], we indicated that a famous theorem of Kuratowski about space curves, in the case of a complex, can be restated in the following manner: $\Phi^2(K) = 0$ is a necessary and sufficient condition for a 1-dimensional complex to be imbeddable in the plane. The purpose of this section is to prove the following theorem:

Theorem 1. *For a finite simplicial complex K of dimension $n > 2$ to be semilinearly imbeddable in R^{2n}, it is necessary and sufficient that $\Phi^{2n}(K) = 0$.*

The necessity of the condition $\Phi^{2n}(K) = 0$ has already been stated. We will prove the sufficienc as follows. We will assume that K is a Euclidean complex in a Euclidean space R^N of sufficiently high dimension N.

Let R^{2n-1} be an $(n-1)$-dimensional linear subspace of R^{2n} dividing R^{2n} into two halfspaces, R^{2n}_+ and R^{2n}_-. Let T be a linear imbedding of the $(n-1)$-skeleton K^{n-1} of K into R^{2n-1}. For any n-simplex σ of K, let $\overline{O}_\sigma$ be the barycenter of σ, let O_σ be the barycenter of the n-simplex in R^{2n-1} determined by $T(\dot{\sigma})$, let P_σ be the n-dimensional linear subspace spanned by $T(\dot{\sigma})$, and let S_σ be the unit $(n-1)$-sphere in P_σ with center O_σ. For each $x \in S_\sigma$, construct the halfline $O_\sigma x$ from O_σ to x, intersecting $T(\dot{\sigma})$ at x'. Let the distance between O_σ and x' be $\rho_\sigma(x)$. Then ρ_σ is a continuous function on the differentiable manifold S_σ. Choose arbitrary $\epsilon > 0$, then, by an approximation theorem of Whitney, we may easily construct a continuously differentiable function f on S_σ, such that for any $x \in S_\sigma$, $\epsilon > \rho_\sigma(x) - f(x) > 0$. For any $\alpha > 0$, let $f_{\sigma,\alpha}(x)$ be the point on the halfline $O_\sigma x$ whose distance from O_σ is $\alpha f(x)$. Then for $\alpha > 1$, $f_{\sigma,\alpha}$ is a differentiable topological mapping of S_σ to P_σ.

Now, through O_σ construct a halfline L in R^{2n}_+, orthogonal to R^{2n-1}, and choose an arbitrary point O'_σ on it. Through O'_σ construct the n-dimensional linear space P'_σ parallel to P_σ. Assume the previously choosen ϵ is sufficiently small. For any $x \in S_\sigma$, let the straight line through $f_{\sigma,\epsilon}(x)$ and orthogonal to R^{2n-1} intersect P'_σ at $f'_{\sigma,\epsilon}(x)$, let the line segment connecting $f_{\sigma,1-\epsilon}(x)$ and $f'_{\sigma,\epsilon}(x)$ be l_x, and let π_x be the orthogonal projection onto $x'O_\sigma$ of the broken line with consecutive vertices x', $f_{\sigma,1-\epsilon}(x)$, $f'_{\sigma,\epsilon}(x)$ and O'_σ. Define $T(\overline{K}^{n-1})$ as before. Extend T to the interior of each n-simplex $\sigma \in K$ as follows: for any $\overline{x}' \in \dot{\sigma}$, if $x' = T(\overline{x}')$, x is the intersection of the halfline $O_\sigma x'$ with S_σ, and $T_{\overline{x}}$, is the linear mapping of the line segment $\overline{x}'\overline{O}_\sigma$ to the line segment $x'O_\sigma$, then $T/\overline{x}'\overline{S}_\sigma \equiv \pi_x^{-1} T_{\overline{x}}$. Next, on the plane determined by $O_\sigma x$, l_x and $O'_\sigma f'_{\sigma,\epsilon}(x)$, construct a circular arc C_x, tangent to the straight line $O_\sigma x$ at $f_{\sigma,1-\epsilon/2}(x)$ and tangent to l_x at p_x. Also construct a circular arc C'_x tangent to $O'_\sigma f'_{\sigma,\epsilon}(x)$ at O'_α and tangent to l_x at p'_x. Let l'_x be the segment of l_x between p_x and p'_x, and let π_x be the orthogonal projection of the union of the line segment $x'f_{\sigma,1-\epsilon/2}(x)$, the arc C_x, the segment l'_x and the arc C'_x onto $x'O_\sigma$. Define $T'(\overline{K}^{n-1})$ as $T(\overline{K}^{n-1})$, and extend T' to the interior of each n-simplex $\sigma \in K$, so that for any $\overline{x}' \in \dot{\sigma}$, if x', x and $T_{\overline{x}}$, are as previously defined, then $T'/\overline{x}'\overline{O} = \pi_x'^{-1} T_{\overline{x}}$. It is obvious that by choosing ϵ sufficiently small and suitably choosing O'_σ, we may make T into a semilinear pseudoimbedding through some subdivision of K, with no triple points, and any double points must be in the interior of the segments l'_x. Similarly, T' has no triple points, its double points agree with those of T, its restriction to the interior of every n-simplex $\sigma \in K$ is a differentiable topological mapping, T' is sufficiently close to T, and in a sufficiently small neighborhood of $\overline{K}^{n-1}$, T' coincides with T.

In the following, $\overline{x} \in \overline{K}$ will be called a singularity of T or T' if its image, $T(\overline{x}) = T'(\overline{x})$ is a double point of both T and T'.

Choose an arbitrary orientation of R^{2n} and let ϕ be the integer intersection number with respect to this orientation. Let σ_i, $i = 1, \cdots, r$ be all the n-simplices of K, and τ_α, and let $\alpha = 1, \cdots, s$ be all the $(n-1)$-simplices of K, each arbitrarily oriented, so that for any $\sigma_i * \sigma_j$ or $\sigma_i * \tau_\alpha \in K^*$, we have

$$\left.\begin{aligned} \sigma_i * \sigma_j &= (-1)^{\dim \sigma_i \dim \sigma_j} \sigma_j * \sigma_i = (-1)^n \sigma_j * \sigma_i, \\ \sigma_i * \tau_\alpha &= (-1)^{\dim \sigma_i \dim \tau_\alpha} \tau_\alpha * \sigma_i = \tau_\alpha * \sigma_i. \end{aligned}\right\} \tag{1}$$

According to the definition in [1], for any arbitrary $2n$-cell $\sigma_i * \sigma_j \in K^*$,

$$\varphi_T(\sigma_i * \sigma_j) = (-1)^n \phi(T\sigma_i, T\sigma_j) \tag{2}$$

determines the imbedding cochain $\phi_T \in \Phi^{2n}(K) \in H^{2n}(K^*)$ of a semilinear pseudoimbedding T. But by assumption, $\Phi^{2n}(K) = 0$, so there exists a $\chi \in C^{2n-1}$, such that

$$\delta\chi = \varphi_T, \tag{3}$$

where χ is defined as

$$\chi(\sigma_i * \tau_a) = C_{ia},\ \sigma_i * \tau_a \in K^*. \tag{4}$$

Let J denote the set of all pairs of indices (i, α), such that $\sigma_i * \tau_\alpha \in K^*$. Let $a_i = \Sigma_\alpha |C_{i\alpha}|$, $b_\alpha = \Sigma_i |C_{i\alpha}|$, each summed over all α or all i, such that $(i, \alpha) \in J$. In each σ_i, choose a_i distinct interior points, all distinct from any singularities: $\bar{x}_{i\alpha,1}, \cdots, \bar{x}_{i\alpha,|C_{i\alpha}|}$, $(i,\alpha) \in J$. In each τ_α, also choose b_α distinct interior points: $\bar{y}_{i\alpha,1}, \cdots, \bar{y}_{i\alpha,|C_{i\alpha}|}$, $(i, \alpha) \in J$. Let $T(\bar{x}_{i\alpha,k}) = x_{i\alpha,k}$ and $T(\bar{y}_{i\alpha,k}) = y_{i\alpha,k}$. For each $y_{i\alpha,k}$ construct an $(n+1)$-dimensional linear space $P_{i\alpha,k}$ orthogonal to $T(\tau_\alpha)$ and passing through $y_{i\alpha,k}$. Construct an n-sphere $S_{i\alpha,k}$ in $P_{i\alpha,k}$ of sufficiently small radius $\epsilon_{i\alpha,k} > 0$ and center at $y_{i\alpha,k}$, which intersects R^{2n-1} in the $(n-1)$-sphere $S^{(0)}_{i\alpha,k}$, dividing $S_{i\alpha,k}$ into two hemispheres: $S^+_{i\alpha,k} \subset R^{2n}_+$ and $S^-_{i\alpha,k} \subset R^{2n}_-$. Let $z_{i\alpha,k}$ be the intersection of $S^+_{i\alpha,k}$ with the halfline through $y_{i\alpha,k}$, orthogonal to R^{2n-1} and in R^{2n}_+. Let $z^+_{i\alpha,k}$ be a sufficiently small spherical neighborhood of $z_{i\alpha,k}$ in $S^+_{i\alpha,k}$. Let $z^-_{i\alpha,k}$ be what remains of $S_{i\alpha,k}$ after taking away the interior of $z^+_{i\alpha,k}$. Let the common boundary of $z^+_{i\alpha,k}$ be $z^{(0)}_{i\alpha,k}$. We orient $P_{i\alpha k}$ so that

$$\phi(P_{i\alpha,k}, T\tau_\alpha) = (-1)^n \operatorname{sgn} C_{i\alpha}, \tag{5}$$

where $\operatorname{sgn} C_{i\alpha} = +1, -1$ or 0 according as $C_{i\alpha} > 0$, < 0 or $= 0$. Orient $S_{i\alpha,k}$ with respect to the orientation of $P_{i\alpha,k}$. Then, by taking sufficiently small $\epsilon_{i\alpha,k} > 0$, we may construct, in $T(\sigma_i)$, $(n-1)$-spheres $S'_{i\alpha,k}$ with center $x_{i\alpha,k}$ with radius $\epsilon_{i\alpha,k}$ such that the $(S'_{i\alpha,k})$'s are mutually disjoint on $T(\sigma_i)$ and they do not contain any double points in any of their interiors. Let $T^i_{i\alpha,k}$ be the interior of $S'_{i\alpha,k}$, with the orientation induced from that of σ_i. Let $T^e_{i\alpha,k}$ be the intersection of $S'_{i\alpha,k}$ and $\dot{\sigma}_i$, also with the orientation of σ. In R^{2n}_+, construct an (infinitely) differentiable simple arc $C_{i\alpha,k}$, connecting $x_{i\alpha,k}$ and $z_{i\alpha,k}$ so that the arcs mutually do not meet, and meet $T(\bar{K})$, $T'(\bar{K})$ and all the $S_{i\alpha,k}$ only at $x_{i\alpha,k}$ and $z_{i\alpha,k}$. Following construction A of §1, we may construct a differentiable tube $C_{i\alpha,k}$ in a sufficiently small neighborhood of $C_{i\alpha,k}$, connecting $S'_{i\alpha,k}$ and $z^{(0)}_{i\alpha,k}$, such that after suitably orienting $C_{i\alpha,k}$, the orientation on $S'_{i\alpha,k}$ induced by $C_{i\alpha,k}$ and $T^i_{i\alpha,k}$ are the same, while the orientation on $z^{(0)}_{i\alpha,k}$ induced by $C_{i\alpha,k}$ and $z^-_{i\alpha,k}$ are opposite. In other words, $T^e_{i\alpha,k} + C_{i\alpha,k} + z^-_{i\alpha,k}$ is a relative cycle mod $\dot{\sigma}$. Now alter T'/σ_i to T'', mapping $T'^{-1}(T^i_{i\alpha,k}) \subset \sigma_i$ to $C_{i\alpha,k} + z^-_{i\alpha,k}$, $(i, \alpha) \in J$; T'' coincides with T' on the remainder of σ_i. The mapping may be smoothed out near $S'_{i\alpha,k}$ and $z^{(0)}_{i\alpha,k}$ by at most infinitesimal changes, so we may assume that T'' is a differentiable topological mapping in the interior of σ_i. By choosing all the $\epsilon_{i\alpha,k}$'s sufficiently small, we may also assume that all the $C_{i\alpha,k}$ are mutually disjoint, and only meet $T'(\bar{K})$ at $S'_{i\alpha,k}$. Then, $T'': \bar{K} \to R^{2n}$ is continuous and possesses the following properties: T'' is a differentiable topological mapping in the interior of each σ_i; T'' has no triple points, and each of its double points is in the interior of both $T''\sigma_i$ and $T''\sigma_j$; the tangent planes of $T''\sigma_i$ and $T''\sigma_j$ at this double point intersect only at this point. Also, since for each $\sigma_i \in K$, T'' coincides with T' or T in a sufficiently small neighborhood of $\dot{\sigma}_i$, T'' is a linear imbedding on a sufficiently small neighborhood of $\bar{K}^{n-1}$.

Let $\sigma_i * \sigma_j \in K^*$, and

$$\partial\sigma_i = \sum_{\alpha} \eta_{i\alpha}\tau_\alpha, \quad \partial\sigma_j = \sum_{\beta} \eta_{j\beta}\tau_\beta \quad (\eta_{i\alpha}, \eta_{j\beta} = \pm 1 \text{ or } 0),$$

Let $T_{i\beta,k}$ denote the union of the interior of $S_{i\beta,k}$ in $P_{i\beta,k}$ and $S_{i\beta,k}$, so that $T_{i\beta,k}$ is compatible with $S_{i\beta,k}$, and we have $(k = 1, 2, \cdots, |C_{i\beta}|)$

$$\begin{aligned}\phi(S_{i\beta,k}, T'\sigma_j) &= \phi(\partial T_{i\beta,k}, T'\sigma_j) = (-1)^{n+1}\phi(T_{i\beta,k}, \partial T'\sigma_j)\\ &= (-1)^{n+1}\eta_{j\beta}\phi(P_{i\beta,k}, T'\tau_\beta) = (-1)^{n+1}\eta_{j\beta}\phi(P_{i\beta,k}, T\tau_\beta)\\ &= -\eta_{j\beta}\operatorname{sgn} C_{i\beta}.\end{aligned}$$

Similarly, we have $(l = 1, 2, \cdots, |C_{j\alpha}|)$,

$$\begin{aligned}\phi(T''\sigma_i, S_{j\alpha,l}) &= \phi(T''\sigma_i, \partial T_{j\alpha,l}) = (-1)^n\phi(\partial T''\sigma_i, T_{j\alpha,l})\\ &= (-1)^n\eta_{i\alpha}\phi(T''\tau_\alpha, P_{j\alpha,l}) = (-1)^n\eta_{i\alpha}\phi(T\tau_\alpha, P_{j\alpha,l})\\ &= (-1)^n\eta_{i\alpha}\cdot(-1)^{(n-1)(n+1)}\phi(P_{j\alpha,l}, T\tau_\alpha)\\ &= (-1)^{n+1}\eta_{i\alpha}\operatorname{sgn} C_{j\alpha}.\end{aligned}$$

Hence

$$\begin{aligned}&\phi(T''\sigma_i, T''\sigma_j) - \phi(T'\sigma_i, T'\sigma_j)\\ &= \phi(T''\sigma_i - T'\sigma_i, T'\sigma_j) + \phi(T''\sigma_i, T''\sigma_j - T'\sigma_j)\\ &= \sum_{\beta,k}\phi(C_{i\beta,k} - T^i_{i\beta,k} + S_{i\beta,k} - z^+_{i\beta,k}, T'\sigma_j)\\ &\quad + \sum_{\alpha,l}\phi(T''\sigma_i, C_{j\alpha,l} - T^i_{j\alpha,l} + S_{j\alpha,l} - z^+_{j\alpha,l}),\end{aligned}$$

Since $C_{i\beta,k}$, $T^i_{i\beta,k}$ and $z^+_{i\beta,k}$ do not meet $T'\sigma_j$, and $C_{j\alpha,l}$, $T^i_{j\alpha,l}$ and $z^+_{j\alpha,l}$ also do not meet $T''\sigma_i$, we may simplify the above formula to

$$\begin{aligned}&= \sum_{\beta,k}\phi(S_{i\beta,k}, T'\sigma_j) + \sum_{\alpha,l}\phi(T''\sigma_i, S_{j\alpha,l})\\ &= \sum_{\beta} - |C_{i\beta}|\eta_{j\beta}\operatorname{sgn} C_{i\beta} + \sum_{\alpha}(-1)^{n+1}|C_{j\alpha}|\eta_{i\alpha}\operatorname{sgn} C_{j\alpha}\\ &= -\sum_{\beta} C_{i\beta}\eta_{j\beta} + (-1)^{n+1}\sum_{\alpha} C_{j\alpha}\eta_{i\alpha}.\end{aligned} \tag{6}$$

On the other hand, we have

$$\begin{aligned}(-1)^n\phi(T'\sigma_i, T'\sigma_j) &= (-1)^n\phi(T\sigma_i, T\sigma_j) = \varphi_T(\sigma_i * \sigma_j)\\ &= \delta\chi(\sigma_i * \sigma_j) = \chi(\partial\sigma_i * \sigma_j) + (-1)^n\chi(\sigma_i * \partial\sigma_j)\\ &= \sum_{\alpha}\eta_{i\alpha}\chi(\tau_\alpha * \sigma_j) + (-1)^n\sum_{\beta}\eta_{j\beta}\chi(\sigma_i * \tau_\beta)\\ &= \sum_{\alpha}\eta_{i\alpha}\chi(\sigma_j * \tau_\alpha) + (-1)^n\sum_{\beta}\eta_{j\beta}\chi(\sigma_i * \tau_\beta)\\ &= \sum_{\alpha} C_{j\alpha}\eta_{i\alpha} + (-1)^n\sum_{\beta} C_{i\beta}\eta_{j\beta}.\end{aligned} \tag{7}$$

By comparing (6) and (7), we get

$$\phi(T''\sigma_i, T''\sigma_j) = 0, \quad \sigma_i * \sigma_j \in K^*.$$

Since the intersection number of $T''\sigma_i$ and $T''\sigma_j$ at each double point is ± 1, we know that $T''\sigma_i$ and $T''\sigma_j$ must intersect at an even number of points. Let this number be $2n_{ij}$. Then the intersection number at each of n_{ij} double points is $+1$, and the intersection number at each of the remaining n_{ij} double points is -1. Since T'' is a differentiable topological mapping in the interiors of σ_i and σ_j, and at every double point, the tangent planes of $T''\sigma_i$ and $T''\sigma_j$ intersect only at this point, by construction B of §1, we may alter the restriction of T'' on σ_i and σ_j, to obtain a mapping which is still a differentiable topological mapping on σ_i and σ_j, but whose images of σ_i and σ_j will be disjoint; and there are no new double points resulting from the intersection of the images of σ_i and σ_j with the image of any other σ_k. Now apply this construction to every $\sigma_i * \sigma_j \in K^*$ to obtain a mapping T_0: $\overline{K} \to R^{2n}$, with the following properties:

1°. T_0 is a differentiable topological mapping in the interior of every σ_i.

2°. T_0 is a linear imbedding on a sufficiently small neighborhood of $\overline{K}^{n-1}$.

3°. T_0 has no triple points. For every double point, $T_0(x) = T_0(y) = p$, x and y must be in the interior of two separate simplices σ_i and σ_j, and σ_j have a common vertex.

Because of 1°, 2° and a theorem of Cairns (e.g., see [6], Theorem 2), we may construct a sufficiently close approximation T_0' of T_0 such that for any σ_i, T_0' coincides with T_0 in some sufficiently small neighborhood of $\dot{\sigma}_i$. By choosing this approximation sufficiently close, we may obviously still preserve property 3°. So following the Van Kampen construction (§1, construction C), we may alter T_0' to remove all the double points. The resulting mapping h: $\overline{K} \to R^{2n}$ is a semilinear imbedding of K through some division of it.

Thus, we have obtained a semilinear imbedding of K to R^{2n} and proved the sufficiency of the theorem.

For an arbitrary Hausdorff space X, we have introduced a system of topological invariants $\Phi^m(X) \in H^m(X^*, I_{(m)})$, $m > 0$ of X, called the m-dimensional imbedding index of X. Here, X^* is the twofold symmetric product of X. We have also proved that when X is a finite polyhedron and K is a simplicial subdivision of it, under a fixed isomorphism $H^m(X^*, I_{(m)}) \approx H^m(K^*, I_{(m)})$, $\Phi^m(X)$ and $\Phi^m(K)$ are cohomologous. Hence the above theorem has the following corollary:

Theorem 1'. *For a finite polyhedron X of dimension $n > 2$ to be semilinearly imbeddable in R^{2n}, it is necessary and sufficient that $\Phi^m(X) = 0$.*

§3. Some suficient conditions for $K^n \subset R^{2n}$

The purpose of this section is to derive, from the main Theorem 1 of last section, some sufficient conditions for an n-dimensional finite complex K to be semilinearly imbeddable in R^{2n}. These conditions are either determined by the homology of K, or easily derived from the complex structure of K, see Theorems 2–6 below.

Theorem 2. *A finite simplicial complex of dimension $n \neq 0$ is semilinearly imbeddable in R^{2n} if $H_n(K, \bmod 2) = 0$.*

Proof. The theorem is obvious if $n = 1$. So assume $n > 2$. Let $\widetilde{K}^*$ be the subcomplex of $K \times K$ spanned by all cells $\sigma \times \tau$, where σ and $\tau \in K$ have no common vertex. By assumption,

$$H_{2n}(K \times K, \bmod 2) \approx H_n(K, \bmod 2) \otimes H_n(K, \bmod 2) = 0$$
$$H_{2n+1}(K \times K, \tilde{K}^*; \bmod 2) = 0$$

and from the exact sequence of $(K \times K, \widetilde{K}^*)$,

$$\cdots \to H_{2n+1}(K \times K, \widetilde{K}^*; \bmod 2) \to H_{2n}(\widetilde{K}^*, \bmod 2) \to$$
$$\to H_{2n}(K \times K, \bmod 2) \to \cdots$$

it follows that $H_{2n}(\widetilde{K}^*, \bmod 2) = 0$. Since the highest dimension of $\widetilde{K}^*$ is $2n$, this implies that $Z_{2n}(\widetilde{K}^*, \bmod 2) = 0$, where Z_{2n} denotes the group of $2n$-cycles. In other words, there exist no mod 2 $2n$-cycles in $\widetilde{K}^*$.

Obviously, $\widetilde{K}^*$ is a twofold covering complex of K^*, where the covers for $\sigma * \tau \in K^*$ are the cells $\sigma \times \tau$ and $\tau \times \sigma$. If $\rho_2 \Sigma_{i=1}^{s} \sigma_i * \tau_i$ is a mod 2 $2n$-cycle of K^*, where the $\sigma_i * \tau_i$ are mutually disjoint, and ρ_r denotes reduction mod r, then $\rho_2 \Sigma \sigma_i \times \tau_i + \rho_2 \Sigma \tau_i \times \sigma_i$ will be a nonzero mod 2 $2n$-cycle of $\widetilde{K}^*$, contradicting the above. So we must have $Z_{2n}(K^*, \bmod 2) = 0$, and hence $Z_{2n}(K^*, \bmod 2m) = 0$ for any integer $m \geq 0$. So, assuming ϕ to be a cocycle of Φ^{2n}, ϕ mod 2 is orthogonal to all mod $2m$ $2n$-cycles of K^*, i.e. $\phi \cdot z = 0 \bmod 2m$ for all $z \in Z_{2n}(K^*, \bmod 2m) = 0$. Next, from Theorem 8 of [1], $\Phi^{2n} = \frac{1}{2}\delta \Phi^{2n-1}$, so $\phi = \frac{1}{2}\delta\phi'$, where $\rho_2\phi' \in \Phi^{2n-1} \in H^{2n-1}(K^*, \bmod 2)$. Thus, for any arbitrary mod $(2m+1)$ $2n$-cycle z of K^*, if we let z' be an integral chain such that $\rho_{2m+1} z' = z$, we have $\phi \cdot z' = \frac{1}{2}\delta\phi' \cdot z' = \frac{1}{2}\phi' = 0$, mod $2m+1$. Hence, whether m is even or odd, ϕ is orthogonal to any mod m $2n$-cycle of K^*. From a theorem of Whitney [Whitney, *On matrices of integers and combinational topology*, Duke Math. J. 3(1937) 35–45], ϕ is a coboundary, or $\Phi^{2n} = 0$. Since $n > 2$, the theorem follows from Theorem 1.

Theorem 2 has the following corollary.

Theorem 2′. *A finite polyhedron X of dimension $n \neq 2$ can be topologically imbedded in R^{2n} if $H_n(X, \bmod 2) = 0$.*

Theorem 3. *A finite simplicial complex K of dimension $n \neq 2$ can be semilinearly imbedded in R^{2n} if $H^n(K) = 0$.*

This theorem obviously follows from Theorem 2 and the lemma:

Lemma. *If, for an n-dimensional finite complex K, $H^n(K) = 0$, then $H_n(K, \bmod 2) = 0$.*

Proof. From the universal coefficient theorem, we have

$$H_n(K, \bmod 2) \approx H_n(K) \otimes I_2 + \mathrm{Tor}(H_{n-1}(K), I_2)$$
$$H^n(K) \approx \mathrm{Hom}(H_n(K), I) + \mathrm{Ext}(H_{n-1}(K), I)$$

where I is the additive group of integers and I_2 is the mod 2 integer group. By formula (2) and the assumption that $H^n(K) = 0$, it follows that $\mathrm{Hom}(H_n(K), I) = 0$ and $\mathrm{Ext}(H_{n-1}(K), I) = 0$. From the former, it follows that $H_n(K)$ must be a finite group. Since K is of dimension n, there is no torsion in the nth dimension, so $H_n(K) = 0$. From the latter, it follows that $H_{n-1}(K)$ has no element of finite order, hence $\mathrm{Tor}(H_{n-1}(K), I_2) = 0$. From formula (1), we have $H_n(K, \bmod 2) = 0$.

Theorem 3′. *A finite polyhedron X of dimension $n \neq 2$ can be topologically imbedded in R^{2n} if $H^n(X) = 0$.*

In *Topologie* I of Alexandroff and Hopf (AH for brevity), Chapter 7, §1, there was defined the so-called closed complex and irreducible closed complex (irreduzible geschlossene Komplexe). We will prove

Theorem 4. *If a finite simplicial complex K of dimension $n \neq 2$ is an irreducible closed complex, then K can be semilinearly imbedded in R^{2n}.*

Proof. If $n = 1$, then from p. 284 of AH, Theorem 12, K is a simple closed polygon, so the theorem is obvious. So we may assume $n > 2$. It follows from AH, Chapter 7, §1 No. 4 Theorem 5 and Chapter 7, §1 No. 5 that K has a natural modulus m, $m = 0$ or $m \geq 2$. When $m \geq 2$, there exist a mod m n-cycle $\rho_m z$ in K, where $z = \Sigma a_i \sigma_i$, the a_i's are nonzero integers, a_i and m are relatively prime and Σ is summed over all n-simplices σ_i of K; so that for any coefficient group G, $H_n(K, G) = Z_n(K, G)$ consists of all the cycles gz, where $g \in G$ and $mg = 0$. So when m is odd, $H_n(K, \bmod 2) = 0$, and it follows from Theorem 3 that K can be semilinearly imbedded in R^{2n}. If $m \geq 2$ is even, then $Z_n(K, \bmod 2)$ has only one nonzero mod 2 cycle, $z_2 = \rho_2 \Sigma a_i \sigma_i = \rho_2 \Sigma \sigma_i$. From the Künneth Theorem, we know that $Z_{2n}(K \times K, \bmod 2) = H_{2n}(K \times K, \bmod 2)$ also has only one nonzero mod 2 cycle, $z_2 \otimes z_2 = \rho_2 \Sigma \sigma_i \times \sigma_j$. Now let $\tilde{z}^* = \rho_2 \Sigma' a_{ij} \sigma_i \times \sigma_j$ be a mod 2 $2n$-cycle of $\tilde{K}^*$, where $a_{ij} = 0$ or 1 and Σ' is summed over all pairs of indices (i, j) such that σ_i and σ_j have no common vertices. Then, viewing $\tilde{z}^*$ as a mod 2 cycle of $K \times K$, we should have $\tilde{z}^* = a(z_2 \otimes z_2)$, where $a = 0$ or 1. But $z_2 \otimes z_2$ has terms of the form $\sigma_i \times \sigma_i$, while $\tilde{z}^*$ cannot have such terms, so a must be 0, and $H_{2n}(\tilde{K}^*, \bmod 2) = Z_{2n}(\tilde{K}^*, \bmod 2) = 0$. From reasoning similar to that in the proof of Theorem 2, we have $\Phi^{2n}(K) = 0$. Hence, from Theorem 1, K may be semilinearly imbedded in R^{2n}.

Next, assume $m = 0$. Then from Theorem 4 of AH, Chapter 7, §1 No. 4, K has an n-cycle with integral coefficient, $z = \Sigma \sigma_i$, where the Σ is summed over all n-simplices σ_i of K, each with suitable orientation, such that for any coefficient group G, $H_n(K, G) = Z_n(K, G)$ consists of all cycles gz, where $g \in G$ is arbitrary. In particular, $Z_n(K, \bmod 2)$ has only one nonzero mod 2 cycle, $\rho_2 z = \rho_2 \Sigma \sigma_i$. By the same reasoning, we have $H_{2n}(\tilde{K}^*, \bmod 2) = 0$, hence $\Phi^{2n}(K) = 0$, and so K can be semilinearly imbedded in R^{2n}. The theorem is now completely proved.

From AH, Chapter 10, §3 No. 5 and Theorem 4 of AH, Chapter 13, §4 No. 4, whether a finite complex K is a closed complex or not is a topological invariant of $\bar{K}$. Similarly, from AH, Chapter 8, §4 No. 7, whether a finite complex K is an irreducible closed complex or not, is also a topological invariant $\bar{K}$. Let us call the space underlying a closed complex or an irreducible closed complex, a closed polyhedron or an irreducible closed polyhedron. The Theorem 4 has the following corollary:

Theorem 4′. *If X is an irreducible closed polyhedron of dimension $n \neq 2$, then X can be topologically imbedded in R^{2n}.*

Theorem 5 (Van Kampen). *If any $(n-1)$-simplex of the n-dimensional complex K is at most the face of two n-simplices, then K can be semilinearly imbedded in R^{2n}.*

Proof. The theorem is obvious if $n < 2$.

Now assume $n = 2$. Also assume first that K is a 2-dimensional homogeneous complex. Then K must be constructed as follows: let K_i, $i = 1, \cdots, r$ be some complexes, obtained by suitably subdividing some connected surfaces, with or without boundary; then K is obtained by identifying some vertices of the (K_i)'s. From the well-known result of the classification of surfaces, we know that there exist semilinear imbeddings $T_i: \bar{K}_i \to R^4$. We may assume that the $T_i(\bar{K}_i)$'s are mutually disjoint. Then the T_i's together determine a semilinear imbedding $T: \Sigma \bar{K}_i \to R^4$, where $T/\bar{K}_i \equiv T_i$. Now let $\bar{a}_j$, $j = 1, 2, \cdots, N$, be all the vertices of the (K_i)'s that will become the vertices of K only after identification. Assume also that all the $\bar{a}_j$ are divided into several families $(\bar{a}_{j_1}, \cdots, \bar{a}_{j_t})$, $j_1 < j_2 < \cdots < j_t$, where the vertices in the same family will be identified, while those of different families will not be identified. Let $T(\bar{a}_j) = a_j$. For each family $(\bar{a}_{j_1}, \cdots, \bar{a}_{j_t})$, we may construct a family of simple broken lines $l_{j_2}, \cdots, l_{j_t}$, connecting $a_{j_1}, a_{j_2}; \cdots; a_{j_1}, a_{j_t}$ respectively, such that these broken lines and $T(\bar{K}_i)$ are mutually disjoint, except for the common endpoints. Let $j = j_2, \cdots,$ or j_t, $\bar{a}_j$ be a vertex of K_{ij} and V_j be a sufficiently small linear neighborhood of a_j in $T(\bar{K}_{ij})$. Then, in a sufficiently small neighborhood of each broken line l_{j_r}, we may construct a "linear conical surface" C_{j_r} with "vertex" a_{j_1}, "base" the boundary of V_{j_r}, "axis" the broken line l_{j_r}, and we can assume that these conical surfaces are mutually disjoint except for the vertices, and only meet $T(\Sigma \overline{K}_i)$ at the vertices and the bases. Alter T to T': $\Sigma \overline{K}_i \to R^4$ so that T' coincides with T on $\Sigma K_i - \Sigma \operatorname{Int} V_j$, and T' maps V_j to C_j. Then T' can be viewed as a semilinear imbedding of K into R^4 as required. Next, still assume $n = 2$, but let K be arbitrary. Let K' be the subcomplex of K, consisting of all 2-simplices of K and their faces. From the preceding, we may construct a semilinear imbedding T: $\overline{K}' \to R^4$. Obviously, this imbedding can be extended to a semilinear imbedding of K. Hence the theorem is proved for $n = 2$.

Now we prove the theorem, assuming $n > 2$.

First, assume K is a regular connected (regulärer zusammenhängender, see AH, Chapter 4) n-dimensional homogeneous complex. Then, for any two n-simplices σ and $\sigma' \in K$ with no common vertices, there exists a chain of n-simplices in K, $\sigma_1 = \sigma, \sigma_2, \cdots, \sigma_r = \sigma'$ such that σ_i and σ_{i+1} $(i = 1, \cdots, r-1)$ have only an $(n-1)$-simplex τ_i as their common face. Among $\sigma_1, \cdots, \sigma_{r-1}$, let σ_s be the last simplex having no common vertex with σ_r, $1 \leq s < r-1$. Then after suitably orienting the σ's and τ's, in K^*, we have

$$\begin{aligned}
\{\sigma_1 * \sigma_r\} - \{\sigma_2 * \sigma_r\} &= \delta\{\tau_1 * \sigma_r\},\\
\{\sigma_2 * \sigma_r\} - \{\sigma_3 * \sigma_r\} &= \delta\{\tau_2 * \sigma_r\},\\
\{\sigma_{s-1} * \sigma_r\} - \{\sigma_s * \sigma_r\} &= \delta\{\tau_{s-1} * \sigma_r\},\\
\{\sigma_s * \sigma_r\} \qquad\qquad &= \delta\{\tau_s * \sigma_r\},
\end{aligned}$$

where $\{\xi * \eta\}$ $(\xi, \eta \in K, \xi * \eta \in K^*)$ represents the integral cochain wtth value 1 on $\xi * \eta$, but value 0 on any other cell of K^*. Adding these formulas, we see that the cochain $\{\sigma_1 * \sigma_r\} = \{\sigma * \sigma'\}$ is a coboundary. Hence $H^{2n}(K^*) = 0$, and in particular, $\Phi^{2n}(K) = 0$. By the assumption that $n > 2$ and Theorem 1, K can be semilinearly imbedded in R^{2n}.

Next, consider the general case for $n > 2$. Let K' be the homogeneous complex consisting of all the n-simplices of K and their faces. Let L be the subcomplex consisting of all the r-simplices, $r \leq n-1$, which are not faces of any n-simplices. According to AH, Chapter 4, §5 No. 8, K' can be decomposed into regular connected regular components (regulärer Komponenten) $K_1, \cdots, K_s$, such that

$$K = K_1 + \cdots + K_s + L,$$

where the common portion of any two subcomplexes on the right-hand side is a subcomplex of dimension at most $n-2$. Thus

$$K^* = \sum_{i<j} K^*_{ij} + \sum_i L^*_i + L^*,$$

where K^*_{ij}, L^*_i and L^* are formed by cells of the form $\sigma * \tau$, σ and τ have no common vertices, $\sigma \in K_i$, $\tau \in K_j$, or $\sigma \in K_i$ and $\tau \in L$, or $\sigma, \tau \in L$. It is easy to see that L^* is a subcomplex of dimension at most $2n-2$, each L^*_i is a subcomplex of dimension at most $2n-1$ and the common portion of any two subcomplexes among the (K^*_{ij})'s and (L^*_i)'s is a subcomplex of dimension at most $2n-2$. Hence, any mod $2m$ $(m \geq 0)$, $2n$-cycle z of K^* can be written as

$$z = \sum_{i<j} z_{ij},$$

where each z_{ij} is a mod $2m$ chain of K_i^*, indeed it is a mod $2m$ cycle.

Now order the vertices of K in a sequence $a_0 < a_1 < \cdots < a_n$, such that the vertices of K_i $(i = 1, \cdots, s-1)$ are all in front of those of K_{i+1}, but not in front of those of $K_1, \cdots,$ or K_{i-1}. At the end of the sequence are the vertices that are in L, but not in any of the K_i's. Otherwise, the ordering is arbitrary. It follows from Theorem 11 of [1], that, with respect to this ordering, there exists a cocycle in $\Phi^{2n}(K)$,

$$\varphi^{2n} = \Sigma\{(a_{i_0} \cdots a_{i_n}) \cdot (a_{j_0} \cdots a_{j_n})\},$$

where Σ is summed over all possible groups of indices (i, j) such that $i_0 < j_0 < i_1 < \cdots < i_n < j_n$. Let $\sigma_i = (a_{i_0}, \cdots, a_{i_n}) \in K_i$, $\sigma_j = (a_{j_0}, \cdots, a_{j_n}) \in K_j$, $i_0 < \cdots < i_n$, $j_0 < \cdots < j_n$ and $\sigma_i * \sigma_j \in K_{ij}^*$. If $i < j$, then, since K_i and K_j has at most an $(n-2)$-simplex in common, by choice of the order of the vertices, we cannot have either $i_0 < j_0 < i_1 < \cdots < i_n < j_n$ or $j_0 < i_0 < j_1 < \cdots < j_n < i_n$. So $\phi^{2n}(\sigma_i * \sigma_j) = 0$. Therefore, by formula (3), we have $\rho_{2m}\phi^{2n}(z) = \Sigma\rho_{2m}\phi^{2n}(z_{ii})$. Now let λ_i be the inclusion mapping of K_{ii}^* into K^*. Then obviously, $\lambda_i^{\#}\phi^{2n} \in \Phi^{2n}(K_i) \in H^{2n}(K_{ii}^*)$. But from the last part of this proof, $\Phi^{2n}(K_i) = 0$. Hence $\lambda_i^{\#}\phi^{2n} \sim 0$, and $\rho_{2m}\phi^{2n}(z_{ii}) = \lambda_i^{\#}\rho_{2m}\phi^{2n}(z_{ii}) = 0 \bmod 2m$. Thus $\rho_{2m}\phi^{2n}(z) = 0$, i.e. $\phi^{2n} \bmod 2m$ is orthogonal to any mod $2m$ $2n$-cycles of K^*. Analogous to the last part of the proof of Theorem 2, we have, for any integer $m = 0$ or ≥ 2, that $\phi^{2n} \bmod m$ is orthogonal to any mod m $2n$-cycles of K^*. Thus, $\phi^{2n} \sim 0$, or $\Phi^{2n}(K) = 0$. By the assumption that $n > 2$ and Theorem 1, K can be semilinearly imbedded in R^{2n}. Thus Theorem 5 is completely proved.

Remark. Theorem 5 above is Theorem 4 of Van Kampen's original paper [3]. His corrected proof (see Lemma 6 of [4], uses a deformation theorem of [3] ([3], Theorem 2), so is different from the proof given here.

Theorem 5 has the following corollary:

Theorem 6. *Every combinatorial manifold (with or without boundary) of dimension n can be semilinearly imbedded in* R^{2n}.

BIBLIOGRAPHY

[1] Wu Wen-jun (Wu Wen-chün), *On the realization of complexes in euclidean spaces.* I, Acta Math. Sinica 5 (1955), 505–552. MR 17, 883.

[2] ——, *On the realization of complexes in euclidean spaces.* II, Acta. Math. Sinica 7 (1957), 79–101. MR 20 #3536.

[3] E. R. Van Kampen, *Komplexe in euklidischen Räumen*, Abh. Math. Sem. Hamburg 9 (1932), 72–78.

[4] ——, Supplement to the preceding, ibid. 152–153.

[5] H. Whitney, *The self-intersections of a smooth n-manifold in 2n-space*, Ann. of Math. 45 (1944), 220–246. MR 5, 273.

[6] J. H. C. Whitehead, *On C^1-complexes*, Ann. of Math. 41 (1940), 809–824. MR 2, 73.

Translated by:
V. Kraines

Acta Mathematica Sinica, 263–273
Vol. 18, December 1975

On Universal Invariant Forms

Wu Wen-tsun

Institute of Mathematics
Academia Sinica

(Received May 23, 1974)

1. Introduction

The concept of integral invariant of invariant form has already had a fruitful effect on the theory and application of mechanical system. In 1947, Mr. H.C. Lee [6] in our country introduced also the concept of universal integral invariant or universal invariant form for the Hamilton system,and proved that besides those already discussed by Poincare and E.Cartan, there are no other such invariant forms. Those universal invariant forms under Lee's meaning, can be generalized to following more general understanding: let M be a space of n variables, G be an infinite transformation group on M under E.Cartan's meaning, a vector field X on M will be called belonging to G, if the transformation in the local single parametric group produced by X only needs to be sufficiently closed to the identity transformation then it belongs to G. We denote the collection of these vector fields as $\mathscr{L}G$. Accordingly an exterior differential from θ on M will be defined as an universal invariant form of G (in the following we abbreviate it as the universal invariant form). If for any vector field $x \in \mathscr{L}G$, the $\mathscr{L}_{ie}$ derivative of θ along X: $\mathscr{L}_x\theta = 0$. When M is a symplectic manifold and G is an infinite transformation group formed by all symplectic transformations on M, the universal invariant form of G is similar to that defined by H.C. Lee.

E.Cartan has pointed out that there are six classes of primitive infinite transformation groups, where four classes are single(see $1°, 2°, 4°, 6°$ in the following). Cartan's results up to now still have not been proved, we list them as follows:

1° The group G_n^I formed by all transformations on n variables.

2° The group G_n^{II} formed by all transformations preserving the volume element

$$\Theta = dx_1 \wedge \cdots \wedge dx_2$$

invariant on n variables $x_1, \ldots, x_n$.

3° The group G_n^{III} formed by all transformations which varies only a non-zero constant factor of the above mentioned volume element Θ on n variables $x_1, \ldots, x_n$.

4° The group G_n^{IV} formed by all regular transformations preserving the form

$$\Omega = dp_i \wedge dq_i (= dp_1 \wedge dq_1 + \cdots + dp_n \wedge dq_n)$$

invariant on $2n$ variables $p_1, \ldots, p_n, q_1, \ldots, q_n (n > 2)$.

5° The group G_n^V formed by all transformations on $2n$ variables $p_i, q_i (i = 1, \ldots, n; n \geq 2$ which vary only a non-zero constant factor of the above mentioned form Ω.

6° The group G_n^{VI} formed by all transformations preserving the form

$$w = dt + p_i dq_i - q_i dp_i$$
$$(= dt + p_i dq_i + \cdots + p_n dq_n - q_1 dp_1 - \cdots - q_n dp_n)$$

invariant on 2n+1 variables $t, p_i, q_i (i = 1, \cdots, n)$.

This paper will determine the universal invariant form of these above mentioned infinite groups. The group G_n^{IV} in 4° is what H.C. Lee exploring. The case of 1° is insignificant, the group in 3° and 5° are subgroups of the groups in 2° and 4° respectively, therefore its exploration may be concluded trivially to the latter. Hence we need to explore only G_n^{II} in 2° and G_n^{VI} in 6° (see Section 2 and Section 4).

Many conservative laws in mechanics reflect a certain symmetry of the mechanical system. They can be expressed by using the concept of universal invariant forms of certain subgroups in G_n^{IV}, these subgroups are formed by all transformations preserving a group of function that is the so-called "moved constants". The complete determination of the universal invariant forms of these subgroups is equivalent to the determination of the corresponding conservative law of the system, (see Section 5).

The function, vector field and form etc. mentioned in this paper belong to C^∞. More accurately, they should be treated as sprout rundle section of function, vector field, the form. Similarly, the so-called transformation also implies the local homeomorphism C^∞ transformation, the transformation group is the so-called pseudo-group. But since we consider only the problems of local property on the whole, hence sprout bundle, pseudo-group such vocabularies are not very necessary and have not been used in this paper.

In many formulas, in accordance with the custom of differential geometry overlapping exponent indicate to take sum, we only write the sigma sign $\sum$ and its range of taking sum clearly when the range of indices my cause confusions, otherwise neglected.

2. Infinite Group of Volume-preserving Transformations

Let G_n^{II} be the single infinite group of type II formed by all transformations preserving the volume element

$$\Theta = dx_1 \wedge \cdots \wedge dx_n$$

Invariant on n variables $x_1, \ldots, x_n$. For any vector field $X = X^i \frac{\partial}{\partial x_i}$ we have

$$\mathscr{L}_x \Theta = \frac{\partial X^i}{\partial x_i} \Theta$$

Therefore the necessary and sufficient condition for $X \in \mathscr{L} G_n^{II}$ is

$$div X = \frac{\partial X^i}{\partial x_i} = 0.$$

Hence $\mathscr{L} G_n^{II}$ contains the following $\frac{1}{2}n(n+1)$ special vector fields:

$$A_i = \frac{\partial}{\partial x_i}, i = 1, \ldots, n.$$

$$A_{ij} = x_j \frac{\partial}{\partial x_i}, i, j = 1, \ldots, n; i \neq j.$$

Now let

$$\theta = a_{i_1 \cdots i_r} dx_{i_1} \wedge \cdots \wedge dx_{i_r}$$

be an universal invariant form of G_n^{II}, where $a_{i_1 \cdots i_r}$ are functions of $x_1, \ldots, x_n$, anti-symmetric for the lower index $i_1, \ldots, i_r$. Accordingly for any $X \in G_n^{II}$, particularly for $X = A_i$ or A_{ij} we have $\mathscr{L}_X \theta = 0$. We have computed

$$\mathscr{L}_{A_i} \theta = \frac{\partial a_{i_1 \cdots i_r}}{\partial x_i} dx_{i_1} \wedge \cdots \wedge dx_{i_r}$$

$$\mathscr{L}_{A_{ij}} \theta = x_j \frac{\partial a_{i_1 \cdots i_r}}{\partial x_i} dx_{i_1} \wedge \cdots \wedge dx_{i_r} + r a_{i i_i \cdots i_{r-1}} dx_{i_j} \wedge dx_{i_1} \wedge \cdots \wedge dx_{i_{r-1}}.$$

By $\mathscr{L}_{A_i} \theta = 0$ we obtain $\frac{\partial a_{i_1 \cdots i_r}}{\partial x_i} = 0$, hence $a_{i_1 \cdots i_r}$ are constants. If $r = n$ then $\theta = n! a_{1 \cdots n} dx_1 \wedge \cdots \wedge dx_n$ is a constant multiple of Θ. Let $r < n$, then for any r indices $(i, i_1, \ldots, i_{r-1})$ such that $i < i_1 < \cdots < i_{r-1}$ we can take any index j not equal to i and $i_1, \ldots, i_{r-1}$, for this pair (i, j) the condition $\mathscr{L}_{A_{ij}} \theta = 0$ gives $a_{i i_1 \cdots i_{r-1}} = 0$. By this we obtain $\theta = 0$ and have the following

Theorem *The unique universal invariant form of the infinite group G_N^{II} of type II is a constant multiple of the volume variable Θ.*

3. Regular Transformation Infinite Group— H.C. Lee's Theorem

Let G_n^{IV} be the single infinite group of type IV formed by all regular transformations preserving the symplectic form

$$\Omega = dp_i \wedge dq_i \tag{1}$$

invariant on $2n$ variables $p_i, q_i (i = 1, \ldots, n)$. H.C. Lee has proved the following described

Theorem ([6] 1947) *The unique universal invariant form of the infinite group G_n^{IV} is a constant multiple of Ω and its outer power $\Omega^2 = \Omega \wedge \Omega, \Omega^3 = \Omega^2 \wedge \Omega, \ldots, \Omega^n = \Omega^{n-1} \wedge \Omega$.*

Since this theorem and the computation in its proof is needed to use in future, therefore we repeat it according to the form a bit different from the original paper as follows.

Showing the indices of range $1, \ldots, n$ with the Latin letters $i, j, k, \ldots$ and showing the indices of range $1, 2, \ldots, 2n$ with the Greek letters $\alpha, \beta, \lambda, \mu, \ldots$. We introduce the new variable as follows

$$x^i = q_i, \quad x^{n+i} = p_i, \tag{2}$$

also we show the anti-symmetric matrix which are the inverses of each others with $\varepsilon^*, \varepsilon_*$:

$$\varepsilon^* = (\varepsilon^{\alpha\beta}) = \begin{pmatrix} 0 & I \\ -I & 0 \end{pmatrix}, \tag{3*}$$

$$\varepsilon_* = (\varepsilon_{\alpha\beta}) = \begin{pmatrix} 0 & -I \\ I & 0 \end{pmatrix}, \tag{3_*}$$

where 0 and I are the zero matrix and identity matrices of order n respectively, accordingly Ω becomes

$$\Omega = \frac{1}{2}\varepsilon_{\alpha\beta}dx^\alpha \wedge dx^\beta. \tag{4}$$

With respect to any vector field

$$X = X^\lambda \frac{\partial}{\partial x^\lambda},$$

where X^λ is a function of x^α, there is

$$\mathscr{L}_X\Omega = \tfrac{1}{2}\varepsilon_{\alpha\beta}(dX^\alpha \wedge dx^\beta + dx^\alpha \wedge dX^\beta) = -d(\varepsilon_{\alpha\beta}X^\beta dx^\alpha).$$

Therefore the necessary and sufficient condition of $X \in \mathscr{L}G_n^{IV}$ is

$$d(\varepsilon_{\alpha\beta}X^\beta dx^\alpha) = 0.$$

Since we only consider within a local range, therefore from Poincare's lemma the condition becomes that there is function H determined to a constant such that

$$\varepsilon_{\alpha\beta}X^\beta dx^\alpha = dH,$$

or $\varepsilon_{\alpha\beta}X^\beta = \frac{\partial H}{\partial x^\alpha}$, $X^\lambda = \varepsilon^{\lambda\mu}\frac{\partial H}{\partial x^\mu}$. By this we obtain the following

Lemma *The necessary and sufficient condition for $X \in \mathscr{L}G_n^{IV}$ is that there is a function H such that*

$$X = \varepsilon^{\lambda\mu}\frac{\partial H}{\partial x^\mu}\frac{\partial}{\partial x^\lambda}. \tag{5}$$

X is determined in this lemma by H uniquely, in the future we will denote it as X_H. Conversely, H determines a constant by $x \in \mathscr{L}G_n^{IV}$ uniquely, in future we will denote a difference of constants as H_X.

Now let any form of degree r $(1 \leq r \leq 2n)$:

$$\theta = A_{\alpha_1\cdots\alpha_r}dx^{\alpha_1} \wedge \cdots \wedge dx^{\alpha_r}. \tag{6}$$

Where $A_{\alpha_1\cdots\alpha_r}$ are functions of x^λ and anti-symmetric for the lower indices $\alpha_1, \ldots, \alpha_r$. For any

$$X = \varepsilon^{\lambda\mu}\frac{\partial H}{\partial x^\mu}\frac{\partial}{\partial x^\lambda} \in \mathscr{L}G_n^{IV}$$

there is

$$\mathscr{L}_X\theta = [\varepsilon^{\lambda\mu}\frac{\partial A_{\alpha_1\cdots\alpha_r}}{\partial x^\lambda}\frac{\partial H}{\partial x^\mu} + \delta^\nu_{\alpha_i}\varepsilon^{\lambda\mu}J^i_\lambda A_{\alpha_1\cdots\alpha_r}\frac{\partial^2 H}{\partial x^\mu \partial x^\nu}]dx^{\alpha_1} \wedge \cdots \wedge dx^{\alpha_r},$$

where J^i_λ means to change the ith index α_i in $A_{\alpha_1\cdots\alpha_r}$ to the operator of λ:

$$J^i_\lambda A_{\alpha_1\cdots\alpha_i\cdots\alpha_r} = A_{\alpha_1\cdots\alpha_{i-1}\lambda\alpha_{i+1}\cdots\alpha_r}. \tag{7}$$

The necessary and sufficient condition for the form θ is to an universal invariant form is that for any H hence for any $\frac{\partial H}{\partial x^\mu}$ and $\frac{\partial^2 H}{\partial x^\mu \partial x^\nu}$ we should have $\mathscr{L}_X\theta = 0$. By this we obtain the following H.C. Lee's system of equations:

$$\begin{cases} \varepsilon^{\lambda\mu}\dfrac{\partial A_{\alpha_1\cdots\alpha_r}}{\partial x^\lambda} = 0 \\ (\delta^\nu_{\alpha_i}\varepsilon^{\lambda\mu} + \delta^\mu_{\alpha_i}\varepsilon^{\lambda\nu})J^i_\lambda A_{\alpha_1\cdots\alpha_r} = 0. \end{cases} \tag{8}$$

By the front part of the system of equations, all $A_{\alpha_1\cdots\alpha_r}$ are constants. By the rear part (where $\mu, \nu, \alpha_1, \ldots, \alpha_r$ are arbitrary), H.C. Lee uses pure algebraic method to obtain that when r is an even number $2s$,

$$\theta = c \cdot \underbrace{\Omega \wedge \cdots \wedge \Omega}_{s},$$

where c is a constant, and when $r =$ odd number

$$\theta = 0.$$

By this Lee has proved his theorem.

4. Tangential Transformation Infinite group

Let G_n^{VI} be the single infinite group of type VI formed by all tangential transformation preserving the form

$$\omega = dt + p_i dq_i - q_i dp_i$$

invariant on $2n+1$ variables $t, p_i, q_i (i = 1, \ldots, n)$. Applying the similar symbols in Section 3 we may write ω as

$$\omega = dt + \varepsilon_{\alpha\beta} x^\alpha dx^\beta, \tag{1}$$

hence

$$d\omega = \varepsilon_{\alpha\beta} dx^\alpha \wedge dx^\beta = 2\Omega,$$

here Ω is same as (4) of Section 3.

Consider any vector field

$$X = T\frac{\partial}{\partial t} + X^\lambda \frac{\partial}{\partial x^\lambda},$$

where T, X^λ are functions of x^α and t, setting

$$K = \frac{1}{2}(T + \varepsilon_{\alpha\beta} x^\alpha X^\beta), \tag{2}$$

$$K_\lambda = \frac{\partial K}{\partial x^\lambda}, \quad K_{\lambda\mu} = K_{\mu\lambda} = \frac{\partial^2 K}{\partial x^\lambda \partial x^\mu}, \tag{3}$$

Then we have

$$\mathscr{L}_X \omega = 2\frac{\partial K}{\partial t} t + 2(\frac{\partial K}{\partial x^\lambda} + \varepsilon_{\alpha\lambda} X^\alpha) dx^\lambda.$$

By $\mathscr{L}_X\omega = 0$ we obtain $\frac{\partial K}{\partial t} = 0$ or K is independent of t and $X^\alpha = \varepsilon^{\alpha\lambda}K_\lambda$ hence

$$X = (2K - x^\alpha K_\alpha)\frac{\partial}{\partial t} + \varepsilon^{\alpha\lambda}K_\lambda\frac{\partial}{\partial x^\alpha}. \tag{4}$$

Its inverse is obviously true therefore we obtain the following

Lemma *The necessary and sufficient condition for $X \in \mathscr{L}G_n^{IV}$ is that X possesses the representation of (4),where K is any function of x^λ but independent of t.*

Now we write any form of degree r as

$$\theta = \varphi + \psi \wedge dt, \tag{5}$$

$$\varphi = A_{\alpha_1\cdots\alpha_r}dx^{\alpha_1} \wedge \cdots \wedge dx^{\alpha_r} \tag{5$'$}$$

$$\psi = B_{\beta_1\cdots\beta_{r-1}}dx^{\beta_1} \wedge \cdots \wedge dx^{\beta_{r-1}}, \tag{5$''$}$$

where all A and B are functions of x^λ and t, and anti-symmetric for the lower index, for the X determined by (4),we directly compute and obtain:

$$\mathscr{L}_x\theta = \tilde{A}_{\alpha_1\cdots\alpha_r}dx^{\alpha_1} \wedge \cdots \wedge dx^{\alpha_r} + \tilde{B}_{\beta_1\cdots\beta_{r-1}}dx^{\beta_1} \wedge \cdots \wedge dx^{\beta_{r-1}} \wedge dt, \tag{6}$$

$$\begin{aligned}\tilde{A}_{\alpha_1\cdots\alpha_r} =& 2K\frac{\partial A_{\alpha_1\cdots\alpha_r}}{\partial t} \\ &- (x^\lambda\frac{\partial A_{\alpha_1\cdots\alpha_r}}{\partial t} + \varepsilon^{\lambda\mu}\frac{\partial A_{\alpha_1\cdots\alpha_r}}{\partial x^\mu} - (-1)^{r-j}\delta^\lambda_{\alpha_j}B_{\alpha_1\cdots\hat{\alpha}_j\cdots\alpha_r})K_\lambda \\ &+ (\delta^\nu_{\alpha_i}\varepsilon^{\lambda\mu}J^i_\lambda A_{\alpha_1\cdots\alpha_r} - (-1)^{r-j}\delta^\nu_{\alpha_j}x^\mu B_{\alpha_1\cdots\hat{\alpha}_j\cdots\alpha_r})K_{\mu\nu},\end{aligned} \tag{6$'$}$$

$$\begin{aligned}\tilde{B}_{\beta_1\cdots\beta_{r-1}} =& 2K\frac{\partial B_{\beta_1\cdots\beta_{r-1}}}{\partial t} - (x^\lambda\frac{\partial B_{\beta_1\cdots\beta_{r-1}}}{\partial t} + \varepsilon^{\lambda\mu}\frac{\partial B_{\beta_1\cdots\beta_{r-1}}}{\partial x^\mu})K_\lambda \\ &+ \delta^\nu_{\beta_j}\varepsilon^{\lambda\mu}J^j_\lambda B_{\beta_1\cdots\beta_{r-1}}K_{\mu\nu}.\end{aligned} \tag{6$''$}$$

Because of K, $K_\lambda$and$K_{\mu\nu}$ can be selected arbitrarily, therefore the necessary and sufficient condition for θ to be the universal invariant form of group G_n^{VI} is that the following equalities hold:

$$\frac{\partial B_{\beta_1\cdots\beta_{r-1}}}{\partial t} = 0, \quad \varepsilon^{\lambda\mu}\frac{\partial B_{\beta_1\cdots\beta_{r-1}}}{\partial x^\mu} = 0, \tag{7}$$

$$(\delta^\nu_{\beta_j}\varepsilon^{\lambda\mu} + \delta^\mu_{\beta_j}\varepsilon^{\lambda\nu})J^j_\lambda B_{\beta_1\cdots\beta_{r-1}} = 0 \tag{7$'$}$$

$$\frac{\partial A_{\alpha_1\cdots\alpha_r}}{\partial t} = 0 \tag{8}$$

$$\varepsilon^{\lambda\mu}\frac{\partial A_{\alpha_1\cdots\alpha_r}}{\partial x^\mu} - (-1)^{r-j}\delta^\lambda_{\alpha_j}B_{\alpha_1\cdots\hat{\alpha}_j\cdots\alpha_r} = 0, \tag{8$'$}$$

$$(\delta^\nu_{\alpha_i}\varepsilon^{\lambda\mu} + \delta^\mu_{\alpha_i}\varepsilon^{\lambda\nu})J^i_\lambda A_{\alpha_1\cdots\alpha_r} - (-1)^{r-j}(\delta^\nu_{\alpha_j}x^\mu + \delta^\mu_{\alpha_j}x^\nu)B_{\alpha_1\cdots\hat{\alpha}_j\cdots\alpha_r} = 0. \tag{8$''$}$$

By (7) we know all B are constants and by (7$'$) such as the proof of Lee's theorem in Section 3 we know (b =constant):

$$\psi = \begin{cases} b\cdot(d\omega)^s, & r = \text{odd } 2s+1, \\ 0, & r = \text{even}. \end{cases} \tag{9}$$

Now set

$$A'_{\alpha_1\cdots\alpha_r} = A_{\alpha_1\cdots\alpha_r} - (-1)^{r-j}\varepsilon_{\gamma\alpha_j}x^\gamma B_{\alpha_1\cdots\hat{\alpha}_j\cdots\alpha_r}. \tag{10}$$

By direct computation we know that (8),(8′),(8″) can be changed to the following equalities respectively:

$$\frac{\partial A'_{\alpha_1\cdots\alpha_r}}{\partial t} = 0, \tag{11}$$

$$\varepsilon^{\lambda\mu}\frac{\partial A'_{\alpha_1\cdots\alpha_r}}{\partial x^\mu} = 0, \tag{11$'$}$$

$$(\delta^\nu_{\alpha_i}\varepsilon^{\lambda\mu} + \delta^\mu_{\alpha_i}\varepsilon^{\lambda\nu})J^i_\lambda A'_{\alpha_1\cdots\alpha_r} = 0. \tag{11$''$}$$

Still similar to the proof of Lee's theorem in §3 we know that there is a constant a such that:

$$\varphi' = A'_{\alpha_1\cdots\alpha_r}dx^{\alpha_1}\wedge\cdots\wedge dx^{\alpha_r} = \begin{cases} a\cdot(d\omega)^s, & r = \text{even } 2s, \\ 0, & r = \text{odd}. \end{cases} \tag{12}$$

First let $r = 2s$ be an even number, then by (9) we obtain $\psi = 0$ and (10) becomes $A'_{\alpha_1\cdots\alpha_r} = A_{\alpha_1\cdots\alpha_r}$. Hence by (12) we obtain:

$$\theta = \varphi' = a\cdot(d\omega)^s, \quad r = 2s. \tag{13}$$

Next let $r = 2s+1$ be an odd number. Then $\varphi' = 0$ and (10) gives

$$A_{\alpha_1\cdots\alpha_r} = (-1)^{j+1}\varepsilon_{\gamma\alpha_j}x^\gamma B_{\alpha_1\cdots\hat{\alpha}_j\cdots\alpha_r}$$

By (5′),(5″) and (1) we obtain

$$\begin{aligned}\varphi &= A_{\alpha_1\cdots\alpha_r}dx^{\alpha_1}\wedge\cdots\wedge dx^{\alpha_r} \\ &= (-1)^{j+1}\varepsilon_{\gamma\alpha_j}x^\gamma B_{\alpha_1\cdots\hat{\alpha}_j\cdots\alpha_r}dx^{\alpha_1}\wedge\cdots\wedge dx^{\alpha_r} \\ &= \varepsilon_{\gamma\alpha_j}x^\gamma dx^{\alpha_j}\wedge B_{\alpha_1\cdots\hat{\alpha}_j\cdots\alpha_r}dx^{\alpha_1}\wedge\cdots\wedge\widehat{dx^{\alpha_j}}\wedge\cdots\wedge dx^{\alpha_r} \\ &= (\omega - dt)\wedge\psi.\end{aligned}$$

By (5) and (9) we obtain:

$$\begin{aligned}\theta &= (\omega - dt)\wedge\psi + \psi\wedge dt = \omega\wedge\psi \\ &= b\cdot\omega\wedge(d\omega)^s, \quad r = 2s+1.\end{aligned}$$

To sum up, we have the following

Theorem *The unique universal invariant form of the infinite group G_n^{VI} is a constant multiple of $\omega\wedge(d\omega)^s$ and $(d\omega)^s$ $(s = 0, 1, 2, \ldots, n)$.*

5. Regular Transformation Infinite Group Possessing Definite Symmetry

Still using the notations in Section 3 we consider the infinite group G_n^{IV} formed by all regular transformations preserving the form

$$\Omega = dp_i \wedge dq_i = \frac{1}{2}\varepsilon_{\alpha\beta} dx^\alpha \wedge dx^\beta \tag{1}$$

invariant on $2n$ variables $p_i, q_i \quad (i = 1, \ldots, n)$. Let $\mathscr{G}$ be a given Lie's group acting on the phase space M on the right, that is there is a mapping $\Phi : M \times \mathscr{G} \to M$ such that for any $x \in M, g \in \mathscr{G}$, setting $\Phi(x, g) = \Phi_x(g) = \Phi_g(x) \in M$, we have $\Phi_g \in G_n^{IV}$ and $\Phi_{gg'} = \Phi_{g'}\Phi_g(g, g' \in \mathscr{G})$. We also let Φ_g be non-degenerate when g is not an identity element e in $\mathscr{G}$. Denote the Lie algebra of $\mathscr{G}$ as g, then for any $a \in g$, regarding a as a left invariant vector field on $\mathscr{G}$, $Y_a(x) = \Phi_{x*}a(e)$ defines a vector field Y_a on M. By $\Phi_g \in G_n^{IV}(g \in \mathscr{G})$ we know $Y_a \in \mathscr{L}G_n^{IV}$ or $\mathscr{L}_{Y_a}\Omega = 0$. Similarly it is easy to prove that corresponding $a \to Y_a$ is a Lie's homomorphism from g to $\mathscr{L}G_n^{IV}$. We call the collection denoting $Y_a(a \in g)$ as the Lie algebra $\mathscr{Y}$. It is easily known that for any function H, if H is invariant under $\mathscr{G}$, that is for any $g \in \mathscr{G}$, there is $\Phi_g^* H = H$, then for any $Y \in \mathscr{Y}$, we have $\mathscr{L}_Y H = 0$,or H is constant of the motion produced by Y, or H takes similar values on each integral curve produced by Y(refer to, for example, [1]).

For any two functions H, K define the poisson bracket to be

$$(H, K) = \frac{\partial(H, K)}{\partial(p_i, q_i)} = \frac{1}{2}\varepsilon_{\alpha\beta}\frac{\partial(H, K)}{\partial(x^\alpha, x^\beta)} = \varepsilon_{\alpha\beta}H_\alpha K_\beta, \tag{2}$$

Here $H_\alpha = \dfrac{\partial H}{\partial x^\alpha}$, similar for the others. According to the lemma in Section 3 from H, K we can determine two vector fields in G_n^{IV}

$$X_H = \varepsilon^{\lambda\mu}H_\mu \frac{\partial}{\partial x^\lambda}, X_K = \varepsilon_{\alpha\beta}K_\beta \frac{\partial}{\partial x^\alpha}.$$

Accordingly

$$\begin{aligned}[X_H, X_K] &= [\varepsilon^{\lambda\mu}H_\mu \frac{\partial}{\partial x^\lambda}, \varepsilon^{\alpha\beta}K_\beta \frac{\partial}{\partial x^\alpha}] \\ &= \varepsilon^{\alpha\beta}\varepsilon^{\lambda\mu}(H_\mu K_{\beta\lambda}\frac{\partial}{\partial x^\alpha} - K_\beta H_{\alpha\mu}\frac{\partial}{\partial x^\lambda}).\end{aligned}$$

interchanging (α, β) and (λ, μ) in the first team on the right hand side, and α and β in the second team,we then obtain

$$\begin{aligned}[X_H, X_K] &= \varepsilon^{\lambda\mu}(\varepsilon^{\alpha\beta}H_\beta K_{\mu\alpha} - \varepsilon^{\beta\alpha}K_\alpha H_{\beta\mu})\frac{\partial}{\partial x^\lambda} \\ &= \varepsilon^{\lambda\mu}(\varepsilon_{\beta\alpha}H_\beta K_\alpha)_\mu \frac{\partial}{\partial X^\lambda},\end{aligned}$$

or

$$[X_H, X_K] = X_{(H,K)}.$$

Hence the function under the Poisson bracket and the vector field under the Lee bracket possess certain dualities.

Now we denote the collection of invariant functions under $\mathscr{G}$ as $\mathscr{H}$. Since the Poisson bracket is invariant under regular transformations. Therefore for arbitrary $g \in \mathscr{G}, \Phi_g \in G_n^{IV}$, we have $(\Phi_g^* H, \Phi_g^* K) = \Phi_g^*(H, K)$, hence when $H, K \in \mathscr{H}$,we also have $(H, K) \in \mathscr{H}$. Also by the known properties of the Poisson bracket, we know that $\mathscr{H}$ becomes a Lie algebra under this bracket, if according to the lemma in Section 3 we denote the collection of all vector fields X_H corresponding to $H \in \mathscr{H}$ as $\mathscr{X} \subset \mathscr{L}G_n^{IV}$. Then the Lie algebra formed by $\mathscr{X}$ under the Lie's bracket and the Lie algebra formed by $\mathscr{H}$ under the Poisson bracket possess the previous described dual property. We will denote the infinite group generated by the regular transformations produced by the vector field X in $\mathscr{X}$ with G_*^{IV}, and denote $\mathscr{X}$ as $\mathscr{L}G_*^{IV}$.

Our purpose is to determine those forms which are invariant forms under the usual meaning for all H in $\mathscr{H}$, or that is the universal invariant form of G_*^{IV}. Every such universal invariant form corresponds to a conservative law possessing relative symmetry with $\mathscr{G}$ in physics.

Because of this for any vector field $Y_a, a \in g$ in $\mathscr{Y}$, according to the lemma in Section 3 take a relative function f_a (determine to a constant) such that $Y_a = X_{f_a}$. For $a, b \in g$ by $[Y_a, Y_b] = Y[a, b]$ and $[X_{f_a}, X_{f_b}] = X_{(f_a, f_b)}$ we obtain $(f_a, f_b) = f_{(a,b)}$ (differs by a constant). Let the whole group of $f_a (a \in g)$ be $\mathscr{C}$, also let the collection after adding all arbitrary functions $F(f_{a_1}, \ldots, f_{a_k}), (a_i \in g)$ in $\mathscr{C}$ be $\overline{\mathscr{C}}$. Then it is easy to see that $\overline{\mathscr{C}}$ is the smallest function set possessing the following two properties and $\supset \mathscr{C}$:

1° There is a function basis $f_1, \ldots, f_m$, the rank of its Jacobi expression $= m$.

2° For any $f', f'' \in$ we also have $(f', f'') \in \overline{\mathscr{C}}$.

Proof. take a basis $a_i, i = 1, \ldots, m$ ($m = dim\mathscr{G}$) for g, and set $f_{a_i} = f_i$, then f_i satisfies 1° and $(f_i, f_j) = c_{ij}^k f_k$ (differ by a constant). Here c_{ij}^k is the structure constant of g, accordingly for any $f', f'' \in \overline{\mathscr{C}}, f', f''$ are functions of f_i and has $(f', f'') = \frac{\partial f'}{\partial f_i} \cdot \frac{\partial f''}{\partial f_j} \cdot (f_i, f_j) \in \overline{\mathscr{C}}$. That is what we want to prove.

The function set possessing the two properties of $1°, 2°$ is called a function group (see [2] chapter 9 or [5] §69). By the theory of function group we know that we can take a standard function basis $p_1, \ldots, \bar{p}_{r+s}, \bar{q}_1, \ldots, q_r$ in $\overline{\mathscr{C}}$ and it can be spanned into a function basis $p_1, \ldots, p_n, q_1, \ldots, q_n$ on M such that it possesses the following relations (as above, particularly see [5] theorem 69.6):

$$\begin{cases} (\bar{q}_i, \bar{q}_j) = 0, \\ (\bar{p}_i, \bar{p}_j) = 0, \quad (i, j = 1, \ldots, n), \\ (\bar{p}_i, \bar{q}_j) = \delta_{ij}. \end{cases} \tag{3}$$

According $(\bar{p}_i, \bar{q}_i)$ can be treated as a group of new coordinate on the $(p_i; q_i)$ phase space and the transformations form (p_i, q_i) to $(\bar{p}_i, \bar{q}_i) \in G_n^{IV}$; and (in the expression i is from 1 to n):

$$\Omega = d\bar{p}_i \wedge d\bar{q}_i. \tag{4}$$

For any $H \in \mathscr{H}$ and $F = F(f_1, \ldots, f_m) \in \overline{\mathscr{C}}$, we have

$$(H, F) = \frac{\partial F}{\partial f_i}(H, f_i) = -\frac{\partial F}{\partial f_i}\mathscr{L}_{Y_{a_i}} H = 0.$$

Conversely, if the function H such that for arbitrary $F \in \overline{\mathscr{C}}$ we have $(H, F) = 0$, then for arbitrary $a \in g$, we have $\mathscr{L}_{Y_a} H = \mathscr{L}_{X_{f_a}} H = (f_a, H) = 0$, hence H is invariant under $\mathscr{G}$ or $H \in \mathscr{H}$. By this we know that $\mathscr{H}$ and $\overline{\mathscr{C}}$ form two function groups which are the inverse of each others, and $\mathscr{H}$ has a standard function basis $\bar{p}_{r+1}, \ldots, \bar{p}_n, \bar{q}_{r+s+1}, \ldots, \bar{q}_n$, refer to [2] and [5].

If we write any vector field X in $\mathscr{L}G_n^{IV}$ as

$$X = \sum_{i=1}^{n} \frac{\partial H}{\partial \bar{p}_i}\frac{\partial}{\partial \bar{q}_i} - \sum_{i=1}^{n} \frac{\partial H}{\partial \bar{q}_i}\frac{\partial}{\partial \bar{p}_i},$$

where H is a certain function of $\bar{p}_i, \bar{q}_i$, then when $X \in \mathscr{X} = \mathscr{L}G_*^{IV}$, we should have

$$\mathscr{L}_x \bar{p}_i = -\frac{\partial H}{\partial \bar{q}_i} = 0, \quad i = 1, \ldots, r+s,$$

$$\mathscr{L}_x \bar{q}_i = \frac{\partial H}{\partial \bar{p}_i} = 0, \quad i = 1, \ldots, r.$$

Summary the above mentioned, we obtain:

Lemma *The necessary and sufficient condition for $X \in \mathscr{X} = \mathscr{L}G_*^{IV}$ is to have a function H which only depends on*

$$\bar{p}_{r+1}, \ldots, \bar{p}_n, \bar{q}_{r+s+1}, \ldots, \bar{q}_n \tag{5}$$

such that

$$X = \sum_{i=r+1}^{n} \frac{\partial H}{\partial \bar{p}_i}\frac{\partial}{\partial \bar{q}_i} - \sum_{i=r+s+1}^{n} \frac{\partial H}{\partial \bar{q}_i}\frac{\partial}{\partial \bar{p}_i}. \tag{6}$$

Introduce the notations

$$\left\{\begin{array}{rcll} t &=& n - r - s & \\ y^i &=& \bar{q}_{r+i}, & \\ y^{s+i} &=& \bar{p}_{r+i}, & (i = 1, \ldots, s) \\ z^i &=& \bar{q}_{r+s+i}, & \\ z^{t+i} &=& \bar{p}_{r+s+i}. & (i = 1, \ldots, t) \end{array}\right. \tag{7}$$

Take the following range of indices:

$$\left\{\begin{array}{l} a, b, c, \ldots = 1, 2, \ldots, s \\ A, B, C, \ldots = 1, 2, \ldots, 2t. \end{array}\right. \tag{8}$$

Also take matrix of order $2t$, which are the inverses of each others

$$\eta^* = (\eta^{AB}) = \begin{pmatrix} 0 & I \\ -I & 0 \end{pmatrix},$$
$$\eta_* = (\eta_{AB}) = \begin{pmatrix} 0 & -I \\ I & 0 \end{pmatrix}, \tag{9}$$

where 0 and I are the zero matrix and identity matrices of order t.

Accordingly any $X \in \mathscr{X}$ of expression (6) can be rewritten

$$X = \eta^{AB}\frac{\partial H}{\partial z^B}\frac{\partial}{\partial z^A} + \frac{\partial}{\partial y^{s+a}}\frac{\partial}{\partial y^a}, \tag{10}$$

where H is an arbitrary function only depending on z^A and y^{s+a}.

Now let θ be any universal invariant form of degree m of G_*^{IV}. Write θ as

$$\theta = \sum_{k\geq 0}\sum_{(i)} du_{i_1} \wedge \cdots \wedge du_{i_k} \wedge \theta_{i_1\cdots i_k},$$

where all u_i represent one of $\bar{p}_1, \ldots, \bar{p}_r, \bar{q}_1, \ldots, \bar{q}_r$ and $\theta_{i_1\cdots i_k}$ is a form of dz^A and dy^a, dy^{s+a}, their coefficients are all functions of the variables $\bar{p}_i, \bar{q}_i$, for X in expression (10) we obviously have

$$\mathscr{L}_X\theta = \sum_{k\geq 0}\sum_{(i)} du_{i_1} \wedge \cdots \wedge du_{i_k} \wedge \mathscr{L}_X\theta_{i_1\cdots i_k},$$

and $\mathscr{L}_X\theta_{i_1\cdots i_k}$ does not contain any differential du, hence by $\mathscr{L}_X\theta = 0$ we obtain $\mathscr{L}_X\theta_{i_1\cdots i_k} = 0$ that is all $\theta_{i_1\cdots i_k}$ are universal invariant forms.

By this the problem is concluded as determining the universal invariant forms of the following shape (let it be of degree $\bar{m}$)

$$\bar{\theta} = \sum_{k\geq 0}\sum_{(i)} d\bar{u}_{i_1} \wedge \cdots \wedge d\bar{u}_{i_k} \wedge \bar{\theta}_{i_1\cdots i_k},$$

where $\bar{u}_i$ shows $\bar{p}_{r+1}, \ldots, \bar{p}_{r+s}, \bar{q}_{r+1}, \ldots, \bar{q}_{r+s}$ that is one of y^a, y^{s+a} and $\bar{\theta}_{i_1\cdots i_k}$ is a form of degree $\bar{m} - k$ of dz^A, its coefficients are all functions of the variables $\bar{p}_i, \bar{q}_i$.

Now we first consider a special case such as $X = X' \in \mathscr{X}$ of expression (10), where H is an arbitrary function of z^A, but independent of y^{s+a}:

$$X' = \eta^{AB}\frac{\partial H}{\partial z^B}\frac{\partial}{\partial z^A}. \tag{11}$$

Accordingly

$$\mathscr{L}_{X'}\bar{\theta} = \sum_{k\geq 0}\sum_{(i)} d\bar{u}_{i_1} \wedge \cdots \wedge d\bar{u}_{i_k} \wedge \mathscr{L}_{X'}\bar{\theta}_{i_1\cdots i_k},$$

and from $\mathscr{L}_{X'}\bar{\theta} = 0$, similar to the above, we obtain $\mathscr{L}_{X'}\bar{\theta}_{i_1\cdots i_k} = 0$, here X' is to show as in expression (11), besides it is arbitrary. By this we know that $\bar{\theta}_{i_1\cdots i_k}$ is the universal invariant form of all regular transformation infinite groups preserving the forms

$$\Omega_z = \frac{1}{2}\eta_{AB}dz^A \wedge dz^B = \sum_{i=r+s+1}^{n} d\bar{p}^i \wedge d\bar{q}^i \tag{12}$$

invariant on $2t$ variables z^A (the variables $\bar{p}_1, \ldots, \bar{p}_{r+s}, \bar{q}_1, \ldots, \bar{q}_{r+s}$ are regarded as constant parameters here). By H.C. Lee's theorem we obtain:

$$\bar{\theta}_{i_1\cdots i_k} = \begin{cases} 0, & \bar{m}-k = \text{odd}, \\ f_{i_1\cdots i_k}(\Omega_z)^{(m-k)/2}, & \bar{m}-k = \text{even}, \end{cases}$$

where $f_{i_1\cdots i_k}$ is a function only depending on $\bar{p}_1, \ldots, \bar{p}_{r+s}, \bar{q}_1, \ldots, \bar{q}_{r+s}$.

Up to now the original problem is concluded further as the problem of determining the universal invariant form of the following shape

$$\varphi = \sum_{k\geq 0}\sum_{(a,b)} g_{a_1\cdots a_k b_1\cdots b_l} dy^{a_1} \wedge \cdots \wedge dy^{a_k} \wedge dy^{s+b_1} \wedge \cdots \wedge dy^{s+b_l} \wedge (\Omega_z)^h$$

where all g are functions of $\bar{p}_1, \ldots, \bar{p}_{r+s}, \bar{q}_1, \ldots, \bar{q}_{r+s}$, anti-symmetric for the lower index a, also anti-symmetric for the lower index b, and $2h+k+l$ is a fixed integer, that is the degree of φ.

Now take any fixed c among the indices $l, \ldots, s$ also take a special $X = X_c \in \mathscr{X}$ shape as $X_c = \frac{\partial H}{\partial y^{s+c}}\frac{\partial}{\partial y^c}$. Where H is an arbitrary function depending only on y^{s+c}, accordingly $\mathscr{L}_{x_c}\Omega_z = 0, \mathscr{L}_{x_c}(dy^{s+b}) = 0$, also when $a \neq c, \mathscr{L}_{x_c}(dy^a) = 0$, and $\mathscr{L}_{x_c}(dy^c) = \frac{\partial^2 H}{\partial (y^{s+c})^2} dy^{s+c}$. By $\mathscr{L}_{x_c}\varphi = 0$ (H is arbitrary) it is easy to know that in φ any term whenever contains dy^c it must at the same time contains dy^{s+c} and has the factor $dy^c \wedge dy^{s+c}$, also g is a function independent of y^c. Since this holds for any exponential c in $1, \ldots, s$, therefore we can write φ as the following shape:

$$\varphi = \sum_{l\geq 0}\sum_{(b)} \psi_{b_1\cdots b_l},$$

where every $\psi_{b_1\cdots b_l}$ has the following shape ($a_i \neq b_i$):

$$\psi = dy^{s+b_1} \wedge \cdots \wedge dy^{s+b_l} \wedge \psi',$$

$$\psi' = \sum_{k\geq 0} g_{a_1\cdots a_k} dy^{a_1} \wedge dy^{s+a_1} \wedge \cdots \wedge dy^{a_k} \wedge dy^{s+a_k} \wedge (\Omega_z)^{e-k},$$

here all g are functions depending only on $\bar{p}_1, \ldots, \bar{p}_{r+s}, \bar{q}_1, \ldots, \bar{q}_r$, also symmetric for the lower index a, and the degree of φ' is then an even number, let it be $2e$. When $k = 0$, the relative coefficient g in the expression of φ has denote as g_0.

Now take a general $X \in \mathscr{X}$ given by expression (10), where H are all arbitrary functions of y^{s+a}, and z^A, by computation we obtain:

$$\mathscr{L}_X\psi = dy^{s+b_1} \wedge \cdots \wedge dy^{s+b_e} \wedge \mathscr{L}_X\psi',$$

$$\mathscr{L}_X\psi' = \sum \tilde{g}_{a_1\cdots a_k,c,A} dy^{a_1} \wedge dy^{s+a_1} \wedge \cdots \wedge dy^{a_k} \wedge dy^{s+a_k} \wedge dy^{s+c} \wedge dz^A \wedge (\Omega_z)^{e-k-1} + \cdots,$$

$$\tilde{g}_{a_1\cdots a_k,c,A} = -[(k+1)g_{ca_1\cdots a_k} + (e-k)g_{a_1\cdots a_k}]\frac{\partial^2 H}{\partial z^A \partial y^{s+c}}.$$

Since for any H we should have $\mathscr{L}_X\varphi = 0$, therefore we obtain (c, a_i, b_j are mutually not equal)

$$g_{ca_1\cdots a_k} = -\frac{e-k}{k+1}\cdot g_{a_1\cdots a_k}.$$

Hence

$$g_{a_1\cdots a_k} = (-1)^k \left(\begin{array}{c} e \\ k \end{array}\right)\cdot g_0.$$

By this we obtain

$$\begin{aligned}\psi &= g_0\cdot dy^B \wedge \sum_{a\neq b}(-1)^k \left(\begin{array}{c} e \\ k \end{array}\right) dy^{a_1}\wedge dy^{s+a_1}\wedge\cdots\wedge dy^{ak}\wedge dy^{s+a_k}\wedge(\Omega_z)^{e-k}\\ &= g_0\cdot dy^B\wedge(\sum_{a\neq b} dy^{s+a}\wedge dy^a + \Omega_z)^e\\ &= g_0\cdot dy^B\wedge(\Omega - \sum_{i=1}^{r} d\bar{p}_i\wedge d\bar{q}_i)^e.\end{aligned}$$

In the above expression $dy^B = dy^{s+b_1}\wedge\cdots\wedge dy^{s+b_l}$.

From the above we know that any universal invariant form θ of G_*^{IV} must be a sum formula, where each term is the outer product of some $d\bar{p}_1,\ldots,d\bar{p}_{r+s},d\bar{q}_1,\ldots,d\bar{q}_r$ and the quadratic form Ω, and with arbitrary functions depending only on $\bar{p}_1,\ldots,\bar{p}_{r+s},\bar{q}_1,\ldots,\bar{q}_r$ as its coefficient, in other words, since $\bar{p}_1,\ldots,\bar{p}_{r+s},\bar{q}_1,\ldots,\bar{q}_r$ is a function basis of $\overline{\mathscr{C}}$, we have already proved the following

Theorem *The universal invariant form of the infinite group G_*^{IV} forms a differential ring produced by functions in $\overline{\mathscr{C}}$ and the form Ω*

Notice that the function in $\overline{\mathscr{C}}$ can completely be determined through simple operations and integrations from $\mathscr{G}$ of the effect to phase space (p_i, q_i), hence the conservative law corresponding to a known symmetric group $\mathscr{G}$ can completely be determined through computations.

References

[1] R. Abraham-J. E. Marsden, Foundations of mechanics, 1967.

[2] C. Caratheodony, Variationsrechnung und partielle differential Gleihungen erster Ordnung, Bd. I, 1956.

[3] E. Cartan, Leçons sur les invariants intégraux, 1958.

[4] S. S. Chern, Pseudo-groups contimus infinis, Géomètrie Différentielle (Colloque CNRS 1953), 119-136.

[5] L.P. Eisenhart, Continuous groups of transformations, 1961.

[6] H. C. Lee, The universal integral invariants of Hamiltonian systems and application to the theory of canonical transformations, Proc. Roy. Soc. Edinburgh, Sect. A, 72 (1947) 237-246.

Vol. XIX No. 5 SCIENTIA SINICA Sep.-Oct. 1976

THEORY OF I^*-FUNCTOR IN ALGEBRAIC TOPOLOGY

EFFECTIVE CALCULATION AND AXIOMATIZATION OF I^*-FUNCTOR ON COMPLEXES

WU WEN-TSÜN (吴文俊)
(*Institute of Mathematics, Academia Sinica*)

Received April 25, 1975.

ABSTRACT

According to a classical definition due to Engels, the pure mathematics has space forms and quantitative relations in the exterior world as its objects of study. These two fundamental notions of mathematics are, however, not to be considered as unrelated, but are often interconnected by "measures." Previously we have introduced the concept of I^* which serves as a measure of space forms by means of quantitative relations. This measure is called a "functor" to follow the current terminology in algebraic topology. This I^*-functor or I^*-measure has the advantage over other known functors of being in general "calculable" to be understood roughly in the following sense: If a new space form is constructed geometrically from some given space forms, the I^*-functor of this new space form is completely determined by the I^*-functors of the given space forms. We have given illustrations of this point in various papers. The aims of the present paper are twofold. First, we not only show the calculability of this functor in principle, but also give a method of effective calculations for practical purposes in the case of finite complexes. Secondly, we have listed a set of representative properties of I^* which are sufficient to characterize it completely, forming thus a so-called axiomatic system in the current terminology. The case of infinite complexes is also considered.

In papers [5, 6, 7], the author basing himself on Sullivan's theory of minimal models[2, 4], has introduced the notion of I^*-functor of spaces and has pointed out that in many cases the I^*-functor is "calculable" while the usual H- and π-functors are often "non-calculable," even restricted to the real field domain. The present paper makes further studies to explain this point. Moreover, we give for the category $\mathscr{K}_0$ of connected, simply-connected finite complexes methods to calculate effectively the I^*-functor from the combinatorial structure of the complex and establish also axiomatic system for this functor.

The notations in the present paper as those in the preceding papers[5, 6, 7], are to be understood here.

1. I^*-FUNCTOR OF K/L AND $K' \cup K''$

Let $K \in \mathscr{K}$, $L \in \mathscr{K}_0$, and f: $L \subset K$. Let C_L be the cone over L. Denote the union $K \cup C_L$ by $K/L \in \mathscr{K}$. Then we have a commutative diagram of simplicial maps:

$$\begin{array}{ccc} C_L & \xrightarrow{\bar{j}} & K/L \\ {\scriptstyle j}\uparrow & & \uparrow{\scriptstyle \bar{f}} \\ L & \xrightarrow{f} & K. \end{array}$$

Set

$$C_f = \sum_{n \geq 0} C_n,$$

$$C_0 = \{x^{\epsilon A^*(K)}/x \text{ takes constant value } \epsilon R \text{ on the simplexes of } L\},$$

$$C_n = \text{Ker}\,[f_n^A\colon A^n(K) \to A^n(L)], \quad n > 0.$$

It is easy to see that C_f is a DGA-algebra $\epsilon \mathscr{A}$ and there is a natural DGA-morphism $i\colon C_f \to A^*(K)$ with the following sequence exact for $n > 0$:

$$0 \to C_n \xrightarrow{i} A^n(K) \xrightarrow{f^A} A^n(L) \to 0.$$

Proposition 1. Min $C_f \approx I^*(K/L)$.

Proof. Let $x \in C_f$ and define $\tau x = \bar{x} \in A^*(K/L)$ as follows. If $\deg x = 0$ with x taking on the constant value $c \in R$ by all simplexes in L, then set $\bar{x} \in A^0(K/L)$ to take on the same constant value c on all simplexes in K/L. If $\deg x > 0$, then $f^A ix = 0$. Hence we can take $\bar{x} \in A^*(K/L)$ with $\bar{f}^A\bar{x} = ix$ and $\bar{j}^A\bar{x} = 0$. Clearly $x \to \bar{x}$ is a DGA-morphism $\tau : C_f \to A^*(K/L)$ and the diagram below is commutative:

$$\begin{array}{ccccc} & \overset{\tau}{\nearrow} & A^*(K/L) & \xrightarrow{\bar{j}^A} & A^*(C_L) \\ & & \downarrow{\scriptstyle \bar{f}^A} & & \downarrow{\scriptstyle j^A} \\ C_f & \xrightarrow{i} & A^*(K) & \xrightarrow{f^A} & A^*(L) \end{array}$$

We prove now

$$\tau_* : H(C_f) \approx H(A^*(K/L)) = H^*(K/L), \tag{1}$$

from which the proposition follows immediately.

To see this, let $\bar{x} \in A^*(K/L)$ with $d\bar{x} = 0$. In case $\deg \bar{x} = 0$ with $\bar{x}$ taking constant value $c \in R$ on all simplexes of K/L, let us set $x \in A^0(K)$ to take the same constant value c on all simplexes of K. Then $x \in C_0$, $dx = 0$ and $\tau x = \bar{x}$. If $\deg \bar{x} > 0$, then in C_L we have $d\bar{j}^A\bar{x} = 0$ so that there exists $y \in A^*(C_L)$ with $dy = \bar{j}^A\bar{x}$. As $\bar{j}^A$ is an epimorphism, we can take $\bar{y} \in A^*(K/L)$ with $\bar{j}^A\bar{y} = y$. Set $\bar{z} = \bar{x} - d\bar{y}$, then $\bar{j}^A\bar{z} = 0$. Hence $f^A\bar{f}^A\bar{z} = j^A\bar{j}^A\bar{z} = 0$ or $\bar{f}^A\bar{z} \in C_f$. Now $\tau\bar{f}^A\bar{z} = \bar{z} = \bar{x} - d\bar{y} \sim \bar{x}$ and $d\bar{f}^A\bar{z} = 0$. From these we see that τ_* is an epimorphism.

Next suppose $z \in C_f$ with $dz = 0$ and $\tau z = d\bar{a}$, $\bar{a} \in A^*(K/L)$. As $d\bar{j}^A\bar{a} = 0$, C_L is contractible and $\bar{j}^A$ is an epimorphism, there exists $\bar{b} \in A^*(K/L)$ with $\bar{j}^A\bar{a} = d\bar{j}^A\bar{b}$. Then $\bar{f}^A(\bar{a} - d\bar{b}) \in C_f$ and $z = d\bar{f}^A(\bar{a} - d\bar{b}) \sim 0$. Hence τ_* is a monomorphism.

The isomorphism (1) is thus proved. Hence the Proposition 1.

Theorem 1. *$I^*(K/L)$ is completely determined by the natural DGA-morphism*

$$g = f^I\colon I^*(K) \to I^*(L). \tag{2}$$

Proof. Set $I^*(K) = M, I^*(L) = N$. From §§ 15—16 of [2], we see easily that there exist $K' \in \mathscr{K}$, $L' \in \mathscr{K}_0$, f': $L' \subset K'$ with $M \approx I^*(K')$, $N \approx I^*(L')$ and the following diagram is homotopically commutative (ρ being the canonical homomorphisms):

$$\begin{array}{ccc} M & \xrightarrow{g} & N \\ \rho \downarrow & & \downarrow \rho \\ A^*(K') & \xrightarrow{f'^A} & A^*(L') \end{array}$$

Moreover, we have $K' \simeq K$, $L' \simeq L$, $f' \simeq f$, $K'/L' \simeq K/L$ so that Proposition 1 gives

$$I^*(K/L) \approx I^*(K'/L') \approx \operatorname{Min} C_{f'}.$$

As K', L', f' are constructed from (2), so $I^*(K/L)$ is completely determined by (2), as to be proved.

The determination of $I^*(K/L)$ (to be denoted by J_g) and the natural DGA-morphisms $I^*(K/L) \to I^*(K)$ (to be denoted by j_g) from (2) or

$$g: M \to N$$

will be called the *J-construction*. The above gives only an existence proof of such J-construction. In the next section, we shall give some explicit constructions of J_g and j_g in the special case $L = S^n$.

Entirely analogous to Proposition 1 and Theorem 1, the same method can be applied to the study of union of complexes as in the Mayer-Vietoris sequence. Thus, let K, K', K'', $L \in \mathscr{K}$, $K = K' \cup K''$, $L = K' \cap K''$ and f': $L \subset K'$, f'': $L \subset K''$. Set $D_{f', f''} = \sum_{n \geq 0} D_n$, where

$$D_0 = \{(a', a'')^{\in A^0(K') \times A^0(K'')} / a', a'' \text{ take the same constant values } \in R \text{ on simplexes of } L\},$$

$$D_n = \{(a', a'')^{\in A^n(K') \times A^n(K'')} / f'^A a' = f''^A a''\}, \quad n > 0.$$

Then $D_{f', f''}$ forms naturally a DGA-algebra easily seen to be $\in \mathscr{A}$. We have then (proof omitted):

Proposition 2. $\operatorname{Min} D_{f', f''} \approx I^*(K' \cup K'')$.

Theorem 2. *$I^*(K' \cup K'')$ is completely determined by*

$$f'^I: I^*(K') \to I^*(L),$$

and

$$f''^I: I^*(K'') \to I^*(L).$$

II. J-construction of DGA-morphism $g: M \to N$ in Case $N \approx I^*(S^n)$

Let M, $N \in \mathscr{M}$ and $N \approx I^*(S^n)$, $n \geq 2$. The purpose of this section is to construct explicitly $J_g \in \mathscr{M}$ and the DGA-morphism

$$j_g: J_g \to M$$

from a given DGA-morphism $g: M \to N$. This is the J-construction of I. For this, let

$$N \approx I^*(S^n) = \begin{cases} \Lambda(s), & n = \text{odd}. \\ \Lambda(s, t), & n = \text{even}, \end{cases}$$

in which

$$\deg s = n, \quad ds = 0,$$

and also

$$\deg t = 2n - 1, \quad dt = s^2,$$

for even n. Let $\bar{N} \in \mathscr{A}$ be the DGA-algebra as follows:

$$\bar{N} = R \oplus R\bar{s},$$

the degrees, multiplications and differentiations in $\bar{N}$ being given by:

$$\deg \bar{s} = n, \quad \bar{s}^2 = 0, \quad d\bar{s} = 0.$$

Let $\tilde{S} \in \mathscr{A}$ be the DGA-algebra

$$\tilde{S} = R \oplus R\tilde{s},$$

with degrees, multiplications and differentiations as follows:

$$\deg \tilde{s} = n + 1, \quad \tilde{s}^2 = 0, \quad d\tilde{s} = 0.$$

Denote by γ the natural DGA-morphism from N to $\bar{N}$ which maps s to $\bar{s}$, and also t to 0 in case n is even. Set

$$\gamma g = \bar{g}: M \to \bar{N}.$$

Define now $K_{\bar{g}}$ according as $\bar{s} \in \operatorname{Im} \bar{g}$ or not:

$$K_{\bar{g}} = \begin{cases} R \oplus \sum_{r>0} \bar{g}_r^{-1}(0), & \text{for } \bar{s} \in \operatorname{Im} \bar{g}, \\ M \otimes \tilde{S}, & \text{for } \bar{s} \notin \operatorname{Im} \bar{g}. \end{cases}$$

Clearly $K_{\bar{g}}$ under operations in M, is a DGA-algebra of $\mathscr{A}$. Define also a DGA-morphism

$$k_{\bar{g}}: K_{\bar{g}} \to M$$

as follows. In case $\bar{s} \in \operatorname{Im} \bar{g}$, $k_{\bar{g}}$ is the natural inclusion, while in case $\bar{s} \notin \operatorname{Im} \bar{g}$, we set

$$k_{\bar{g}}(a \otimes 1) = a, \quad k_{\bar{g}}(a \otimes \tilde{s}) = 0 \quad (a \in M).$$

In what follows we shall construct $J_{\bar{g}} \in \mathscr{M}$ and DGA-morphism $j_{\bar{g}}: J_{\bar{g}} \to M$ from $K_{\bar{g}}$ and $k_{\bar{g}}$, and $J_{\bar{g}}$, $j_{\bar{g}}$ will be taken to be the $J_{\bar{g}}$ and $j_{\bar{g}}$ in the beginning of this section, which will be discussed in two seperate cases.

Case I. $\bar{s} \in \operatorname{Im} \bar{g}$.

Define now

$$J_{\bar{g}} = \operatorname{Min} K_{\bar{g}} \in \mathscr{M},$$

and also

$$j_{\bar{g}} = k_{\bar{g}}\rho_{\bar{g}}\colon\ J_{\bar{g}} \to M,$$

in which $\rho_{\bar{g}}\colon\ J_{\bar{g}} \to K_{\bar{g}}$ is any canonical homomorphism.

Case II. $\bar{s} \notin \operatorname{Im} \bar{g}$.

Take any set $x_i \in Z(M)$ forming an additive homology basis of $H(M)$. Then $H(K_{\bar{g}})$ has an additive homology basis:

$$x_i,\ \bar{s},\ x_i \otimes \bar{s}.$$

In the set $x_i \otimes \bar{s}$, let those of the lowest degree (say $m_{\bar{g}} \geqslant n+3$) be denoted by y_j, while the others with degree $> m_{\bar{g}}$ be denoted by z_k. Introduce η_j and construct the DGA-algebra

$$K^1_{\bar{g}} = K_{\bar{g}} \otimes \Lambda(\eta_j),$$

in which

$$\deg \eta_j = m_{\bar{g}} - 1,$$

$$d\eta_j = y_j.$$

As $k_{\bar{g}}(y_j) \sim 0$ (in fact $= 0$), we may take $a_j \in M$ with $k_{\bar{g}}(y_j) = da_j$ (in fact a_j may be taken to be 0). Define the DGA-morphism

$$k^1_{\bar{g}}\colon\ K^1_{\bar{g}} \to M,$$

by

$$k^1_{\bar{g}}(a) = k_{\bar{g}}(a),\quad a \in K_{\bar{g}},$$

$$k^1_{\bar{g}}(\eta_j) = a_j.$$

It is easy to see that $K^1_{\bar{g}}$ has an additive homology basis:

$$x_i,\ \bar{s},\ y^1_j,\ z^1_k,$$

in which

$$\deg y^1_j = m^1_{\bar{g}} > m_{\bar{g}},$$

$$\deg z^1_k > m^1_{\bar{g}},$$

and under $k^1_{\bar{g}}$,

$$k^1_{\bar{g}}(y^1_j) \sim 0,\quad k^1_{\bar{g}}(z^1_k) \sim 0.$$

Now introduce η^1_j and construct the DGA-algebra

$$K^2_{\bar{g}} = K^1_{\bar{g}} \otimes \Lambda(\eta^1_j),$$

in which

$$\deg \eta^1_j = m^1_{\bar{g}} - 1,\quad d\eta^1_j = y^1_j.$$

Define also a DGA-morphism

$$k^2_{\bar{g}}\colon\ K^2_{\bar{g}} \to M,$$

by

$$k^2_{\bar{g}}(a) = k^1_{\bar{g}}(a),\quad a \in K^1_{\bar{g}},$$

$$k^2_{\bar{g}}(\eta^1_j) = a^1_j,$$

in which

$$a^1_j \in M,\quad \deg a^1_j = m^1_{\bar{g}} - 1,\quad da^1_j = k^1_{\bar{g}}(y^1_j).$$

In $K_{\bar{g}}^2$, there is additive homology basis as before and the preceding constructions can be continued to get

$$K_{\bar{g}} \subset K_{\bar{g}}^1 \subset K_{\bar{g}}^2 \subset \cdots.$$

In this manner, we get a DGA-algebra

$$K_{\bar{g}}^{\infty} = K_{\bar{g}} \otimes \Lambda(\eta_j, \eta_j^1, \eta_j^2, \cdots),$$

and a DGA-morphism

$$k_{\bar{g}}^{\infty}: K_{\bar{g}}^{\infty} \to M,$$

with

$$k_{\bar{g}}^{\infty}(a) = k_{\bar{g}}(a), \quad a \in K_{\bar{g}},$$

$$k_{\bar{g}}^{\infty}(\eta_j) = a_j \ , k_{\bar{g}}^{\infty}(\eta_j^r) = a_j^r, \quad r = 1, 2, \cdots.$$

Moreover $H(K_{\bar{g}}^{\infty}) \approx H(M) \oplus H(\tilde{S})$ has an additive homology basis

$$x_i, \bar{s}.$$

Define now the DGA-algebra

$$J_{\bar{g}} = \text{Min}\, K_{\bar{g}}^{\infty} \in \mathscr{M},$$

and the DGA-morphism

$$j_{\bar{g}}: J_{\bar{g}} \to M$$

by

$$j_{\bar{g}} = k_{\bar{g}}^{\infty} \rho_{\bar{g}}^{\infty},$$

in which

$$\rho_{\bar{g}}^{\infty}: J_{\bar{g}} \to K_{\bar{g}}^{\infty}$$

is any canonical homomorphism. Then we have

$$H_q(J_{\bar{g}}) \approx H_q(M) \oplus H_q(\tilde{S}), \quad q > 0,$$

and

$$j_{\bar{g}}{}^*: H(J_{\bar{g}}) \to H(M)$$

is an isomorphism on $H(M)$, and is 0 on $H(\tilde{S})$.

III. Privileged Morphisms of Minimal Models

Let A, $B \in \mathscr{A}$, f: $A \to B$ be a DGA-morphism, $M = \text{Min}\, A$, $N = \text{Min}\, B$, and ρ_A: $M \to A$, ρ_B: $N \to B$ be canonical homomorphisms. The collection of all DGA-morphisms g: $M \to N$ induced from f will be denoted by $G(f)$. In general, the diagram

$$\begin{array}{ccc} A & \xrightarrow{f} & B \\ {\scriptstyle\rho_A}\uparrow & & \uparrow{\scriptstyle\rho_B} \\ M & \xrightarrow{g} & N \end{array} \tag{1}$$

is only homotopically commutative. It is easy to give examples with f given for which no ρ_A, ρ_B and g can be chosen to make (1) commutative. However, when f is an epimorphism, e. g. when in the case $A = A^*(K)$, $B = A^*(L)$ and the DGA-

morphism $f = f^A\colon A \to B$ induced from a simplicial map $f\colon L \subset K$, then we have the following:

Theorem 3. *Let $A, B \in \mathscr{A}$ and $f\colon A \to B$ be a DGA-epimorphism. Then for $M = \operatorname{Min} A$, $N = \operatorname{Min} B$ and a given canonical homomorphism $\rho_B\colon N \to B$, there are a canonical homomorphism $\rho_A\colon M \to A$ and a DGA-morphism $g\colon M \to N$ such that (1) is commutative.*

Definition and notation. The morphisms $g\colon M \to N$ in the theorem will be called *privileged morphisms* associated with $f\colon A \to B$ whose collection will be denoted by $G^0(f) \subset G(f)$. In case $f\colon L \subset K$ and $f = f^A\colon A^*(K) \to A^*(L)$, $G^0(f)$ and $G(f)$ will also be denoted by $G^0(K, L)$ and $G(K, L)$ respectively.

Proof of Theorem 3. We shall go into detail of the proof only in the case $N \approx I^*(S^n)$ and $B_n = 0$ for $m > n$. The proof of the general case is similar, but more complicate.

For this, let us take s as the generator of degree n in N, and set $\rho_B s = c$. As f is an epimorphism, we have $a \in A$ with $fa = c$. The choice of such an a will be explained below.

Denote by $M^{(m)}$ the minimal DGA-algebra generated by generators of M of degree $\leqslant m$, with $M^{(0)} = M^{(1)} = R$. We shall extend $M^{(m)}$ successively as

$$M^{(0)} = M^{(1)}(= R) \subset M^{(2)} \subset \cdots \subset M^{(m)} \subset M^{(m+1)} \subset \cdots \subset M,$$

and define DGA-morphisms

$$\rho_A^{(m)}\colon M^{(m)} \to A,$$
$$g^{(m)}\colon M^{(m)} \to N,$$

such that the following induction hypothesis is observed:

$H\,1^0_m$. $\rho_A^{(m)}/M^{(m-1)} = \rho_A^{(m-1)}$, $g^{(m)}/M^{(m-1)} = g^{(m-1)}$.

$H\,2^0_m$. $\rho_{A*}^{(m)}\colon \begin{cases} H_q(M^{(m)}) \approx H_q(A), & q \leqslant m, \\ H_{m+1}(M^{(m)}) \subset H_{m+1}(A). \end{cases}$

$H\,3^0_m$. The diagram below is commutative:

$$\begin{array}{ccc} A & \xrightarrow{\;f\;} & B \\ {\scriptstyle \rho_A^{(m)}}\uparrow & & \uparrow{\scriptstyle \rho_B} \\ M^{(m)} & \xrightarrow{\;g^{(m)}\;} & N. \end{array}$$

We shall construct successively from

$$G^{(m)} = \{M^{(m)}, \rho_A^{(m)}, g^{(m)}\},$$

which satisfies $H1^0_m - 3^0_m$ to the set

$$G^{(m+1)} = \{M^{(m+1)}, \rho_A^{(m+1)}, g^{(m+1)}\},$$

satisfying $H1^0_{m+1} - 3^0_{m+1}$ as follows.

The construction of $G^{(2)}$ is easy. Suppose that $G^{(m)}$ has been constructed with $m \leqslant n-2$.

From the induction hypothesis $H1^0_m - 2^0_m$, we have exact sequences

$$\begin{cases} 0 \to H_{m+1}(M^{(m)}) \xrightarrow{\rho^{(m)}_{A*}} H_{m+1}(A) \to \operatorname{Coker}_{m+1}\rho^{(m)}_{A*} \to 0, \\ 0 \to \operatorname{Ker}_{m+2}\rho^{(m)}_{A*} \to H_{m+2}(M^{(m)}) \xrightarrow{\rho^{(m)}_{A*}} H_{m+2}(A). \end{cases} \tag{2$_m$}$$

Take now

$$\begin{cases} e^{(m)}_i \in A_{m+1}, & de^{(m)}_i = 0, \\ \xi^{(m)}_j \in M^{(m)}_{m+2}, & d\xi^{(m)}_j = 0, \\ x^{(m)}_j \in A_{m+1}, & \rho^{(m)}_A \xi^{(m)}_j = dx^{(m)}_j, \end{cases} \tag{3$_m$}$$

such that the $e^{(m)}_i \in Z_{m+1}(A)$ form an additive homology basis of $\operatorname{Coker}_{m+1}\rho^{(m)}_{A*}$, and the $\xi^{(m)}_j \in Z_{m+2}(M^{(m)})$ form one of $\operatorname{Ker}_{m+2}\rho^{(m)}_{A*}$.

Since $H(B) \approx H(N) \approx H(S^n)$ and $m \leqslant n-2$, we have $fe^{(m)}_i \sim 0$. As f is an epimorphism, we have $h^{(m)}_i \in A_m$ with

$$fe^{(m)}_i = dfh^{(m)}_i. \tag{4$_m$}$$

From $m \leqslant n-2$ we have further

$$\rho_B g^{(m)} \xi^{(m)}_j = 0. \tag{5$_m$}$$

From $H3^0_m$ we get

$$dfx^{(m)}_j = fdx^{(m)}_j = f\rho^{(m)}_A \xi^{(m)}_j = \rho_B g^{(m)} \xi^{(m)}_j = 0,$$

so that $fx^{(m)}_j \in Z_{m+1}(B)$. As before, we have $y^{(m)}_j \in A_m$ with

$$fx^{(m)}_j = dfy^{(m)}_j. \tag{6$_m$}$$

Define now

$$M^{(m+1)} = M^{(m)} \otimes \Lambda(\varepsilon^{(m+1)}_i, \zeta^{(m+1)}_j), \tag{7$_m$}$$

in which

$$\begin{cases} \deg \varepsilon^{(m+1)}_i = \deg \xi^{(m+1)}_j = m+1, \\ d\varepsilon^{(m+1)}_i = 0, \\ d\zeta^{(m+1)}_i = \xi^{(m)}_j \in M^{(m)}_{m+2}. \end{cases} \tag{8$_m$}$$

Then $M^{(m+1)} \in \mathscr{M}$. Define also the morphism

$$\rho^{(m+1)}_A \colon M^{(m+1)} \to A$$

by

$$\begin{cases} \rho^{(m+1)}_A / M^{(m)} = \rho^{(m)}_A, \\ \rho^{(m+1)}_A \varepsilon^{(m+1)}_i = e^{(m)}_i - dh^{(m)}_i, \\ \rho^{(m+1)}_A \zeta^{(m+1)}_j = x^{(m)}_j - dy^{(m)}_j. \end{cases} \tag{9$_m$}$$

Clearly $d\rho^{(m+1)}_A \varepsilon^{(m+1)}_i = \rho^{(m+1)}_A d\varepsilon^{(m+1)}_i = 0$ and

$$d\rho^{(m+1)}_A \zeta^{(m+1)}_j = dx^{(m)}_j = \rho^{(m)}_A \xi^{(m)}_j = \rho^{(m+1)}_A \xi^{(m)}_j = \rho^{(m+1)}_A d\zeta^{(m+1)}_j.$$

Hence $\rho_A^{(m+1)}$ can be uniquely extended to a DGA-morphism from $M^{(m+1)}$ to A which will still be denoted by $\rho_A^{(m+1)}$. If we define

$$g^{(m+1)} = 0, \tag{10$_m$}$$

then from $(9)_m$, $(10)_m$ we see that $G^{(m+1)}$ so obtained will satisfy $H1_{m+1}^0$ and $H3_{m+1}^0$. Besides, $H2_{m+1}^0$ is also easily verified.

Suppose now $m = n-1$ and construct $G^{(n)}$ from $G^{(n-1)}$ as follows.

Take as before $e_i^{(n-1)} \in A_n$, $\xi_j^{(n-1)} \in M_{n+1}^{(n-1)}$, $x_j^{(n-1)} \in A_n$ to satisfy $(3)_{n-1}$ such that $e_i^{(n-1)}$ form an additive homology basis of $\mathrm{Coker}_n \rho_{A*}^{(n-1)}$, and $\xi_j^{(n-1)}$, one of $\mathrm{Ker}_{n+1}\rho_{A*}^{(n-1)}$.

As $fe_i^{(n-1)} \in Z_n(B)$ and $H_n(B) \approx R$ is generated by c, there are $r_i \in R$ and $h_i^{(n-1)} \in A_{n-1}$ such that

$$fe_i^{(n-1)} = r_i c + dfh_i^{(n-1)}. \tag{4$_{n-1}$}$$

If some $r_i \neq 0$, then $f_*: H_n(A) \to H_n(B)$ is an epimorphism and a will be chosen with $da = 0$. Otherwise a will be chosen arbitrarily. Furthermore, as $g^{(n-1)} = 0$, we have as before $fx_j^{(n-1)} \in Z_n(B)$ so that there are $y_j^{(n-1)} \in A_{n-1}$ and $r_j' \in R$ such that

$$fx_j^{(n-1)} = r_j' c + dfy_j^{(n-1)}. \tag{6$_{n-1}$}$$

Now we define $G^{(n)}$ by $(7)_{n-1} - (10)_{n-1}$, where $(9)_{n-1}$ and $(10)_{n-1}$ in view of $(4)_{n-1}$ and $(6)_{n-1}$ are, however, replaced by formulas below:

$$\begin{cases} \rho_A^{(n)}/M^{(n-1)} = \rho_A^{(n-1)}, \\ \rho_A^{(n)}\varepsilon_i^{(n)} = e_i^{(n-1)} - dh_i^{(n-1)} - r_i a, \\ \rho_A^{(n)}\zeta_j^{(n)} = x_j^{(n-1)} - dy_j^{(n-1)}, \end{cases} \tag{9$_{n-1}$}$$

$$\begin{cases} g^{(n)}/M^{(n-1)} = g^{(n-1)} = 0, \\ g^{(n)}\varepsilon_i^{(n)} = 0, \\ g^{(n)}\zeta_j^{(n)} = r_j' s. \end{cases} \tag{10$_{n-1}$}$$

It is easy to see that $\rho_A^{(n)}$ can be extended to a DGA-morphism with the obtained $G^{(n)}$ satisfying $H1_n^0 - 3_n^0$.

Let $m \geqslant n$. Then $(5)_n$ holds still since $B_{m+2} = 0$. Hence we can construct $G^{(m+1)}$ from $G^{(m)}$ as before.

Set now

$$M = \bigcup_m M^{(m)},$$

$$\rho_A\colon\ M \to A,$$

$$g\colon\ M \to N,$$

such that

$$\rho_A/M^{(m)} = \rho_A^{(m)},$$

$$g/M^{(m)} = g^{(m)}.$$

Then $\{M, \rho_A, g\}$ thus obtained, meets the requirement of the theorem.

IV. J-construction Determined by Combinatorial Spheres in K

Let $K \in \mathscr{K}$, f: $L \subset K$, and L be a combinatorial sphere of dimension $n \geqslant 2$, C_L be a combinatorial cell with L as boundary, and K/L be the complex $K \cup C_L \in \mathscr{K}$. Denote for simplicity

$$A = A^*(K), \quad B = A^*(L), \quad \tilde{A} = A^*(K/L),$$

and let $f = f^A$: $A \to B$, $C = C_f$, i: $C \subset A$, τ: $C \to \tilde{A}$, $\tilde{f} = \tilde{f}^A$: $\tilde{A} \to A$ as in I. Set also $M \approx \operatorname{Min} A \approx I^*(K)$, $N \approx \operatorname{Min} B \approx I^*(L) \approx I^*(S^n)$, then by III there are DGA-morphisms g: $M \to N$, $g \in G^0(K, L)$ with the diagram below commutative (ρ being the canonical homomorphisms):

$$\begin{array}{ccc} A & \xrightarrow{\ f\ } & B \\ {\scriptstyle\rho}\uparrow & & \uparrow{\scriptstyle\rho} \\ M & \xrightarrow{\ g\ } & N. \end{array}$$

Let $\bar{N} = R \oplus R\bar{s}$, γ: $N \to \bar{N}$, and $\bar{g} = \gamma g$: $M \to \bar{N}$ as in II. As $\bar{g}_n$: $M_n \to \bar{N}_n$ is nothing but the homomorphism f^*: $\pi_n^*(K) \to \pi_n^*(L)$ induced by f: $L \to K$, the morphism $\bar{g}$ is completely determined by $L \subset K$.

Definition. The unique DGA-morphism

$$\bar{g}\colon\ M \to \bar{N}$$

determined by the combinatorial sphere $L \subset K (L, K \in \mathscr{K})$ will be called the *characteristic homomorphism* of L *w. r. t.* K.

The present section then aims at proving the following:

Theorem 4. *From the J-construction (II) of the characteristic homomorphism* $\bar{g}$: $M \to \bar{N}$, *we get*

$$J_{\bar{g}} \approx \operatorname{Min} \tilde{A} \approx I^*(K/L),$$
$$j_{\bar{g}} \in G^0(K/L, K),$$

which make the diagram below commutative (ρ, $\tilde{\rho}$ being canonical homomorphisms):

$$\begin{array}{ccc} \tilde{A} & \xrightarrow{\ \tilde{f}\ } & A \\ {\scriptstyle\tilde{\rho}}\uparrow & & \uparrow{\scriptstyle\rho} \\ J_{\bar{g}} & \xrightarrow{\ j_{\bar{g}}\ } & M. \end{array}$$

Proof. We shall distinguish two cases whether $\bar{s} \in \operatorname{Im} \bar{g}$ or not.

Case 1. $\bar{s} \in \operatorname{Im} \bar{g}$.

In this case we have $J_{\bar{g}} = \operatorname{Min} K_{\bar{g}}$, $K_{\bar{g}} = R \oplus \sum_{r>0} \bar{g}_r^{-1}(0)$, $k_{\bar{g}}$: $K_{\bar{g}} \subset M$, $j_{\bar{g}} = k_{\bar{g}} \rho_{\bar{g}}$: $J_{\bar{g}} \to M$, and $\rho_{\bar{g}}$: $J_{\bar{g}} \to K_{\bar{g}}$ is some canonical homomorphism. Consider now the following diagram

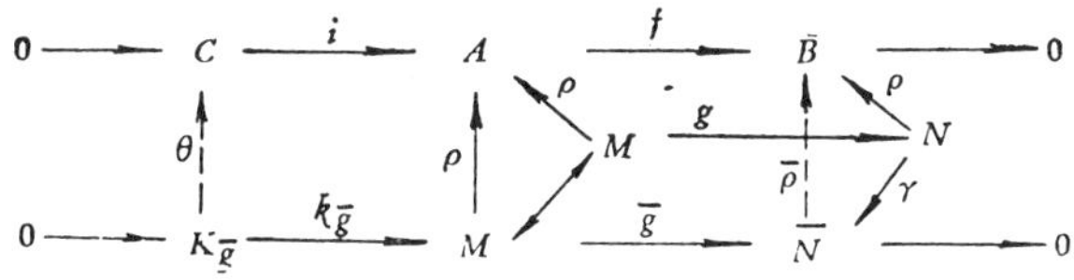

in which the two rows are exact in each positive degree, while the morphism $\bar{\rho}$: $\bar{N} \to B$ is determined by $\bar{\rho}(\bar{s}) = \rho s$. Since the degrees of elements of $B \approx A^*(L)$ are all $\leqslant n$, $\bar{\rho}$ is naturally a DGA-morphism. Then θ: $K_{\bar{g}} \to C$ is the DGA-morphism uniquely determined to make the above diagram commutative. From ρ_*: $H(M) \approx H(A)$, $\bar{\rho}_*$: $H(\bar{N}) \approx H(B)$ and the 5-Lemma we get

$$\theta_*: \ H_r(K_{\bar{g}}) \approx H_r(C), \quad r > 0.$$

Moreover, θ_* is clearly an isomorphism also for $r = 0$.

Define now

$$\tilde{\rho}: \ J_{\bar{g}} \to \tilde{A}$$

by

$$\tilde{\rho} = \tau\theta\rho_{\bar{g}}.$$

Then the following diagram is commutative:

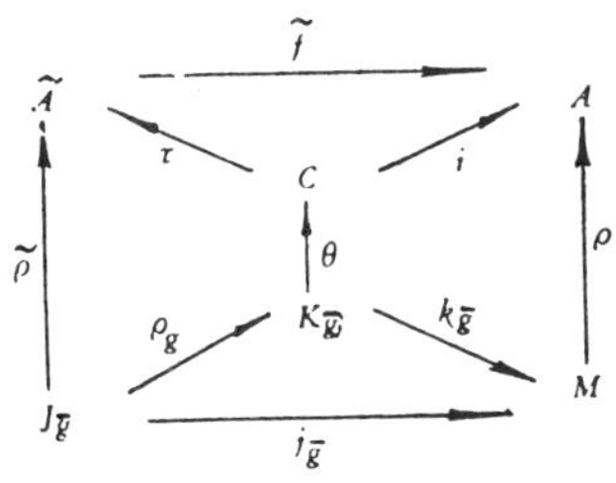

By Proposition 1 of I, τ_*: $H(C) \approx H(\tilde{A})$. Hence we get

$$\tilde{\rho}_*: \ H(J_{\bar{g}}) \approx H(\tilde{A}).$$

It follows that $J_{\bar{g}} \approx \operatorname{Min} \tilde{A} \approx I^*(K/L)$, $\tilde{\rho}$ is a canonical homomorphism, and $j_{\bar{g}} \in G^0(K/L, K)$.

Case II. $\bar{s} \notin \operatorname{Im} \bar{g}$.

This time $\rho M \subset C$ so that ρ: $M \to A$ determines a DGA-morphism ρ^C: $M \to C$ with $i\rho^C = \rho$. We have then

$$H_q(C) = \rho^C_* H_q(M), \quad q \neq n+1,$$

$$H_{n+1}(C) = \rho^C_* H_{n+1}(M) \oplus \delta_* H_n(B),$$

$$\rho^C_*: \ H(M) \subset H(C),$$

$$\delta_*: \ H_n(B) \subset H_{n+1}(C).$$

The generator of $\delta_* H_n(B)$ is given as follows. Take $a \in A$ with $fa = \rho s$ owing to the epimorphism $f\colon A \twoheadrightarrow B$. Then $da \in C$, and $da \in Z(C)$ forms an additive homology basis of $\delta_* H_n(B) \subset H_{n+1}(C)$. Take now any additive homology basis $x_i \in Z(M)$ of $H(M)$, then $H(C)$ has an additive homology basis, viz.

$$\rho^C x_i, \quad da.$$

As $\tilde{f}\colon \tilde{A} \twoheadrightarrow A$ is an epimorphism, there exist $\tilde{a} \in \tilde{A}$ with $\tilde{f}\tilde{a} = a$. In what follows we shall suppose that a, $\tilde{a}$ have been so chosen and taken to be fixed.

Define now

$$\varphi\colon M \otimes \tilde{S} \to \tilde{A},$$

such that

$$\varphi(x) = \tau\rho^C(x), \quad x \in M,$$

$$\varphi(\bar{s}) = \tau da - d\tilde{a}.$$

As C_L is a combinatorial cell of dimension $n+1$, $(\tau da - d\tilde{a})^2$ is 0 on C_L and *a fortiori* also 0 on K, so $(\tau da - d\tilde{a})^2 = 0$. Hence the above two expressions determine φ to be a multiplicative homomorphism. It is easy to see that φ is a DGA-morphism and the diagram below is commutative:

$$\begin{array}{ccc} \tilde{A} & \xrightarrow{\tilde{f}} & A \\ \varphi \uparrow & & \uparrow \rho \\ M \otimes \tilde{S} & \xrightarrow{k_{\bar{g}}} & M. \end{array}$$

Since $\tau da - d\tilde{a} \sim \tau da$ (in $\tilde{A}$) and $\tau_*\colon H(C) \approx H(\tilde{A})$, $H(\tilde{A})$ has an additive homology basis, viz.

$$\tau\rho^C x_i, \quad \tau da - d\tilde{a},$$

or

$$\varphi(x_i), \quad \varphi(\bar{s}),$$

in which the x_i form an additive homology basis of $H(M)$ as before.

Starting from the additive homology basis

$$x_i, \quad \bar{s}, \quad x_i \otimes \bar{s}$$

of $H(M \otimes \tilde{S})$, let us now construct successively according to II

$$M \otimes \tilde{S} = K_{\bar{g}} \subset K_{\bar{g}}^1 \subset K_{\bar{g}}^2 \subset \cdots,$$

$$K_{\bar{g}}^\infty = K_{\bar{g}} \otimes \Lambda(\eta_j, \eta_j^1, \eta_j^2, \cdots),$$

$$J_{\bar{g}} = \operatorname{Min} K_{\bar{g}}^\infty,$$

and

$$k_{\bar{g}}^i\colon K_{\bar{g}}^i \twoheadrightarrow M, \quad k_{\bar{g}}^\infty\colon K_{\bar{g}}^\infty \to M,$$

$$\rho_{\bar{g}}^\infty\colon J_{\bar{g}} \to K_{\bar{g}}^\infty, \quad j_{\bar{g}} = k_{\bar{g}}^\infty \rho_{\bar{g}}^\infty\colon J_{\bar{g}} \to M.$$

Prove now $\varphi\colon K_{\bar{g}} \to \tilde{A}$ can be successively extended to DGA-morphisms

$$\varphi^i\colon K_{\bar{g}}^i \to \tilde{A},$$

with following diagram commutative

$$\begin{array}{ccc} \tilde{A} & \xrightarrow{\tilde{f}} & A \\ {\scriptstyle \varphi^i}\uparrow & & \uparrow{\scriptstyle \rho} \\ K_{\bar{g}}^i & \xrightarrow[k_{\bar{g}}^i]{} & M. \end{array}$$

To see this, suppose that φ^{i-1}: $K_{\bar{g}}^{i-1} \to \tilde{A}$ has been already defined ($\varphi^0 = \varphi$, $K_{\bar{g}}^0 = K_{\bar{g}}$), and define φ^i as follows.

According to the construction of Π, we have

$$\begin{aligned} K_{\bar{g}}^i &= K_{\bar{g}}^{i-1} \otimes \Lambda(\eta_j^{i-1}), \\ d\eta_j^{i-1} &= y_j^{i-1} \in K_{\bar{g}}^{i-1}, \\ k_{\bar{g}}^i(\eta_j^{i-1}) &= a_j^{i-1} \in M, \\ da_j^{i-1} &= k_{\bar{g}}^{i-1}(y_j^{i-1}). \end{aligned}$$

Then φ^i: $K_{\bar{g}}^i \to \tilde{A}$ will be defined as the DGA-morphism determined by the following expressions:

$$\begin{aligned} \varphi^i(x) &= \varphi^{i-1}(x), \quad x \in K_{\bar{g}}^{i-1}, \\ \varphi^i(\eta_j^{i-1}) &= \tau\rho^C(a_j^{i-1}). \end{aligned}$$

From φ^i we get then a DGA-morphism

$$\varphi^\infty: K_{\bar{g}}^\infty \to \tilde{A},$$

with

$$\varphi^\infty / K_{\bar{g}}^i = \varphi^i.$$

Define now

$$\tilde{\rho}: J_{\bar{g}} \to \tilde{A}$$

by

$$\tilde{\rho} = \varphi^\infty \rho_{\bar{g}}^\infty,$$

then we have a commutative diagram:

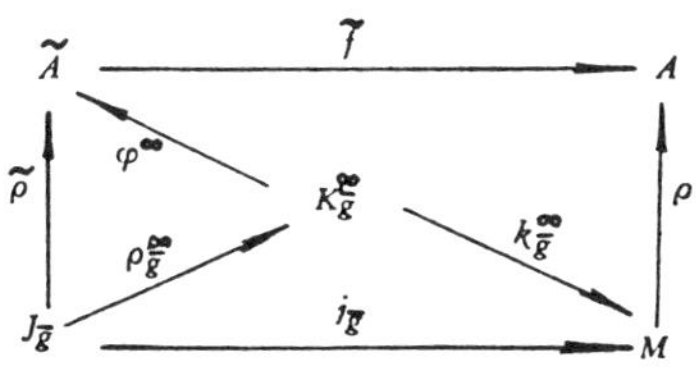

From the construction we know that $\tilde{\rho}_*$: $H(J_{\bar{g}}) \approx H(\tilde{A})$. Consequently $J_{\bar{g}} \approx \operatorname{Min} \tilde{A} \approx I^*(K/L)$, $\tilde{\rho}$ is a canonical homomorphism, and $j_{\bar{g}} \in G^0(K/L, K)$.

The theorem is now completely proved.

Remark. The construction of $J_{\bar{g}}$ depends on the choice of ρ: $M \to A$, the additive homology basis x^i, and $a \in A$, $\tilde{a} \in \tilde{A}$, $a_j^i \in M$, etc. However, the theorem shows that $J_{\bar{g}}$ is independent of such choice and is completely determined by the characteristic homomorphism $\bar{g}$: $M \to \bar{N}$.

From the J-construction, we get also easily the following:

Corollary.

1°. *Let* $L' \in \mathscr{K}$ *be any combinatorial sphere in* K. *Denote the characteristic homomorphism of* K *w.r.t.* L' *by*

$$\bar{g}': M \to \overline{I^*(L')},$$

then

$$\bar{g}' j_{\bar{g}}: J_{\bar{g}} \to \overline{I^*(L')}$$

is the characteristic homomorphism of K/L *w.r.t.* L'.

2°. *If in* K/L *there exists combinatorial sphere* L' *of dimension* $n+1$ *which contains* C_L, *then the characteristic homomorphism of* K *w.r.t.* L

$$\bar{g}: I^*(K) \to \overline{I^*(L)}$$

is $\bar{g} = 0^{(*)}$. *Denote the combinatorial cell* $L' \cap K$ *of dimension* $n+1$ *in* K *by* K', *or* $L' = K' \cup C_L = K'/L$, *and the characteristic homomorphism* $M' = I^*(K') = R \to \bar{N}$ *by* $\bar{g}' = 0$. *Then the characteristic homomorphism* $h_{\bar{g}}: J_{\bar{g}} \to J_{\bar{g}'}$ *is completely determined by* $\bar{g}$ *which makes the following diagram commutative:*

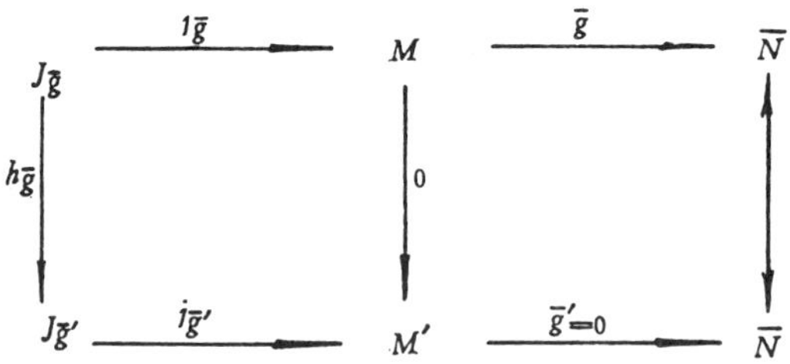

V. Effective Calculations and Axiomatic System of I-functor on $\mathscr{K}^0$

Any $K \in \mathscr{K}_0$ can be represented as

$$K = K_m \supset K_{m-1} \supset \cdots \supset K_1 \supset K_0, \tag{1}$$

in which K_0 is the 2-dimensional squelette, and K_r the union of K_{r-1} with an additional simplex Δ_r, the boundary of Δ_r being

$$\dot{\Delta}_r = L_{r-1} \subset K_{r-1}, \tag{2}$$

so that

$$K_r = K_{r-1} \cup \Delta_r = K_{r-1}/L_{r-1}. \tag{3}$$

Let

$$f_{r-1}: L_{r-1} \subset K_{r-1},$$

and

$$g_{r-1} = f^I_{r-1}: I^*(K_{r-1}) \to I^*(L_{r-1})$$

(*) If $M, N \in \mathscr{M}$, then $h = 0: M \to N$ will denote the DGA-morphism with $h_0: M_0 \approx N_0 (\approx R)$ and $h_r = 0: M_r \to N_r$ for $r > 0$.

be any DGA-morphism of $G(K_{r-1}, L_{r-1})$. By I, $I^*(K_r)$ is determined from g_{r-1} by J-construction, and from this it is easy to establish an axiomatic system of I^*-functor on $\mathscr{K}_0$ by induction *w.r.t.* (1).

However, the J-construction of I is not explicit, while all L_{r-1} are combinatorial spheres, so that we shall rather establish axiomatic system of I^*-functor by means of II—IV which permits to furnish at the same time an effective method of calculations of $I^*(K)$ for $K \in \mathscr{K}_0$.

First of all, to any $K \in \mathscr{K}_0$ we have

$$I^*(K) \in \mathscr{M}$$

unique up to DGA-isomorphism, and to any pair of $K \in \mathscr{K}_0$ and combinatorial subsphere $L \in \mathscr{K}_0$ of K, a characteristic DGA-morphism

$$\bar{g}\colon\ I^*(K) \to \overline{I^*(L)}.$$

We know also that I^* and $\bar{g}$ possess the following properties:

1°. I^* is a homotopic functor, or more precisely, for $K, K' \in \mathscr{K}_0$ with $K \simeq K'$, we have $I^*(K) \approx I^*(K')$.

2°. Let $K \in \mathscr{K}_0$ and $L \in \mathscr{K}_0$ be a combinatorial subsphere of K; $\bar{g}\colon\ I^*(K) \to \overline{I^*(L)}$ be the characteristic morphism. Construct now by the J-construction of II

$$J_{\bar{g}} \in \mathscr{M}$$

and DGA-morphism

$$j_{\bar{g}}\colon\ J_{\bar{g}} \to I^*(K),$$

then

$$J_{\bar{g}} \approx I^*(K/L).$$

3°. Let K, L, $\bar{g}$ be as in 2°. If $L' \in \mathscr{K}_0$ is any combinatorial subsphere of K and let $\bar{g}'\colon\ I^*(K) \to \overline{I^*(L')}$ be the corresponding characteristic morphism, then

$$\bar{g}' j_{\bar{g}}\colon\ I^*(K/L) \to \overline{I^*(L')}$$

is the corresponding characteristic morphism.

4°. Let $K \in \mathscr{K}_0$ and $L \in \mathscr{K}_0$ be a combinatorial subsphere of K. If L is the boundary of some combinatorial cell of K, then the characteristic morphism $\bar{g}\colon\ I^*(K) \to \overline{I^*(L)}$ is given by $\bar{g} = 0$.

5°. Let K, L, $\bar{g}$ be as in 2°. If L is the boundary of some combinatorial cell K' of K, then the DGA-morphism

$$h_{\bar{g}}\colon\ I^*(K/L) \to \overline{I^*(K'/L)}$$

constructed by IV from the characteristic morphism $\bar{g} = 0\colon\ I^*(K) \to I^*(L)$ of 4° is the corresponding characteristic morphism.

6°. For $K', K'' \in \mathscr{K}_0$ we have

$$I^*(K' \vee K'') \approx I^*(K') \vee I^*(K'').$$

7°. If $K \in \mathscr{K}_0$ is an n-dimensional combinatorial sphere, then

$$I^*(K) \approx \begin{cases} E(x), & n = \text{odd}, \\ E(y) \otimes R[x], & n = \text{even}, \end{cases}$$

in which

$$\deg x = n, \quad dx = 0,$$

and also

$$\deg y = 2n - 1, \quad dy = x^2,$$

for even n.

Remark. Owing to 7°, $\overline{I^*(L)}$ and characteristic morphism $\bar{g}: I^*(K) \to \overline{I^*(L)}$ are both meaningful for any combinatorial subsphere $L \in \mathscr{K}_0$ of K.

It is now easy to prove that the above properties 1°—7° form an axiomatic system for the I^*-functor over the category $\mathscr{K}_0$. In other words, we have the following

Theorem 5. *Let* ${}^0I^*$ *be any functor from* $\mathscr{K}_0$ *to* $\mathscr{M}$ *such that to any* $K \in \mathscr{K}_0$ *we have a* ${}^0I^*(K) \in \mathscr{M}$ *and to any pair of* $K \in \mathscr{K}_0$ *and a combinatorial subsphere* $L \in \mathscr{K}_0$ *of* K *we have a characteristic morphism*

$${}^0\bar{g}: \ {}^0I^*(K) \to \overline{{}^0I^*(L)},$$

which satisfies the axioms corresponding to 1°—7°. *Then to any* $K \in \mathscr{K}_0$ *we have*

$${}^0I^*(K) \approx I^*(K), \tag{I}$$

and to any combinatorial subsphere $L' \in \mathscr{K}_0$ *of* K, *there exists a commutative diagram between the above isomorphisms and the various characteristic morphisms, viz.*

$$\begin{array}{ccc} {}^0I^*(K) & \xrightarrow{\ {}^0\bar{g}\ } & \overline{{}^0I^*(L')} \\ \updownarrow & & \updownarrow \\ I^*(K) & \xrightarrow{\ \bar{g}\ } & \overline{I^*(L')} \end{array} \tag{II}$$

Proof. Let $K \in \mathscr{K}_0$ and K be represented by (1)—(3). Denote the expression (I) by $(\mathrm{I})_r$ in case $K = K_r$, and the expression (II) by $(\mathrm{II})_r$ in case $K = K_r$, and $L' = L'_r \in \mathscr{K}_0$ being any combinatorial subsphere of K_r. As $K_0 \simeq S^2 \vee \cdots \vee S^2$, we know by 1°, 6°, 7° that $(\mathrm{I})_0$ and $(\mathrm{II})_0$ hold true. Suppose that $(\mathrm{I})_{r-1}$ and $(\mathrm{II})_{r-1}$ have been proved and proceed to prove $(\mathrm{I})_r$ and $(\mathrm{II})_r$ as follows.

Let us first prove $(\mathrm{I})_r$. Consider the diagram below

$$\begin{array}{ccccc} {}^0I^*(K_r) & \xrightarrow{\ {}^0\bar{g}_r\ } & {}^0I^*(K_{r-1}) & \xrightarrow{\ {}^0\bar{g}_{r-1}\ } & \overline{{}^0I^*(L_{r-1})} \\ \updownarrow & & \updownarrow & & \updownarrow \\ I^*(K_r) & \xrightarrow{\ \bar{g}_r\ } & I^*(K_{r-1}) & \xrightarrow{\ \bar{g}_{r-1}\ } & \overline{I^*(L_{r-1})} \end{array}$$

By induction hypothesis $(\mathrm{I})_{r-1}$ and $(\mathrm{II})_{r-1}$, the two vertical arrows on the right are both DGA-isomorphisms, $\bar{g}_{r-1}$, ${}^0\bar{g}_{r-1}$ are both characteristic homomorphisms, and the right square is commutative. By Theorem 4 of IV we have $I^*(K_r) = I^*(K_{r-1}/L_{r-1}) \approx$

$J_{\bar{g}_{r-1}}$ and $\bar{g}_r = j_{\bar{g}_{r-1}}$: $J_{\bar{g}_{r-1}} \to I^*(K_{r-1})$. By Axiom 2° we have also ${}^0I^*(K_r) = {}^0I^*(K_{r-1}/L_{r-1}) \approx J_{{}^0\bar{g}_{r-1}}$ and ${}^0\bar{g}_r = j_{{}^0\bar{g}_{r-1}}$: $J_{{}^0\bar{g}_{r-1}} \to {}^0I^*(K_{r-1})$. From the construction of J and j, we see then ${}^0I^*(K_r) \approx I^*(K_r)$, i. e., $(I)_r$. Moreover, under these isomorphisms, the left-hand square in the above diagram is also commutative.

Prove next $(\text{II})_r$. For this, let $L'_r \in \mathscr{K}_0$ be a combinatorial subsphere of K_r and consider two cases separately.

Case I. $\Delta_r \notin L'_r$.

This time $L'_r \subset K_{r-1}$.

Case II. $\Delta_r \in L'_r$.

This time

$$\dim L'_r = \dim \Delta_r,$$

$$L'_r = K'_{r-1} \cup \Delta_r,$$

$$K'_{r-1} = \text{combinatorial cell in } K_{r-1}, \text{ with boundary } L_{r-1}.$$

Consider first the Case I. We have then the diagram below:

$$\begin{array}{ccccc}
{}^0I^*(K_r) & \xrightarrow{{}^0\bar{g}_r} & {}^0I^*(K_{r-1}) & \xrightarrow{{}^0\bar{g}'_{r-1}} & {}^0I^*(L'_r) \\
\updownarrow & & \updownarrow & & \updownarrow \\
I^*(K_r) & \xrightarrow{\bar{g}_r} & I^*(K_{r-1}) & \xrightarrow{\bar{g}'_{r-1}} & I^*(L'_r)
\end{array}$$

By the induction hypothesis $(\text{II})_{r-1}$, the two right vertical homomorphisms are both DGA-isomorphisms, ${}^0\bar{g}'_{r-1}$, $\bar{g}'_{r-1}$ are characteristic homomorphisms, and the left square is also commutative (with the same symbols as before). Now by Corollary 1° at the end of IV, $\bar{g}' = \bar{g}_r \bar{g}'_{r-1}$ is the characteristic homomorphism, so is ${}^0\bar{g}' = {}^0\bar{g}_r {}^0\bar{g}'_{r-1}$ by Axiom 3°. It follows that the diagram is commutative or we have $(\text{II})_r$.

For the Case II, we have also $(\text{II})_r$ by both Corollary 2° of IV and Axioms 4°, 5°. The theorem is now proved.

Remark. The Axiom 7° can also be slightly weakened to '7°. If $K \in \mathscr{K}_0$ is a 2-sphere, then

$$I^*(K) \approx E(y) \otimes R[x],$$

in which

$$\deg x = 2, \quad dx = 0,$$

$$\deg y = 3, \quad dy = x^2.$$

VI. I^*-FUNCTOR OF COUNTABLY INFINITE COMPLEXES

Any countably infinite complex $\tilde{K} \in \mathscr{K}$ is the union of finite subcomplexes $K_i \in \mathscr{K}_0$:

$$K_1 \subset K_2 \subset \cdots (\subset \tilde{K}). \tag{1}$$

Write for the inclusion f_i: $K_i \subset K_{i+1}$, then we have a sequence of DGA-morphisms:

$$A^*(K_1) \xleftarrow{f_1^A} A^*(K_2) \longleftarrow \cdots \longleftarrow A^*(K_i) \xleftarrow{f_i^A} A^*(K_{i+1}) \longleftarrow \cdots.$$

Clearly $A^*(\widetilde{K}) = \varprojlim A^*(K_i)$. As each f_i^A is an epimorphism, we can construct successively according to Theorem 3 of III $\{I^*(K_i), \rho_i, g_{i-1}\}$ starting from $\{I^*(K_1), \rho_1\}$, such that the following diagram is commutative in which the ρ_i are canonical homomorphisms, and $g_i \in G^0(f_i^A)$ are privileged homomorphisms:

$$\begin{array}{ccccccccc} A^*(K_1) & \xleftarrow{f_1^A} & A^*(K_2) & \longleftarrow \cdots \longleftarrow & A^*(K_i) & \xleftarrow{f_i^A} & A^*(K_{i+1}) & \longleftarrow & \cdots \\ \uparrow{\scriptstyle \rho_1} & & \uparrow{\scriptstyle \rho_2} & & \uparrow{\scriptstyle \rho_i} & & \uparrow{\scriptstyle \rho_{i+1}} & & \\ I^*(K_1) & \xleftarrow{g_1} & I^*(K_2) & \longleftarrow \cdots \longleftarrow & I^*(K_i) & \xleftarrow{g_i} & I^*(K_{i+1}) & \longleftarrow & \cdots. \end{array} \quad (2)$$

We shall call (g_i) a *sequence of privileged morphisms.*

Theorem 6. *Represent arbitrarily a countably infinite complex $\widetilde{K} \in \mathscr{K}$ as* (1) *and construct a sequence of privileged morphisms*

$$I^*(K_1) \xleftarrow{g_1} I^*(K_2) \longleftarrow \cdots \longleftarrow I^*(K_i) \xleftarrow{g_i} I^*(K_{i+1}) \longleftarrow \cdots. \quad (3)$$

Then

$$I^*(\widetilde{K}) \approx \mathrm{Min} \varprojlim I^*(K_i). \quad (4)$$

Proof. To any $(\xi_i) \in \varprojlim I^*(K_i)$ for which $\xi_i \in I^*(K_i)$, $g_{i-1}\xi_i = \xi_{i-1}$, let us set

$$\tilde{\rho}(\xi_i) = (\rho_i \xi_i) \in \varprojlim A^*(K_i) = A^*(\widetilde{K}).$$

Then we have the DGA-morphism

$$\tilde{\rho}\colon \varprojlim I^*(K_i) \to A^*(\widetilde{K}).$$

From this we get a commutative diagram:

$$\begin{array}{ccc} H(\varprojlim I^*(K_i)) & \xrightarrow{\tilde{\rho}_*} & H(\varprojlim A^*(K_i)) = H^*(\widetilde{K}) \\ \downarrow{\scriptstyle F^I} & & \downarrow{\scriptstyle F^A} \\ \varprojlim H(I^*(K_i)) & \xrightarrow{(\rho_{i*})} & \varprojlim H(A^*(K_i)) \end{array}$$

In the diagram F^I and F^A are both natural morphisms. Now F^I is an isomorphism by [1] Chap. VIII and F^A is one by [3]. Moreover ρ_{i*}: $H(I^*(K_i)) \approx H(A^*(K_i)) = H^*(K_i)$ for all i so that $\tilde{\rho}_*$ is also an isomorphism. It follows that $\tilde{\rho}$ is a canonical homomorphism and we have (4) as to be proved.

References

[1] Eilenberg, S. & Steenrod, N. E.: *Foundations of Algebraic Topology I* (1952).

[2] Friedlander, E., Griffiths, P. A. & Morgan, J.: *Homotopy Theory and Differential Forms*, Mimeog. Notes, (1972).

[3] Kahn, D. W.: The existence and applications of anticommutative cochain algebras, ILL. *J. Math.*, 7 (1963), 376—395.

[4] Sullivan, D.: Differential forms and topology of manifolds, *Symp. Tokyo on Topology* (1973), 37—49.

[5] 吴文俊: 代数拓朴的一个新函子, «科学通报», 7 (1975), 311—312.

[6] 吴文俊: 代数拓朴 I^* 函子论——齐性空间的实拓朴, «数学学报», 18 (1975), 162—172.

[7] Wu, Wen-tsün: Theory of I^*-functors in algebraic topology —— Real topology of fibre squares, *Scientia Sinica*, 18 (1975), 464—482.

Contemporary Mathematics
Volume 29, 1984

Reprinted with permission from the American Mathematical Society.

ON THE DECISION PROBLEM AND THE MECHANIZATION OF THEOREM-PROVING IN ELEMENTARY GEOMETRY[1]

Wu Wen-Tsün[2]

Abstract

The idea of proving theorems mechanically may be dated back to Leibniz in the 17th century and has been formulated in precise mathematical forms in this century through the school of Hilbert as well as his followers on mathematical logic. The problem consists in essence in replacing qualitative difficulties inherited in usual mathematical proofs by quantitative complexities of calculations on standardizing the proof procedures in an algorithmic manner. Such quantitative complexities of calculations, formerly far beyond the reach of human abilities, have become more and more trivial owing to the occurrence and rapid development of computers. In spite of vigorous efforts, however, researches in this direction give rise quite often to negative results in the form of undecidable mathematical theories. To cite a notable positive result, we may mention Tarski's method of proving theorems mechanically in elementary geometry and elementary algebra. The methods of Tarski as well as later ones are largely based on a generalization of Sturm theorem and are still too complicated to be feasible, even with the use of computers. The present paper, restricted to theorems with betweenness out of consideration and based on an

[1] Reprinted from *Scientia Sinica 21* (2), 1978.

[2] Institute of Systems-Sciences, Academia Sinica, Beijing 100080, People's Republic of China

entirely different principle, aims at giving a mechanical procedure which permits to prove quite non-trivial theorems in elementary geometry even by hands.

I. Formulation of the problem

A. Tarski in a classic paper [14] of 1948 has settled the decision problem of real closed field with one of its main aims to give mechanical proofs of theorems in elementary geometry. Alternative proofs of Tarski's result have later been given by Seidenberg, A. Robinson and P.J. Cohen, cf [12,9,2]. These authors have even suggested construction of certain decision machines to carry out such mechanical proofs. However, such a procedure seems to be far from being realized. In fact, only proofs of very trivial theorems in elementary geometry have actually been carried out on computers, cf. e.g. [6,7]. The purpose of the present paper is, leaving aside questions involving betweenness of points, to give an alternative solution of the decision problem of elementary geometry based on a principle entirely different from those employed by the authors above-mentioned. Our method permits to furnish mechanical proofs of quite difficult geometrical theorems which can be practiced even by hands, i.e., by means of papers and pencils only. The programming on a computer, based on such a method, though has not yet been done, will present no actual difficulties at all.

We shall restrict our considerations wholly to plane elementary geometry, though our method may be applied to the consideration of various other kinds of geometry. The first step of our method consists in the algebraic formalization of the geometrical problems involved. Points in the plane are to be defined as ordered pairs of numbers in a fixed field, say the field of rational numbers R. A dictionary is then set up turning geometrical relations into algebraic expressions which may be considered as either definitions or axioms. For example, for points $A_i = (x_i, y_i)$, distinct or not, we shall say:

A_1A_2 is parallel to A_3A_4

if $(x_1-x_2)(y_3-y_4) - (x_3-x_4)(y_1-y_2) = 0$,

A_1A_2 is orthogonal to A_3A_4

if $(x_1-x_2)(x_3-x_4) + (y_1-y_2)(y_3-y_4) = 0$,

the length-square of A_1A_2 is $r^2 = (x_1-x_2)^2 + (y_1-y_2)^2$, etc.

We may replace the basic field R by other fields, make correspondence the points to other kinds of number-sets, or modify the algebraic expressions in the axioms, e.g., instead of the length-square function r^2 given above, we consider the function $r^4 = (x_1-x_2)^4 + (y_1-y_2)^4$. We then go to other realms of geometry, non-euclidean geometry, real or complex projective geometry, finite geometry, etc. We shall however stick ourselves in what follows to plane elementary geometry only which has some representative character.

To illustrate our method of treatment, let us cite first a simple example. Consider the following statement:

(S_n) Let $A_0A_1A_2$ be a right-angled triangle with right angle at A_0. If x_1, x_2 denote the lengths of sides A_0A_1, A_0A_2 and x_3 is the length of the hypotenuse, then

$$x_1^n + x_2^n = x_3^n.$$

The problem is to decide whether the statement (S_n) is true or not and to give an algorithmic procedure of proving or disproving (S_n) which holds good for all statements alike in elementary geometry (with betweenness out of consideration).

To solve this problem, let us remark first that the points, etc., occurring in the statement have a generic character subjected to the conditions implied in the hypothesis of the statement. Thus, if we represent the points in question in coordinates with $A_0 = (x_0, v_0)$, $A_1 = (u_1, v_1)$, $A_2 = (u_2, v_2)$, the coordinates v_0, u_1, v_1, u_2, v_2 can be considered as indeterminates. On the other hand, the other coordinates and geometric entities x_0, x_1, x_2, x_3 are then algebraically dependent on these indeterminates, being restricted by following algebraic equations according to hypothesis of the statement (S_n):

$$f_0 \equiv (u_1-x_0)(u_2-x_0) + (v_1-v_0)(v_2-v_0) = 0,$$

$$f_1 \equiv x_1^2 - (u_1-x_0)^2 - (v_1-v_0)^2 = 0,$$

$$f_2 \equiv x_2^2 - (u_2-x_0)^2 - (v_2-v_0)^2 = 0,$$

$$f_3 \equiv x_3^2 - (u_1-u_2)^2 - (v_1-v_2)^2 = 0.$$

The conclusion in the statement (S_n) is equivalent to

$$g_n \equiv x_3^n - x_1^n - x_2^n = 0.$$

Let us now take once and for all the rational number field R as the base field. Let A^9 be the affine space on R with coordinates $(v_0, u_1, v_1, u_2, v_2, x_0, x_1, x_2, x_3)$ arranged in that definite order. Then the above equations $f_i = 0$ define an algebraic variety V of dimension 5, in the present case irreducible over R, with some generic point $(v_0, u_1, v_1, u_2, v_2, \bar{x}_0, \bar{x}_1, \bar{x}_2, \bar{x}_3)$ of which v_0, u_1, v_1, u_2, v_2 are indeterminates. The truth or untruth of the statement (S_n) amounts to $g_n \equiv 0$ or $g_n \not\equiv 0$ on V respectively.

It turns out that the general decision problem can be formulated in the following manner.

Problem. In a certain affine space A^n of dimension $n = r + d$ over R with coordinates $(u_1, \ldots, u_d, x_1, \ldots, x_r)$, consider an algebraic variety V with defining equations (u_i being independent indeterminates)

$$
\text{(I)} \qquad \begin{cases} f_1(u_1, \cdots, u_d, x_1) = 0, \\ f_2(u_1, \cdots, u_d, x_1, x_2) = 0, \\ \quad \cdots \\ f_r(u_1, \cdots, u_d, x_1, x_2, \cdots, x_r) = 0. \end{cases}
$$

The variety V may eventually split into irreducible components, all of (real) dimension $\leq d$. Those of dimension $= d$, with generic points of the form $(u_1, \ldots, u_d, \bar{x}_1, \ldots, \bar{x}_r)$ for which $\bar{x}_j$ are algebraic over the field $K = R(u_1, \ldots, u_d)$ will have a union V^* usually coincident with V. Let a polynomial $g(u_1, \ldots, u_d, x_1, \ldots, x_r)$ (or a set of such polynomials g_k) in $R[u_1, \ldots, u_d, x_1, \ldots, x_r]$ be given. It is to decide in an algorithmic manner whether

$$g \equiv 0$$

(or all $g_k \equiv 0$) on V^* or not.

In the above formulation the algebraic variety V, or preferably V^* reflects the hypothesis of the geometric statement considered. Either V or V^* will be called the *associated variety* of the statement in question. The variety V^* considered as one defined on the field $K = R(u_1, \ldots, u_d)$ is of dimension 0. The form of equations (I) shows that the algebraically dependent variables $x_1, \ldots, x_r$ are to be adjoined to K *successively* which reflects the geometrical fact that, starting from some generic points on certain generic lines, circles, etc., new points are to be *successively* adjoined in an algebraic manner by various geometric operations of joining points, drawing parallels, perpendiculars or circles, forming intersections of lines and circles, etc. These geometrical constructions give rise to algebraic equations involving x and u which can easily be turned into the form (I) be simple elimination procedure. In fact, the starting equations in x and u are rarely higher than 2. We shall call the variables u as the *parameters* and x as the *dependents*. It is also to be remarked that the condition for all components of V to be of dimension $\leq d$ over R reflects just the *determinate* character of the geometric statements to be considered, and the restriction to V^* reflects the depriving of degeneracies in our consideration. Both of these are, in reality, implicitly implied in the hypothesis of ordinary geometrical theorems. On the other hand, the equation $g=0$ on V^* (or set of $g_k=0$ on V^*) is the algebraic equivalent of the conclusion of the statement to be proved or disproved. We shall call g or set of g_k in what follows the *deciding polynomial(s)* of the geometrical statement in question.

Theoretically, the methods given by Hermann in [4] permit already to solve the above decision problem in an algorithmic manner. However, his methods are so complicated to give rise to astronomical expansions that even the simplest geometrical theorems can hardly be proved. On the contrary the decision procedure given below takes advantage of the particular character of the equations (I) and permits to prove mechanically quite non-trivial theorems even by hands, i.e. by means of pencil and paper only.

Our method of decision procedure is based on the following three theorems:

Theorem 1. *There is an algorithmic procedure permitting us to split the associated variety V^* of any determinative geometric statement defined by (I) into subvarieties V' irreducible over R each of which, considered as defined over the field $K = R(u_1, \ldots, u_d)$ has a representative basis (i.e. basis of the* *associated prime ideal) of the form*

$$(p_1, \cdots, p_r),$$

having the following properties:

(T1)$_1$. Each p_i is a polynomial in $R[u_1, \ldots, u_d, x_1, \ldots, x_i]$ of some degree $m_i > 0$ in x_i.

(T1)$_2$. The coefficients of p_i, considered as a polynomial in x_i, are polynomials in $R[u_1, \ldots, u_d, x_1, \ldots, x_{i-1}]$ having no common factor and with degree in x_j less than m_j for $j = 1, \ldots, i-1$.

(T1)$_3$. The leading coefficient of p_i, considered as a polynomial in x_i, is a polynomial $\neq 0$ in $R[u_1, \ldots, u_d]$ free of all x.

(T1)$_4$. p_1, as a polynomial in x_1, is irreducible in the field $K = R(u_1, \ldots, u_d)$, and for each $i > 1$, p_i, as a polynomial in x_i, is irreducible in the field obtained by adjoining $x_1, \ldots, x_{i-1}$ to K by the algebraic equations $p_1 = 0, \cdots, p_{i-1} = 0$.

It is clear that the polynomials $p_1, \ldots, p_r$ are uniquely determined by V' up to multipliers in R and will be said to form a *privileged basis* of V' (more exactly, of the prime ideal associated to V' over K), with respect to the given order $x_1, \ldots, x_r$ of the dependents. Remark that the notion is in reality due to Gröbner under the name of prime basis, cf. e.g. [3] and cf. also [8] for the intimately related concept of characteristic sets introduced by R. F. Ritt.

Theorem 2. *Let $(p_1, \ldots, p_r)$ be a privileged basis of any irreducible component V' of the associated variety V^*. There is an algorithmic procedure which permits us to determine, for any polynomial h in $R[u_1, \ldots, u_d, x_1, \ldots, x_r]$, an equation of the form*

$$Dh = \sum_{i_s=0}^{m_s-1} h_{i_1 \cdots i_r} x^{i_1} \cdots x^{i_r} + \sum_{i=1}^{r} A_i p_i,$$

verifying the following conditions:

$(T2)_1$. $D, h_{i_1 \dots i_r}$ are all polynomials in $R[u_1, \dots, u_d]$ and $D \neq 0$,

$(T2)_2$. A_i are all polynomials in $R[u_1, \dots, u_d, x_1, \dots, x_r]$.

The polynomials $h_{i_1, \dots, i_r}$ which are uniquely determined up to multipliers in R by the algorithmic procedure, will be called the ***remainder constituents*** of the polynomial h with respect to the privileged basis $(p_1, \dots, p_r)$ of V', or, by abuse of language, simply the remainder constituents of h with respect to V'.

Theorem 3. *For a geometrical statement with associated variety V^* and deciding polynomial g (or a set g_k of deciding polynomials) to be true, it is necessary and sufficient that for any irreducible component V' of V^*, all remainder constituents of g (or of all g_k) should be identically zero.*

II. Examples

Before giving proofs of these theorems in IV, we shall illustrate their use by some examples below.

Ex. 1. For the geometrical statement (S_n) about right-angled triangles as cited in the beginning of the present paper, we see readily that the associated variety V in parameters v_0, u_1, v_1, u_2, v_2 and dependents x_0, x_1, x_2, x_3, is already irreducible over R and possesses a privileged basis (p_1, p_2, p_3, p_4), where

$$p_1 = x_0^2 - (u_1+u_2)x_0 + u_1u_2 + (v_1-v_0)(v_2-v_0),$$

$$p_2 = x_1^2 + (u_1-u_2)x_0 - u_1^2 + u_1u_2 - (v_1-v_0)(v_1-v_2),$$

$$p_3 = x_2^2 - (u_1-u_2)x_0 - u_2^2 + u_1u_2 + (v_2-v_0)(v_1-v_2),$$

$$p_4 = x_3^2 - (u_1-u_2)^2 - (v_1-v_2)^2.$$

It is readily verified that

$$g_2 = p_4 - p_2 - p_3,$$

while for $n > 2$, we have

$$g_n \not\equiv 0 \mod (p_1, p_2, p_3, p_4).$$

It follows that (S_n) is true only for $n=2$ which corresponds to the Keu-Kou Theorem.

Ex. 2. Our decision procedure can also be applied to give mechanical proofs of trigonometric identities. Consider, e.g. the following statements:

(S) If $A_1 + A_2 + A_3 = 180°$, then

$$\sin 2A_1 + \sin 2A_2 + \sin 2A_3 = 4 \sin A_1 \sin A_2 \sin A_3.$$

(C) If $A_1 + A_2 + A_3 = 180°$, then

$$\cos 2A_1 + \cos 2A_2 + \cos 2A_3 + 4 \cos A_1 \cos A_2 \cos A_3 = 0.$$

To decide whether (S) or (C) is true, let us set

$$\sin A_i = s_i, \quad \cos A_i = c_i, \quad (i = 1,2),$$

$$\sin 2A_i = x_i, \quad \cos 2A_i = y_i, \quad (i = 1,2,3),$$

$$\sin A_3 = z_1, \quad \cos A_3 = z_2.$$

Take c_1 and c_2 as parameters and

$$s_1, s_2, z_1, z_2, x_1, x_2, x_3, y_1, y_2, y_3$$

as dependents (in this order), then the associated variety of (S) or (C) is already irreducible in R and possesses a privileged basis $(p_1, \ldots, p_{10})$ given by

$$p_1 = s_1^2 + c_1^2 - 1,$$

$$p_2 = s_2^2 + c_2^2 - 1,$$

$$p_3 = z_1 - c_2 s_1 - c_1 s_2,$$

$$p_4 = z_2 - s_1 s_2 + c_1 c_2,$$

$$p_5 = x_1 - 2c_1 s_1,$$

$$p_6 = x_2 - 2c_2 s_2,$$

$$p_7 = x_3 - 2z_1 z_2,$$

$$p_8 = y_1 - 2c_1^2 + 1,$$

$$p_9 = y_2 - 2c_2^2 + 1,$$

$$p_{10} = y_3 + 4c_1 c_2 s_1 s_2 - 4c_1^2 c_2^2 + 2c_1^2 + 2c_2^2 - 1.$$

The deciding polynomials of the statements (S) and (C) are given respectively by

$$g_s = x_1 + x_2 + x_3 - 4s_1 s_2 z_1,$$

$$g_c = y_1 + y_2 + y_3 + 4c_1 c_2 z_2.$$

We verify readily that

$$g_s = -2c_2 s_2 p_1 - 2c_1 s_1 p_2 - 2p_3(s_1 s_2 + c_1 c_2)$$
$$+ 2z_1 p_4 + p_5 + p_6 + p_7$$
$$\equiv 0 \mod(p_1, \ldots, p_{10}),$$

$$g_c = -1 + 4c_1 c_2 p_4 + p_8 + p_9 + p_{10}$$
$$\not\equiv 0 \mod(p_1, \ldots, p_{10}).$$

Hence (S) gives a true identity while (C) does not.

Ex. 3. Let us consider the Simson-Line Theorem which corresponds to the following statement:

(S_s). From a point A_4 on the circumscribed circle of a triangle $A_1A_2A_3$, perpendiculars are drawn to the sides of the triangle. Then the feet of the perpendiculars are in a line.

To prove this, let us take for simplicity the center of the circumscribed circle as the point (0,0), while the radius is r. Let the points A_i $(i = 1,2,3,4)$ be (x_i, u_i) and the feet of perpendiculars A_j be (x_j, y_j), $j = 5,6,7$. Consider r, u_1, u_2, u_3, u_4 as parameters and

$$x_1, x_2, x_3, x_4, y_5, y_6, y_7, x_5, x_6, x_7$$

as the dependents in the order indicated.

A prime basis $(p_1, \ldots, p_{10})$ of the associated Simson variety which is irreducible is readily given as follows:

$$p_1 = x_1^2 + u_1^2 - r^2,$$

$$p_2 = x_2^2 + u_2^2 - r^2,$$

$$p_3 = x_3^2 + u_3^2 - r^2,$$

$$p_4 = x_4^2 + u_4^2 - r^2,$$

$$p_5 = 2r^2 y_5 - h_5,$$

$$p_6 = 2r^2 y_6 - h_6,$$

$$p_7 = 2r^2 y_7 - h_7,$$

$$p_8 = u_{23}x_5 - x_{23}y_5 + u_3x_2 - u_2x_3,$$

$$p_9 = u_{31}x_6 - x_{31}y_6 + u_1x_3 - u_3x_1,$$

$$p_{10} = u_{12}x_7 - x_{12}y_7 + u_2x_1 - u_1x_2.$$

In the above equations we have set for simplicity:

$$u_{ij} = u_i - u_j, \quad x_{ij} = x_i - x_j, \quad (i,j = 1,2,3),$$

$$h_5 = u_4 x_2 x_3 - (u_2 x_3 + u_3 x_2) x_4 + r^2(u_2 + u_3 + u_4) - u_2 u_3 u_4.$$

Similarly for h_6 and h_7.

The deciding polynomial is given by

$$g = x_5(y_6 - y_7) + x_6(y_7 - y_5) + x_7(y_5 - y_6).$$

Straightforward calculation shows again

$$g \equiv 0 \mod(p_1, \ldots, p_{10}),$$

which proves the truth of Simson statement (S_s).

Ex. 4. For a less trivial example, let us consider the Feuerbach theorem which corresponds to the following statement:

(S_F) The 9-point circle of a triangle is tangent to the four inscribed and the escribed circles of the triangle.

Let us take the three vertices of the triangle as $(2u_i, 2v_i)$, the center and radius of the 9-point circle as (x_1, y_1) and r_1, and the center and radius of either the inscribed or any of the escribed circles as (x_2, y_2) and r_2. Introduce also variables z_1, z_2, z_3 corresponding to lengths of the sides of the triangle. Then with $u_1, v_1, u_2, v_2, u_3, v_3$ as the parameters and $z_1, z_2, z_3, x_1, y_1, x_2, y_2, r_1, r_2$ as dependents in this order, the associated Feuerbach variety V_F splits into irreducible ones having privileged basis $(p_1, \ldots, p_9)$ given by:

$$p_1 = z_1^2 - u_{23}^2 - v_{23}^2,$$

$$p_2 = z_2^2 - u_{31}^2 - v_{31}^2,$$

$$p_3 = z_3^2 - u_{12}^2 - v_{12}^2,$$

$$p_4 = 4\Delta x_1 - \alpha_1,$$

$$p_5 = 4\Delta y_1 - \beta_1,$$

$$p_6 = 2\Delta x_2 - \alpha_2,$$

$$p_7 = 2\Delta y_2 - \beta_2,$$

$$p_8 = 4\Delta r_1 - z_1 z_2 z_3,$$

$$p_9 = 2\Delta r_2 + \gamma.$$

In the above formulas we have put for simplicity

$$u_{ij} = u_i - u_j, \quad v_{ij} = v_i - v_j, \qquad (i,j = 1,2,3),$$

$$2\Delta = \begin{vmatrix} u_1, & v_1, & 1 \\ u_2, & v_2, & 1 \\ u_3, & v_3, & 1 \end{vmatrix},$$

$$\alpha_1 = v_{23}v_{31}v_{12} + u_1^2 v_{23} + u_2^2 v_{31} + u_3^2 v_{12}$$
$$- 2(u_2u_3v_{23} + u_3u_1v_{31} + u_1u_2v_{12}),$$

$$\beta_1 = - u_{23}u_{31}u_{12} - v_1^2 u_{23} - v_2^2 u_{31} - v_3^2 u_{12}$$
$$+ 2(v_2v_3u_{23} + v_3v_1u_{31} + v_1v_2u_{12}),$$

$$\alpha_2 = -[v_{23}v_{31}v_{12} + v_1(u_2^2 - u_3^2) + v_2(u_3^2 - u_1^2) + v_3(u_1^2 - u_2^2)$$
$$+ \varepsilon_2\varepsilon_3 v_{23}z_1z_3 + \varepsilon_3\varepsilon_1 v_{31}z_3z_1 + \varepsilon_1\varepsilon_2 v_{12}z_1z_2],$$

$$\beta_2 = u_{23}u_{31}u_{12} + u_1(v_2^2 - v_3^2) + u_2(v_3^2 - v_1^2) + u_3(v_1^2 - v_2^2)$$
$$+ \varepsilon_2\varepsilon_3 u_{23}z_2z_3 + \varepsilon_3\varepsilon_1 u_{31}z_3z_1 + \varepsilon_1\varepsilon_2 u_{12}z_1z_2,$$

$$\gamma = \varepsilon_1(u_{31}u_{12} + v_{31}v_{12})z_1 + \varepsilon_2(u_{12}u_{23} + v_{12}v_{23})z_2$$
$$+ \varepsilon_3(u_{23}u_{31} + v_{23}v_{31})z_3 + \varepsilon_1\varepsilon_2\varepsilon_3 z_1z_2z_3.$$

Remark that the choice of $\varepsilon_i = +1$ or -1 corresponds to the 4 inscribed or escribed circles and reflects the reducibility of the

Feuerbach variety.

Now the deciding polynomial of statement (S_F) is of the form

$$g_\eta = (\eta r_1 - r_2)^2 - (x_1 - x_2)^2 - (y_1 - y_2)^2,$$

in which $\eta = +1$ or -1. By a straightforward calculation, however lengthy and tedious, we verify that g_η will be a linear combination of $p_1, \ldots, p_9$ for $\eta = +\varepsilon_1\varepsilon_2\varepsilon_3$ but not so for $\eta = -\varepsilon_1\varepsilon_2\varepsilon_3$. Thus the Feuerbach statement is a true theorem and $\eta = +\varepsilon_1\varepsilon_2\varepsilon_3$ reflects the manner of contact between the 9-point circle and the respective in- or es-cribed circle. Remark that in our formulation we have left aside the question of betweenness and the manner of contact of circles is not of interest to us. In this way we can take $g = g_{+1} \cdot g_{-1}$ as the deciding polynomial if we like. We may also take the Feuerbach variety in the affine space of u_i, v_i, x_j, y_j only without the introduction of z_1, z_2, z_3 so that it is irreducible at the outset.

III. Some lemmas

To make some preparations we shall consider a field $R(u_1, \ldots, u_d, x_1, \ldots, x_r)$ in which $u_1, \ldots, u_d$ are transcendental while $x_1, \ldots, x_r$ are algebraic over the base field R. The algebraic extensions to $x_1, \ldots, x_r$ are defined successively by the following equations

$$\text{(II)} \quad \begin{cases} p_1(x_1) \equiv p_{10}x_1^{m_1} + p_{11}x_1^{m_1-1} + \cdots + p_{1m_1} = 0, \\ p_2(x_1, x_2) \equiv p_{20}x_2^{m_2} + p_{21}x_2^{m_2-1} + \cdots + p_{2m_2} = 0, \\ \cdots \\ p_r(x_1, \cdots, x_r) \equiv p_{r0}x_r^{m_r} + p_{r1}x_r^{m_r-1} + \cdots + p_{rm_r} = 0. \end{cases}$$

It is assumed that for $1 \leq i \leq r$, the p_{ij}'s are polynomials in the ring $P_{i-1} = R[u_1, \ldots, u_d, x_1, \ldots, x_{i-1}]$, that p_{i0} are polynomials $\neq 0$ in $P_0 = R[u_1, \ldots, u_d]$, and that p_i considered as polynomials in x_i are irreducible over the field $K_{i-1} = R(u_1, \ldots, u_d, x_1, \ldots, x_{i-1})$ defined by the equations $p_1(x_1) = 0, \cdots, p_{i-1}(x_1, \ldots, x_{i-1}) = 0$. We put here

$$K_0 = K = R(u_1, \cdots, u_d).$$

Let Υ be the collection of sets of indices $I = (i_1, \ldots, i_r)$ with $0 \leq i_j \leq m_j - 1$. For such an I we shall write symbolically

$$x^I = x_1^{i_1} \cdots x_r^{i_r}.$$

Any polynomial of the form .
with coefficients a_I in a certain ring or field F will be called then a *normalized* one in $F[x_1, \ldots, x_r]$.

Lemma 1. *There is an algorithmic procedure which permits to determine uniquely for any polynomial A in P_r, a set of integers $s_1, \ldots, s_r \geq 0$ and a set of polynomials A_I in P_0 for $I \in \Upsilon$ verifying the following conditions:*

$(L1)_1$. Modulo some linear combination of p_i over P_r, we have

$$p_{10}^{s_1} \cdots p_{r0}^{s_r} A \equiv \sum A_I x^I.$$

$(L1)_2$. A_I are polynomials in P_0 with coefficients linear in those of A, considered as polynomial in P_r.

$(L1)_3$. $s_1, \ldots, s_r$ are the least integers ≥ 0 to make $(L1)_1$ and $(L1)_2$ possible.

Proof. Considering both A and p_r as polynomials in x_r with coefficient in p_{r-1}, we get by division for some integer $s_r \geq 0$ taken to be least possible

$$p_{r0}^{s_r} A = Q_r p_r + R_{r-1},$$

with Q_r, R_{r-1} polynomials in P_r for which the degree of R_{r-1} in x_r is $< m_r$. Considering now p_{r-1} and R_{r-1} as polynomials in x_{r-1} with coefficients in $R[u_1, \ldots, u_d, x_1, \ldots, x_{r-2}, x_r]$, we get by division for some integer $s_{r-1} \geq 0$ taken to be least possible

$$p_{r-1,0}^{s_{r-1}} R_{r-1} = Q_{r-1} p_{r-1} + R_{r-2},$$

with Q_{r-1}, R_{r-2} polynomials in P_r for which the degrees of R_{r-2} in x_r and x_{r-1} are $< m_r$ and $< m_{r-1}$ respectively. Proceeding in this manner, we get successively

$$p_{r-2,0}^{s_{r-2}} R_{r-2} = Q_{r-2} p_{r-2} + R_{r-3},$$

$$\ldots$$

$$p_{10}^{s_1} R_1 = Q_1 p_1 + R_0,$$

with R_0 as a polynomial in P_r, for which the degree in x_i is $< m_i$, $1 \leq i \leq r$. We may then write R_0 as $\sum A_I x^I$ and get the expression verifying all the conditions in the Lemma as required.

Lemma 2. *There is an algorithmic procedure which permits us to determine for any polynomial in some indeterminate y of the form*

$$A = A_0 y^m + A_1 y^{m-1} + \cdots + A_m$$

with each A_i in P_r and $A_0 \neq 0$ in K_r expressions of the form

$$HA = B + C_1 p_1 + \cdots + C_r p_r,$$

with

$$B = B_0 y^m + B_1 y^{m-1} + \cdots + B_m,$$

verifying the following conditions:

(L2)$_1$. *All* B_i *are normalized polynomials in* P_r *and* B_0 *is a polynomial* $\neq 0$ *in* P_0.

(L2)$_2$. C_i *are all polynomials in* $P_r[y]$ *and* H *is one in* P_r.

(L2)$_3$. $H \not\equiv 0 \mod (p_1, \ldots, p_r)$.

Any such polynomial B *satisfying (L2)*$_1$ *will then be said to be a normalized polynomial over* P_r *in the indeterminate* y.

Proof. By Lemma 1, we have for some integers $s_1, \ldots, s_r \geq 0$ an expression of the form

$$p_{10}^{s_1} \cdots p_{r0}^{s_r} A = A'_0 y^m + A'_1 y^{m-1} + \cdots + A'_m \mod (p_1, \cdots, p_r)$$

with all A'_i normalized in P_r and $A'_0 \neq 0$ in K_r. Suppose that A'_0 is free of $x_{i+1}, \ldots, x_r$ but not so for x_i. As A'_0 has a degree $< m_i$ in x_i while p_i is irreducible in K_{i-1} and of degree m_i in x_i, we find by the usual division algorithm polynomials h, k in P_i and A'_{00} in P_{i-1} such that

$$hA'_0 + kp_i = A'_{00},$$

in which h is $\neq 0$ in K_i and $A'_{00} \neq 0$ in K_{i-1}. We have then an expression of the form

$$hp_{10}^{s_1} \cdots p_{r0}^{s_r} A + kp_i y^m = A'_{00} y^m + hA'_1 y^{m-1} + \cdots + hA'_m \mod (p_1, \cdots, p_r).$$

Applying Lemma 1 again to hA'_i and A'_{00}, we get then some expression

$$h'' A = A''_0 y^m + A''_1 y^{m-1} + \cdots + A''_m \mod (p_1, \cdots, p_r),$$

for which all A''_i are normalized in P_r with $A''_0 \neq 0$ in K_r and not containing any x_j for $j \geq i$, and h'' is some polynomial in P_r which is $\neq 0$ in K_r. If A''_0 is free of all $x_1, \ldots, x_r$, we may take

$$B = A''_0 y^m + A''_1 y^{m-1} + \cdots + A''_m$$

as the polynomial B required or otherwise we proceed as before.

Lemma 3. *There is an algorithmic procedure which permits to factorize in K_r any polynomial in $P_r[y]$*

$$A = A_0 y^m + A_1 y^{m-1} + \cdots + A_m$$

with A_i in P_r and $A_0 \neq 0$ in K_r, $m \geq 2$. More precisely, it permits to find expression of the form

$$HA = A' + \sum C_i p_i,$$

with

$$A' = \bar{A}_1 \cdots \bar{A}_t,$$

verifying the following conditions:

$(L3)_1$. Each $\bar{A}_j$ is a normalized polynomial in $P_r[y]$ and is irreducible in K_r.

$(L3)_2$. H is a polynomial in P_r which is $\neq 0$ in K_r.

$(L3)_3$. C_i are polynomials in $P_r[y]$.

Proof. The method of Hermann in [4] permits us to give, in an algorithmic manner, a factorization of A into irreducible ones in K_r, so that after clearing of fractions we have an expression of the form

$$DA = B_1 \cdots B_t + \sum B'_i p_i,$$

with B_i, B'_i polynomials in $P_r[y]$, B_i irreducible in K_r and D a polynomial in P_r which is $\neq 0$ in K_r. Applying now Lemma 2 to each B_i, we get then the expression required. Consider also the reference [15], p. 130.

IV. Proofs of the theorems

We are now in a position to give proofs of Theorems 1-3 quite simply as follows:

Proof of Theorem 1. For the defining system of equations (I) of the associated variety V, let us consider $f_1(u_1, \ldots, u_d, x_1)$ as a polynomial in x_1 with coefficients in $R[u_1, \ldots, u_d]$ and factorize f_1 into ones irreducible in $K = R(u_1, \ldots, u_d)$ by applying the algorithm in Lemma 3, with x_1 as y and 0 as r there. Take any such irreducible factor as $p_1(x_1)$. Let K_1 be the field obtained from K by adjoining x_1 defined by the equation

$$p_1(x_1) = 0$$

Now f_2, considered as a polynomial in the indeterminate x_2, cannot be identical with 0 in the field K_1, for otherwise the variety V would be of dimension $> d$, contrary to the determinancy hypothesis of our geometric statement. Applying Lemma 3 to f_2 with x_2 as y and 1 as r there, we get a certain polynomial f'_2 in $P_2 = R[u_1, \ldots, u_d, x_1, x_2]$ with an expression of the form

$$h_2 f_2 = f'_2 + C_{21} p_1,$$

in which f'_2, as a polynomial in the indeterminate x_2, is a product of normalized factors irreducible and $\neq 0$ in the field K_1. Take any such factor as $p_2(x_1, x_2)$. Let K_2 be the field obtained from K_1 by adjoining x_2 defined by the equation

$$p_2(x_1, x_2) = 0.$$

Then f_3 is a polynomial in the indeterminate x_3 over P_2 which cannot be identical with 0. Applying Lemma 3, we get an expression of the form

$$h_3 f_3 = f'_3 + C_{31} p_1 + C_{32} p_2,$$

in which the polynomial f'_3 in x_3 is a product of normalized factors irreducible and $\neq 0$ in K_2. Take any factor as $p_3(x_1,x_2,x_3)$, adjoin x_3 to K_2 by the equation

$$p_3(x_1,x_2,x_3)=0,$$

and proceed further as before. In this manner, we get finally system of subvarieties V' irreducible over R, each defined on K by systems of equations of the type

$$\text{(III)} \qquad \begin{cases} p_1(x_1)=0, \\ p_2(x_1,x_2)=0, \\ \quad \cdots \\ p_r(x_1,x_2,\cdots,x_r)=0, \end{cases}$$

verifying some obvious conditions as described in III.

It is easy to see that the collection of all such subvarieties exhausts the given variety V^*. In fact, consider an irreducible component in R of V^* with a generic point, say $(u_1,\ldots,u_d, \bar{x}_1,\ldots,\bar{x}_r)$, in which $u_1,\ldots,u_d$, are independent indeterminates while $\bar{x}_1,\ldots,\bar{x}_r$ depend algebraically on them. As $(u_1,\ldots,u_d,\bar{x}_1)$ should satisfy the system of equations (I), in particular the equation $f_1=0$, they should annul some one of the irreducible factors of f_1, say p_1 before. Now, $(u_1,\ldots,u_d,\bar{x}_1,\bar{x}_2)$ should satisfy the other equations in the system (I) as well as the equation $p_1(x_1)=0$. It follows from the expression about f'_2 given above that they should also satisfy the equation $f'_2=0$ and hence should annul one of the irreducible factors of f'_2, say $p_2(x_1,x_2)$ above. Proceeding in the same manner we see that $(u_l,\ldots,u_d, \bar{x}_1,\ldots,\bar{x}_r)$ should satisfy a system of equations of the type (III) and hence is a generic point of an irreducible subvariety among the collection found above. This completes the proof of the theorem.

Proofs of Theorems 2 and 3.

Theorem 2 follows immediately from the algorithm in Lemma 1. The truth of Theorem 3 is quite clear from Theorems 1 and 2.

Final Remark. Our algorithm for the mechanization of theorem-proving in elementary geometry involves mainly such polynomial manipulations as arithmetic operations and simple eliminations, which were all originated and quite developed in 12-14c. Chinese mathematics, cf. e.g. the book of late Chien [1] for the explanations. In fact, the algebrization of geometrical problems and systematic method of their solutions by algebraic tools were some of the main achievements of Chinese mathematicians at that time, much earlier than the appearance of analytic geometry in 17c.

Added in Proof. (Dec. 1977).

The same principle has been applied to the mechanization of theorem-proving in elementary differential geometry with the aid of Ritt's theory of differential algebra, of which the details will be given later.

Bibliography

[1] Chien, Baozong. *The history of Chinese mathematics*, Science Press, 1964.

[2] Cohen, P.J. Decision procedures for real and p-adic fields, *Comm. Pure Appl. Math*, **22** (1969), 131-152.

[3] Gröbner, W. *Moderne algebraische Geometrie* (1949), Wien.

[4] Hermann, G. Die Frage der endlich vielen Schritte in der Theorie der Polynomideale, *Math. Ann.*, **95** (1926), 736-788.

[5] Hilbert, D. *Grundlagen der Geometrie*.

[6] McCharen, J.D. *et al.* Problems and experiments for and with automated theorem-proving programs, *IEEE Trans. on Computers*, **C-25** (1976), 773-782.

[7] Reiter, R. A semantically guided deductive system for automatic theorem proving, *IEEE Trans. on Computers*, **C-25** (1976), 328-334.

[8] Ritt, R.F. *Differential Algebra* (1950).

[9] Robinson, A. *Introduction to Model Theory and to the Metamathematics of Algebra* (1963), Amsterdam.

[10] Robinson, A. Algorithms in algebra, in model theory and algebra, *Lect. Notes in Math.*, No. 498 (1975), 14-40.

[11] Robinson, A. A decision method for elementary algebra and geometry-revisited, *Proc. of Tarski Symposium* (1974), 139-152.

[12] Seidenberg, A. A new decision method for elementary algebra, *Annals of Math.*, **60** (1954), 365-374.

[13] Tarski, A. What is elementary geometry? *The Axiomatic Method with Special Reference to Geometry and Physics* (1959), 16-29, Amsterdam.

[14] Tarski, A. and McKinsey, J.C.C. *A Decision Method for Elementary Algebra and Geometry*, 2nd ed., Berkeley and Los Angeles (1948-1951).

[15] Waerden, Van der. *Moderne Algebra*, Bd. 1 (1930).

[16] Waerden, Van der. *Einführung in die algebraischen Geometrie* (1945).

Vol.2 No.2 1982

数学物理学报 Acta Mathematica Scientia

TOWARD MECHANIZATION OF GEOMETRY —SOME COMMENTS ON HILBERT'S "GRUNDLAGEN DER GEOMETRIE"

Wu Wen-tsün (吴文俊)

Institute of Systems—Science, Academia Sinica

The great merit of Hilbert's classic "Grundlagen der Geometrie" of 1899 is universally recognized as being representative for axiomatization of mathematics, laying in particular a rigorous foundation of the euclidean geometry. However, another great merit (perhaps greater in the opinion of the present author) of this classic seems hardly to be noticed up to the present. In fact, this classic is also representative for the mechanization of geometry, showing clearly at the same time the way to achieve it. The present paper has the object of trying to clarify these points.

First of all let us remark that in the statement of theorems (or even axioms) there are usually some implicit assumptions about *genericity* or *non-degeneracy* of figures involved without which the theorems may be meaningless or even fall into fallacies. The following simple examples may serve as illustrations.

Ex.1 The three altitudes of a triangle are concurrent.

It is implicitely assumed here that the triangle in question should be *generic* in the sense that it does not degenerate into one with vertices collinear or with two vertices coincident. In the first case the theorem is not true and in the second case it is meaningless.

Ex.2 The opposite sides of a parallelogram are congruent.

*Received — Nov., 1981

The theorem will not be true if the parallelogram in question degenerates into one with vertices lying on the same line.

Ex.3 Desargues Theorem. If the three pairs of corresponding sides of two triangles are all parallel to each other, then the three lines joining the corresponding vertices of these triangles are either concurrent or are parallel to each other.

The theorem will not be true if the two triangles in question degenerate into ones with collinear vertices.

Ex.4 Desargues Theorem. If the three lines joining the corresponding vertices of two triangles are either concurrent or parallel, and two pairs of corresponding sides of these triangles are both parallel to each other, then the same will be true for the third pair of corresponding sides.

Again the theorem will not be true if one of the pairs of corresponding sides parallel to each other degenerates into a pair lying on the same line.

In view of these examples we see that theorems of elementary geometry are usually true only in the *generic* or *non-degenerate* case which are implicitely assumed as hypothesis but usually not clearly expressed in the statements of the theorems. In each degenerate case we have to investigate separately whether the theorem is meaningful or not and if it is so whether the theorem remains hold true or not.

Now to prove theorems in the usual euclidean fashion one should incessantly make resort to previously proved theorems considered to be already known. As these known theorems are only true under certain non-degeneracy conditions one should, each time when these theorems are to be applied, verify whether these non-degeneracy conditions are observed or not. One should consider different cases to deprive off one by one each of these degeneracy situations. This renders the proof of a theorem very cumbersome, the more so because such non-degeneracy conditions of theorems to be applied are usually not clearly stated. In fact, even for a theorem of moderate complexity, it would be quite impossible to take care of all these non-degenerate cases occured in the known theorems to be applied in order to make the proof meeting the usual standard of rigor. A scrutiny of the proofs of theorems concerning the establishment of rules of number systems in a Desarguesian geometry, in which Desargues theorem should be applied over and over, as described in the classic of Hilbert, may well illustrate this point.

On the other hand Hilbert, in his classic, after laying down the foundation for the algebraization and coordinatization of a geometry involving Desargues theorem and Pascal theorem, has stated a theorem (numbered Th.62 in the later editions), which is in essence equivalent to a mechanized procedure for the proofs of a certain kind of theorems. Let us call for short a plane geometry to be *pascalian* if the planar axioms of incidence, the axioms of order, the axiom of parallelism in the strengthened form, as well as the so-called Pascal theorem all hold true. Then the original statement of the theorem in question runs somewhat as follows.

Th.62 Each *pure intersection-point theorem*, in a Pascalian geometry, if true, can always be proved by the aid of suitably constructing auxiliary points and lines, as consequence of a combination of a finite number of Pascal configurations.

For the meaning of pure intersection-point theorem (abb. PIP-Th) Hilbert has given two explanations which are in fact not equivalent. For the first one Hilbert defined a PIP-Th as one in which only incidence of points and lines as well as parallelism of lines are involved. Then, he argued that every such PIP-Th. may be described in more details in the following form:

Select first a system of finite number of points and lines arbitrarily. Then successively in a definite prescribed manner draw some parallels to some of the lines get, choose some points on some of the lines get, and draw some lines through some of the points get. If, by constructing in this prescribed manner the joining lines, intersection points as well as parallels through points already constructed, one arrives finally to a finite set of lines, then the theorem will assert that these lines will be either concurrent or parallel to each other.

To distinguish between these two concepts of PIP-Th, in fact not equivalent, we shall call the later one the PIP-Th *of constructive type*. Now the explanations given just before the statement of Theorem 62 in the classic furnish an idea of a proof of the following theorem, which may be considered as an alternative version of Th. 62 and will be called

Mechanization Theorem of Hilbert. In a planar Pascalian geometry there is a mechanical procudure which permits to prove or disprove in a finite number of steps any pure intersection-point theorem of constructive type under certain subsidiary non-degeneracy conditions also generated mechanically during the procedure.

As the intended proof of Theorem 62 or the above Mechanization Theorem as given in Hilbert's classic is not only vague but also somewhat inexact in various respects, we shall rewrite the proof in what follows which consists in giving a mechanical procedure meeting the purpose as described in the Mechanization Theorem.

First of all we shall take in the plane a coordinate system which may be chosen arbitrarily. To save the labor of computations we may choose, if we want, the coordinate system in a convenient manner which is however immaterial to the reasonings to follow. The points and lines involved in any pure intersection-point theorem will then be represented by number pairs or linear equations. To fix the ideas, we shall avoid the use of linear equations for lines and restrict ourselves to considerations of points or number pairs alone as described below. As a point may be either chosen in an arbitrary manner or constructed in a definite manner from prescribed geometric conditions, we shall distinguish two types of coordinates: parametric ones and geometrically-bounded ones, which will be denoted by u_i and x_j respectively. They will be denoted by a_k if no distinction for u or x is necessary. We shall now represent the points occuring successively in the PIP-Th. of constructive type to be proved by number pairs (a_i, a_j) one after the other as follows.

First let us remark that the construction of a point occuring in the theorem in question will be one of the following 10 types:

1. A point is arbitrarily given or chosen.

The point will then be represented as (u_i, u_j) with u_i, u_j as parameters.

2. A line is arbitrarily given or chosen.

Instead of representing the line by a linear equation of current coordinates, we shall take arbitrarily two points represented by (u_i, u_j), (u_k, u_l) respectively with the line in question as their joining line.

3. Construct an arbitrary line through a point (a_i, a_j) already constructed.

As in 2, we shall take an arbitrary point (u_k, u_l) and represent the line as the joining line of this point and the point (a_i, a_j).

4. Construct the joining line of two points already constructed.

As the line has been determined by the two points thereon, this construction is no more necessary.

5. Choose a point arbitrarily from a line already constructed.

If the line is determined by two points (a_i, a_j), (a_k, a_l) already constructed, then the arbitrary point chosen thereon may be either represented by (u_r, x_s) or (x_s, u_r) satisfying the following equation,

$$(a_j - a_i)u_r - (a_i - a_k)x_s + a_i a_l - a_j a_k = 0,$$

or

$$(a_j - a_l)x_s - (a_i - a_k)u_r + a_i a_l - a_j a_k = 0.$$

6. Construct arbitrarily a parallel to a line already constructed. As the line is determined by two points thereon already cons-

tructed say (a_i, a_j), (a_k, a_l), the parallel to be constructed will be determined as follows. First take an arbitrary point (u_m, u_n), and then a point (u_r, x_s) or (x_s, u_r) such that the joining line of this point to the point (u_m, u_n) will be parallel to the line determined by (a_i, a_j) and (a_k, a_l), so that x_s will satisfy the following equation:

$$(a_k - a_i)(x_s - u_n) - (a_l - a_j)(u_r - u_m) = 0,$$

or

$$(a_l - a_j)(x_s - u_m) - (a_k - a_i)(u_r - u_n) = 0.$$

7. Construct a line through a point (a_m, a_n) already constructed parallel to a line already constructed.

Let the line already constructed be determined by the points (a_i, a_j) and (a_k, a_l), then the line to be constructed will be taken to be one determined by (a_m, a_n) and a further point (u_r, x_s) or (x_s, u_r) satisfying the following equation:

$$(a_k - a_i)(x_s - a_n) - (a_l - a_j)(u_r - a_m) = 0,$$

or

$$(a_l - a_j)(x_s - x_m) - (a_k - a_i)(u_r - a_m) = 0,$$

8. Construct the intersection-point of two intersecting lines already constructed.

Let the two lines be determined respectively by pairs of points (a_i, a_j), (a_k, a_l) and (a_p, a_q), (a_r, a_s) already constucted. Then the intersection-point will be taken to be (x_g, x_h), satisfying the following system of equations:

$$(a_j - a_l)x_g - (a_i - a_k)x_h + a_i a_l - a_j a_k = 0,$$

and

$$(a_q - a_s)x_g - (a_p - a_r)x_h + a_p a_s - a_q a_r = 0.$$

9. Construct the intersection-point of a line already constructed and a line through a point (a_m, a_n) already constructed and parallel to a second line already constructed.

Let the two lines already constructed be determined respectively by (a_i, a_j), (a_k, a_l) and (a_p, a_q), (a_r, a_s). Represent the point to be constructed by (x_g, x_h), then x_g, x_h will satisfy the following system of equations:

$$(a_r - a_p)(x_h - a_n) - (a_s - a_q)(x_g - a_m) = 0,$$

$$(a_j - a_l)x_g - (a_i - a_k)x_h + a_i a_l - a_j a_k = 0.$$

10. Construct the intersection-point of two lines passing through each of two points already Constructed and parallel respectively to each of two lines already constructed.

This is similar to 8 and 9 and may be similarly treated.

Of course we may reduce the constructions 9 and 10 to the previous ones by introducing more coordinates u or x.

We are now in a position to describe the mechanical procedure to be followed. As the intersection-point theorem to be proved is of constructive type, the points and lines as described in the theorem will occur one after the other in a definite order of succession. It follows that the coordinates of the points involved (with lines replaced by two points thereon), whether parametric or geometrically-bounded, can be arranged in a definite order in accordance with their ordering of appearance in the construction as follows:

$$u_1 \prec u_2 \prec \cdots\cdots \prec u_m,$$

$$x_1 \prec x_2 \prec \cdots\cdots \prec x_n.$$

In particular, whenever a new point is introduced by the construction 8,9 or 10, its coordinates attributed will be two x's in succession, say (x_g, x_{g+1}).

Let us introduce now some sets as follows.

A set of parametric coordinates

$$U = \{u_1, \cdots, u_M\}.$$

A set of geometrically-bounded coordinates

$$X = \{x_1, \cdots, x_N).$$

A set of *degeneracy-polynomials*

$$\Delta = \{D_1, \cdots, D_R\}.$$

with each D a polynomial $\neq 0$ in $u_1, \cdots_M$ alone.

A set of *solutions*

$$S = \left\{\frac{Q_1}{P_1}, \cdots, \frac{Q_N}{P_N}\right\},$$

with all P_i's, Q_i's polynomials in $u_1, \cdots, u_M$ alone and each P_i a non-zero power product of the D's in Δ.

The meaning of the sets U and X are already clear. The meaning of the sets Δ and S is this:

Under the non-degeneracy conditions

$$D_1 \neq 0, \cdots, D_R \neq 0 \qquad (D_i \text{ in } \Delta)$$

the coordinates x_i are given by

$$x_1 = \frac{Q_1}{P_1}, \cdots, \quad x_N = \frac{Q_N}{P_N}.$$

with x_i in X, Q_i/P_i in S.

We begin now by setting all the sets X, Δ, and S to be empty ones, U to te $\{u_1, \cdots, u_m\}$, and proceed by enlarging these sets in following in steps the successive constructions as described in the theorem to be proved in question.

Suppose that we have proceeded to a certain step of construction but not yet finished. We distinguish two cases according as the next step of construction is one of the types 1-7 or of the types 8-10. In the first case we get a single equation

$$A_{N+} x_{N+1} + B_{N+} = 0$$

with a new bounded coordinate x_{N+1} and with A_{N+1}, B_{N+1} both polynomials in $u_1, \cdots, u_M$ and $x_1, \cdots, x_N$ already occured in the sets U and X. Replace now x_i's in A_{N+1} and B_{N+1} by Q_i/P_i given already in S, with $P_i \neq 0$ due to the already introduced non-degeneracy-conditions $D_i \neq 0$ with D_i in Δ already determined. Let the new fractional expressions thus obtained in $u_1, \cdots, u_N$ alone be denoted by A^*_{N+1} and B^*_{N+1} respectively so that the equation in x_{N+1} becomes

$$A^*_{N+1} x_{N+1} + B^*_{N+1} = 0.$$

Several possibilities may then occur.

If $A^*_{N+1} \neq 0$ then we write the numerator of A^*_{N+1} as D_{R+1} and put D_{R+1} into the set Δ to turn it into a new Δ, and introduce as a new non-degeneracy condition $D_{R+1} \neq 0$. Next we solve for x_{N+1} in the form

$$x_{N+1} = -B^*_{N+1}/A^*_{N+1} = Q_{N+1}/P_{N+1},$$

and put x_{N+1} into X and Q_{N+}/P_{N+1} into S to turn X and S into new set X and new solution set S. We note that P_{N+1} is again a power product of the polynomials D_j's in the new set Δ.

We proceed then to the next step of construction if there remains any.

If instead $A^*_{N+1} \equiv 0$, but $B^*_{N+1} \neq 0$, then we set again D_{R+1} = numerator of B^*_{N+1} and put it into Δ. The hypothesis of the theorem in question is now itself contradictory at least under the subsidiary conditions

$$D_j \neq 0$$

for D_j in the new Δ. The whole procedure will then be stopped.

If finally both A^*_{N+1} and $B^*_{N+1} \equiv 0$, then x_{N+1} undergoes in reality no restrictions and we may introduce a new parameter u_{M+1} and set

$$x_{N+1} = u_{M+1} = Q_{N+1}/P_{N+1}$$

(with $Q_{N+1} = u_{M+1}$, $P_{N+1} = 1$, the same in what follows).
We put then u_{M+1} into U, x_{N+1} into X, and Q_{N+1}/P_{N+1} into S with Δ unchanged. With this new system of sets U, X, Δ and S we then proceed to the next step of construction if there remains any.

This ends the treatment in the first case.

Suppose now we are in the second case of a construction of the types 8, 9, or 10. For any such type we will get a system of two equations say

$$A_{11} x_{N+1} + A_{12} x_{N+2} + B_1 = 0,$$

$$A_{21}x_{N+1}+A_{22}x_{N+2}+B_2=0,$$

with two new bounded coordinates x_{N+1}, x_{N+2} introduced atthe same time. The polynomials $A's$ and $B's$ are all in the variables $u_1,\cdots,u_n$ of U and $x_1\cdots,x_N$ in X. Replace now these $x_i's$ by Q_i/P_i in S for $i=1,\cdots,N$ we get for the $A's$ and $B's$ some fractional expressions in the $u's$ of U alone to be denoted by $A^{*\prime}s$ and $B^{*\prime}s$. In this way the equations above become then the following ones:

$$A^*_{11}x_{N+1}+A^*_{12}x_{N+2}+B^*_1 \;=0,$$
$$A^*_{21}x_{N+1}+A^*_{22}x_{N+2}+B^*_2 \;=0,$$

Various possibilities may now occur.

First suppose that the determinant of the coefficients of these equations in x_{N+1}, x_{N+2} is not identically zero:

$$E=\begin{vmatrix} A^*_{11} & A^*_{12} \\ A^*_{21} & A^*_{22} \end{vmatrix} \neq 0.$$

In this case let us express E in the form of a fraction of which the denominator is necessarily a power-product of $D_j's$ already present in Δ. We set now the numerator of E to be a new D_{R+1} and put it into Δ to enlarge it into a new one still denoted by Δ.Solve the two equations above we get then two expressions of the form

$$x_{N+1}=Q_{N+1}/P_{N+1}, \qquad x_{N+2}=Q_{N+2}/P_{N+2},$$

in which both P_{N+1} and P_{N+2} are power-products of $D_j's$ in the new Δ We put now x_{N+1}, x_{N+2} into X, and $\frac{Q_{N+1}}{P_{N+1}}, \frac{Q_{N+2}}{P_{N+2}}$ into S, denote the sets thus enlarged still by X and S, and then proceed to the next step of construction if there remains any.

Suppose now E is identically 0 but not all $A^{*\prime}s$ are so. In this case there will exist polynomialsa_1, a_2 not both 0 such that

$$a_1A^*_{11}+a_2A^*_{21}\equiv 0,$$
$$a_1A^*_{12}+a_2A^*_{22}\equiv 0.$$

From the two equations above we get

$$a_1B^*_1 \;+\; a_2B^*_2 \;=0.$$

If $a_1B^*_1 \;+\; a_2B^*_2 \;\neq 0$, then we set

$$D_{R+1}=\text{numerator of } a_1B^*_1 \;+\; a_2B^*_2$$

and put it into Δ. In this case the hypothesis of the theorem in question is contradictory in itself under the non-degeneracy conditions

$$D_j\neq 0,$$

with D_j in the new Δ. We stop then and proceed no more.

In case $a_1B^*_1 \;+\; a_2B^*_2$ is$\equiv 0$, the two equations will reduce to a

single one, say the first one to fix the ideas. Not both A^*_{11} and A^*_{12} can be $\equiv 0$, say $A^*_{12} \neq 0$ to fix the ideas.
Set then

$$D_{R+1} = \text{Numerator of } A^*_{12}$$

and put it into Δ. Introduce also a new u_{M+1} to be put in U, set

$$x_{N+1} = u_{M+1} = Q_{N+1}/P_{N+1}$$

and solve for x_{N+2} in the form

$$x_{N+2} = Q_{N+2}/P_{N+2}$$

with P_{N+2} a power-product of D_j's in the new Δ. We put now x_{N+1}, x_{N+2} into X and Q_{N+1}/P_{N+1}, Q_{N+2}/P_{N+2} into S. All the sets Δ, U, X, S thus enlarged will again be denoted by the same letters and then proceed to the next step of construction if there remains any.

Finally let us suppose that all the A^*'s are identically 0. If at least one of B^*, B^*_2 is not identically 0 then we set the numerators of these non-zero B^*'s as new D_{R+1} or D_{R+1} and D_{R+2}, put this one or both into Δ with the new Δ still denoted by the same letter. We stop then the whole procedure with the conclusion that the hypothesis of the theorem is in contradiction under the non-degeneracy conditions $D_j \neq 0$ with D_j in the new Δ.

If finally both B^*_1 and B^*_2 are identically zero, then we introduce two new u_{M+1}, u_{M+2} to be put into U, put x_{N+1}, x_{N+2} into X, and set

$$x_{N+1} = u_{M+1} = Q_{N+1}/P_{N+1},$$
$$x_{N+2} = u_{M+2} = Q_{N+2}/P_{N+2},$$

with Q_{N+1}/P_{N+1}, Q_{N+2}/P_{N+2} to be put into S. The new sets U, etc. will then again be denoted by the same letters. We proceed now to the next step of construction if there remains any.

This procedure will be stopped with a final system of sets,

$$U = \{u_1, \cdots, u_M\},$$
$$X = \{x_1, \cdots, x_N\},$$
$$\Delta = \{D_1, \cdots, D_R\},$$

and

$$S = \left\{\frac{Q_1}{P_1}, \cdots, \frac{Q_N}{P_N}\right\}.$$

Now two cases may occur.

Case 1 The hypothesis of theorem in question is contradictory under the subsidiary non-degeneracy conditions

$$D_j \neq 0, \quad (D_j \text{ in } \Delta).$$

In this case we have already achieved our aim of proof.

Case 2 In this contrary case we have to consider the conclusion of the theorem in question which amounts to say that

$$G = 0,$$

where G is a certain polynomial in the variables $u_1, \cdots, u_m$ and $x_1, \cdots, x_n$. Now N is necessarily $= n$ and $M \geq m$ in the present case and we may thus put

$$x_i = Q_i/P_i, \quad i=1,\cdots,n,$$

and substitute these expressions into G. The polynomial G becomes then a fraction with denominator a power-product in the D_j's occuring in Δ. Under the non-degeneracy conditions

$$D_j \neq 0 \quad (D_j \text{ in } \Delta),$$

we verify by usual computations whether G is identically zero or not and arrive at the two possible conclusions below.

1^0 G becomes identically zero so that the theorem is true under the above non- degeneracy conditions. The whole procedure constitutes then an actual proof of this theorem with restrictions clearly exhibited by the above non-degeneracy conditions.

2^0 G is not identically zero so that the theorem is not true at least under the above non-degeneracy conditions.

This terminates the whole mechanical procedure with precise conclusions under precise subsidiary conditions also mechanically generated.

The mechanization theorem of Hilbert is thus completely proved.

As illustrations of the above mechanical proving procedure for intersection-point theorems let us consider some examples shown below.

Ex.5 Consider the Desargues theorem of Ex. 3 stated in the following form: Let ABC and $A'B'C'$ be two triangles with three pairs of corresponding sides mutually parallel to each other. Let the line $l_1 = AA'$ and $l_2 = BB'$ meet at a point 0. Then CC' should also pass through 0.

Let us turn the statement of the theorem into a constructive form as follows. To simplify the calculations we shall take l_1, l_2 as the two coordinate axes. We shall take first two arbitrary points A, A' on l_1 with coordinates

$$A = (u_1, 0), \quad A' = (u_2, 0)$$

Then take an arbitrary point B on l_2 and an arbitrary point C on the plane with coordinates

$$B = (0, u_3), \quad C = (u_4, u_5).$$

Draw now through A' a parallel of AB meeting l_2 in B', and then through A' and B' draw parallels to AC and BC respectively to meet together at C'. The coordinates of the new points will be taken to be

$$B' = (0, x_1), \quad C' = (x_2, x_3).$$

Then the theorem asserts that the points O, C, C' are collinear.

Now the hypothesis of the theorem consists of the following equations

$$A'B' \mathbin{/\!/} AB \Longleftrightarrow u_1x_1 - u_2x_3 = 0,$$
$$A'C' \mathbin{/\!/} AC \Longleftrightarrow (u_4 - u_1)x_3 - u_5(x_2 - u_2) = 0,$$
$$B'C' \mathbin{/\!/} BC \Longleftrightarrow u_4(x_3 - x_1) - (u_5 - u_3)x_2 = 0.$$

The conclusion of the theorem is given by:

$$O, C, C' \text{ are collinear} \Longleftrightarrow G = 0,$$

with

$$G = u_4x_3 - u_5x_2.$$

Following the mechanical procedure indicated above we get

$$U = \{u_1, u_2, u_3, u_4, u_5\},$$
$$X = \{x_1, x_2, x_3\},$$
$$\Delta = \{u_1, u_1u_3 - u_1u_5 - u_3u_4\},$$
$$S = \left\{\frac{u_2u_3}{u_1}, \frac{u_2u_4}{u_1}, \frac{u_2u_5}{u_1}\right\}.$$

It follows that the non-degeneracy conditions to be observed are

$$u_1 \neq 0,$$

and

$$u_1u_3 - u_1u_5 - u_3u_4 \neq 0,$$

Under these conditions we verify that

$$G = u_4x_3 - u_5x_2 = u_4 \cdot \frac{u_2u_5}{u_1} - u_5 \cdot \frac{u_2u_4}{u_1} = 0,$$

so that the Desargues theorem in question is true in this non-degenerate case.

The first degeneracy case $u_1 = 0$ means that A coincides with O. It is of no geometrical interest and may be left out of consideration at all. If we want we may take $A = O$ as one of the hypothesis of the theorem and proceed in the same mechanical manner as before with the result that the theorem will then be trivially true under however some further non-degeneracy conditions.

The second degeneracy condition

$$u_1u_3 - u_1u_5 - u_3u_4 = 0$$

means that the points A, B, C are collinear. Take this as one of the hypothesis we verify as before that the theorem will then be no more true at least under certain further non-degeneracy conditions. This accounts for the fallacies of the Desargues theorem as indicated in Ex. 3. The same may be said about the other examples 1, 2 and 4. What is important for us is that the degeneracy conditions which may cause the fallacies of theorems present *automatically* during a mechanical procedure and may be treated alternatively and systematically also in a mechanical way which is actually impossible for the usual euclidean fashion proofs.

Ex. 6 To give a less trivial example let us consider the general Pascal theorem in projective geometry. Let us call for short a hexagon to be *Pascalian* if the three pairs of opposite sides intersect in collinear points. Now the necessary and sufficient condition for a hexagon to be Pascalian is that the six vertices of the hexagon should lie on the same conic (or *co-conic* for short). It follows that if a hexagon $A_1A_2A_3A_4A_5A_6$ is Pascalian, then, the six vertices being co-conic, any hexagon $A_{i_1}A_{i_2}A_{i_3}A_{i_4}A_{i_5}A_{i_6}$ arising from reordering of the vertices, is also Pascalian. In this way we may announce theorems in the form of pure intersection-point theorems in a Pascalian geometry with the mention of the notion of conics completely avoided. The intersection-point theorems thus arrived may be divided into various types and, to fix the ideas, let us consider for example the following one.

If the hexagon $A_1A_2A_3A_4A_5A_6$ is Pascalian then so is $A_1A_4A_3$ $A_2A_5A_6$.

In more details, the theorem states that:

If the points of intersection

$$P=A_1A_2\wedge A_4A_5,\quad Q=A_2A_3\wedge A_5A_6,\quad R=A_3A_4\wedge A_6A_1$$

are collinear, then so are the points of intersection

$$P'=A_1A_4\wedge A_2A_5,\quad Q'=A_3A_4\wedge A_5A_6,\quad R'=A_2A_3\wedge A_6A_1$$

(Here $\wedge$ stands for intersection).

For the proof let us first turn the above statement into one of constructive type as follows.

Take first an arbitrary point A_6 and then two arbitrary lines l_1, l_2 through A_6. For the mere sake of simplifying the calculations we shall take A_6 as the origin 0 and the two lines as the coordinate axes.

Take on l_1 an arbitrary point $A_1=(u_1,0)$

Through A_1 construct an arbitrary line and take thereon an arbitrary point $A_2=(u_2,u_3)$.

Through A_2 construct an arbitrary line and take thereon an arbitrary point $A_3=(u_4,u_5)$.

Through A_3 construct an arbitrary line and take thereon an arbitrary point $A_4=(u_6,u_7)$.

Let the line A_3A_4 meet l_1 in the point $R=(x_1,0)$.

Let the line A_2A_3 meet l_2 in the point $Q=(0,x_2)$.

Let the line A_3A_4 meet l_2 in the point $Q' = (0, x_3)$.

Let the line A_2A_3 meet l_1 in the point $R' = (x_4, 0)$.

Join the line QR and let it meet A_1A_2 in the point $P = (x_5, x_6)$.

Join PA_4 and let it meet l_2 in the point $A_5 = (0, x_7)$.

Join A_1A_4 as well as A_2A_5 meeting in the point $P' = (x_8, x_9)$,

With these hypothesis the conclusion is now: P', Q', R' are collinear.

Now the hypothesis of the theorem in question reads as follows:

R lies on $A_3A_4 \Longleftrightarrow (u_5 - u_7)x_1 + u_4u_7 - u_5u_6 = 0$,

Q lies on $A_2A_3 \Longleftrightarrow (u_2 - u_4)x_2 - u_2u_5 + u_3u_4 = 0$,

Q' lies on $A_3A_4 \Longleftrightarrow (u_4 - u_6)x_3 - u_4u_7 + u_5u_6 = 0$,

R' lies on $A_2A_3 \Longleftrightarrow (u_3 - u_5)x_4 + u_2u_5 - u_3u_4 = 0$,

P lies on $A_1A_2 \Longleftrightarrow u_3x_5 + (u_1 - u_2)x_6 - u_1u_3 = 0$,

P lies on $QR \Longleftrightarrow x_2x_5 + x_1x_6 - x_1x_2 = 0$,

$A_5 = PA_4 \wedge l_2 \Longleftrightarrow (u_6 - x_5)x_7 + u_7x_5 - u_6x_6 = 0$,

P' lies on $A_1A_4 \Longleftrightarrow u_7x_8 + (u_1 - u_6)x_9 - u_1u_7 = 0$,

P' lies on $A_2A_5 \Longleftrightarrow (u_3 - x_7)x_8 - u_2x_9 + u_2x_7 = 0$.

The conclusion becomes:

P', Q, R' are collinear $\Longleftrightarrow G \equiv x_3x_8 + x_4x_9 - x_3x_4 = 0$.

Fowllowing the mechanical procedure given in the Mechanization Theorem of Hilbert we see that the theorem is true after a long and tedious but easy and mecahnical computations under certain subsidiary non-degeneracy conditions uninteresting to be explicitely given. The computations constitute then automatically a proof of the theorem in the *generic* or *non-degeneracy* case.

Ex.7 Let us consider the previous example again with however the theorem in question not turned into constructive type. Thus, let us take A_6A_1, A_6A_5 still as the coordinate axes but with coordinates of the various points as follows.

$$
\begin{aligned}
A_1 &= (u_1, 0), & A_2 &= (u_2, u_3), \\
A_3 &= (u_4, u_5), & A_5 &= (0, u_6), \\
A_4 &= (u_7, x_1), & Q &= (0, x_2), \\
R &= (x_3, 0), & R' &= (x_4, 0), \\
P &= (x_5, x_6), & Q' &= (0, x_7), \\
P' &= (x_8, x_9).
\end{aligned}
$$

The hypothesis of the theorem becomes then

A_4, A_5, P are collinear $\Longleftrightarrow (x_1 - u_6)x_5 - u_7x_6 + u_6u_7 = 0$,

P,Q,R are collinear $\Longleftrightarrow x_2x_5+x_3x_6-x_2x_3=0,$

A_3,A_4,R are collinear $\Longleftrightarrow (x_1-u_5)x_3-u_4x_1+u_5u_7=0,$

A_1,A_2,P are collinear $\Longleftrightarrow u_3x_5+(u_1-u_2)x_6-u_1u_3=0,$

A_2,A_3,Q are collinear $\Longleftrightarrow (u_2-u_4)x_2-u_2u_5+u_3u_4=0,$

etc.

From these we get an equation of the form

$$Ax_1^2+Bx_1+C=0,$$

with A,B,C all polynomials in the $u's$ alone and A not identically zero. It follows that under the non-degeneracy condition $A\neq 0$ the previous mechanical procedure does not work.

The last example shows clearly that we should distinguish between two types of pure intersection-point theorems, those of constructive type and those not, and that the idea of the mechanical procedure as indicated by Hilbert in his classic works only for theorems of constructive type. For pure intersection-point theorems which cannot be turned into constructive form or for theorems which involve non-linear equations as are usual for various kinds of geometries, we have to device other mechanical procedures to give mechanical proofs. Now such mechanical procedures do exist for a large class of theorems so far no order relations are involved and the procedures are even *feasible* in the sense that quite difficult theorems may be proved in this manner on a computer of moderate size in a reasonable period of time. The same can even be done for (local) differential geometry. We shall not enter into this more which has been sketched in some original papers of the present author. A book with details is now also in preparation.

The Out-In Complementary Principle

Wu Wenchun

Ancient Chinese geometry with its long history, rich content and many achievements forms a school of thought peculiar in style and systematically different from Euclidean geometry. Much of its history remains to be explored. However, the "out-in complementary principle" pervades it and is clearly defined in the following major classics handed down to date:

Zhou Bi Suan Jing (*The Arithmetical Classic of the Gnomon and the Circular Paths*), or *Zhou Bi* for short;

Jiu Zhang Suan Shu (*Nine Chapters on the Mathematical Art*), or *Jiu Zhang* for short;

Jiu Zhang Suan Shu Zhu (*Annotation on the Nine Chapters on the Mathematical Art*) by Liu Hui, or *Liu Zhu* for short;

Hai Dao Suan Jing (*Sea Island Mathematical Manual*), or *Hai Dao* for short;

Ri Gao Tu Shuo (*Theory with Diagrams of the Sun's Altitude*), or *Ri Gao Shuo* for short; and *Gou Gu Yuan Fang Tu Shuo* (*Theory with Diagrams of the Right Triangle Making Use of Circles or Squares*), or *Gou Gu Shuo* for short, both by Zhao Shuang.

As everywhere else, geometry in China arises from land mensuration and astronomical observation. These practices in ancient times gave rise to the calculation of planar areas and methods of surveying based on the properties of the right triangle. Later, solid figures were involved in earthwork, etc., leading to a theory of volumes. One of the characteristics of ancient Chinese geometry is its fairly high power of abstraction in formulating the seemingly most commonplace out-in complementary principle which arose from diverse experiences. It has, however, been applied successfully to solving problems of extreme diversity.

Simple Applications and the Theory of Proportion

The essence of the so-called out-in complementary principle is the assumption of the following obvious facts: 1) The area of a planar figure remains the same when the figure is rigidly shifted to another place on the plane. 2) If a planar figure is cut into several sections, the sum of the areas of the sections is equal to the area of the original figure. It follows that the areas of the various sections involved before and after the out-in procedures possess simple arithmetic relations. The principle also applies to solid figures in space.

It is easy to apply this principle to obtaining the ordinary formula that the area of any triangle is equal to half the product of one side and the associated altitude. From this the area of any polygon can be calculated.

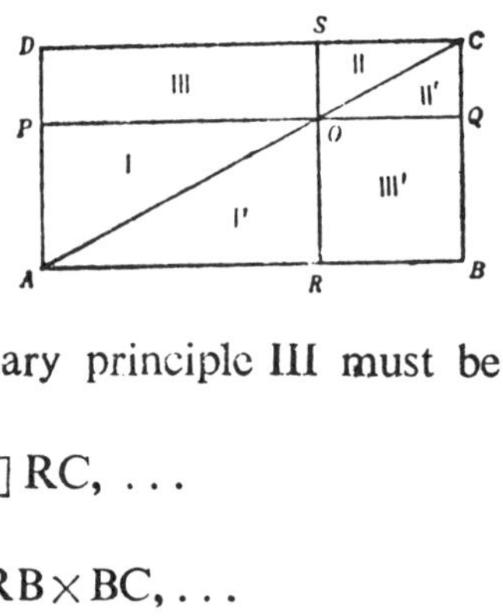

Another simple application is diagrammed as follows:

If $\triangle ACB$ is considered as $\triangle ACD$ shifted, and I' and II' as I and II shifted, then according to the out-in complementary principle III must be equal to III' in area, too.

$$\text{Likewise, } \square PC = \square RC, \ldots$$

From this we know

$OP \times OS = OR \times OQ$, $PQ \times QC = RB \times BC, \ldots$

Therefore $AR:OQ = OR:CQ$, $AB:OQ = BC:QC, \ldots$

That is, the corresponding sides of the similar right triangles ARO and OQC and also of ABC and OQC are in proportion. From this we know that certain other corresponding parts are also in proportion.

Though these simple results are not explicitly stated in *Jiu Zhang*, they are time and again manifested in the solution of various practical problems (Ref. *Liu Zhu*).

Gnomon, Shadow and Double Differences

The method of using two gnomons to find the altitude of the sun is given in *Zhou Bi*. The formula appears below:

$$\text{Altitude of the sun} = \frac{\text{height of the gnomon} \times \text{distance between the gnomons}}{\text{difference between the lengths of shadows of the two gnomons}} + \text{height of the gnomon}$$

As shown in the following diagram:

A is the position of the sun, BI represents the ground level, ED and GF are the two gnomons, while DH and FI are the two shadows projected on the ground.

In *Hai Dao* the same method is used to measure the height of an island from the shore. In the same diagram above, AB is the height of the island, H and I are the observer's positions where the observer's eye, the tops of the gnomons and the top of the island are in line. The formula then becomes:

$$\text{Height of the island} = \frac{\text{height of the gnomon} \times \text{distance between the two gnomons}}{\text{difference between the distances of observer from the gnomons}} + \text{height of the gnomon}$$

Liu Hui's original proof and diagram have been lost. But we have pieced these together drawing inspiration from other sources as well as extant fragments of diagrams in *Ri Gao Shuo* to be roughly as follows:

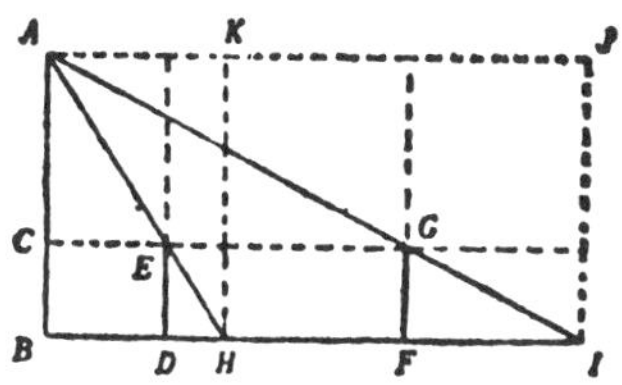

According to the out-in complementary principle, we know

$$\square\ JG = \square\ GB \quad (1)$$

$$\square\ KE = \square\ EB \quad (2)$$

(1)-(2) $\square\,JG - \square\,KE = \square\,GD$,

Therefore $(FI - DH) \times AC = ED \times DF$,

That is

difference between the distance of observer from the two gnomons $\times$ (height of island $-$ height of gnomon) $=$ height of gnomon $\times$ distance between the two gnomons

From this we arrive at the formula for the sea island.

In *Hai Dao* altogether nine practical problems are listed, all having to do with the measurement of heights and distances. In all the nine formulae given, differences occurring from two observations are usually taken to be the denominator. Probably this is where the term "double differences" comes in. The other eight formulae can all be proved likewise on the out-in complementary principle.

Some of the problems carried in *Si Yuan Yu Jian* (*Precious Mirror of the Four Elements*), written by Zhu Shijie of the Yuan Dynasty 1,100 years later than *Hai Dao*, are essentially the same as the nine posed in *Hai Dao*. Zhu must have drawn heavily upon his predecessors' work. Careful analysis of Zhu's method as shown in the *tianyuanshu* brings us to the conclusion that Liu's proof of the

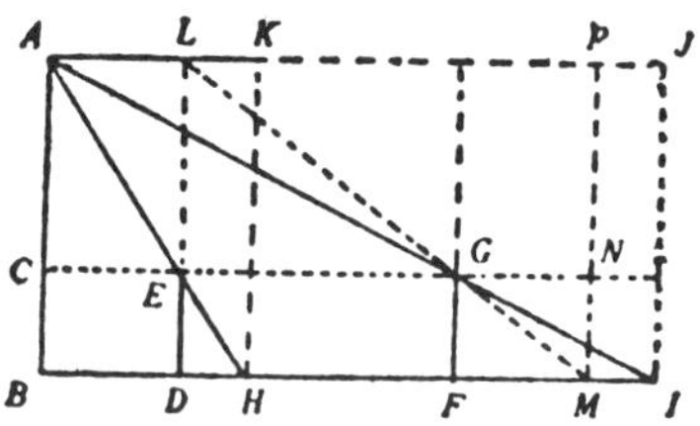

sea island formula is possibly somewhat more sophisticated than that given above. Accordingly, we suggest the following alternative proof to be considered as Liu's "original":

By the out-in complementary principle we have besides (1), (2) also

$\square\,PG = \square\,GD$ in the diagram above. (3)

From (1), (2) and (3) we get

□ JN = □ EB = □ KE,

Therefore IM = DH, (4)

FM = FI − IM = FI − DH = difference between the distances from observer to the two gnomons

From (3) we arrive at the formula for the sea island.

If done in the usual manner according to Euclidean geometry, an auxiliary line GM' should naturally be drawn parallel to AH to make the proving plain, as shown in the diagram on the right. The rest can then be proved by making use of the similar triangles and the theory of proportion. In fact the proving of the formula

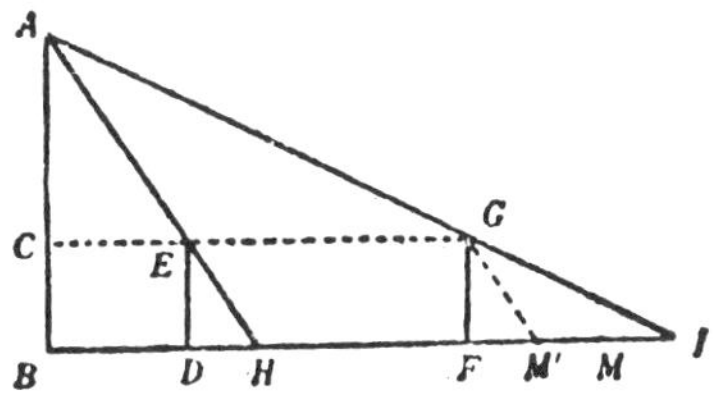

has been so traced by historians of mathematics in China and elsewhere in recent times, including Li Huang of the Qing Dynasty (1644-1911). But this is surely not the original method of Liu Hui; it is in fact totally out of accord with the spirit of ancient Chinese geometry. Note GM' parallel to AH makes FM' = DH. The constructed point M' here and the M point taken for equation (4) are quite different, each being typical of an independent school of geometry.

The Italian priest Matteo Ricci who came to China near the end of the Ming Dynasty (1368-1644) took the teaching of Euclidean geometry as one of his academic missions. In the book *Method and Theory of Surveying* dictated by him there appears a problem almost identical with the sea island problem. However, instead of proving it according to the Euclidean method he takes without reason a point M on FI to meet the requirement of (4) above, then goes on to prove the formula by proportions. This runs counter to Euclidean geometry but coincides with the Chinese tradition. Why Matteo Ricci should have done so is quite puzzling.

The *Gougu* Theorem

The Pythagorean theorem is called the *gougu* theorem in traditional Chinese geometry, and in both *Zhou Bi* and *Jiu Zhang* it is clearly prescribed in the written texts: Multiply the shorter and longer arms enclosing the right angle by their own values respectively and add up the squares; the sum is equal to the hypotenuse multiplied by its own value; i.e., $gou^2 + gu^2 = xuan^2$. Though the original proof has long been lost, we can still trace it from the texts of *Gou Gu Shuo*, *Liu Zhu*, and especially from the few diagrams left from Zhao Shuang. It is clearly stated that the proof is based on the out-in complementary principle; therefore it can be something like this:

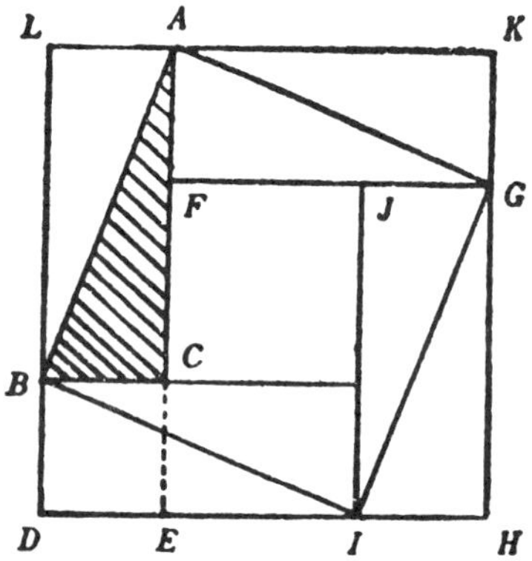

In the diagram on the left, ABC is the right triangle. BCDE is the square on the *gou* (the shorter arm), while EFGH is equal to the square on the *gu* (the longer arm). In the planar shape DBCFGH, cut off the triangle △BDI and shift it to the position of △ABC; cut off △ GHI and move it to the position of △ AFG. We then have ABIG equal to the square of the hypotenuse AB, and hence the *gougu* theorem.

In Euclid's *Elements of Geometry* the Pythagorean theorem is proved as illustrated in the diagram below:

It is clear that before the Pythagorean theorem is tackled, a lot of preparatory work must be done. First, a few theorems with regard to identical triangles and triangular areas must be established. That is why the Pythagorean theorem does not appear in the first volume of *Elements of Geometry* until near the end of the book. Euclid's book gives practically no applications of the theorem, but in ancient China the *gougu* theorem was widely employed

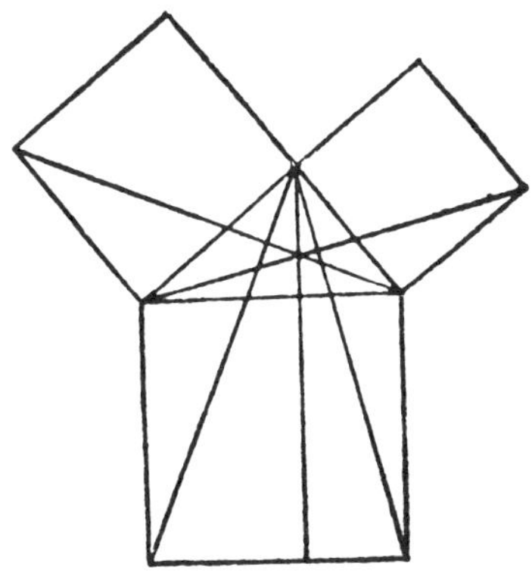

in diverse applications as early as in *Jiu Zhang*. It was a source of development over more than 2,000 years of Chinese mathematics (cf. the diagrams at the end of this article). The same theorem played quite a different role in the Eastern and Western systems of ancient geometry.

Gou, Gu, Xuan, Their Sums and Differences and Methods of Finding One from the Others

Gou, gu and *xuan*, the sum of and the difference between any two of the three, give out nine values. One can find the unknown from two knowns. Any one of the three sides can be found provided the other two are given. This is mainly a problem of extracting a square root. But the sum of or the difference between two sides is more often employed in solving practical problems such as those listed in the *gougu* chapter of *Jiu Zhang*:

1. Given the difference between *xuan* (the hypotenuse) and

gu (the longer arm), and *gou* (the shorter arm); find *xuan* and *gou*. Five problems are listed.

2. Given the difference between *gou* and *gu*, and *xuan*; find *gou* and *gu*. One problem.

3. Given the difference between *xuan* and *gou*, and *gu* respectively; find *gou*, *gu* and *xuan*. One problem.

4. Given the sum of *xuan* and *gu*, and *gou*; find *gu* and *xuan*. One problem.

Formulae are given for the problems in *Jiu Zhang*. The propositions in *Gou Gu Shuo* are of the same nature. In *Liu Zhu* proofs of the formulae are worked out, making use of the out-in complementary principle; sometimes also the theory of proportion. Take Problem No. 13 in the *gougu* chapter, the problem of the "broken bamboo", for example:

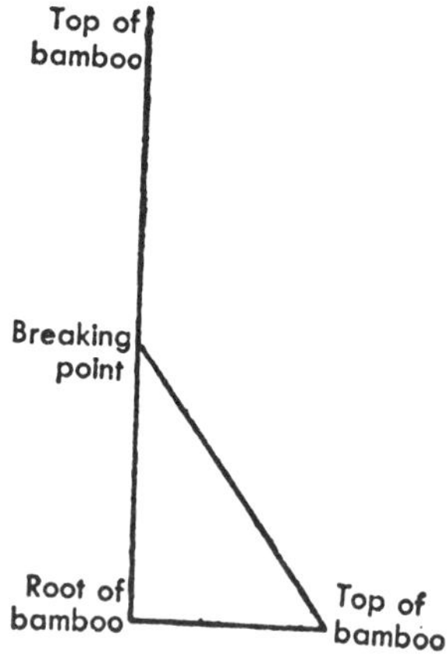

The height of the bamboo (*gu* plus *xuan*) is known. When bent the top touches the ground at a known distance from the stem (*gou*). Find the height of the break (*gu*).

The formula is given as follows:

$$xuan - gu = \frac{gou^2}{xuan + gu};$$

$$xuan, gu = \frac{(\text{sum of } xuan \text{ and } gu \pm \text{ difference between } xuan \text{ and } gu)}{2}$$

Liu Zhu provides another formula: $gu = \dfrac{(\text{sum of } xuan \text{ and } gu)^2 - gou^2}{2 \times \text{ sum of } xuan \text{ and } gu.}$

To prove the former formula, see in the diagram below:

The side of the squares ABCD or AEFG is equal respectively to the *xuan* or *gu* of the right triangle. According to the *gou gu* theorem the area of EBCDGF is equal to gou^2. Shift □ FD to the position of □ CH, then according to the out-in complementary principle, the area of □ BH is equal to gou^2, while the longer and shorter sides of this rectangle are equal to the sum of *xuan* and *gu* and the difference between them respectively. From this we get the former formula.

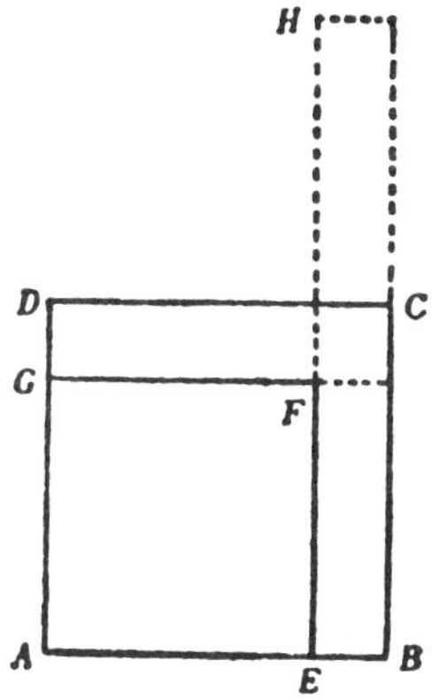

Liu Hui's proof for the other formula is done likewise. In the diagram below:

The area of the reversed L-shaped figure in the lower right corner is equal to gou^2 by the *gougu* theorem. The area bordered by the bold lines is thus equal to $(xuan + gu)^2 - gou^2$. Shift I to the position

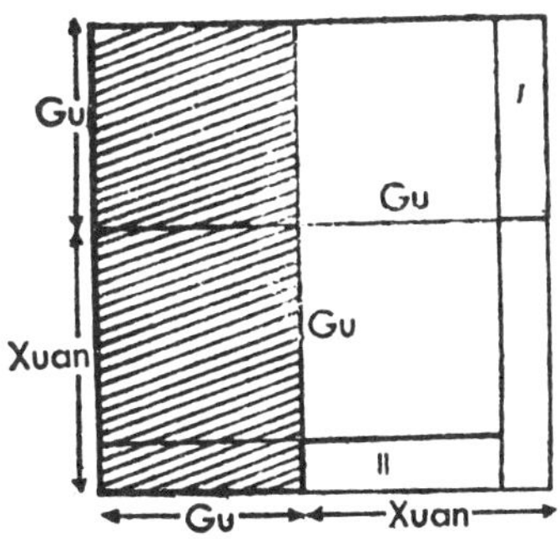

of II and we see according to the out-in complementary principle that this area is two times the shaded area; i.e., $2 \times gu \times (xuan + gu)$. The formula is therefore proved.

Qin Jiushao's Formula[1]

In Qin Jiushao's *Shu Shu Jiu Zhang*[2] (*Mathematical Treatise in Nine Sections*) 1247, there is a problem of finding the area of a scalene triangular plot. Given the three unequal sides of the triangle *da*, *zhong*, *xiao*, (the longest side, the medium side and the shortest side). Qin Jiushao's solution can be formulated as follows:

$$\text{Area}^2 = \frac{1}{4}\left[xiao^2 \cdot da^2 - \left(\frac{da^2 + xiao^2 - zhong^2}{2}\right)\right]^2.$$

Qin says nothing about the source of this formula. The proof of the formula has also been lost. Making use of the results and methods in *Liu Zhu*, we may infer the lost proof to be somewhat as follows:

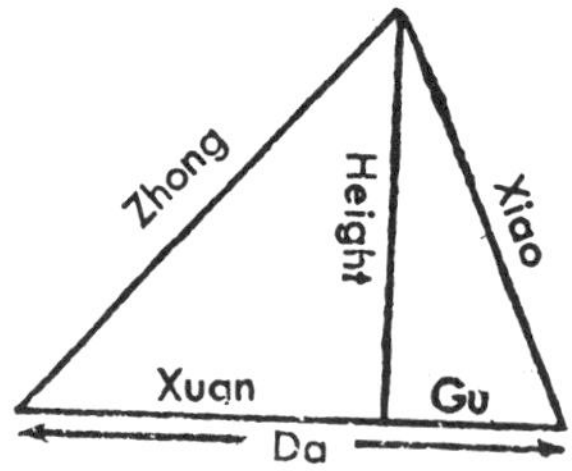

Draw an altitude of the triangle perpendicular to *da*, dividing *da* into two parts. Let the longer and the shorter parts be the *xuan* and *gu* of a right triangle. From *Jiu Zhang* we know the area of a triangle to be $^1/_2 \times$ altitude $\times$ *da*, therefore our problem becomes one of finding the altitude, then further boils down to finding the *gu* of that right triangle. Since

$$xuan + gu = da,$$
$$gou^2 = xuan^2 - gu^2 = zhong^2 - xiao^2,$$

our problem is the same as that of finding *gu*, given *gou* and the sum

---

[1] Qin Jiushao was one of the greatest Chinese mathematicians of the 13th century.

[2] A very important mathematical classic written by Qin, known especially for its treatments of numerical equations of higher degree and indeterminate analysis.

of *xuan* and *gu*. From Liu Hui's formula we have:

$$gu = \frac{(xuan + gu)^2 - gou^2}{2 \times (xuan + gu)} = \frac{da^2 - (zhong^2 - xiao^2)}{2 \times da}.$$

$$\text{Altitude}^2 = xiao^2 - gu^2 = xiao^2 - \frac{(da^2 + xiao^2 - zhong^2)^2}{2 \times da}.$$

From this we get Qin's formula.

Qin's formula looks rather odd. But the proof traced above is quite natural and perfectly in line with ancient Chinese mathematical tradition. We may even regard it as the original proof.

Heron's formula in Western geometry, however, is neat in form and good-looking:

Area of a triangle $= \frac{1}{4}\sqrt{(a+b+c)(b+c-a)(c+a-b)(a+b-c)}$,
where a, b, c are the three sides of the triangle.

Qin's formula is not likely to have been derived from Heron's, and we may conclude that it has its indigenous origin independent of Heron's influence.

Extracting the Square or Cubic Root

To find the hypotenuse from the two arms enclosing the right angle in a right triangle, we add up the two squares on the arms and extract the square of the sum. Thus the application of the *gougu* theorem inevitably leads to the extraction of the square root. In fact, in the ancient mathematical classic *Zhou Bi* the square roots of many concrete numbers are provided. Detailed steps in extracting square roots are stated in *Jiu Zhang*. The method is geometric, based on the out-in complementary principle. Suppose the task is to find the square root of the number 55,225. In geometry this is to find the side of a square the area of which is 55,225. Note the decimal system has long been in use in China. First we must decide on how many digits the root is going to have. The square root of a five-digit number has three digits. So our task is to ascertain the first, second and third digits successively. Since our number 55,225 lies between 40,000 and 90,000, its square root must lie between 200 and 300. Our first digit is therefore a 2. (In *Jiu Zhang* this process of ascertainment is called *yi*,[1] or "to suggest".) In the diagram let ABCD be the square the area of which is 55,225. On one side AB we take a point E and let AE be equal to 200. Draw the square AEFG. Cut off AEFG from ABCD. The area of the

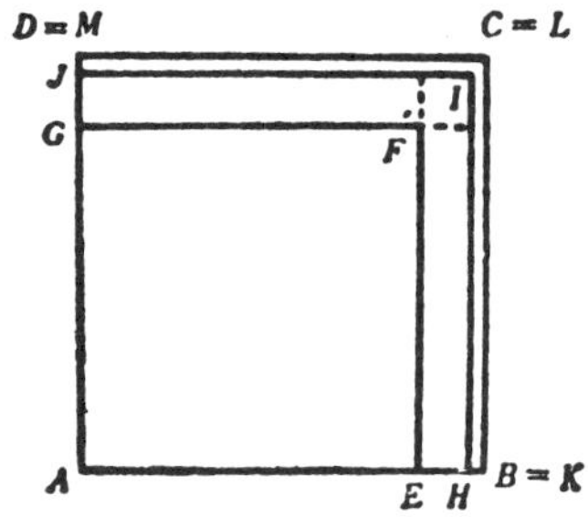

remaining inverted L-shaped figure is therefore $55{,}225-200^2=15{,}225$. We then suggest that the second digit be a 3. On EB we again take a point H making EH equal to 30. Draw the square AHIJ. Cut the inverted L-shaped figure into three parts: □ FH, □ FJ, □ FI. Their areas are respectively $30\times$ EF, $30\times$FG, 30^2. But EF = FG = 200, so the area of the remaining inverted L-shaped figure is equal to

$$15{,}225-(2\times 30\times 200+30^2)=2{,}325.$$

Let us then suggest that the third digit be a 5, and on HB we take a point K making HK equal to 5. Draw the square AKLM. The area of the remaining inverted L-shaped figure, if any, must be

$$2{,}325-(2\times 5\times 230+5^2)=0$$

In that case K and B must coincide, and the square root of 55,225 is 235.

The same method is used in extracting the cubic root. It will of course be more complicated to dissect a cube but the principle is still geometric and still that of out-in complementation. The method is described in detail in *Jiu Zhang*.

These methods of extracting the square and cubic roots date back to very ancient times in China. They are clearly geometric

1 Some say the character means "to discuss".

and display a superiority in the decimal place-value system of numeration employed.

By the middle of the 11th century Chinese mathematicians had already improved the methods of extracting the square and cubic roots to the solution of equations of higher degree. This is called *zeng zheng kai fang fa* (the method of extracting equational roots by successive additions and multiplications). A diagram illustrating the different coefficients of the various terms in the expansion formulae of binomial powers of high degrees had also appeared and was called *kai fang zuo fa ben yuan tu* (diagram illustrating the origin and method of extracting equational roots). The geometric nature and the high degree notion involved in *zeng zheng kai fang fa* show that Chinese mathematicians in ancient times might already have had primitive ideas about hypercubes and hypergeometry.

Quadratic Equations

In extracting the square root, we make use of the diagram on p.74. $2 \times EF$ in the diagram is called the *dingfa*. Having obtained AE, we come to find EB from the known area of the inverted L-shaped figure EBCDGF. Shift □ DF to the position of □ CH, the area of □ BH is the same as that of the inverted L-shaped figure according to the out-in complementary principle. Note that the difference between the longer and shorter sides of □ BH is equal to $2 \times EF$ (*dingfa*), which is also known. The problem of finding EB is therefore a problem of

(A) finding the longer and shorter sides of a rectangle, given its area and the difference between the two sides.

Conversely, the solution of problem (A) can be reduced to one as from the second step onwards in the method of square-root extraction, which in *Jiu Zhang* is called *kai dai cong ping fang fa*. The solution of (A) in *Jiu Zhang* is stated in the following words:

(B) "Take [the area of the rectangle] as *shi* and [the difference between the length and width] as *congfa*, then *kai fang chu zhi* (literally "to extract the square root" which means here *kai dai cong ping fang*) and the root is the [width]."

The term *congfa* comes from *dingfa* in extracting the square root. The term *kaifang* (root extraction) shows its origin.

The following problem is taken from *Jiu Zhang*. In the diagram on the right, ABCD is a square walled city. At point G there is a big tree of known distance in terms of human steps northward from the north gate (north steps for short). A man takes a definite number of steps southward out of the south gate (south steps for short), then turns west and also counts his steps till he is just able to see the tree (west steps for short). Find the length of each side of the square city. The answer given in *Liu Zhu* is obtained on the out-in complementary principle as follows: □ EJ = 2 □ EG = 2 □ KG = 2 × north steps × west steps. In □ EJ the difference between the length and the width is equal to the sum of north steps and south steps. The problem is thus reduced to one in the form of (A) above. According to *Jiu Zhang* its solution is as follows: Take 2 × north steps × west steps as *shi*, and the sum of north steps and south steps as *congfa*, *kai ping fang chu zhi* and we find the length of one side of the city as represented by EI in the diagram.

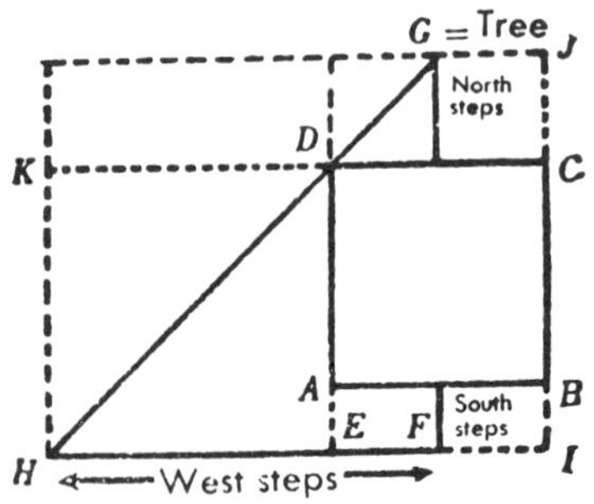

Not only the numerical value of problem (A) can be found by means of the *kai dai cong ping fang fa* method, but also a precise expression of the solution of (A) may be obtained on the out-in complementary principle. In fact, if in the rectangle we take the width as the *gou* and the length as the *gu* of a right triangle, then problem (A) becomes the following:

(C) Given the product of *gou* and *gu*, and the difference between them in a right triangle, find *gou* and *gu*.

Let us examine a diagram left by Zhao Shuang in which there are two squares the sides of which are equal to the sum of, and the

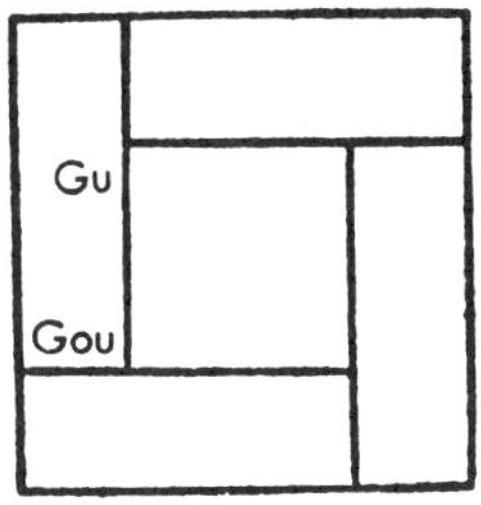

difference between, *gou* and *gu* of the right triangle respectively. We therefore have

$$(gou+gu)^2 = 4\,(gou \times gu)-(gu-gou)^2.$$

From this we get the sum of *gou* and *gu*, and *gou* and *gu* consequently. Similarly, *gou* and *gu* can be found given their sum and their product. Reference can be made to the last proposition in *Gou Gu Shuo*.

In the Song and Yuan dynasties (10th to 14th century) the notion of the unknown was explicitly and clearly introduced into traditional Chinese mathematics. If x (called *tianyuanyi*[1] then; while the *tianyuan* notation is one used by the Song algebraists for the expression of numerical equations of high degree. It is a way of arraying counting rods on counting boards. The array is of a "matrix" character. Different terms are used for distinguishing figures on different "storeys", with the constant term on the lowest, and the coefficient of the highest degree term on the highest storey above;) stands for the width of the rectangle, our problem (A) is equivalent to solving a quadratic equation of the form

$$x^2+bx = c, \text{ with } b \text{ as } \textit{congfa} \text{ and } c \text{ as } \textit{shi}.$$

Ancient Chinese mathematicians furnished both numerical and accurate solutions to quadratic equations of the above type (with b

1 *Tianyuanyi* has different meanings in the works of Song and Yuan dynasty mathematicians.

and c positive). During the Song and Yuan dynasties the *kaifangshu* (method of root extraction) was extended to solving numerical equations of high degree. As for the method of accurate solution of equations of higher degree, historical traces have long been lost. Judging from what Wang Xiaotong wrote in the early years of the Tang Dynasty (618-907) and from historical comments on Zu Chongzhi (429-500), we cannot totally rule out the possibility that geometrical approaches have been attempted with some success in accurate solution of cubic equations.

In other countries, the Arab mathematician Al-Khowārizmi in his well-known classic on algebra (A.D. 829) gives accurate solutions for quadratic equations of various types. His method was geometrical in spirit, similar to ours on the out-in complementary principle. Later, Italian mathematicians in the 16th century worked out solutions for cubic equations. Their methods were also geometrical.

Theory of Volumes and Liu Hui's Principle

Since the area of a rectangle is the product of its length and width, it is easy to infer on the out-in complementary principle that

(1) the area of a triangle = $^1/_2$ × its height × its base.

It is also easy to derive further the formulae for areas of polygons. All these fall within the category of plane geometry.

In solid geometry, however, although we know that the volume of a rectangular parallelepiped must be equal to its length × its width × its height, it is by no means definite whether we can on the out-in complementary principle reason that

(2) the volume of a tetrahedron = $^1/_3$ × its altitude × the area of its base surface, and hence form a theory for volumes of polyhedra. In fact this constitutes a most difficult problem in geometry which was presented as one of the 23 unsolved problems at the International Congress of Mathematicians in 1900 by the celebrated David Hilbert. This problem has been solved by Max Dehn who proved that besides being of equal volumes certain conditions must further be satis-

fied before two polyhedra can be cut into a number of mutually congruent smaller ones. These conditions have since been called Dehn's conditions. In 1965 the Swiss mathematician Sydler proved that Dehn's conditions are also sufficient. Even so, it appears that the problem may still be regarded as not yet satisfactorily settled. Dehn's conditions are too complicated to be accepted as final.

A probe into how the problem was dealt with by ancient Chinese mathematicians would probably provide us with some food for thought.

In both *Jiu Zhang* and *Liu Zhu* the starting point from which problems of polyhedra volumes are solved is to cut some regular

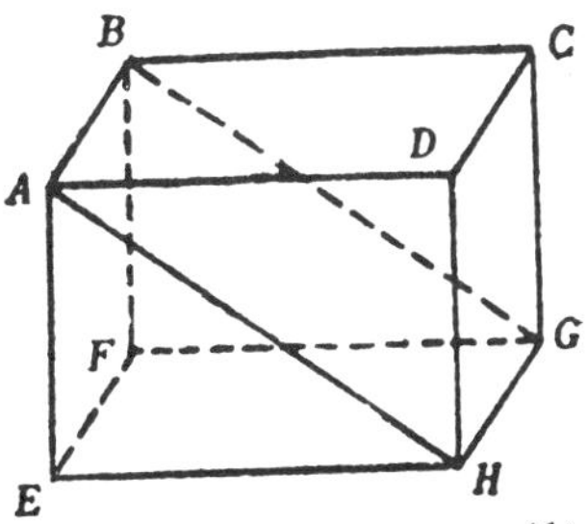

(1)

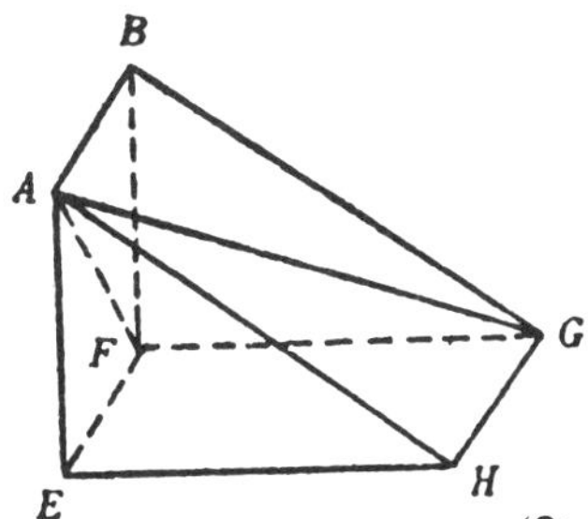

(2)

polyhedra into several basic solid figures which will be helpful in analysis. A rectangular parallelepiped can be cut diagonally (passing through two diagonally opposite edges) into two *qiandu* (right triangular prisms), as shown in diagrams (1) and (2). A *qiandu* in turn can be cut into a *yangma* (pyramid) and a *bienao* (tetrahedral wedge) as shown in (3) and (4). The basic features of a *bienao* are that it has AB perpendicular to the plane BFG, and FG perpendicular to the

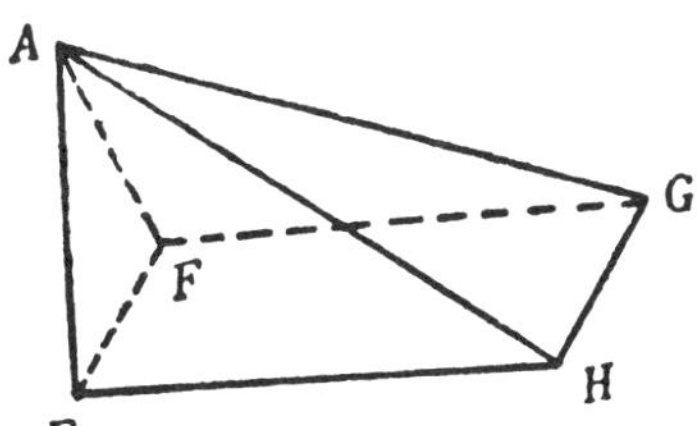

(3)

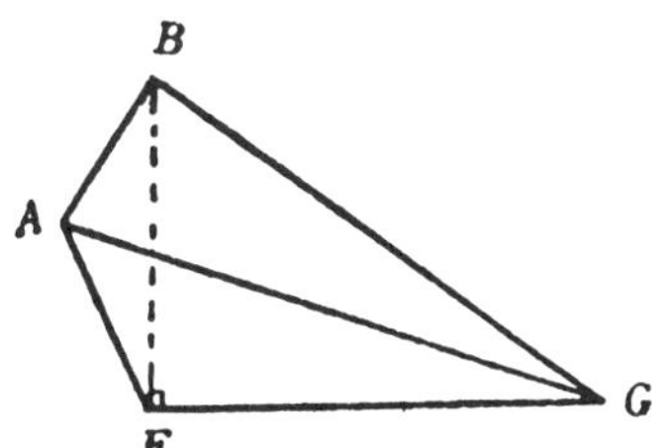

(4)

plane ABF as shown in the diagram. Since any polyhedron can be cut into tetrahedra and any tetrahedron can be cut into six *bienao* as shown in the diagram below, the whole problem boils down to finding the volumes of the *bienao* (and the *yangma*) so produced.

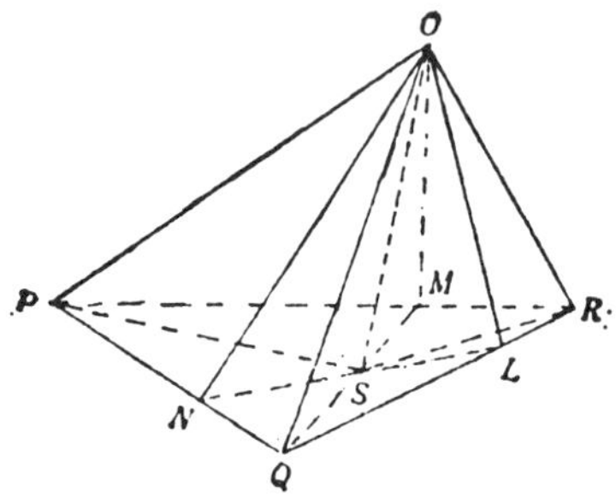

In Liu Hui's own words *yangma* and *bienao* are the "basic figures for the whole theory and practice involving volumes of polyhedra".

We then come to the problem of finding the volumes of *yangma* and *bienao*. If our parallelepiped is simplified into a cube, it will be easy to see that the volume of the pyramid cut from the prism is twice that of the tetrahedral wedge. Liu Hui proved in a long dissertation that this is the case not only in the *qiandu* from a cube, but in all *qiandu* alike. In Liu Hui's words, "In a *qiandu* the volume of the *yangma* is always twice that of the *bienao*." We may well call this statement Liu Hui's principle. In modern language, "If any rectangular parallelepiped is cut diagonally into two prisms, and the prisms are further cut into pyramids and tetrahedra, the ratio between the volumes of the pyramid and tetrahedron so produced is always 2:1."

From this principle it will be easy to arrive at the formulae for volumes of *yangma* and *bienao*. It is then no problem to prove formula (2) above. The whole theory for volumes of polyhedra may then be based on the principles of Liu Hui and of out-in complementation.

Liu Hui's long and detailed dissertation is proof of his principle, proof based on some limit considerations. What has been made clear by Hilbert and his followers can be construed as that volumes are different from planar areas in that the mere out-in complementary principle is insufficient for a satisfactory theory. In fact, it must be supplemented by some axiom or principle of continuity. Though in 1903 Shatunovsky argued that the principle of continuity could be omitted and that the foundation of the theory of polyhedra volumes could be built on formula (2), it nevertheless requires a proof of the independence of the choice of altitude and base which is neither plain nor trivial at all. In comparison with the method of exhaustion of the ancient Greeks and the method employed in Legendre's *Elèments*, Liu Hui's treatment of polyhedra volumes based on his principle and the out-in complementary principle can be safely regarded as the most natural one surpassing all others in simplicity and elegance.

It seems that much yet remains to be proved in the field of the polyhedra. It might be an aid if the conceptions and methods in ancient Chinese geometrical approaches were duly taken into account.

The *Xianchu* Theorem

The term *xianchu* (a wedge with trapezoid base and both sides sloping, see the diagram below) as well as other strange terms for polyhedra have come down from ancient Chinese architecture and earthwork.

In *Jiu Zhang*, volumes of polyhedra are calculated on the out-in complementary principle and by the *yangma* and *bienao* formulae.

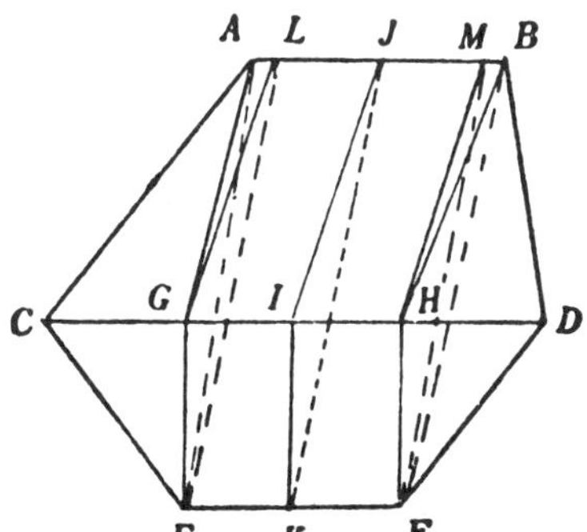

Take the *xianchu* in the diagram for instance. ABCD form a trapezoid on the ground surface. CDEF is another trapezoid in a plane perpendicular to the ground. ABEF is a slope. The whole solid ABCDEF in the form of a tunnel is *xianchu*. Plane IJK is perpendicular both to the ground and plane CDEF. It bisects *xianchu* into two symmetrical parts. EG, FH and KI show the depth of *xianchu*. IJ is the length of *xianchu* on the ground. CD, EF and AB are called the upper width, the lower width, and the hind width of *xianchu*. The formula for the volume of *xianchu* given in *Jiu Zhang* is as follows:

$$\text{Volume of } \textit{xianchu} = \frac{1}{6}\left(\begin{matrix}\text{upper}\\ \text{width}\end{matrix} + \begin{matrix}\text{lower}\\ \text{width}\end{matrix} + \begin{matrix}\text{hind}\\ \text{width}\end{matrix}\right) \times \text{depth} \times \text{length}.$$

To prove this, Liu Hui in his book *Liu Zhu* cuts *xianchu* into several parts, and supposes CD> AB> EF as in the diagram above. *Xianchu* is therefore regarded as composed of a *qiandu* EFGHLM, two small *bienao* AGEL and BFHM, and two big irregular *bienao* ACEG and BDFH. From formula (2) above and the formulae for *qiandu* and *bienao*, the formula for the volume of *xianchu* is therefore obtained. The same method is employed in *Jiu Zhang* in calculating the volumes of *chumeng* (wedge with rectangular base and both sides sloping), *chutong*, *panchi*, *minggu* (three variations of a frustum of pyramid with rectangular base of unequal sides), and other polyhedra.

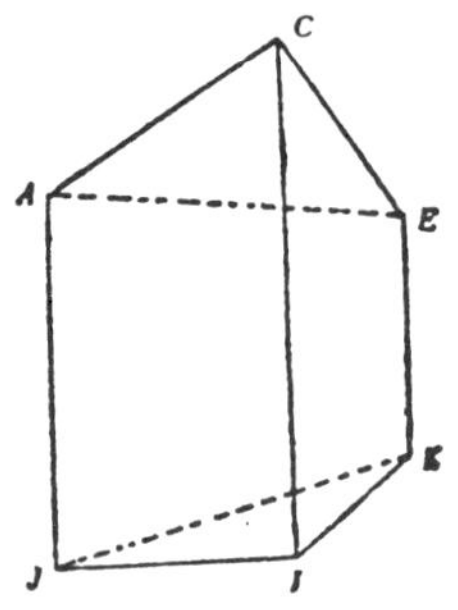

The formula of the *xianchu* volume is of special importance in that half of the *xianchu* standing erect on the right triangular base IJK will be equal to a right-angled prism cut slantwise at the upper end. Its volume will simply be the product of the average height and the right triangular (*gougu* form) base. Now a pillar bounded at the top by any curved surface can be regarded as composed of such slant-topped prisms approximately. Therefore the integral approximate formula of a function f(x, y) can be obtained analogous to Simpson's integral approximate formula in the case of an area under a curve. This shows the particular significance of the *xianchu* formula.

In Western mathematics, the earliest formula for the volume of a pillar cut slantwise at the top appeared in 1794 in Legendre's *Elèments de géométrie*, and has since been called Legendre's formula. Legendre's book is the earliest work to take the place of Euclid's *Elements*. Legendre's proof of his own formula is also based on the volume of the tetrahedron but with different method of dissection from that in *Liu Zhu*. Reference can be made to both for comparison.

Volume of the Sphere and the Principle of Zu Geng

Within the 300 years or so between the writing of *Jiu Zhang* and that of *Liu Zhu* a fairly complete theoretical system with regard to volumes of polyhedra had arisen. Yet ancient Chinese mathematicians at that time stopped short at bodies bounded by curved surfaces, especially spheres, the volume of which remained unsolved till Zu Geng of the 5th-6th centuries put forward a famous principle named after him. In Zu Geng's own words the principle is as follows:

"If the *mi* (cross-sections, areas) are the same on the same *shi* (level), the *ji* (whole volumes) cannot be different."

The same principle appeared in Europe in the 16th century by the name of Cavalieri's principle, which was an important step towards the invention of calculus.

Zu Geng's proof for his formula of spherical volumes is described in detail in an annotation by Li Chunfeng (in about 656) to *Jiu Zhang*. The arguments are very clear in three successive steps:

1. Within a cube draw two inscribed cylinders at cross direc-

tions. That part in common of the two cylinders is called *mou he fang gai* (literally "the common square cover"). Cut a small cube $^1/_8$ of the original cube. According to Zu Geng's principle, the following proportion is obtained:

$$^1/_8 \text{ volume of sphere} : {}^1/_8\, fanggai = \pi:4.$$

2. That part of *fanggai* within the $^1/_8$ cube is the inner *qi*, and those three parts within the small cube but left out of *fanggai* are the outer *qi*.

From the small cube cut an inverted *yangma*. Prove by the *gougu* theorem that if we cut the *yangma* horizontally at a certain level from the base, the cross-section of the *yangma* is equal to the total cross-sections of the outer *qi* cut at the same level in area.

3. Prove by Zu Geng's principle that the total volume of the outer *qi* is equal to that of the *yangma*.

From these the formula for the volume of the sphere is immediate.

The idea of *mou he fang gai* was first introduced by Liu Hui. The first step of Zu Geng had actually been worked out by Liu also. In fact, in *Liu Zhu* he had time and again made use of what was later called Zu Geng's principle to find the volumes of solid bodies bounded by curved surfaces, such as the volumes of the cylinder from the polygonal pillar, of the cone from the pyramid, of the frustum of cone from the frustum of pyramid, etc. Zu Geng's merits not only consist in actual solution of volumes of *mou he fang gai* and the sphere, but also in his summing up of practical experiences and objective facts in the form of a general principle. Whether the principle should be called the Liu-Zu principle to give Liu Hui his due is a matter that deserves discussion.

Other Applications

Jiu Zhang is so comprehensive that, leaving other topics aside, the out-in complementary principle is by no means applied merely to the various problems above. The problem of the inscribed circle in a right triangle in *Jiu Zhang* treated on this principle has since been further developed. It is fully treated in *Ce Yuan Hai Jing*

(*Sea Mirror of Circle Measurement*, 1248) by Li Ye. In the works of Qin Jiushao and Li Ye, the problem of the "square city" above has been replaced by a problem of "circular walled city" which was beyond the masters of older times. The invention of such methods as *tianyuanshu* in the Song and Yuan dynasties not only solves heretofore unsolvable problems but also largely simplifies old problems. Compared with the older methods, the new methods give results with far less effort. The essence of the new methods and new theories lies in the algebraization of geometry, which blazed the trail for both analytical geometry and modern algebra.

Conclusion

The out-in complementary principle together with the principles of Liu Hui and Zu Geng demonstrated the considerable abilities of ancient Chinese masters in scientific abstraction. Drawing intrinsic conclusions from objective facts, they summed up the conclusions into succinct principles. These principles, plain in reasoning and extensive in application, form a unique character of ancient Chinese mathematics. The emphasis has always been on the tackling of concrete problems and on simple, seemingly plausible principles and general methods. The same spirit permeates even such outstanding achievements as the algebraization of geometry and the place-value decimal system of numbers. Western mathematics, in contrast, lays emphasis on conceptions and the logical relationships between them.

The majority of the ancient Chinese mathematical classics have sunk into oblivion because of feudal obscurantism — a most deplorable loss in human society. Zu Geng's contributions would also have been lost had it not been for the rather casual entry by Li Chunfeng in his annotation to *Jiu Zhang*. However, judging from what is still available, the historical facts that ancient Chinese mathematics had its origin in human productive activities and had thrived in its own, independent way before the 15th century are still clear, as pithily shown in the following two diagrams:

Diagram I

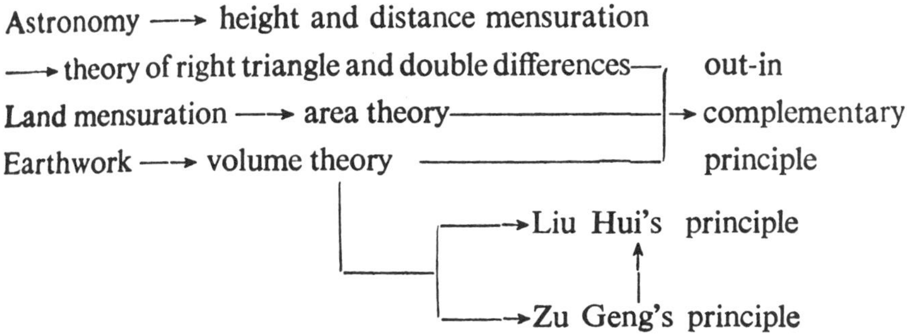

Diagram II

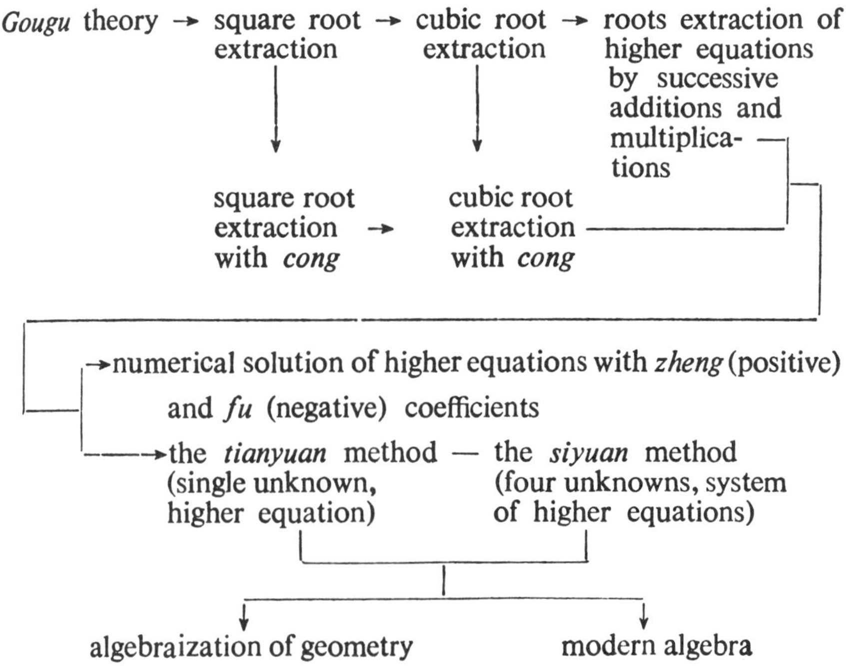

A CONSTRUCTIVE THEORY OF DIFFERENTIAL ALGEBRAIC GEOMETRY BASED ON WORKS OF J.F.RITT WITH PARTICULAR APPLICATIONS TO MECHANICAL THEOREM-PROVING OF DIFFERENTIAL GEOMETRIES

Wu Wen-tsun (吴文俊)

Institute of Systems-Science, Academia Sinica

The nowadays algebraic geometry is mainly of an EXISTENTIAL character. For example, it is proved that any algebraic variety is the union of a finite number of irrecucible ones but with no indications at all how such a decomposition can actually be carried out. There are even no indications how a given variety, defined by a set of equations, is known to be irreducible or not. The criteria for the irreducibility of a variety is non-CONSTRUCTIVE and cannot be applied to arrive at final result except rare cases by means of special devices. On the contrary, J. F. Ritt has already established a theory of algebraic varieties which is in the main CONSTRUCTIVE. He has even established such a theory for the more general case of a variety defined over a field possessing a further operation of differentiation. We shall call such a variety a DIFFERENTIAL-ALGEBRAIC VARIETY and the geometry thus founded the DIFFERENTIAL-ALGEBRAIC GEOMETRY. The present note is the simplified version of a paper bearing the same title to be published elsewhere which has the aim of giving an exposition of this theory of Ritt with emphasis on its CONSTRUCTIVE character. The concepts, and most of the results too, are all due to Ritt as may be found in the two books [R1,R2] of Ritt. We remark only that, while the original aim of Ritt is to establish a theory of differential equations from

Reprinted from *Differential Geometry and Differential Equations*, Lecture Notes in Mathematics, No. 1255 (Springer, 1984), pp. 173–189. With kind permission of Springer Science and Business Media.

an algebraic standpoint, our interest lies rather in its applications to mechanical theorem proving of differential geometries in particular. For this purpose we have suppressed all notions of IDEAL and its alike, being not appropriate for mechanical theorem proving.

For the sake of simplicity we shall restrict ourselves to the case of functions of only one independent variable. Let us recall first some definitions of Ritt, naturally with due modifications.

A DIFFERENTIAL FIELD [abbr. d-FIELD], say F, is a field of characteristic 0 which has, besides the usual arithmetic operations, a further operation of DIFFERENTIATION such that any element A of F has a DERIVATIVE DiA verifying the usual rules. We write for simplicity

DiA = D1...D1A

with D1 i times and call DiA the i-th DERIVATIVE of A. The element A itself is also considered as 0-th DERIVATIVE of A: A=D0A.

In what follows the d-FIELD F will be fixed in advance. An INDETERMINATE Y is just a symbol having an infinity of DERIVATIVES DiY none of which is zero. A DIFFERENTIAL POLYNOMIAL (abbr. d-POL), say P, in INDETERMINATES Y1, Y2, ..., Yn over F is a polynomial in DiYj(i>=0,1<=j<=n) with coefficients in F. For P we can then form its successive DERIVATIVES DiP as well as various PARTIAL DERIVATIVES dP/d(DiYj) in the usual formal manner.

To any d-POL P<>0 is associated three characteristic numbers, viz.,

(a) The CLASS cls(P) which is the greatest p such that some DjYp is actually present in P. If no such DjYp is present in P for any p>0 so that P is itself an element of F, then the CLASS cls(P) will be set to be 0.

(b) The ORDER ord(P) which is the greatest m such that the m-th DERIVATIVE DmYp with p=cls(P) is actually present in P. In case cls(P)=0 we define the ORDER ord(P) to be 0.

(c) The DEGREE deg(p) which is the highest degree in DmYp present in P where p=cls(p) and m=ord(p). In case cls(P)=0 we define the DEGREE deg(P) to be 0.

More generally, for any i with 1<=i<=n we shall denote by ord(i,P) the greatest m such that the m-th DERIVATIVE DmYi is actually present in P and then by deg(i,P) the highest degree in DmYi actually present in P. If neither Yi nor any of its DERIVATIVEs is present in P, then we just set ord(i,P)=-1.In particular, ord(p,Q)=ord(p,P), deg(P)=deg(p,P) if p=cls(P)>0. A d-POL Q is then said to be REDUCED with respect to a d-POL P of CLASS p>0 if either ord(p,Q)<ord(P), or ord(p,Q)=dor(P), but deg(p,Q)<deg(P).

Any d-POL P of CLASS p>0, ORDER m, and DEGREE d can now be written in the form

P = CO*DmYp^d + Cl*DmYp^(d-i) + ... + Cd,

with cls(Ci)<p, or cls(Ci)=p and ord (p,Ci)<m for i=0,1,...,d. The leading coefficient CO, which is itself a d-POL, is then called the INITIAL of P and dP/dDmYp is called the SEPARANT of P.

A finite sequence of non-zero d-POLs

Pl, P2, ... , Pr (ASC)

is called an ASCENDING SET (abbr. ASC-SET) if either

(a) r=1, or

(b) r>1, cls(Pl)>0, and for any j>i, cls(Pj)>cls(Pi) and Pj is REDUCED with respect to Pi.

For the ASC-SET (ASC) as above let Si and Ii be respectively the SEPARANT and INITIAL of Pi, i=1,2,...r. A d-POL G will be said to be REDUCED with respect to (ASC) if it is REDUCED with respect to each Pi in (ASC). In particular all SEPARANTs Si and INITIALs Ii are REDUCED with respect to (ASC).

For any system (DP) of d-POLs and any two d-POLs Q1, Q2 we shall write for simplicity

Q1 :=: Q2 d-mod (DP)

if Q1-Q2 is a linear combination of a finite number of d-POLs of (DP) or their DERIVATIVEs with coefficients themselves d-POls. The following lemma is then fundamental in the theory of Ritt:

LEMMA. Let (ASC) be an ASCENDING SET as given above. For any d-POL G there exist then unique non-negative integers Ki, Li such that, setting J the power product of SEPARANTs Si and INITIALs Ii of Pi in (ASC) as given below

Si^Ki*...*Sn^Kn*Ii^Li*...*In^Kn = J,

we shall have an equation of the form

J*G :=: R d-mod (ASC) (REM)

in which the d-POL R is REDUCED with respect to (ASC).

The d-POL R uniquely determined is called the REMAINDER of G with respect to (ASC). The procedure in passing from G to R as described in the LEMMA is then called the REDUCTION of G with respect to (ASC). The corresponding formula (REM) will then be called the REMAINDER FORMULA of G with respect to (ASC). The REMAINDER R will also be denoted as Rem(G/ASC).

Let a d-FIELD F be given. A d-FIELD F1 will be said to be an EXTENSION of F if, as an algebraic field, it is an extension field of F in the ordinary sense, and moreover any element A of F1 which is also in F will have the same p-th DERIVATIVE for any p>0 whether it is considered as an element of F or of F1.

Let the d-FIELD F and INDETERMINATEs Y1, Y2,..., Yn be now fixed in advance. Consider any finite or infinite system (DP) of d-POLs in Y1, ..., Yn over F. The system of equations P=0 for all P in (DP) will be represented symbolically by (DP)=0.

Suppose that there exists a certain EXTENSION F1 of F and a set of n elements Z1,...,Zn in F1, such that when each Yi is replaced by Zi in the d-POLs of (DP), these d-POLs all reduce to 0. Then we call the set (Z1,...,Zn) a ZERO of (DP) or alternatively a SOLUTION of (DP)=0. The totality of all ZEROs of (DP), for all possible extensions F1 of F, will be called the DIFFERENTIAL-ALGEBRAIC VARIETY (abbr. DIFF-ALG VARIETY) of (DP) or (DP)=0, to be denoted in what follows by Zero(DP). A ZERO of (DP) will also be called a POINT of the DIFF-ALG VARIETY Zero(DP). If Zero(DP) is an empty set, then we shall say that (DP) is a CONTRADICTORY system. Furthermore, if a certain d-POL G is given, then the totality of ZEROs of (DP) which are not ZEROs of G will be denoted by Zero(DP/G).

Consider now a finite system (DP) of non-zero d-POLs in the INDETERMINATEs Y1,...,Yn over F. The following theorem plays an important role in the theory of Ritt which we shall call the Ritt Well Ordering Principle or simply the

RITT PRINCIPLE. There is a mechanical procedure which permits to decide in a finite number of steps for a given finite system (DP) or non-zero d-POLs, either (DP) is CONTRADICTORY and possesses no ZEROs at all or there is some enlarged system (DP)' of (DP) and a particular ASC-SET (CS)

C1, C2, ... , Cr (CS)

consisting of d-POLs Ci in (DP)' having the following properties:

(1) cls(C1)>0.

(2) (DP)' has the same DIFF-ALG VARIETY of ZEROs as that of (DP).

(3) Any d-POL in (DP)' has its REMAINDER 0 with respect to (CS).

More precisely, we have in fact the following explicit formula for the structure of the DIFF-ALG VARIETY Zero(DP):

Zero(DP) = Zero(CS/J) + SUMi Zero(DPi')
+ SUMi Zero(DPi"). (RITT)

In the formula (RITT) the d-POL J is the product of all INITIALs Ii and SEPARANTs Si of Ci in (CS). Each DPi' is the enlarged system of (DP) with i-th INITIAL Ii adjoined to it and each DPi" is the one with i-th SEPARANT Si adjoined to it. The ASC-SET (CS) occuring in the formula (RITT) is of particular importance and is called a CHARACTERISTIC SET (abbr. CHAR-SET) of the given system (DP) of d-POLs. Remark that this terminology of CHAR-SET used here is a little different from that one used by Ritt.

The formula (RITT) above gives a decomposition of set of ZEROs of a system (DP) of d-POLs into several parts. It will be decomposed further to an ultimate form which will lead to some fundamental facts about DIFF-ALG VARIETYs. For this purpose, let us consider an ASC-SET (ASC)

P1, P2, ... , Pr (ASC)

with steadily increasing CLASSes

(0 <) cls(P1) < cls(P2) < ... < cls(Pr).

For any k with 1<=k<=r let (ASCk) be the ASCENDING SET formed by the first k d-POLs in the sequence (ASC)=(ASCr). Then we lay down the following

DEFINITION. The ASC-SET (ASC) is said to be d-IRREDUCIBLE if the following holds:

For each k>=1 and <=r let h=k-1. Then for any d-POL H REDUCED with respect to (ASCh), which is of CLASS either < cls(Pk), or of CLASS=cls(Pk) but of ORDER < ord(Pk), there can exist no relations of the form

H*Pk :=: P'*P" d-mod (ASCh),

in which P' and P" are both of the same CLASS and ORDER as Pk.

According to Ritt we can reduce the problem of deciding whether an ASC-SET of d-POLs is d-IRREDUCIBLE or not to a problem involving ordinary polynomials over ordinary fields which we shall not enter.

Consider now a system (DP) consisting of a finite number of non-zero d-POLs and also a d-POL G. For the structure of the set of ZEROs Zero(DP/G) we have then the following

ZERO DECOMPOSITION THEOREM. There is a mechanical procedure

which permits to decide in a finite number of steps whether Zero(DP/G) is an empty set and in the non-empty case to furnish a decomposition of the following form:

Zero(DP/G) = SUMi Zero(ASCi/Ri). (Z-DEC)

In this decomposition formula each ASCi is a d-IRREDUCIBLE ASC-SET and Ri is the non-zero REMAINDER of Ji*Gi with respect to (ASCi), where Ji is the product of INITIALs and SEPARANTs of d-POLs in (ASCi), and Gi is certain non-zero d-POL.

The proof consists of giving such a mechanical procedure as described below.

Step 1. Form, as in the RITT PRINCIPLE, a CHAR-SET (CS) of (DP). If (CS) is consisting of a single d-POL which is a non-zero element of the basic d-FIELD F, then Zero(DP), a fortiori Zero(DP/G), is empty and the procedure stops. In the contrary case let the INITIALs and SEPARANTs of the d-POLs in (CS) be respectively Ii and Si. Then the RITT PRINCIPLE will give rise to a decomposition of the form

Zero(DP/G) = Zero(CS/J*G) + SUMi Zero(DPi'/G)
+ SUMi Zero(DPi"/G),

in which J is the product of all INITIALs Ii and SEPARANTs Si of (CS), while each (DPi') resp. (DPi") is the enlarged system of (DP) with Ii resp. Si adjoined to it.

Step 2. Consider the set Zero(CS/J*G).

Suppose first that (CS) is d-IRREDUCIBLE. Form the REMAINDER R of J*G with respect to (CS). By the REMAINDER FORMULA we have clearly

Zero(CS/J*G) = Zero(CS/G).

If R=0 then Zero(CS/J*G) is empty and should be removed in the above decomposition. Otherwise we just replace Zero(CS/J*G) in the decomposition by Zero(CS/R). In any case we proceed to the next step.

Suppose now (CS) is d REDUCIBLE. Let (CS) consist of d-POLs

C1, C2, ... , Cr (CS)

of CLASSes

(0<) cls(C1) < cls(C2) < ... < cls(Cr).

There will then be some k<=r, h=k-1, and some d-POLs H and P', P" such that

H*Ck :=: P'*P" d-mod (CSh)

with corresponding properties observed. Let (CS') and (CS") be now the system of d-POLs obtained from (CS) in replacing Ck by P', P" respectively, and (CSO) the enlarged one obtained from (CS) by adjoining H to it. Then it is clear that

Zero(CS/J*G) = Zero(CSO/J*G) + Zero(CS'/H*J*G)
+ Zero(CS"/H*J*G).

Replace now Zero(CS/J*G) in the decomposition of Step 1 by the above union of sets of ZEROs and proceed to the next step.

Step 3. Treat now in turn each set of ZEROs occuring in the decomposition of Step 1 or Step 2 in returning to Step 1, to be considered as the new (DP), removing any empty set of ZEROs if it appears, and proceeding as before.

It can be proved that we have to stop after a finite number of steps. We have thus finally arrived at either an empty set or a decomposition in the form as described in the theorem.

The ZERO DECOMPOSITION THEOREM furnishes us with a complete description of the structure of the set of ZEROs of a finite system of d-POLs. It can be applied to give a CONSRUCTIVE proof of HILBERT ZERO THEOREM which, even in the case of ordinary polynomials, is usually proved in a mere EXISTENTIAL manner. It can also be applied to give a CONSTRUCTIVE proof of the decomposition of a DIFF-ALG VARIETY into IRREDUCIBLE ones in the following way.

Let a d-IRREDUCIBLE ASC-SET (ASC) be given as above with INITIALs Ii and SEPARANTs Si. Construct now a ZERO of (ASC) as follows. For any two d-POLs P, Q the relation that P-Q has its REMAINDER 0 with respect to (ASC) is easily seen to be an equivalence relation and we shall say accordingly that P, Q belong to the same REMAINDER CLASS with respect to (ASC). It is also easy to see that the original algebraic operations pass naturally to these REMAINDER CLASSes so that these CLASSes form a ring. The d-IRREDUCIBILITY of (ASC) shows that this is an integral domain so that we may form its quotient field. Furthermore, differentiation in the given d-FIELD F will also induce one in the above quotient field to turn it into a d-FIELD, to be denoted as d-FIELD(ASC). Let us identify the INDETERMINATEs Yi to its CLASS, denoted however by Zi, then the above d-FIELD becomes an EXTENSION d-FIELD of the d-FIELD F. It is clear that Z=(Z1,...,Zj) is a ZERO of (ASC).

DEFINITION. The ZERO Z=(Z1,...,Zn) of the d-IRREDUCIBLE ASC-SET (ASC) with Zi the CLASS of Yi in the d-FIELD(ASC) is called the GENERIC ZERO of (ASC).

The importance of this notion lies in the following

THEOREM G. A d-POL P has its REMAINDER 0 with respect to the d-IRREDUCIBLE ASC-SET (ASC) if and only if P has the GENERIC ZERO Z of (ASC) as a ZERO.

The system of all d-POLs P which has the above GENERIC ZERO as a ZERO, or what is the same, those having REMAINDER 0 with respect to (ASC), forms thus a prime ideal closed under a further operation of DIFFERENTIATION and will be denoted by d-IDEAL(ASC). The ZEROs of this system form then a DIFF-ALG VARIETY which will be denoted by d VAR(ASC). Remark that d-VAR(ASC) is in general different from Zero(ASC).

From the ZERO-DECOMPOSITION FORMULA (Z-DEC) of (DP) given above we see that Zero(DP) is non-empty if and only if terms actually present in the right hand side since each (ASCi/Ri) has non-empty ZEROs, say the GENERIC ZERO of (ASCi) which cannot be ZERO of the non-zero d-POL Ri, known to be REDUCED with respect to (ASCi).

It is also clear that each d-VAR(ASCi) is a non-empty DIFF-ALG VARIETY which is d-IRREDUCIBLE. We deduce easily the following VARIETY DECOMPOSITION THEOREM. There is a mechanical procedure which permits to decompose any DIFF-ALG VARIETY, when non-empty, into a finite sum of d-IRREDUCIBLE DIFF-ALG VARIETYs, viz.,

Zero(DP) = SUMi d-VAR(ASCi) (V-DEC)

We shall now apply the above ZERO DECOMPOSITION THEOREM to the mechanical proving of theorems in differential geometries. For this purpose we shall restrict ourselves in this note to the case of differential geometry of curves which involves functions of only one independent variable. We shall also restrict ourselves to such theorems for which both hypothesis and conclusion are expressed in the form of P=0 with P certain d-POL in a number of INDETERMINATEs Y1,...,Yn over a certain d-FIELD F (e.g. the field of all reals with trivial differentiation). Thus the hypothesis is, say, (HYP) = 0 where (HYP) is a finite system of such d-POLs and the conclusion is, say, CONC=0 with CONC another d-POL. A ZERO of (HYP) is then just a geometrical confiquration (over possibly certain extended field, e.g. complex field extension of the real field) verifying the hypothesis of the given theorem. To prove a theorem to be true seems thus equivalent to the following problem

(A) To decide whether CONC=0 follows from (HYP)=0 or not, i.e. to decide whether

Zero(HYP/CONC) = empty (Z-CONC)

or not.

Mathematically we can give a complete answer to the above problem (A). In fact, so far the hypothesis are not CONTRADICTORY in themselves or Zero(HYP) is non-empty, we shall get by some mechanical procedure the decomposition below:

Zero(HYP) = SUMi Zero(ASCi/Ri) (H-DEC)

In the formula (H-DEC) each (ASCi) is a d-IRRECUCIBLE ASC-SET and

Ri = Rem(Ji*Gi/ASCi) <> 0

for some d-POL Gi with Ji the product of all INITIALs and SEPARANTs of (ASCi). The answer of the above-mentioned problem (A) is then implied by the following

PROPOSITION. For (Z-CONC) to be true it is necessary and sufficient that for each i we have

Rem(CONC/ASCi) = 0 (R-CONCi)

which can be verified by direct computations.

This PROPOSITION, which in appearance completely settles the problem (A), does not however meet the REALITY of geometrical situations, and in this sense it cannot be accepted as a CORRECT solution to the problem of mechanical proving of geometrical theorems. In fact, if we define a THEOREM to be TRUE by (Z-CONC), then actually no THEOREM will be TRUE by this definition. The reason is this: The THEOREMs which one encounters in all kinds of geometries are usually TRUE only in a certain GENERIC sense, i.e., TRUE only if certain subsidiary NON-DEGENERACY conditions are observed. Examples are too many to be cited here. One may just take any one in the elementary plane geometry to see the point.

To lay down a CORRECT formulation of how a THEOREM is TRUE is to be defined,we shall first introduce the concept of DIMENSION as follows.

DFFINITION. For a non-CONTRADICTORY d-IRREDUCIBLE ASC-SET (ASC) consisting of r d-POLs we define the integer n-r as the DIMENSION of (ASC) and will denote it by dim(ASC). This integer is also defined as the DIMENSION of the DIFF-ALG VARIETY d-VAR(ASC) ASSOCIATED to (ASC), to be denoted as dim(d-VAR(ASC)). If for a system (DP) of d-POLs the set Zero(DP) is decomposed as the union of Zero(ASCi/Ri) as in the formula (Z-DEC) before, we shall define the DIMENSION of the DIFF-ALG VARIETY Zero(DP) by dim Zero(DP) = MAXi (dim(ASCi))

= MAXi (dim(d-VAR(ASCi))).

We shall leave aside the legitimacy of the definition. Admitting this, we shall lay down the following

DEFINITION. A THEOREM with hypotheses system (HYP) and conclusion CONC is said to be GENERICALLY TRUE if for the decomposition formula (H-DEC), (ASCi/Ri*CONC)=empty is true for all indices i for which dim(ASCi)=dim Zero(HYP).

As Zero(ASCi/Ri*CONC)=empty is equivalent to (R-CONCi) which can by verified by direct computations, we have the following theorem which meets the REALITY of geometrical situations and forms the underlying principle of our method of mechanical theorem proving of differential geometries.

THEOREM T. There is a mechanical procedure which permits to decide in a finite number of steps whether the hypothesis system of a THEOREM is CONTRADICTORY or not, and if not so, whether the THEOREM is GENERICALLY TRUE or not. In case that the THEOREM is GENERICALLY TRUE, then the procedure itself gives a PROOF of the THEOREM.

As the procedure in getting a formula of the form (H-DEC) requires factorization which is usually quite complicate and is thus not so convenient to use in practice, we shall adopt an alternative definition for a THEOREM to be TRUE, viz.,

DEFINITION. Let (HYP) and CONC be as before. Let N be a d-POL such that

Zero(HYP/N*CONC) = empty.

Then we say that the THEOREM is TRUE GENERICALLY under the subsidiary NON-DEGENERACY CONDITION

N <> 0.

The CONDITION is said to be REASONABLE if

dim Zero(HYP-N) < dim Zero(HYP)

in which (HYP-N) is the enlarged system of (HYP) with N adjoined to it.

The difficulty of applying the above definition to mechanical theorem proving lies clearly in the finding of such a d-POL N. However, we have discovered a method to meet this difficulty in the following way.

Let us form a CHAR-SET (CS) of the system (HYP) by the RITT PRINCIPLE so that we get the formula

Zero(HYP) = Zero(CS/J) + SUMi Zero(HYPi')
+ SUMi Zero(HYPi").

In the formula J is the product of all INITIALs Ii and SEPARANTs Si of (CS), and (HYPi'), (HYPi") are respectively the enlarged systems of (HYP) with Ii and Si adjoined to it. Form now the REMAINDER R of CONC with respect to (CS) to get a formula of the form

J' * CONC :=: R   d-mod (CS)

in which J' is some power product of the INITIALs Ii and the SEPARANTs Si of (CS). Now if R=0, then we see that

CONC = 0

will follow from (HYP)=0 so far J<>0. On the other hand if (CS) is d-IRREDUCIBLE, and if CONC=0 follows from (HYP)=0 so far J<>0, then CONC will have the GENERIC ZERO of (CS) as a ZERO and by THEOREM G we would have R=0. We have thus the following

THEOREM N. If the REMAINDER R of CONC with respect to the CHAR SET (CS) of (HYP) is 0, then the THEOREM in question is TRUE GENERICALLY under the subsidiary NON-DEGENERACY CONDITION J<>0. If (CS) is d-IRREDUCIBLE then the converse is also true.

This THEOREM N is at the basis of our method of mechanical theoem proving and has been accordingly programmed. It turns out that it will meet our purposes in general and has been proved to be quite efficient in practice. In fact, based on the last THEOREM we have programmed on some small computers like HP9835A and HP1000. Several differential-geometrical theorems have accordingly be proved on the computers in this way. The following is one of such examples.

Let C, C' be a pair of curves in the ordinary 3-space which are in 1-1 correspondece. Suppose that the lines joining pairs of corresponding points p, p' be common principal normals to the curves. Let k, t be the curvature and torsion of C, similarly let k', t' be those of C'. Then the following conclusions are known to be true:

(a) The distance r between the corresponding points p, p' is constant.

(b) The angle alpha between the tangents to the curves at corresponding points is constant.

(c) The curvature k and torsion t of C satisfy some linear relation with constant coefficients, i.e., there are constants a, b, c not all 0 such that a*k+b*t=c. The same is true for C'.

Less well-known are the following conclusions:

(d) The product t*t' of torsions of the curves C, C' at corresponding points is constant.

(e) Let z, z' be the centers of curvature of C, C' on the common principal normal at corresponding points p, p', then the cross ratio (p,z,p',z') is constant.

To prove these theorems let us choose the arc lengths s, s' on the curves as parameters. Then s', r, alpha, k, t, etc. are all functions of s. Remark that a computer can treat only rational entities so we have to replace the transcendental functions of alpha by rational ones, viz.

u = cos alpha,   v = sin alpha,

connected by the rational relation

u^2 + v^2 = 1.

Let us take now the usual moving frames (e1,e2,e3) and (e1',e2',e3') at the corresponding points p and p' with e1, e1' the tangent vectors, e2 and e2'=+ or -e2 the principal normal vectors, and e3, e3' the binormal vectors. To fix the ideas, let us suppose e.g. e2'=+e2. Then we have:

$$
\begin{aligned}
p' &= p + r*e2,\\
e1' &= u*e1 + \quad\quad + v*e3,\\
e2' &= \quad\quad + e2,\\
e3' &= -v*e1 + \quad\quad + u*e3,
\end{aligned}
\qquad \text{(FRAME)}
$$

Treat now the functions

$$r,\ u,\ v,\ ds'/ds,\ k,\ t,\ k',\ t'$$

as indeterminates and replace them by symbols Y1, Y2,..., Y8 in the above order. Then the hypothesis system (HYP), by comparing the FRENET-DARBOUX EQUATIONs of the two curves, will be found to be consisting of d-POLs P1,...,P10 listed below:

$$
\begin{aligned}
P1 &= D1Y1,\\
P2 &= D1Y2,\\
P3 &= D1Y3,\\
P4 &= Y2^2 + Y3^2 - 1,\\
P5 &= Y1*Y5 + Y2*Y4 - 1,\\
P6 &= Y1*Y6 - Y3*Y4,\\
P7 &= Y4*Y7 + Y3*Y6 - Y2*Y5,\\
P8 &= Y3*Y4*Y8 + Y2*Y4*Y7 - Y5,\\
P9 &= Y2*Y4*Y8 - Y3*Y4*Y7 - Y6,\\
P10 &= Y4*Y8 - Y2*Y6 - Y3*Y5.
\end{aligned}
$$

It follows that the conclusions (a) and (b) are already seen to be true from the equations P1=0, P2=0, and P3=0. The other conclusions are however not so evident and we have to resort to our program based on the last THEOREM N. We find thus a CHAR-SET (CS) of the hypothesis system (HYP) after the removal of some simple factors of Y1, Y2, Y3, to be consisting of d-POLs C1,...,C7 as given below:

C1 = D1Y1,

C2 = D1Y2,

C3 = Y3^2 + Y2^2 - 1,

C4 = Y1*Y5 + Y2*Y4 - 1,

C5 = Y1*Y6 - Y3*Y4,

C6 = Y1*Y4*Y7 + Y4 - Y2,

C7 = Y1*Y4*Y8 - Y3.

The NON-DEGENERACY CONDITION is then seen to be given automatically as some power product of the INITIALLs and SEPARANTs of (CS), viz. for certain Ki>0,

N = Y1^K1*Y2^K2*Y3^K3*Y4^K4 <> 0.

Alternatively we may replace the NON-DEGENERACY CONDITION N<>0 by a set of CONDITIONs

Y1 <> 0, Y2 <> 0, Y3 <> 0, Y4 <> 0,   (COND)

of which the geometrical meanings are quite clear.

The conclusions (c), (d), (e) may now be replaced by CONCi=0, i=1,2,3 with the d-POLs CONCi as given below,

CONC1 = D1Y5*D2Y6 - D2Y5*D1Y6,

CONC2 = D1Y6*Y8 + Y6*D1Y8,

CONC3 = Y1*Y5*D1Y7 + Y1*D1Y5*Y7 + D1Y5 - D1Y7.

We find on the computer that all the d-POLs CONCi have their REMAINDER 0 with respect to (CS). It follows that the conclusions (a),...,(e) are all GENERICALLY TRUE and are proved under the NON-DEGENERACY CONDITION N<>0. If we like we can proceed in the same way as before to test in turn whether the conclusions remain true in the respective degenerate case afforded by adjoining each of the conditions in (COND) in turn to (HYP).

We add a final remark to our method. As our method of proving is purely algebraic in character which has nothing to do with the real nature of the curves of being analytic or not, we come to the following

PRINCIPLE. If a THEOREM is TRUE for all geometrical entities which are analytic, then the THEOREM will remain TRUE for all geometrical entities merely differentiable up to certain degree sufficiently high.

This PRINCIPLE was already anounced by the author in some of the previous works. It shows that the analyticity or high differentiability plays actually no role in the truth of a THEOREM.

REFERENCES

[R1] J.F.Ritt, Differential equations from the algebraic standpoint, Amer. Math. Soc., (1932).

[R2] J.F.Ritt, Differential algebra, Amer. Math. Soc., (1950).

[WU1] Wu Wen-tsun, On the mechanization of theorem-proving in elementary differential geometry, Scientia Sinica, Math. Supplement (I), 94-102 (1979).

[WU2] Wu Wen-tsun, Mechanical theorem proving in elementary geometry and differential geometry, in Proc. 1980 Beijing DD-Symposium, Beijing, V.2, 1073-1092 (1982).

系 统 科 学 与 数 学
J. Sys. Sci. & Math. Scis.
4 (3) (1984), 207—235

BASIC PRINCIPLES OF MECHANICAL THEOREM PROVING IN ELEMENTARY GEOMETRIES

WU WENJUN (WU WEN-TSÜN)
(*Institute of Systems Science, Academia Sinica, Beijing*)

Dedicated to Professor Kwan Chao-chih, the late Director of the Institute of Systems Science

§ 1. INTRODUCTION.

By *elementary geometry* we shall mean the one described in Hilbert's *Grundlagen der Geometrie* in which no notion of differentiation is involved, as a contrast to *differential geometry*. It is well known by the theorem of Tarski that the ordinary Euclidean geometry, as one of such elementary geometries, is *decidable*, or in our terminology, *mechanizable* in the following sense: There exists an algorithmic method by which any "theorem" or a geometrical statement meaningful in the geometry in question can be shown, in a *finite* number of steps, to be either true as a real *theorem*, or false so that it is not a *theorem* at all. Any elementary geometry possessing such an algorithmic method will be said to be *mechanizable*, and the theorem in asserting that the geometry in question does possess such an algorithmic method will be called a *Mechanization Theorem*. In the mechanizable case we may program according to the algorithm shown to exist and practise on a computer so that the proof (or disproof) of a theorem in that geometry may be carried out on the computer. This method will be called *mechanical theorem proving* for short. It will lead to what may be called *mechanical theorem discovering* of new theorems. We remark that all these notions can be naturally extended to the case of a given class of theorems or meaningful statements in the geometry in question, not necessarily to the geometry as a whole. In this sense the Theorem of Tarski mentioned may be called the Mechanization Theorem of ordinary Euclidean geometry. However, the algorithmic procedure given by Tarski, even with the great simplifications due to Seidenberg, is too complicated to be feasible. In fact, no theorems of any geometrical interest seem to have been proved in this way up to the present day. On the other hand, the author discovered in 1977 an algorithmic method which leads to Mechanization Theorems of many kinds of elementary geometries including the ordinary Euclidean geometry, as long as we restrict ourselves to the class of theorems involving no order relations. What is important to us is that our method is very efficient. In fact, in the past years we have programed on some small computers and arrived at the proof and discovery of quite nontrivial theorems. Mr. S. C. Chou, now at University of Texas at Austin, USA, has also practised on some computer there, on the basis of our algorithm, and proved some interesting new theorems. The present paper is aimed at explaining the basic principles underlying our method with some illustrative examples about the theorems proved or discovered in this way.

Consider a certain kind of geometry in the sense of Hilbert. As shown in the classical

Received March 12, 1984,

Grundlagen of Hilbert, in starting from the defining axiomatic system of the geometry we may introduce some number system intrinsically associated to that geometry and then to coordinate systems which will turn any geometrical entities and relations into algebraic ones. Let us restrict ourselves to the case that the geometry admits some axiom of infinity as well as some Pascalian axiom so that the number system is a commutative field of characteristic 0. The algebraic relations corresponding to the geometrical relations occuring in a theorem will then be polynomial equations, polynomial inequations, or polynomial inequalities, with coefficients in the associated field, or even with rational or integer coefficients, as is usually the case. Now let us restrict ourselves further to the case that no order relations and axioms are involved in the geometry in question or to a restricted class of theorems in which no order relations are involved. In the algebraic relations above there will appear only polynomial equations and inequations but not any polynomial inequalities. Remark further that all theorems in geometries are actually only *generically* true, or true only under some *non-degeneracy conditions* which are usually not easy to be made explicit and thus only implicitly assumed in the statement of theorems. It turns out that the problem of mechanical theorem proving in the restricted cases mentioned above is algebraically equivalent to the following one:

Problem. Given a system Σ of polynomial equations (or equivalently, system of polynomials) as well as another polynomial g, all in the same finite set of variables $x, y, \cdots$, decide in a finite number of steps either of the two cases below:

Case 1. A finite set of polynomials D_α is determined such that $g=0$ is a consequence of the system Σ under the non-degeneracy conditions $D_\alpha \neq 0$ such that $D_\alpha=0$ are themselves not consequences of the system Σ.

Case 2. No such set $S=\{D_\alpha\}$ can exist so that Case 1 holds.

In the above formulated problem in the algebraic form the polynomials in Σ correspond to the hypotheses and g the conclusion of the theorem in question whose truth is to be decided. The theorem is seen to be generically true in Case 1 under the non-degeneracy conditions $D_\alpha \neq 0$ found during the procedure but not so in Case 2. The polynomials naturally have their coefficients in the field intrinsically associated to the geometry considered. A solution of the above problem constitutes the Mechanization Theorem of geometries in the algebraic form. The algorithm in furnishing such a solution as well as the proof will be given in Section 4. In Sections 2 and 3 we shall make some preparations. All these depend heavily on the works of J. F. Ritt as exhibited in his two books [2, 3], which seem to be however undeservedly little known in the present days.

§ 2. Well-ordering of a Polynomial Set.

In what follows K will be a fixed basic field of characteristic 0. Consider two sets of variables

$$u_1, \cdots, u_e \text{ and } x_1, \cdots, x_N,$$

arranged in a fixed order

$$u_1 \prec \cdots \prec u_e \prec x_1 \prec \cdots \prec x_N.$$

We shall consider a linear space K^{e+N} of dimension $e+N$ over the field K, with a basis corresponding to $u_1, \cdots, u_e, x_1, \cdots, x_N$. In what follows by a polynomial we shall always mean one in the variables $u_1, \cdots, u_e, x_1, \cdots, x_N$ with coefficients in K, i.e., an element in the ring $K[u_1, \cdots, u_e, x_1, \cdots, x_N]$.

A monomial

$$\mu = a u_1^{i_1} \cdots u_e^{i_e} x_1^{m_1} \cdots x_N^{m_N} \qquad (a \in K),$$

will sometimes be written in the simple form

$$\mu = a U^I X^M, \quad I = (i_1, \cdots, i_e), \quad M = (m_1, \cdots, m_N)$$

or

$$\mu = a z^{\alpha}, \alpha = (I, M) = (i_1, \cdots, i_e, m_1, \cdots, m_N).$$

If $a \neq 0$, and the last one $\neq 0$ in the N-tuple $(m_1, \cdots, m_N)$ is m_p, then we say that the monomial is of *class* p; otherwise we say that the class of the monomial $\mu \neq 0$ is 0 .In that case in μ there occurs at most u but not x.

For two sets of non-negative integers

$$\alpha = (a_1, \cdots, a_s), \qquad \beta = (b_1, \cdots, b_s)$$

we say that α *precedes* β, or β *follows* α, which is denoted as

$$\alpha \prec \beta \quad \text{or} \quad \beta \succ \alpha,$$

if there is some k such that

$$a_{k+1} = b_{k+1}, \quad \cdots, \quad a_s = b_s$$

while $a_k < b_k$. For two non-zero monomials

$$\lambda = a u_1^{i_1} \cdots u_e^{i_e} x_1^{l_1} \cdots x_N^{l_N}, \quad a \neq 0,$$

$$\mu = b u_1^{j_1} \cdots u_e^{j_e} x_1^{m_1} \cdots x_N^{m_N}, \quad b \neq 0,$$

we say that λ *precedes* μ or μ *follows* λ, which is denoted as

$$\lambda \prec \mu \quad \text{or} \quad \mu \succ \lambda$$

if

$$(i_1, \cdots, i_e, l_1, \cdots, l_N) \prec (j_1, \cdots, j_e, m_1, \cdots, m_N).$$

Any non-zero polynomial F can be written in the form

$$F = a_1 z^{\alpha_1} + a_2 z^{\alpha_2} + \cdots + a_s z^{\alpha_s}$$

in which

$$a_i \in K, \; a_1 \neq 0, \; \cdots, \; a_s \neq 0,$$

$$\alpha_1 \succ \alpha_2 \succ \cdots \succ \alpha_s.$$

In that case we say that $a_1 z^{\alpha_1}$ is the *leading term* of F, and the class of z^{α_1} will be called the *class* of F.

If a non-zero polynomial F has its class $= p \succ 0$, and the leading term $a_1 z^{\alpha_1}$ of F has its degree in $x_p = m$, then F can be written in the form

$$F = C_0 x_p^m + C_1 x_p^{m-1} + \cdots + C_m,$$

in which the C's are all polynomials in u and $x_1, \cdots, x_{p-1}$, containing none of $x_p, x_{p+1}, \cdots, x_N$, with $C_0 \neq 0$ in particular. The polynomial C_0 will then be called the *initial* of F. If the leading term of C_0 is c_0, then the leading term of F is clearly $c_0 x_p^m$.

Consider two non-zero polynomials F and G and any variable x_p. If the highest degree of x_p appearing in F is less than that in G, then we say that F has a *lower rank* than G

or G has a *higher rank* than F *with respect to* x_p. We say that F and G have the *same rank with respect to* x_p when neither is of higher rank than the other.

For two non-zero polynomials F and G we say that F has a *lower rank* than G or G a higher rank than F, which is denoted as

$$F \prec G \quad \text{or} \quad G \succ F,$$

if one of the following two cases occurs:

1. class $F <$ class G;

2. class $F =$ class G, say, $= p > 0$, while the degree of x_p in F is less than that of x_p in G, or in other words, F has lower rank than G with respect to x_p.

In the case neither of F and G is of higher or lower rank than the other, F and G will be said to be of the *same rank*, denoted as

$$F \sim G.$$

For example, two non-zero polynomials are of the same rank if both are of class $= 0$. Let F be a polynomial of class $p > 0$. Any polynomial G of rank lower than F with respect to x_p will be said to be *reduced* with respect to F. Clearly the initial of F is of class $< p$ and is already reduced with respect to F.

Let F be of class $p > 0$ written in the form

$$F = f_0 x_p^m + f_1 x_p^{m-1} + \cdots f_m,$$

in which

$$f_i \in K[u_1, \cdots, u_e, x_1, \cdots, x_{p-1}], \quad f_0 \neq 0.$$

Any non-zero polynomial G which has not been reduced with respect to F can then always be written in the form

$$G = g_0 x_p^M + g_1 x_p^{M-1} + \cdots + g_M,$$

in which

$$g_i \in K[u_1, \cdots, u_e, x_1, \cdots, x_{p-1}, x_{p+1}, \cdots, x_N],$$

and

$$g_0 \neq 0, \quad M \geqslant m.$$

By the division algorithm of polynomials, we would get, in dividing G by F, an expression of the form

$$f_0^s G = QF + R,$$

where Q, R are both polynomials with, in the case $R \neq 0$, the degree of x_p in $R < m$ so that R is already reduced with respect to F. The integer s will be determined as the smallest to make possible such an expression that s is unique and is $\leqslant M - m$. If G is already reduced with respect to F, then we can simply take $s = 0$, $Q = 0$, $R = G$ so that the above expression holds true still. In any way, the polynomial R will be called the *remainder* of G with respect to F. The procedure to get the remainder R from G will then be called the *reduction* of G with respect to F.

In what follows we shall consider sequences formed by a finite number of polynomials A_i like the one below.

$$\mathscr{A}: A_1, A_2, \cdots, A_r.$$

Such a sequence will be called an *ascending set* if either (a) or (b) below holds true:

(a) $r = 1$ and $A_1 \neq 0$;

(b) $r > 1$, $0 < \text{class}\ (A_1) < \text{class}\ (A_2) < \cdots < \text{class}\ (A_r)$, and moreover A_j is reduced with respect to A_i for each pair $j > i$.

It is clear that for an ascending set one always has $r \leqslant N$.

An ascending set will be said to be *contradictory* if $r = 1$, $A_1 \neq 0$ with class $(A_1) = 0$.

Given a second ascending set

$$\mathscr{B}: B_1, B_2, \cdots, B_s,$$

we say that $\mathscr{A}$ has a *higher rank* than $\mathscr{B}$ or $\mathscr{B}$ a *lower rank* than $\mathscr{A}$, which is denoted as

$$\mathscr{A} \succ \mathscr{B} \text{ or } \mathscr{B} \prec \mathscr{A},$$

if either (a)′ or (b)′ below holds true:

(a)′ There is some $j \leqslant \min(r, s)$ such that

$$A_1 \sim B_1, \cdots, A_{j-1} \sim B_{j-1}, \text{ while } A_j \succ B_j;$$

(b)′ $s > r$ and

$$A_1 \sim B_1, \cdots, A_r \sim B_r.$$

If neither of the ascending sets $\mathscr{A}$ and $\mathscr{B}$ is of higher rank than the other, then we say that $\mathscr{A}, \mathscr{B}$ are of the *same rank*, denoted as $\mathscr{A} \sim \mathscr{B}$. In that case we have

$$r = s, \text{ and } A_1 \sim B_1, \cdots, A_r \sim B_r.$$

It is clear that the collection of all ascending sets is partially ordered by the rank. Hence for any set of ascending sets we can speak of the notion of *minimal ascending set*, if it exists. The following lemma, simple as it is, will play an important role in the whole theory.

Lemma 1. *Let*

$$\Phi_1, \Phi_2, \cdots, \Phi_q, \cdots$$

be a sequence of ascending sets Φ_q for which the rank never increases, or for any q we have either $\Phi_{q+1} \prec \Phi_q$ or $\Phi_{q+1} \sim \Phi_q$. Then there is an index q' such that for any $q > q'$ we have

$$\Phi_q \sim \Phi_{q'}.$$

In other words, there is some q' such that any Φ_q for which $q \geqslant q'$ is a minimal ascending set of the above sequence.

Proof. For the ascending set Φ_q let us denote by r_q its number of polynomials and by A_q the first polynomial in the set. Then

$$A_1, A_2, \cdots, A_q, \cdots$$

is a sequence of polynomials for which the rank never increases, or for any q we have either $A_{q+1} \prec A_q$ or $A_{q+1} \sim A_q$. Consequently for any q we have class $(A_{q+1}) \leqslant$ class (A_q) and in the case class (A_{q+1}) = class (A_q), say, $=p>0$ the degree in x_p of A_{q+1} should be $\leqslant$ the corresponding degree in x_p of A_q. As both class and degree are non-negative integers, there should be some index q_1 such that all A_q are of the same rank for $q \geqslant q_1$.

If there is some $q_1' \geqslant q_1$ such that all $r_q = 1$ for any $q \geqslant q_1'$, then the lemma is clearly true. Suppose the contrary. Then there should be some $q_1' \geqslant q_1$ such that all $r_q \geqslant 2$ for any $q \geqslant q_1'$. Denote the second polynomial in such Φ_q by $A_q^{(1)}$. Then

$$A_{q_1'}^{(1)}, A_{q_1'+1}^{(1)}, \cdots, A_q^{(1)}, \cdots$$

will be a sequence of polynomials with non-increasing ranks. As before there will then be some $q_2 \geqslant q_1'$ such that all $A_q^{(1)}$ are of the same rank for any $q \geqslant q_2 \geqslant q_1' \geqslant q_1$.

If all $r_q \leqslant 2$, then the lemma is proved already. Suppose the contrary. Then there will be some $q_2' \geqslant q_2$ such that all $r_q \geqslant 3$ for any $q \geqslant q_2'$ and we may take the third polynomial $A_q^{(2)}$ in such Φ_q's to form a sequence of polynomials with non-increasing ranks. As for all q we have $r_q \leqslant n$, so proceeding in this way we should stop at some r and some q' such that for all $q \geqslant q'$ we have $r_q = r$ and the r-th polynomials taken from such Φ_q will all have the same rank. It follows that all such Φ_q's will have the same rank and the lemma is proved.

From this lemma we get the following

Lemma 1′. *If in a sequence of ascending sets the ranks are steadily decreasing, then such a sequence can only be composed of a finite number of ascending sets.*

Suppose now we have a non-empty collection $\Sigma = \{F_\alpha\}$ of non-zero polynomials F_α. An ascending set $\mathscr{A}$ of polynomials will be said to *belong* to Σ if each polynomial in $\mathscr{A}$ belongs to Σ. Since each single $F_\alpha \neq 0$ forms by itself an ascending set, such ascending sets belonging to Σ exit naturally. Any minimal ascending set of the collection of all ascending sets belonging to Σ will then be called a *basic set* of Σ.

The following lemma points out not only the existence of such basic sets but also some constructive method of arriving at such basic sets.

Lemma 2. *Let Σ be a finite set of non-zero polynomials. Then Σ has necessarily basic sets and there is a mechanical method in getting such a basic set in a finite number of steps.*

Proof. As Σ is finite, the existence of basic sets is quite evident. So the problem reduces to the mechanical generation of such a basic set.

To show this let us find at the outset a polynomial, say A_1, of lowest rank from $\Sigma = \Sigma_1$. This can clearly be done in a mechanical manner. If class $(A_1)=0$, then A_1 alone will form already a basic set. Suppose therefore class $(A_1)>0$. Check whether each polynomial except A_1 in Σ_1 is already reduced with respect to A_1. If no such polynomial exists in Σ_1, then A_1 by itself forms already a basic set of Σ_1. Otherwise let Σ_2 be the subset of Σ_1 formed by all such polynomials except A_1 already reduced with respect to A_1. From the choice of A_1 all polynomials in Σ_2 will have a rank higher than that of A_1. Now let A_2 be a polynomial in Σ_2 of lowest rank. If Σ_2 has not any polynomial which is different from A_2 and is already reduced with respect to A_2, then A_1, A_2 will form a basic set of Σ. Otherwise let Σ_3 be the subset of Σ_2 consisting of all polynomials except A_2 which have already been reduced with respect to A_2. Choose from Σ_3 a polynomial A_3 of lowest rank and proceed as before. As the classes of the polynomials $A_1, A_2, A_3, \cdots$ are steadily increasing and unlikely to become $> N$, we have to stop in a finite number of steps and get finally a basic set in a mechanical manner, Q. E. D.

Lemma 3. *Let Σ be a finite set of non-zero polynomials with a basic set*

$$\mathscr{A}: A_1, A_2, \cdots, A_r$$

of which class $(A_1) > 0$. Let B be a non-zero polynomial reduced with respect to all A's. Then the set Σ' obtained from Σ by adjunction of B will have a basic set of rank lower than that of $\mathscr{A}$.

Proof. If class $(B) = 0$, then B alone will form a basic set of Σ' of rank lower than that of $\mathscr{A}$. Suppose therefore class $(B) = p > 0$. As B is already reduced with respect to all A's, there should be some $i \geqslant 0$ and $\leqslant r$ such that $p >$ class (A_{i-1}) and $p \leqslant$ class (A_i). Moreover, in the case $p =$ class (A_i), the degree of x_p in B will be less than that of x_p in A_i. Hence

$$A_1, A_2, \cdots, A_{i-1}, B$$

will be an ascending set of Σ' with a rank lower than that of $\mathscr{A}$. The basic set of Σ' will have therefore *à fortiori* a rank lower than that of $\mathscr{A}$, Q. E. D.

Remark. The above lemmas are clearly also true for any infinite set of polynomials and the proofs remain essentially the same as long as the axiom of choice is applied. As the use of axiom of choice will be in opposition to the mechanical thought, the main theme of the whole theory, we have deliberately restrict ourselves to the case of finite sets of polynomials.

Consider now an ascending set

$$\mathscr{A}: A_1, A_2, \cdots, A_r$$

as before with class $(A_1) > 0$. Let class $(A_i) = p_i$ and let the initial of A_i be I_i. Then

$$0 < p_1 < p_2 < \cdots < p_r$$

and for each i we have

$$\text{class } (I_i) < p_i,$$

$$I_i \text{ reduced with respect to } A_1, \cdots, A_{i-1}.$$

Let B be an arbitrary polynomial. Set $B = R_r$. With respect to the polynomials in $\mathscr{A}$ starting from A_r to A_1 we can form successively the remainders $R_{r-1}, \cdots, R_0$ of R_r so that we get $(s_i \geqslant 0)$:

$$I_r^{s_r} R_r = Q_r A_r + R_{r-1},$$

$$I_{r-1}^{s_{r-1}} R_{r-1} = Q_{r-1} A_{r-1} + R_{r-2},$$

$$\cdots$$

$$I_1^{s_1} R_1 = Q_1 A_1 + R_0.$$

Set $R_0 = R$. Then we get an expression of the form

$$I_1^{s_1} \cdots I_r^{s_r} B = Q_1' A_1 + \cdots + Q_r' A_r + R,$$

in which Q' are all polynomials. The polynomial R is determined from B and the ascending set $\mathscr{A}$. We shall call R the *remainder* of B with respect to $\mathscr{A}$. We call also the above formula the *Remainder Formula*.

It is clear that any term of R will have the degree in x_{p_i} less than the corresponding degree in x_{p_i} of A_i. In other words, R is reduced with respect to all polynomials A_i in $\mathscr{A}$. We shall say briefly that R is *reduced* with respect to $\mathscr{A}$ and call the above procedure of getting R from B and $\mathscr{A}$ the *reduction* of B with respect to $\mathscr{A}$. As the determination of one polynomial with respect to the other is done mechanically by the division algorithm, we may state the result in the form of the following

Lemma 4. *Given a non-zero polynomial B and an ascending set $\mathscr{A}$ of which the first polynomial is of class > 0, there is an algorithm which permits to determine the remainder of B with respect to $\mathscr{A}$ in a finite number of steps. Denote the i-th polynomial in $\mathscr{A}$ by A_i and its class by p_i. Then any term in the remainder R will have its degree of x_{p_i} in A_i less than the degree of x_{p_i} in A_i for each i.*

Come now to the *well-ordering* of a polynomial set as follows. For this purpose let us review briefly the notion *zero* of such a set.

Consider any polynomial F. Suppose that there is a set of numbers

$$u_1^0, \cdots, u_e^0, x_1^0, \cdots, x_N^0$$

in K which will turn F into 0 when these numbers are substituted for the variables $u_1, \cdots, u_e, x_1, \cdots, x_N$ in F. Then this set of numbers, which may be considered as the coordinates of a point in the linear space K^{e+N}, is called a *zero* of the polynomial F or alternatively a *solution* of the equation $F = 0$. If the various u^0, x^0 are not numbers of K, but of some extension field $\widetilde{K}$ of K, which still turn F into 0 when substituted into it, then, the set of numbers, considered as a point of the linear space $\widetilde{K}^{e+N}$ on $\widetilde{K}$, will be called an *extended zero* of F or an *extended solution* of $F = 0$. In order to make the involved field $\widetilde{K}$ explicit, it will also be called a *$\widetilde{K}$-zero* of F or a *$\widetilde{K}$-solution* of $F = 0$.

Given a set of polynomials Σ, if a set of numbers as given above is a zero (or extended zero, or $\widetilde{K}$-zero) of every polynomial in Σ, then it will be called simply a *zero* (resp. an *extended zero* or a *$\widetilde{K}$-zero*) of Σ or a *solution* (resp. an *extended solution* or a *$\widetilde{K}$-solution*) of $\Sigma = 0$.

Consider now a set $\Sigma = \Sigma_1$ of non-zero polynomials, supposed to be finite in number. By Lemma 2, Σ_1 will have some basic set, say Φ_1. If Φ_1 is a contradictory set, then Φ_1 consists of a single polynomial A_1 belonging to Σ_1 for which class $(A_1) = 0$. Suppose on the contrary that Φ_1 is not contradictory so that the first polynomial in Φ_1 has its class > 0. For polynomials B, which belong to Σ_1 but not to Φ_1, let us form the remainders R_B of B with respect to Φ_1 supposed not all 0. Adjoin all such remainders R_B, whenever non-zero, to the set Σ_1 to get an enlarged set of non-zero polynomials Σ_2. From the formula about remainders each R_B, when non-zero, will be a linear sum of polynomials in Φ_1 as well as the polynomial B, with polynomials as coefficients. It follows that the set Σ_2 will have the same set of zeros (or extended zeros, or $\widetilde{K}$-zeros for any extended field $\widetilde{K}$) as the original set Σ_1. Form now the basic set Φ_2 of Σ_2. By Lemma 3 Φ_2 will have a rank lower than that of Φ_1. If Φ_2 is not a contradictory ascending set then we can proceed as before. In this way we shall get either a contradictory ascending set after a finite number of steps or a sequence of finite sets of polynomials

$$\Sigma_1 \subset \Sigma_2 \subset \cdots \subset \Sigma_q \subset \cdots,$$

where all Σ_i have same set of zeros (or extended zeros or $\tilde{K}$-zeros for any extended field $\tilde{K}$) with the corresponding non-contradictory basic sets Φ_i having steadily decreasing ranks:

$$\Phi_1, \Phi_2, \cdots, \Phi_q, \cdots.$$

Now by Lemma 1, such a sequence can have only a finite number of terms. In other words, if the last one of such a sequence of finite sets of polynomials is Σ_q, with Φ_q as the corresponding basic set, then the remainder of any polynomial of Σ_q not in Φ_q with respect to Φ_q will be equal to 0.

Let Φ_q be

$$\Phi_q: F_1, F_2, \cdots, F_r,$$

in which each F_i is either belonging originally to Φ_{q-1}, or is the non-zero remainder of some polynomial in Σ_{q-1} with respect to Φ_{q-1}. By the remainder formula each F_i is thus a linear sum of polynomials in Φ_{q-1} with polynomials as coefficients. It follows that any zero of Σ_{q-1} and thus any zero of Σ is also a zero of Φ_q.

On the other hand let the initials of polynomials in Φ_q be $I_1, I_2, \cdots, I_r$. From the construction we know that for any polynomial G in Σ_q, there should be non-negative integers $s_i \geqslant 0$ such that

$$I_1^{s_1} \cdots I_r^{s_r} G = Q_1 F_1 + \cdots + Q_r F_r.$$

It follows that any zero of Φ_q, if not a zero of any one of the initials $I_1, \cdots, I_r$, is necessarily also a zero of Σ_q and thus a zero of $\Sigma = \Sigma_1$. The same is clearly true for extended zeros or $\tilde{K}$-zeros for any extended field $\tilde{K}$.

Denote Φ_q by Φ. Then what we have proved may be reformulated as the theorem below:

Theorem (Ritt). *There is an algorithm which permits to get, after mechanically a finite number of steps, either a polynomial A of class 0, i. e. one in variables $u_1, \cdots, u_e$ so that any zero of Σ is also a zero of A, or a non-contradictory ascending set*

$$\Phi: F_1, \cdots, F_r,$$

with initials $I_1, \cdots, I_r$ such that any zero of Σ is also a zero of Φ, and any zero of Φ which is not zero of any of the initials I_i, will also be a zero of Σ. The same is ture for extended zeros and $\tilde{K}$-zeros.

We shall call the mechanical procedure which permits to determine Φ from Σ a *well-ordering* of Σ and the above theorem will be called the *Well-Ordering Theorem*. The theorem is due to Ritt and forms the basis of our method. We shall call the theorem *Ritt Principle* accordingly. The polynomial set Φ in the theorem is called a *characteristic set* of Σ.

§ 3. A Constructive Theory of Algebraic Varieties

As before, let K be the basic field of characteristic 0 and

$$x_1 \prec x_2 \prec \cdots \prec x_N$$

be a set of variables arranged in a definite order with $u_1, \cdots, u_c$ neglected. A polynomial will always be understood as one in $K[x_1, \cdots, x_N]$,

A finite set of non-zero polynomials will simply be called a *polynomial set*. The polynomial set Σ obtained from putting together the polynomials in two polynomial sets Σ_1 and Σ_2 will be denoted as $\Sigma_1 + \Sigma_2$. For polynomials F, G, etc., $\Sigma + \{F\}$ will also be denoted as $\Sigma + F$, and $\Sigma + \{F, G\}$ as $\Sigma + F + G$, etc.

We say that a polynomial set Σ defines an *algebraic variety* or simply *a variety*, to be denoted as $|\Sigma|$, with Σ as its *defining set*. For two polynomial sets Σ_1 and Σ_2, if any extended zero of Σ_1 is also an extended zero of Σ_2, then we say that the algebraic variety defined by Σ_1 is a *subvariety* of that defined by Σ_2, to be denoted as

$$\Sigma_2 = 0 | \Sigma_1, \quad \text{or} \quad |\Sigma_1| \subset |\Sigma_2|.$$

If, further, we have $|\Sigma_2| \subset |\Sigma_1|$ so that Σ_1, Σ_2 have the same set of extended zeros, then we say that Σ_1, Σ_2 are *equivalent*, denoted as

$$\Sigma_1 \approx \Sigma_2, \quad \text{or} \quad |\Sigma_1| = |\Sigma_2|.$$

If $|\Sigma_1| \subset |\Sigma_2|$ but $|\Sigma_1| \neq |\Sigma_2|$, or $|\Sigma_1| \subsetneqq |\Sigma_2|$, then we say that the variety defined by Σ_1 is a *true* subvariety of that defined by Σ_2.

Given a polynomial F, if any extended zero of Σ is also one of F, i.e.

$$\{F\} = 0 | \Sigma \quad \text{or} \quad |\Sigma| \subset |\{F\}|,$$

then we say that $F = 0$ on Σ, denoted as $F = 0 | \Sigma$. Otherwise we denote this as

$$F \neq 0 | \Sigma,$$

Given $k + 1$ polynomial sets $\Sigma, \Sigma_1, \cdots, \Sigma_k (k > 1)$ having the following property: Any extended zero of Σ is also an extended zero of at least one of the sets $\Sigma_1, \cdots, \Sigma_k$, and conversely, any extended zero of any Σ_i is also one of Σ, then we say that $\Sigma_1, \cdots, \Sigma_k$ are a *decomposition* of Σ, or the corresponding algebraic varieties $|\Sigma_1|, \cdots, |\Sigma_k|$ are a *decomposition* of $|\Sigma|$, denoted as

$$|\Sigma| = |\Sigma_1| \cup \cdots \cup |\Sigma_k| \qquad (k > 1).$$

If for any i, $|\Sigma_i|$ cannot be omitted in the above decomposition, then the decomposition is said to be *uncontractible*. In this case the variety defined by each Σ_i is a true subvariety of the variety defined by Σ, but not a subvariety defined by the union of other Σ_j's.

We say that the polynomial set Σ is *reducible* if it has some uncontractible decomposition and the variety defined by it is also said to be *reducible*. In the contrary case we say that Σ as well as the variety defind by it is *irreducible*. If in a certain decomposition of Σ each Σ_i is irreducible, then we say that this decomposition is an *irreducible decomposition* of Σ; the same for the variety defined by Σ. In this case each Σ_i or the variety defined by it is called an *irreducible component* of Σ or the variety defined by it.

We consider now the problem of reducibility of a polynomial set or its defining algebraic variety. The following two lemmas give some well-known criteria for their *irreducibility*.

Lemma 1. *A necessary and sufficient condition for a polynomial set Σ to be irreducible is that there cannot exist two non-zero polynomials G and H such that*

$$GH = 0 | \Sigma,$$

while

$$G \neq 0 | \Sigma, \ H \neq 0 | \Sigma.$$

For the second criterion let us first introduce the important notion of the so-called *generic point* of a variety. Consider two extension fields $\widetilde{K}$ and K' of K and two points $\xi = (\tilde{x}_1, \cdots, \tilde{x}_N)$, $\tilde{x}_i \in \widetilde{K}$, and $\xi' = (x'_1, \cdots, x'_N)$, $x'_i \in K'$, in the N-dimensional linear spaces $\widetilde{K}^N$ and K'^N on $\widetilde{K}$ and K' respectively. Suppose that these two points possess the following property:

For any polynomial $F(x_1, \cdots, x_N)$ in $K[x_1, \cdots, x_N]$, that ξ is an extended zero of F would imply that ξ' is also an extended zero of F; in other words, $F(x'_1, \cdots, x'_N) = 0$ as long as $F(\tilde{x}_1, \cdots, \tilde{x}_N) = 0$.

In this case ξ' will be called a *specialization* of ξ with respect to K, or simply a specialization of ξ if no misunderstanding can occur.

Suppose the polynomial set Σ has a certain extended zero ξ such that any extended zero of Σ is a specialization of ξ with respect to K, then we say that ξ is a *generic point* of the polynomial set Σ or one of the algebraic variety $|\Sigma|$ defined by it. The following lemma gives the second irreducibility criterion of polynomial sets or algebraic varieties:

Lemma 2. *A necessary and sufficient condition for a polynomial set Σ or its variety to be irreducible is that Σ has generic points.*

The two lemmas above give some necessary and sufficient conditions which are however merely existential in character and not constructive at all. Given a polynomial set Σ, there is no means to ascertain in a finite number of steps whether the conditions in the lemmas can be satisfied or not. For the purpose of mechanical theorem proving, we have to devise some mechanical procedure which permits to decide in a finite number of steps whether a given polynomial set is irreducible or not, and in the case it is reducible, to give in a finite number of steps the various irreducible components of the decomposition. Such a mechanization may be considered as constituting a *constructive theory* of algebraic geometry. It was given in details in the two books of J. F. Ritt[2,3] and we shall give some outlines in somewhat revised form of this theory below.

Consider an ascending set

$$\Phi: A_1, A_2, \cdots, A_n$$

in which the class of A_i is p_i with

$$0 < p_1 < p_2 < \cdots < p_n.$$

We shall change the notations in setting

$$x_{p_1} = y_1, \cdots, x_{p_n} = y_n$$

and denote the other x's in the original order as $u_1, \cdots, u_d$. We call

$$d = N - n$$

the *dimension* of the ascending set Φ, denoted as

$$d = \dim \Phi.$$

Write now the polynomials A_i in Φ in the following form:

$$\Phi: \begin{cases} A_1 = C_{10}y_1^{m_1} + C_{11}y_1^{m_1-1} + \cdots + C_{1m_1}, \\ A_2 = C_{20}y_2^{m_2} + C_{21}y_2^{m_2-1} + \cdots + C_{2m_2}, \\ \cdots \\ A_n = C_{n0}y_n^{m_n} + C_{n1}y_n^{m_n-1} + \cdots + C_{nm_n}. \end{cases}$$

In the expressions $C_{i0} \neq 0$ are initials of A_i, and each C_{ij} is a polynomial in $u_1, \cdots, u_d$, $y_1, \cdots, y_{i-1}$ with coefficients in K. Furthermore each A_i has already been reduced with respect to $A_1, \cdots, A_{i-1}$ so that the degrees of $y_1, \cdots, y_{i-1}$ in C_{ij} are less than $m_1, \cdots, m_{i-1}$ respectively. The first problem to be considered is to give conditions for Φ to be the basic set of a certain irreducible polynomial set.

For this problem let us suppose that the ascending set Φ possesses the following property:

Let the transcendental extension field $K(u_1, \cdots, u_d)$ of K got by adjoining $u_1, \cdots, u_d$ be denoted by K_0; then A_1, as a polynomial in $K_0[y_1]$ with coefficients in K_0, is irreducible in $K_0[y_1]$.

Let the algebraic extension field of K_0 got by adjoining an extended zero η_1 of $\tilde{A}_1 = 0$ be denoted by $K_0(\eta_1) = K_1$; then the polynomial $\tilde{A}_2$ in $K_1[y_2]$ obtained by substituting η_1 for y_1 in A_2 is irreducible in $K_1[y_2]$.

Let the algebraic extension field of K_1 got by adjoining an extended zero η_2 of $\tilde{A}_2 = 0$ be denoted by $K_1(\eta_2) = K_2$; then the polynomial $\tilde{A}_3$ in $K_2[y_3]$ obtained by substituting η_1 for y_1 and η_2 for y_2 in A_3 is irreducible in $K_2[y_3]$.

Suppose that proceeding in the same manner we get successively algebraic extensions $K_i = K_{i-1}(\eta_i)$, polynomials $\tilde{A}_i$ obtained by substituting $\eta_1, \cdots, \eta_{i-1}$ for $y_1, \cdots, y_{i-1}$ in A_i, and some extended zeros η_i of $\tilde{A}_i = 0$, where each $\tilde{A}_i$ is irreducible in $K_{i-1}[y_i]$ for $i = 1, 2, \cdots, n$. Under these conditions we say that the ascending set Φ is *irreducible*. By known methods there exist some mechanical procedures which permit to decide in a finite number of steps whether Φ is irreducible or not.

Let Φ be irreducible and satisfy the conditions above. Then u_i, η_j are all elements in $\tilde{K} = K_n$ and $\tilde{\eta} = (u_1, \cdots, u_d, \eta_1, \cdots, \eta_n)$ can be considered as a point of the linear space $\tilde{K}^{d+n} = \tilde{K}^N$. We shall call $\tilde{\eta}$ a *generic point* of Φ and $\tilde{K}$ a *generating field* of Φ.

The following lemma is quite important for the theory.

Lemma 3. *If the ascending set Φ is irreducible with*

$$\tilde{\eta} = (u_1, \cdots, u_d, \eta_1, \cdots, \eta_n)$$

a generic point as above, then for a polynomial $F \in K[u_1, \cdots, u_d, y_1, \cdots, y_n]$ to have the remainder $R = 0$ with respect to Φ, it is necessary and sufficient that $\tilde{\eta}$ is an extended zero of F.

Proof. Denote the ascending set formed by the first k terms in Φ by

$$\Phi_k: A_1, A_2, \cdots, A_k \ (1 \leqslant k \leqslant n).$$

Denote by K_k the $(d+k)$-dimensional linear space over K with basis $u_1, \cdots, u_d, y_1, \cdots,$

y_k. Similarly for the others. Then Φ_k is clearly irreducible, and

$$\tilde{\eta}_k = (u_1, \cdots, u_d, \eta_1, \cdots, \eta_k),$$

when considered as a point in K_k^{d+k}, is a generic point of Φ_k while K_k is the generating field of Φ_k.

We shall prove by induction on k the following two assertions:

1_k. $\tilde{\eta}_{k-1}$ is not an extended zero of C_{k0}.

2_k. If $R_k \in K[u_1, \cdots, u_d, y_1, \cdots, y_k]$ is already reduced with respect to Φ_k and $\tilde{\eta}_k$ is an extended zero of R_k, then R_k is identically 0.

As $C_{k+1,0} \in K[u_1, \cdots, u_d, y_1, \cdots, y_k]$ is known to be reduced with respect to Φ_k and is $\neq 0$, so 1_{k+1} is a consequence of 2_k.

Suppose 2_{k-1} has already been proved. Consider any R_k satisfying the conditions in 2_k. Write R_k as a polynomial in y_k,

$$R_k = S_0 y_k^r + S_1 y_k^{r-1} + \cdots + S_r,$$

in which $S_i \in K[u_1, \cdots, u_d, y_1, \cdots, y_{k-1}]$ with $r < m_k$. Substitute $y_1, \cdots, y_{k-1}$ in S_i by $\eta_1, \cdots, \eta_{k-1}$ with the resulting S_i as $\tilde{S}_i \in K_{k-1}$. Set

$$\tilde{R}_k = \tilde{S}_0 y_k^r + \tilde{S}_1 y_k^{r-1} + \cdots + \tilde{S}_r \in K_{k-1}[y_k].$$

By hypothesis η_k is an extended zero of $\tilde{R}_k = 0$. As $r < m_k$ and η_k is an extended zero of the irreducible polynomial $\tilde{A}_k$ in K_{k-1}, $\tilde{R}_k$ should be identically 0 and so $\tilde{S}_0 = 0, \cdots, \tilde{S}_r = 0$. As R_k is reduced with respect to Φ_k so that each S_i is reduced with respect to Φ_{k-1}, by induction hypothesis 2_{k-1} we have necessarily $S_i = 0$ so that $R_k = 0$, i. e., 2_k holds true. It follows that 1_{k+1} is also true. The above proof is clearly valid for 2_1 while 1_1 is quite evident. Consequently 1_k and 2_k are true for $k = 1, 2, \cdots, n$.

It is now easy to complete the proof of Lemma 3 as follows.

Let the remainder of F with respect to $\Phi_n = \Phi$ be R; then we have the following remainder formula

$$C_{10}^{s_1} \cdots C_{n0}^{s_n} F = Q_1 A_1 + \cdots + Q_n A_n + R.$$

Suppose $R = 0$. Since $\tilde{\eta}$ is an extended zero of all A_i's while by 1_k it is not an extended zero of any C_{k0}, so by the formula above it should be an extended zero of F. Conversely, if $\tilde{\eta}$ is an extended zero of F, then by the same formula $\tilde{\eta}$ should also be an extended zero of R. By 2_n we have necessarily $R = 0$. This completes the proof.

Lemma 4. *Let the ascending set*

$$\Phi: A_1, A_2, \cdots, A_n$$

be irreducible with a generic point

$$\tilde{\eta} = (u_1, \cdots, u_d, \eta_1, \cdots, \eta_n)$$

as before. If the polynomial $F \in K[u_1, \cdots, u_d, y_1, \cdots, y_n]$ has its remainder $\neq 0$ with respect to Φ, then in $K[u_1, \cdots, u_d, y_1, \cdots, y_n]$ there are polynomials G and $Q_i, i = 1, \cdots, n$ such that

$$GF-(Q_1A_1+\cdots+Q_nA_n)\in K[u_1,\cdots,u_d]$$

and that

$$G(\bar\eta)\neq 0.$$

Proof. Omitted.

Given an irreducible set Φ as above, let Ω be the set of all polynomials in $K[u_1,\cdots,u_d,y_1,\cdots,y_n]$ for which the remainder with respect to Φ is 0. By Lemma 3, this set will form clearly a module. By the Hilbert basis theorem, there will be a finite number of polynomials in Ω, such that any polynomial of Ω is a linear combination of these polynomials with polynomial coefficients. We may add the A_i's of Φ into this finite set and denote the enlarged finite set by Ω_Φ. By Lemma 3 this polynomial set will have clearly Φ as its basic set and $\bar\eta$ as an extended zero.

Let G be any polynomial with $\bar\eta$ as an extended zero; then by Lemma 3 G has its remainder $=0$ with respect to Φ. By the construction of Ω_Φ, G is a linear sum of polynomials in Ω_Φ so that $G=0/\Omega_\Phi$. It follows that any extended zero of Ω_Φ is a specialization of $\bar\eta$ or that Ω_Φ is an irreducible polynomial set with $\bar\eta$ as a generic point. We thus get the following

Theorem 1. *Any irreducible ascending set Φ is the basic set of some irreducible polynomial set Ω_Φ,*

The above proof showing how to get an irreducible polynomial set Ω_Φ from a given irreducible ascending set Φ is based on the use of the finite basis theorem of Hilbert. As Ω is transfinite, and the existence of a finite basis depends on the axiom of choice, only the existence of such an irreducible polynomial set Ω_Φ has been actually proved. However, there does exist some mechanical procedure to produce in a finite number of steps such an irreducible polynomial set Ω_Φ consisting of a finite number of polynomials. In other words, we may strengthen the above theorem to the following form:

Theorem 1'. *There exists some mechanical procedure for any irreducible ascending set Φ which will permit to determine in a finite number of steps a finite number of polynomials including those of Φ that form an irreducible polynomial set Ω_Φ with any generic point of Φ as its generic point.*

The proof of the constructive Theorem 1' is not a simple one. As in applications the mere existence of such an irreducible polynomial set Ω_Φ will already be sufficient, as guaranteed by the Hilbert basis theorem, we shall satisfy ourselves in merely stating the theorem while putting aside the proof.

The next problem to be studied is the decomposition of a polynomial set or the corresponding algebraic variety into irreducible components. For this purpose let Φ, $\bar\eta$ and Ω_Φ be as before. we have shown that the irreduciblity of Φ is a sufficient condition for Φ to be the basic set of some irreducible polynomial set Ω_Φ with the same generic point $\bar\eta$ as Φ which can even be determined in a mechanical manner in a finite number of steps. To this we now give the following supplement:

Lemma 5. *Let the basic set Φ of a polynomial set Λ be irreducible with the class of each polynomial A_i in Φ being >0. Denote the initial of A_i by I_i, $i=1,\cdots,n$. If any polynomial in Λ has its remainder 0 with respect to Φ, then Λ has a decomposition*

$$|A| = |\Omega_\Phi| \cup |A + I_1| \cup \cdots \cup |A + I_n|,$$

in which Ω_Φ or the corresponding algebraic variety $|\Omega_\Phi|$ is irreducible.

Proof. For such a polynomial G in A or not with its remainder 0 with respect to Φ we would have, for some $s_i \geqslant 0$ and $Q_i \in K[u_1, \cdots, u_d, y_1, \cdots, y_n]$,

$$I_1^{s_1} \cdots I_n^{s_n} G = Q_1 A_1 + \cdots + Q_n A_n.$$

By the construction of Ω_Φ, G should be a linear sum of polynomials in Ω_Φ so that any extended zero of Ω_Φ should be an extended zero of G and hence an extended zero of A. Conversely, any extended zero of A may be considered an extended zero of A_i's. Hence by the above formula it should be an extended zero of either any such G or some I_i. In other words, it should be an extended zero of Ω_Φ or some $A + I_i$. Thus we have the decomposition as shown in the lemma.

Lemma 6. *Let A, Φ be as in Lemma 5 with A being irreducible. Then*

$$A \approx \Omega_\Phi \text{ or } |A| = |\Omega_\Phi|.$$

Proof. Let the initials of the polynomials in Φ be I_i, $i = 1, \cdots, n$. Then it is clear by definition that

$$|A + I_1| \cup \cdots \cup |A + I_n| = |A + I_1 \cdots I_n|.$$

The decomposition given in Lemma 5 can therefore be written in the form

$$|A| = |\Omega_\Phi| \cup |A + I_1 \cdots I_n|.$$

As the generic point of Φ is also a generic point of Ω_Φ but cannot be any extended zero of $I_1 \cdots I_n$, so $|\Omega_\Phi| \not\subset |A + I_1 \cdots I_n|$. If A has some extended zero which is not an extended zero of Ω_Φ, it should be an extended zero of $A + I_1 \cdots I_n$ so that we shall have $|A + I_1 \cdots I_n| \not\subset |\Omega_\Phi|$. In this way $|A|$ would have an uncontractible decomposition contrary to the irreducibility hypothesis of A. Hence we should have $|A| \subset |\Omega_\Phi|$. As conversely we should have $|\Omega_\Phi| \subset |A|$, so $|A| = |\Omega_\Phi|$, Q. E. D.

Consider now an ascending set Φ as before but with Φ not necessarily irreducible. Then there will be some k such that

$$\Phi_{k-1}: A_1, A_2, \cdots, A_{k-1}$$

is irreducible, with

$$\tilde{\eta}_{k-1} = (u_1, \cdots, u_d, \eta_1, \cdots \eta_{k-1})$$

as a generic point, and that the polynomial $\tilde{A}_k$ got from A_k by substituting $\eta_1, \cdots, \eta_{k-1}$ for $y_1, \cdots, y_{k-1}$ is reducible in $K_{k-1}[y_k]$, where $K_{k-1} = K_0(\eta_1, \cdots, \eta_{k-1})$. Let the irreducible factorization of $\tilde{A}_k$ in $K_{k-1}[y_k]$ be given by

$$\tilde{A}_k = g_1 \cdots g_h,$$

in which each $g_i \in K_{k-1}[y_k]$ is irreducible, and $h \geqslant 2$. As in g_i the coefficients of powers of y_k are all elements of K_{k-1} and can thus be expressed as the quotients of two polynomials in $u_1, \cdots, u_d, \eta_1, \cdots, \eta_{k-1}$, multiplying by a common multiple of the denominators we would get an expression of the form

$$\widetilde{D}\widetilde{A}_k = \widetilde{G}_1 \cdots \widetilde{G}_h,$$

in which $D \in K[u_1, \cdots, u_d, y_1, \cdots, y_{k-1}]$, $G_i \in K[u_1, \cdots, u_d, y_1, \cdots, y_k]$, while $\widetilde{D}$, $\widetilde{G}_i$ are got from D, G_i by substituting $\eta_1, \cdots, \eta_{k-1}$ for $y_1, \cdots, y_{k-1}$ and are polynomials in $K_{k-1}[y_k]$. We may also consider D as already reduced with respect to Φ_{k-1}. Similarly we may consider G_j as already reduced with respect to Φ_k.

Write the polynomial $G_1 \cdots G_h - DA_k$ in a form according to powers of y_k, say,

$$G_1 \cdots G_h - DA_k = \sum_j B_j y_k^j,$$

in which $B_j \in K[u_1, \cdots, u_d, y_1, \cdots, y_{k-1}]$. Denote by b_j the element in $K_{k-1} = K_0(\eta_1, \cdots, \eta_{k-1})$ got from B_j by substituting $\eta_1, \cdots, \eta_{k-1}$ for $y_1, \cdots, y_{k-1}$. Then we have $b_j = 0$ since $\widetilde{D}\widetilde{A}_k = \widetilde{G}_1 \cdots \widetilde{G}_k$. In other words, each B_j will have $\tilde{\eta}_{k-1}$ as an extended zero. It follows from the proof of Lemma 5 that each B_j will have its remainder 0 with respect to the irreducible ascending set Φ_{k-1}, so that there are non-negative integers $s_{j1}, \cdots, s_{j,k-1}$ and polynomials $Q_{ji} \in K[u_1, \cdots, u_d, y_1, \cdots, y_{k-1}]$ verifying the relation ($C_{i_0} = I_i$)

$$I_1^{s_{j1}} \cdots I_{k-1}^{s_{j,k-1}} B_j = \sum_{i=1}^{k-1} Q_{ji} A_i.$$

Set $s_i = \max_j (s_{ji})$; we then get

$$I_1^{s_1} \cdots I_{k-1}^{s_{k-1}} (G_1 \cdots G_h - DA_k) = \sum_{i=1}^{k-1} Q_i A_i$$

or

$$I_1^{s_1} \cdots I_{k-1}^{s_{k-1}} G_1 \cdots G_h = \sum_{i=1}^{k} Q_i A_i,$$

in which Q_i are polynomials in $u_1, \cdots, u_d, y_1, \cdots, y_k$.

From the above it is easy to get the following

Lemma 7. *Let the polynomial set Λ have Φ as basic set, and let the class of term A_i be > 0 and the initial of A_i be I_i, $i = 1, \cdots, n$. Suppose that Φ is reducible, so that there is some k for which the ascending set Φ_{k-1} formed by the first $k-1$ terms of Φ is irreducible with $\tilde{\eta}_{k-1} \in K_{k-1}$ as a generic point, while the polynomial got from A_k by substituting $\tilde{\eta}_{k-1}$ for the corresponding variables is reducible with an irreducible factorization into polynomials $G_1, \cdots, G_h$. Then there is a decomposition of the form*

$$|\Lambda| = |\Lambda + I_1| \cup \cdots \cup |\Lambda + I_{k-1}| \cup |\Lambda + G_1| \cup \cdots \cup |\Lambda + G_h|.$$

Proof. Any extended zero of either a $\Lambda + I_i$ or a $\Lambda + G_j$ on the right-hand side of the above expression is clearly also an extended zero of Λ. Conversely, any extended zero of Λ is also an extended zero of all A_i's. From the expression just before the lemma it is also an extended zero of some I_i or some G_j, i. e. one of some $\Lambda + I_i$ or $\Lambda + G_j$. This proves the decomposition formula.

Lemma 8. *Let Λ be a polynomial set with Φ as basic set as in Lemma 5 or Lemma*

7. Then the basic set of any polynomial set $\Lambda + I_i$ or $\Lambda + G_j$ appearing in the right-hand side of the decompositions of these lemmas will have its rank lower than that of Φ.

Proof. As each I_i is already reduced with respect to Φ and each G_j is assumed to be reduced with respect to Φ_k and hence also reduced with respect to Φ, the present lemma is an immediate consequence of Lemma 3 of Section 2.

Lemma 9. *Let the polynomial set Λ be irreducible with an irreducible ascending set Φ as its basic set. Suppose also that any polynomial in a polynomial set Λ' or Λ has its remainder 0 with respect to Φ. Then*

$$|\Lambda| \cup |\Lambda'| = |\Lambda'|,$$

or the decomposition $|\Lambda| \cup |\Lambda|'$ is contractible.

Proof. By Lemma 6 we have $|\Omega_\Phi| = |\Lambda|$. By hypothesis any polynomial G' in Λ' has its remainder 0 with respect to Φ. It follows therefore that the generic point of Φ, or the generic point of Ω_Φ, is an extended zero of G', whence $G' = 0/\Omega_\Phi$. Consequently $\Lambda' = 0/\Omega_\Phi$, or $|\Omega_\Phi| \subset |\Lambda'|$, or $|\Lambda| \subset |\Lambda'|$. This proves the lemma.

From the above lemmas and also the preceding section we get the following mechanical procedure for getting the uncontractible irreducible decomposition of a polynomial set.

Let the given polynomial set be Σ. By the well-ordering theorem given in the preceding section, we can, in following some mechanical procedure, successively enlarge the given set Σ to get a sequence of polynomial sets steadily increasing as shown below:

$$\Sigma = \Sigma_1 \subset \Sigma_2 \subset \cdots \subset \Sigma_q = \Lambda.$$

These polynomial sets are actually mutually equivalent, viz.

$$\Sigma = \Sigma_1 \approx \Sigma_2 \approx \cdots \approx \Sigma_q = \Lambda.$$

Two cases may appear. In the first case Λ turns out, in a certain step, to be a contradictory set consisting of a single term which is a non-zero element in K. In this case Σ itself is a contradictory set with no extended zeros. Hence it is only necessary to consider the second case. In that case Λ has a basic set

$$\Phi: A_1, A_2, \cdots, A_n,$$

with $I_1, \cdots, I_n$ as initials and class of $A_1 > 0$. Moreover, Λ will possess the following properties: Any polynomial in Λ will have its remainder 0 with respect to Φ, any extended zero of Σ is also one of Φ, and any extended zero of Φ, if not one of any initial I_i, is also an extended zero of Σ.

Now according to the beginning part of this section, there is some mechanical procedure to verify whether Φ is reducible, or whether A_i's are reducible in the successively extended fields K_{i-1}. We have two subcases again.

In the first subcase Φ is irreducible. By Lemma 5 there is a decomposition

$$|\Lambda| = |\Omega_\Phi| \cup |\Lambda + I_1| \cup \cdots \cup |\Lambda + I_n|,$$

in which Ω_Φ is irreducible while all $\Lambda + I_i$ have some basic sets of ranks lower than that of Λ. We may then consider each $\Lambda + I_i$ as a new polynomial set Σ and proceed again as in the beginning.

In the second subcase Φ is reducible. Then we have by Lemma 7 some decomposition

$$|A| = |A + I_1| \cup \cdots \cup |A + I_{k-1}| \cup |A + G_1| \cup \cdots \cup |A + G_h|,$$

in which each $A + I_i$ or $A + G_j$ has some basic set of a rank lower than that of A. We may then consider each $A + I_i$ or $A + G_j$ as a new polynomial set and proceed again as before.

Whatever the subcase may be, we may take each $A + I_i$ or $A + G_j$ as a new polynomial set Σ' in succession and proceed as before to get a sequence

$$\Sigma' = \Sigma'_1 \approx \Sigma'_2 \approx \cdots \Sigma'_{q'} = A'.$$

In the case that A' has a basic set consisting of a single term which is a non-zero element of the field K, we may remove $|A'|$ or the original $|A + G_j|$ or $|A + I_i|$ from the decomposition. In the contrary case $|A'|$ will be decomposed further into several algebraic varieties with basic sets of rank lower than the preceding ones for the corresponding polynomial set, plus possibly one with corresponding irreducible polynomial set $\Omega_{\Phi'}$ having an irreducible ascending set Φ' as a basic set. In this way we will get a further decomposition of $|A|$ or $|\Sigma|$ itself. In the decomposition there will appear irreducible polynomial sets of the form Ω_Φ, $\Omega_{\Phi'}$ as well as those of the form $A' + I'$ or $A' + G'$. For the latter ones we may decompose them further as before.

As in each step for further decomposition the polynomial sets $A' + I'$ or $A' + G'$ involved have their basic sets of ranks lower than the preceding ones, the decomposition should stop in a finite number of steps owing to the well-ordering theorem of Section 2. Consequently, in a finite number of steps we shall arrive at a decomposition of the following form:

$$|\Sigma| = |\Omega_{\Phi_1}| \cup |\Omega_{\Phi_2}| \cup \cdots \cup |\Omega_{\Phi_s}|,$$

in which each Φ_i is an irreducible ascending set, and Ω_{Φ_i} is the irreducible polynomial set got from Φ_i as described in Theorem 1.

According to the above construction, each $|\Omega_{\Phi_i}|$ cannot be a subvariety of any $|\Omega_{\Phi_j}|$, $j > i$, but we cannot say that some $|\Omega_{\Phi_i}|$ cannot be a subvariety of any $|\Omega_{\Phi_j}|$, $j < i$. This is because we apply only Theorem 1 which asserts the mere existence of Ω_{Φ_i} from Φ_i. If we take into account Theorem 1′ which asserts a mechanical procedure for the concrete determination of Ω_{Φ_i} from Φ_i, then we may use Lemma 9 to prove if any $|\Omega_{\Phi_i}|$ is a subvariety of a preceding $|\Omega_{\Phi_j}|$, $j < i$, or not. It follows that, on the basis of Theorem 1′, we can get a noncontractible irreducible decomposition of $|\Sigma|$ in a mechanical manner.

In a word, we get finally the following

Theorem 2. *There is a mechanical procedure which permits to determine for a polynomial set Σ, in a finite number of steps, a noncontractible irreducible decomposition of the form*

$$|\Sigma| = |\Omega_{\Psi_1}| \cup \cdots \cup |\Omega_{\Psi_t}|,$$

in which each Ψ_i is an irreducible ascending set of Ω_{Ψ_i}.

For the application to mechanical theorem proving, it is however actually not necessary to carry out the decomposition up to the end to arrive at a noncontractible one. In fact, it is usually sufficient to have an irreducible decomposition which may be a contractible one.

Hence for the applications the existential Theorem 1, but not necessarily the constructive Theorem 1′, will be quite sufficient to meet the purpose.

§4. Proof of the Algebraic Mechanization Theorem

We give below the proof of the Mechanization Theorem in the algebraic form as described in Section 1. For this we first make some preparations.

Given a set of variables $x_1, \cdots, x_N$ arranged in a definite order:

$$x_1 \prec x_2 \prec \cdots \prec x_N,$$

and given a basic field K of characteristic 0 and an ascending set of polynomials in $K[x_1, \cdots, x_N]$,

$$\Phi: A_1, A_2, \cdots, A_n,$$

for which the classes satisfy the relations

$$0 < p_1 < p_2 < \cdots < p_n,$$

we rewrite each x_{p_i} as y_i and the other x's as $u_1, \cdots, u_d$ with $d = N - n$. Then A_i's can be put in the form

$$A_i = C_{i0}y_i^{m_i} + C_{i1}y_i^{m_i-1} + \cdots + C_{im_i},$$

in which

$$C_{ij} \in K[u_1, \cdots, u_d, y_1, \cdots, y_{i-1}], \quad i = 1, \cdots, n; j = 0, 1, \cdots, m_i.$$

The initials I_i of A_i are then just the polynomials $I_i = C_{i0} \in K[u_1, \cdots, u_d, y_1, \cdots, y_{i-1}]$. We call each inequation

$$I_i \neq 0$$

a *non-degeneracy condition*.

Let a polynomial $G \in K[u_1, \cdots, u_d, y_1, \cdots, y_n]$ be given. Construct the remainder R of G with respect to Φ. Then by the remainder formula we have

$$I_1^{s_1} \cdots I_n^{s_n} G = Q_1 A_1 + \cdots + Q_n A_n + R,$$

for certain non-negative integers $s_i \geq 0$, with each $Q_i \in K[u_1, \cdots, u_d, y_1, \cdots, y_n]$.

We shall investigate the necessary and sufficient conditions such that

$$G = 0$$

may be deduced as a consequence of the equations $A_i = 0$, $i = 1, \cdots, n$. We shall prove that, under the subsidiary non-degeneracy conditions $I_i \neq 0$ and under the hypothesis that Φ is irreducible, the necessary and sufficient condition is just $R = 0$. Whether the set Φ is irreducible or not, the sufficiency of the condition is quite evident from the above remainder formula. So we have the following

Theorem 1. *Let Φ, A_i, I_i, G be as above and $R=0$; then under the non-degeneracy conditions*

$$I_i \neq 0, i = 1, \cdots, n,$$

$G = 0$ is a consequence of $A_i = 0$, $i = 1, \cdots, n$, whether Φ is reducible or not.

If Φ is irreducible, under the non-degeneracy conditions for $G=0$ to be a consequence of $A_i=0$, $i=1,\cdots,n$, the condition $R=0$ is not only necessary but also sufficient, as in the following theorem which follows directly from Lemma 3 in Section 3.

Theorem 2. *Let Φ, A_i, I_i, G be as above and Φ be irreducible. If under the non-degeneracy conditions $I_i \neq 0$ the equation $G=0$ is a consequence of the equations $A_i=0$, $i=1,\cdots,n$ (for a certain extension field of K), then the remainder R of G with respect to Φ is 0.*

Remark. The proofs of these theorems depend very much on the theory developed in Section 3 and are rather involved. If we restrict ourselves to real field as is the case of ordinary Euclidean geometry and pay no attention to the constructive aspects, then the proofs will be much simpler.

We now give the proof of the Mechanization Theorem of unordered geometries in its algebraic form.

Given a geometrical statement (S) in a certain unordered geometry, our object is to give a mechanical method to decide whether (S) is true or not. For this purpose we choose first a coordinate system, express the points involved by coordinates, denote these coordinates by x_i, and arrange them in a certain definite order:

$$x_1 \prec x_2 \prec \cdots \prec x_N.$$

Next we translate the various geometrical relations in the statement (S) into algebraic relations of these coordinates. Then the hypothesis in the statement (S) will be translated into a system of equations

$$F_1=0, \quad \cdots, \quad F_s=0,$$

in which F_i are polynomials in $K[x_1, \cdots, x_N]$, with K the basic field of characteristic 0 associated to the geometry in question. Actually all these polynomials are with rational or even integer coefficients. The conclusions of the statement (S) will then be turned into another system of equations

$$G_1=0, \quad \cdots, \quad G_t=0,$$

with all G_j being polynomials in $K[x_1, \cdots, x_N]$, also with rational or integer coefficients. Without loss of generality we may suppose that there is only one such polynomial G_j, denoted simply by G henceforward. The polynomials F_i are then called *hypothesis polynomials* of the statement (S), and the G_j's or G the *conclusion polynomial(s)* of (S).

The proof of the Mechanization Theorem consists in exhibiting a mechanical procedure which permits to determine first in a finite number of steps a set of polynomials $D_1, \cdots, D_r$ for *non-degeneracy conditions*, with all D_k in $K[x_1, \cdots, x_N]$, which will actually be all with rational or even integer coefficients. Secondly the same mechanical procedure will also permit to decide in a finite number of steps whether under the non-degeneracy conditions

$$D_1 \neq 0, \quad \cdots, \quad D_r \neq 0,$$

the equation $G=0$ will be a consequence of $F_1=0, \cdots, F_s=0$.

With the language of algebraic geometry, the proof of Mechanization Theorem can also be restated in an alternative form in the following manner:

Denote the set of hypothesis polynomials F_i by $\Sigma = \{F_i\}$. The set Σ defines an algebraic variety $|\Sigma|$, with dimension d, viz., the dimension of any characteristic set of Σ. The proof of the Mechanization Theorem consists then in exhibiting a mechanical procedure which permits to determine a set of polynomials $D_1, \cdots, D_r$ such that in adjoining each D_i to Σ, the resulting polynomial set $\Sigma + D_i$ will define an algebraic variety $|\Sigma + D_i|$ of dimension $< d$. Furthermore, the same procedure will permit to decide, under the non-degeneracy conditions $D_1 \neq 0, \cdots, D_r \neq 0$, whether $G = 0$ or not; in other words, whether G will be 0 or not on the remaining part of the algebraic variety $|\Sigma|$ after removal of the true subvarieties $|\Sigma + D_i|$.

As briefly indicated in Section 3, we can decompose the algebraic variety into irreducible components, each of which has an irreducible basic set Φ_i which determines in turn that irreducible component in question, denoted by $|\Omega_{\Phi_i}|$. Furthermore, in the case the dimension d_i of $|\Omega_{\Phi_i}|$ is less than the dimension d of $|\Sigma|$, then this true subvariety is got from a certain previous $|\Omega_{\Phi_j}|$ by adjoining to Φ_j some polynomial D_i which is either an initial I_k or some G_l in the previous notations and $|\Omega_{\Phi_i}|$ is a subvariety of $|\Phi_j + D_i|$. We take each such D_i as a non-degeneracy polynomial. Suppose after removal of all these true subvarieties, the remaining irreducible components of dimension d are

$$|\Omega_{\Phi_1}|, \cdots, |\Omega_{\Phi_t}|.$$

Denote the initials of each Φ_j by $I_{j1}, \cdots, I_{jh}$ and consider them also as non-degeneracy polynomials D_{jk}. Now whether $G = 0$ is a consequence of $F_1 = 0, \cdots, F_s = 0$ under the non-degeneracy conditions $D_i \neq 0$, $D_{jh} \neq 0$, is just the same as whether $G = 0$ on the remaining parts of $|\Omega_{\Phi_1}|, \cdots, |\Omega_{\Phi_t}|$ after removal of the components $|\Phi_j + D_i|$ and those defined by $D_{jk}' = 0$. By Theorems 1, 2 above this can be decided by whether the remainders of G with respect to Φ_j are all 0. It furnishes the mechanical procedure required and thus gives the proof of the Mechanization Theorem in question.

The above mechanical procedure of theorem-proving is theoretically quite simple in appearance. However it would be quite difficult to apply this method to the proof of concrete theorems. The reason is that the irreducible decomposition of algebraic varieties depends on factorization of polynomials which, though theoretically almost self-evident, is a rather difficult problem in practice for which no method of high efficiency is available even up to now. Consequently, the above method is entirely non-feasible in practice. Fortunately, for the theorem-proving in geometries, we usually hope that the theorem in question is really a true theorem and we hope to prove it true in an affirmative manner. For this purpose it is enough to prove, by Theorem 1, that the remainder of the conclusion polynomial G is 0 with respect to some ascending set, whether irreducible or not. Therefore, to each concrete theorem whose truth is to be tested and to be proved in the case it is really true, we may apply Theorem 1 directly. If by computation we know that G has its remainder 0 with respect to the ascending set, then the theorem in question is true and the computation furnishes actually a proof of this theorem. In this case everything is done. Only in the case that the remainder is not 0 should we ask further whether the corresponding ascending set is irreducible or not. For this reason we shall modify the above mechanical procedure of proof to the following form which has been proved to be very efficient in practice (some examples will be given in the next section).

The modified mechanical procedure runs somewhat as follows.

Consider a set $\mathscr{P}$ of polynomial sets and a set Δ of polynomials, where Δ is called the *degeneracy set*. In the outset, $\mathscr{P}$ will consist of a single polynomial set, viz. the set of hypothesis polynomials

$$\Sigma = \{F_1, \cdots, F_s\},$$

and the degeneracy set will be an empty one, viz.

$$\Delta = \emptyset.$$

During the procedure we shall increase or decrease the number of polynomial sets in $\mathscr{P}$ and also adjoin non-degeneracy polynomials into Δ to get the final

$$\Delta = \{D_1, \cdots, D_r\}$$

as required.

Step 1. Let $\mathscr{P}$ be non-empty. Then take arbitrarily a polynomial set Σ from $\mathscr{P}$, and remove it from $\mathscr{P}$ to get a new $\mathscr{P}$. Using the well-ordering theorem in Section 2 to enlarge Σ to successive polynomial sets as shown below:

$$\Sigma = \Sigma_1 \subset \Sigma_2 \subset \cdots \subset \Sigma_q = \Lambda.$$

If Λ has an element which is a non-zero number in K, then Λ is a contradictory set. In this case the hypothesis in the statement (S) is contradictory in itself and the procedure will be stopped. In the contrary case let the basic set of Λ be

$$\Phi: A_1, A_2, \cdots, A_n.$$

The initials of A_i will be denoted by I_i. By construction, any polynomial in Λ except A_i will have its remainder 0 with respect to Φ. In that case we have also

$$\dim|\Sigma| = \dim\Phi = N - n = d.$$

If Step 1 is just the first step from the very beginning of the whole procedure, then the dimension d will be recorded for future reference.

If Step 1 is the successive step from the other ones during the procedure, then we compare the new dimension d with the previous d recorded in the beginning.

If this new $d =$ the previously recorded d, then we adjoin the initials I_i to Δ to get some enlarged new degeneracy set Δ, and proceed to Step 2.

If this new $d <$ the previously recorded d, and the present Σ is obtained as some $\Lambda + I_i$ or $\Lambda + G_j$ during Step 3 below, then we adjoin this I_i or G_j to Δ to get a new Δ. We then return to Step 1 and proceed as before.

Step 2. Find the remainder R of G with respect to Φ.

Suppose $R = 0$. If in $\mathscr{P}$ there is not any more polynomial set, then the statement (S) is true under the non-degeneracy conditions

$$D_k \neq 0 \quad (D_k \in \Delta),$$

and the procedure will be stopped. In this case the theorem is true and is proved under the non-degeneracy conditions. Otherwise we return to Step 1 and proceed again as before.

Suppose $R \neq 0$. Then we proceed to Step 3.

Step 3. Check the irreducibility of the basic set Φ.

Suppose that Φ is irreducible. Then as G has its remainder $\neq 0$ with respect to Φ, by Theorem 2 under the non-degeneracy conditions

$$D_k \neq 0 \qquad (D_k \in \Delta)$$

statement (S) is not true; the procedure will then be stopped. In this case the theorem is not true under the above non-degeneracy conditions.

Suppose that Φ is reducible. Then there will be some decomposition

$$|A| = |A + I_1| \cup \cdots \cup |A + I_{l-1}| \cup |A + G_1| \cup \cdots \cup |A + G_h|.$$

Consider such $A + I_i$ and $A + G_j$ as new polynomial sets Σ, and adjoin all these to $\mathscr{P}$ to get a new enlarged set $\mathscr{P}$. Then return to Step 1 and proceed again as before.

According to the previous sections, the above procedure should stop in a finite number of steps. In this way we get a final degeneracy set

$$\Delta = \{D_k\}$$

and one of the following three conclusions should be true:

1) Under the non-degeneracy conditions

$$D_k \neq 0 \qquad (D_k \in \Delta)$$

the hypotheses in the statement (S) are contradictory in themselves.

2) Under the above non-degeneracy conditions, or under the additional hypothesis $D_k \neq 0$, the statement (S) is true, or, what is the same, the theorem in question is true.

3) Under the above non-degeneracy conditions, or under the additional hypothesis $D_k \neq 0$, the statement (S) is not true, or, what is the same, the theorem is not true.

Generally speaking, the degeneracy conditions

$$D_k = 0$$

are not worth any more consideration. If there is some necessity to consider such a degeneracy condition $D_k = 0$, we may simply take it as a new hypothesis to be adjunct to the statement (S), i.e., we consider $\{F_1, \cdots, F_s, D_k\}$ instead of $\{F_1, \cdots, F_s\}$ and then proceed as above.

The above mechanical procedure is very feasible. We have implemented it on small computers, proving and thus also discovering quite non-trivial theorems in this way. The next section will describe a few illustrative examples.

§5. Programming and Examples.

It is clear how to program according to the procedure described in the preceding sections. In fact, programming has been done and various theorems have been proved on rather small computers. Before we explain certain theorems proved in this way, let us first add some remarks.

First, we may lessen the labour of computation by modifying slightly the definition of the basic set and characteristic set. Thus, we shall define an ascending set

$$\mathscr{A}: A_1, A_2, \cdots, A_r$$

to be one *in loose sense* or *in weak sense* in requiring only that each A_i in the set be reduced merely with respect to the variables occuring in the leading term of A_i alone. The notions of basic set, etc. derived in this way are then also said to be in *loose sense* or *weak sense*. This will not affect the final conclusions but will greatly simplify the programming and the computation Thus. the polynomial set corresponding to the hypothesis of a theorem in the ordinary geometry is usually already in the form of an ascending set and hence also a basic set in the above loose or weak sense. In the worse case a few strokes of simple hand computations may be required. The procedure of well-ordering is not necessary in general because it is quite laboursome.

Secondly, we are only interested in arriving at *true* theorems so that only the sufficiency part of our criterion will be considered in the programming. Thus, if the remainder of the conclusion polynomial with respect to the hypothesis polynomial set, supposed already a basic set in loose sense, is zero, then the theorem is *true* generically under the non-degeneracy conditions furnished by the initials of the hypothesis polynomials and we have achieved our aim. Only in the case of non-zero remainders is the truth of theorem doubtful, and further investigations about the reducibility of the polynomials may then be required.

Finally, we remark that though the hypotheses as well as the conclusion polynomials usually have only a few terms, the polynomials got successively during the reduction in the determination of the remainder may rise up quickly to hundreds and thousands of terms. To avoid the appearance of this phenomenon the following *branching* device has been adopted in our programming. Thus, let some polynomial g of the form (m_p = degree in y_p of A_i of class p in $\mathscr{A}$)

$$g = g_0 y_p^{m_p-1} + g_1 y_p^{m_p-2} + \cdots + g_{m_p-1},$$

in which each g_i is of class $< p$, appear during the successive reduction of the conclusion polynomial. Then, instead of verifying further whether the remainder of g with respect to $\mathscr{A}$ is zero, we may verify this for each g_i in turn. Furthermore, we shall use an *index set* $[TCD]$ to indicate the complexity of a polynomial, where T is the number of terms, C the class, and D the degree in the leading variable y_C of the polynomial. The successive reduction of the conclusion polynomial up to the final remainder which constitutes in fact a *proof* of the theorem in the case of zero remainder may then be clearly shown by a *flowing chart* of the index sets. As a simple example, with suitable coordinates the well-known Pappus Theorem will have 6 hypothesis polynomials already in the form of a basic set in the loose sense whose index sets are:

$$[4\ 7\ 1],\ [3\ 8\ 1],\ [4\ 9\ 1],\ [3\ 10\ 1],\ [4\ 11\ 1],\ [4\ 12\ 1].$$

The conclusion polynomial has an index set [6 12 1] and the flowing chart of the reductions, as done on a computer, runs as follows:

$$[6\ 12\ 1] \longrightarrow [8\ 11\ 1] \longrightarrow [12\ 10\ 1] \longrightarrow [16\ 9\ 1] \longrightarrow [18\ 8\ 1] \longrightarrow [16\ 7\ 1] \longrightarrow 0.$$

The final zero means that the theorem is true (of course generically only) and is proved with the above running chart as a proof. Remark that different choices of coordinates will give rise to different running charts which correspond to different proofs.

We have applied our program to the proof of various famous theorems in the ordinary geometry: theorems of Keukou, Pappus, Pascal, Simson, Feuerbach, Morley, etc. Perhaps the proof of the theorem of Morley is the most difficult and is quite instructive in itself. So let us state the theorem in full below.

Theorem of Morley. *For a triangle $A_1A_2A_3$ the neighbouring trisectors of the three angles of the triangle will intersect to form 27 triangles in all, of which 18 are equilateral.*

In appearance this theorem is out of the reach of our method which works only for unordered geometries without notion of order or only for theorems not involving order relations in an ordered geometry. Thus, in an unordered geometry, there is no notion of rays and an angle cannot be defined in the usual way as two rays emanating from a common point. However, we can define an angle $\angle(l_1, l_2)$ simply as an ordered pair of lines l_1, l_2, and attribute a magnitude $T(l_1, l_2)$ to it corresponding to the tangent function of the angle in the case of ordinary geometry.

We may now define a bisector of the angle $\angle(l_1, l_2)$ in the unordered geometry as a line such that the *reflection* (well-defined in the geometry) of l_1 with respect to t is just l_2. If l_1, l_2 intersect, then t is a line through the intersecting point such that $T(t, l_1) = T(l_2, t)$ corresponding to the ordinary formula $\angle(t, l_1) \equiv \angle(l_2, t)$ or $2\angle(t, l_1) \equiv \angle(l_2, l_1) \bmod \pi$. However, in the unordered geometry there may exist two such bisectors for an angle and there is no means to distinguish these two bisectors.

Similar ambiguity occurs for trisectors of an angle. To fix the ideas, let us call a line t a *primary trisector* of an angle $\angle(l_1, l_2)$ if a formula in T holds which corresponds to the ordinary formula $3\angle(t, l_1) \equiv \angle(l_2, l_1) \bmod \pi$. There are 3 such primary trisectors which there is no means to distinguish from each other. To each such primary trisector t however is uniquely associated a *secondary trisector* t' such that $T(l_2, t') = T(t, l_1)$.

Consider now a triangle $A_1A_2A_3$. Let t_1 be any one of the primary trisectors of the angle $\angle(A_1A_2, A_1A_3)$ at vertex A_1 with associated secondary trisector t_1'. Similarly let t_2, t_2' be a primary and an associated secondary trisector of the angle $\angle(A_2A_3, A_2A_1)$ and t_3, t_3' be those of the angle $\angle(A_3A_1, A_3A_2)$. Let t_1, t_2' intersect at a point A_4, in notation $A_4 = t_1 \wedge t_2'$, Similarly let $A_6 = t_2 \wedge t_3'$, $A_5 = t_3 \wedge t_1'$. The triangles $A_4A_5A_6$ are clearly 27 in all. The Morley theorem asserts that 18 among them are equilateral.

First of all we have to settle the problem how the 18 triangles should be chosen. For this let us denote by θ an angle for which the T-value has square $= 3$. In ordinary geometry this means $\theta \equiv \pm\frac{\pi}{3} \bmod 2\pi$. Remark in passing that in an unordered geometry it is not legitimate to speak about $+\sqrt{3}$ or $-\sqrt{3}$. Now we choose the primary trisectors t_1, t_2, t_3 such that some relation in the T-values corresponding to the ordinary formula

$$\angle(t_1, A_1A_2) + \angle(t_2, A_2A_3) + \angle(t_3, A_3A_1) \equiv \theta \bmod 2\pi$$

holds true. Under this condition the number of triangles $A_4A_5A_6$ is then reduced to 18 which will be proved to be all equilateral.

Adopting now a certain coordinate system with coordinates of various points and the T-

values of various angles involved in the theorem as x_i's arranged in a certain definite order, we shall get a set of hypothesis polynomials H_i, 18 in number, and a certain conclusion polynomial g. Without entering the details we merely list the index sets of various polynomials below:

For hypothesis-polynomials:

[2 3 1], [3 4 1], [4 5 1], [3 7 1], [3 8 1], [4 9 1], [3 10 1],
[2 11 1], [2 12 2], [8 13 1], [4 14 1], [4 15 1], [4 16 1],
[2 17 1], [5 18 1], [3 19 1], [4 20 1], [4 21 1].

For conclusion-polynomial: [4 21 1].

To verify the theorem by means of our program we remark that separation will occur when we come to the point after the reductions with respect to H_9 and H_4. The following is a rough scheme about the successive reductions with index set of successive polynomials indicated.

C_0[4 21 1] $\xrightarrow{(1)}$ [1208 12 1] ⟶ [592 11 2] ⟶ [376 6 15] ⟶ [19 5 3] ⟶ 0

→ C_1[616 11 2] ⟶ [375 6 15] ⟶ [20 5 3] ⟶ 0

→ C_2[357 6 14] → [26 5 3] → 0

→ C_3[355 6 14] ⟶ [23 5 3] ⟶ 0

→ C_5 ⋯ → C_6 ⋯

⋮ ⋮

→ C_{25}[54 6 3] ⟶ [29 5 3] ⟶ 0 → C_{24}[53 6 3] ⟶ [28 5 3] ⟶ 0

→ C_{27}[25 6 2] ⟶ 0 → C_{26}[25 6 2] ⟶ 0

Remark that each arrow in the above scheme consists of a number of successive reductions. For example, the arrow marked (1) is detailed as follows.

C_0[4 21 1] ⟶ [8 20 1] ⟶ [4 19 1] ⟶ [18 18 1] ⟶ [36 17 1]
⟶ [36 16 1] ⟶ [66 15 1] ⟶ [132 14 1] ⟶ [236 13 2] ⟶ [832 13 1]
⟶ [1960 12 3] ⟶ [1208 12 1].

Thus a certain polynomial of 1960 terms occurs in the whole procedure of reductions. If we do not adopt separation devices at convenient places in selecting suitable coordinate systems and coordinates of points, the polynomials during the procedure may quickly grow too large to be admitted even by a big computer. For the present case as all remainders (28 in all) are zero, the Morley theorem is true and the above scheme furnishes such a proof of the theorem.

We add finally that the above scheme shows that we have indeed proved a theorem a little more general than the original one. For the same proof holds also in the case of certain unordered geometries like complex geometries, for example. In such geometries isotropic lines may exist. However, if we restrict our theorem so that no isotropic lines are involved in the statement, then the mechanical proof applies still.

As a further example let us consider the problem of determining all triangles ABC with two equal bisectors t_A and t_B of angles A and B. It is well-known, but is quite non-trivial to prove, that the triangle ABC should be isoceles ($AC = BC$) if the two equal bisectors in question are both *internal* ones. Mr. S. C. Chou has raised the question of proving this fact by the mechanical theorem-proving method. Now it is easy to see that $AC = BC$ would not be true (generically) if one of the bisectors t_A, t_B is an *internal* and the other is an *external* one. Chou and I have tried on the computer and found the rather unexpected result that $AC = BC$ is still not true if the equal bisectors are both *external* ones.

In principle the above problem is again out of reach of our method. However, in view of the nature of the problem that the order relations only enter the hypothesis but not the conclusion at all, our method in combination with that of Seidenberg in reducing inequalities to equalities by introducing new auxiliary variables will lead to some information about the final results to be found. Thus, let us denote by AE and BD the two equal bisectors in question and by I their point of intersection. Take coordinates with

$A = (-1, 0)$, $B = (+1, 0)$, $I = (x_2, x_3)$, $C = (x_{12}, x_{13})$, etc.

Denote also the slopes of AE, BD by x_4, x_5, etc. Introduce a further auxiliary variable x_1 by setting

$$x_4 x_5 = -x_1^2, \tag{1}$$

or

$$x_4 x_5 = +x_1^2. \tag{2}$$

Equation (1) means that AE, BD are either both internal or both external bisectors which will be distinguished by either

$$x_3 x_{13} > 0,$$

or

$$x_3 x_{13} < 0.$$

On the other hand equation (2) means that one of AE, BD is an internal while the other is an external bisector.

Consider e.g. the case of equation (1). From the hypothesis including the equality of bisectors we get on running the program a set of equations, with extraneous factors corresponding to degenerate cases already removed, as follows:

$$x_2 f(x) = 0, \tag{3}$$

with

$$f(x) = (1 - x_2^2)(x_1^2 - 1)^2(x_1^2 - 2) - 4, \tag{4}$$

$$x_3^2 = x_1^2(1 - x_2^2), \tag{5}$$

$$(1 - x_1^2)x_3 x_{13} = 2x_3^2, \tag{6}$$

etc.

Equation (5) shows that in the non-degenerate case we have

$$x_2^2 < 1. \tag{7}$$

Equation (6) shows that we have

$$x_1^2 < 1 \text{ or } > 1$$

according as the two bisectors AE, BD are both internal ones or both external ones.

Suppose first $x_1^2 < 1$. Then from (4) we see that $f(x) < 0$. From (3) it follows that we have necessarily

$$x_2 = 0.$$

This just proves the classical theorem that a triangle with two equal *internal* bisectors is isosceles.

Suppose next $x_1^2 > 1$ so that the two bisectors are both *external* ones. Then $f(x) = 0$ will have positive roots of x_1^2 for $x_2^2 < 1$ so that there are an infinity of *non-isoscelcs* triangles ABC with equal *external* bisectors AE, BD for which the corresponding point I (x_2, x_3) will lie on a certain oval defined by the following equation together with (7):

$$x_3^6 - 4x_3^4(1 - x_2^2) + 5x_3^2(1 - x_2^2)^2 - 2(1 - x_2^2)^3 - 4(1 - x_2^2)^2 = 0.$$

The case of equation (2) or the case of one internal and one external bisector can be treated in entirely the same manner. We find thus infinities of non-isosceles triangles with equal bisectors one internal and one external for which the corresponding points I will lie on two ovals defined by the same equation above with the restriction $x_2^2 > 1$. The problem raised above is thus completely settled.

We have also applied our method to the mechanical theorem discovering of "new" theorems in ordinary geometry. Several theorems have been discovered in this way. We shall illustrate below.

Ex. Pascal-Conic Theorem

Suppose we are given 6 points $A_1, \cdots, A_6$ on the same conic. Let us call any point of intersection $A_iA_j \cap A_kA_l$ (for i, j, k, l mutually unequal) a *Pascal point*. Such Pascal points are 45 in all which lie three by three on 60 so-called *Pascal lines*. These points and lines constitute a configuration which has been much studied by numerous geometers including Steiner, Staudt, Cayley, Kirkmaun. However, most of the interesting theorems found by them are of a *linear* character: collinearity of certain points and concurrency of certain lines. Now we put the following problem: What theorems of a *quadratic* character can be found about this configuration? In particular, we ask what combinations of 6 among the 45 Pascal points will lie on the same conic (*co-conic* for short). Of course we are only interested in such combinations of 6 Pascal points lying on some conic not degenerated into two Pascal lines.

The problem will be studied with further specialization. Consider for example a permutation $s = (123456)$ which will act on the 45 Pascal points. We now ask for what Pascal points P the six points P, sP, $s^2P, \cdots$, s^5P will lie on some non-degenerate conic. By trials we see that the only possible points are $A_1A_3 \wedge A_2A_5$ or the equivalent ones. Assuming that the usual Pascal theorem is known, then this amounts to whether the hexagons formed of the six points s^iP, $i = 0, 1, \cdots, 5$, are Pascalian or not, i.e., whether the three points of intersection of the opposite sides of the hexagons are collinear or not. Formulating the theorem to be proved in this way we verified again on the computer that this is really the case. So we get a number of non-degenerate conics on each of which lie 6 Pascal points. We call these conics the *Pascal conics* and the theorem thus discovered the *Pascal-Conic Theorem*. It was first discovered in 1980 and verified on an HP9835A.

Of course it is very likely that the theorem is known already in the last century. Moreover, simple and elegant proof may also be easily found for this theorem. However, these are neither of any interest nor of any importance to us from the point of view of mechanical theorem proving. The example shown may well indicate the powerfulness in discovering really non-trivial new theorems in various kinds of geometries besides the ordinary geometry, e.g. the non-Euclidean geometries, the circle geometries, or geometries of even more modern nature, in which known interesting theorems are rare. Even in the case of Pascal configurations we may put forward some problems to which our method may give some answer: Are there other conics through at least 6 of the Pascal points or touching at least 6 Pascal lines besides those found above? Are there any interesting geometrical relations between these conics and the various Pascal points, Pascal lines and other known points and lines of significance ? Are there also cubic relations between the 45 Pascal points, i.e., are there non-degenerate cubics passing through at least 9 out of the 45 Pascal points, etc. Of course innumerable problems can be set forth in this way.

References

[1] Hilbert, D., Grundlagen der Geometrie, Teubner, 1899.

[2] Ritt, J. F., Differential equations from the algebraic standpoint, Amer Math. Soc. 1932.

[3] Ritt, J. F., Differential algebra, Amef. Math. Soc., 1950.

[4] Wu Wen-tsün, On the decision problem and the mechanization of theorem proving in elementary geometry, *Scientia Sinica,* 21 (1978), 159—172.

初等几何定理机器证明的基本原理

吴　文　俊

（中国科学院系统科学研究所，北京）

摘　　要

1976 与 1977 之交，我发现了一个初等几何定理证明的机械化方法，见文献 [4]. 这一方法适用于各种无序的但满足 Pascal 公理的初等几何，或各种初等几何中不牵涉次序关系的那类定理. 本文 § 4 叙述了这一方法所依据的基本原理并给出了详细证明. 在 § 2 与 § 3 中则阐述了基本原理所依赖的关于多项式组的整序理论与代数簇的构造性理论. 二者俱源出 Ritt 的著作，见文献 [2，3]. 最后在 § 5 中以 Morley 定理与我所发现的 Pascal 锥线定理为例，说明这一方法在计算机上实施的具体情况.

ON ZEROS OF ALGEBRAIC EQUATIONS

——AN APPLICATION OF RITT PRINCIPLE

Wu Wenjun (Wu Wen-tsün 吴文俊)
(*Institute of Systems Science, Academia Sinica, Beijing*)

Received December 18, 1984.

Let K be a basic field of characteristic 0, and $f_i, i = 1, \cdots, r$, be polynomials in $K[x_1, \cdots x_n]$. Consider the system of algebraic equations

$$f_i = 0, \; i = 1, \cdots, r$$

which defines an algebraic variety V consisting of zeros of the system in an arbitrary extension field of K. The study of the structure of V or that of the set of zeros satisfying these equations is one of the main themes of algebraic geometry. Moreover the actual determination of the zeros in the complex field when they are finite in number for $K = Q, R$, or C is of great importance in applications. The present paper aims at giving a rather complete answer to these questions with a decomposition of V into parts quite different from the usual one in algebraic geometry but is more adapted to applications. The method is based on the so-called Ritt principle as described in a previous paper[1] of the author. Various notations and terminologies are also to be referred to that paper.

Consider thus a polynomial system S consisting of a finite number of polynomials and a polynomial G all in $K[x_1, \cdots, x_n]$. We shall denote the set of zeros of equations $S = 0$ in an arbitrary extension field of K for which G is not 0 by Zero (S/G). If the extension field is prescribed to be $\widetilde{K}$, then the set of $\widetilde{K}$-zeros for which $G \neq 0$ is denoted by $\widetilde{K}$-Zero(S/G). Our main result is then the following

Structure Theorem. *There is an algorithmic procedure which permits to decide in a finite number of steps whether Zero (S/G) is empty and in the contrary case to furnish a decomposition of the following form:*

$$\text{Zero } (S/G) = \text{Union Zero } (A_i/R_i).$$

In this decomposition formula each A_i is an irreducible ascending set and R_i is the non-zero remainder of J_iG_i with respect to A_i, where J_i is the product of initials of polynomials in A_i, and G_i is certain non-zero polynomial.

The proof consists in giving such an algorithmic procedure as described below.

Step 1. Form, as in [5], the characteristic set C of S. If C is contradictory, in other words C is composed of a single polynomial which is a non-zero constant of the basic field K, then Zero (S), in particular Zero (S/G), is empty and the procedure stops. In the contrary case let the initials of the polynomials in C be I_i with product J. Then the Ritt principle as described in [5] will give a decomposition of the form

$$\text{Zero }(S/G) = \text{Zero }(C/JG) + \text{Union Zero }(S_i/G),$$

in which each S_i is the enlarged system of S with I_i adjoined to it. Of course only non-constant I_i is to be considered.

Step 2. Consider the set Zero (C/JG). Suppose first that C is irreducible which may be determined by various known methods. Form now the remainder R of JG with respect to C. By the remainder formula we have clearly

$$\text{Zero }(C/JG) = \text{Zero }(C/R).$$

If $R=0$, then Zero (C/JG) is empty and should be removed in the above decomposition. Otherwise we just replace Zero (C/JG) in the decomposition by Zero (C/R). In any case we proceed to the next step.

Suppose now that C is reducible. Let C be consisting of polynomials

$$g_1, g_2, \cdots, g_r$$

of classes $(0<)\, p_1 < \cdots < p_r$. Then there will be some $s \leqslant r$ enjoying the following properties: First, for each $i<s$, $g_i \in K[x_1, \cdots, x_{p_i}]$ is irreducible as a polynomial of $K_{i-1}[x_{p_i}]$, where K_{i-1} is the field $K(x_1, \cdots, x_{p_i-1})$ which is obtained from K successively by either simple transcendental extension by x_j, $j \neq p_1, \cdots, p_{i-1}$, or a simple algebraic extension of $x_{p_1}, \cdots, x_{p_{i-1}}$ by means of the polynomials $g_1, \cdots, g_{i-1}$ as minimal polynomials. Secondly, g_s, as a polynomial in $K_{s-1}[x_{p_s}]$, with K_{s-1} likewisely defined, is a reducible one. By the known methods of factorization, we get then, after clearing of fractions, an expression in K_{i-1} of the form

$$hg_s = g_s' g_s'',$$

in which g_s', g_s'' are polynomials in $K[x_1, \cdots, x_{p_s}]$ both of degree >0 in x_{p_s}, and h is one in $K[x_1, \cdots, x_{p_s-1}]$ reduced with respect to the irreducible ascending set consisting of $g_1, \cdots, g_{s-1}$. Let C', C'' be the polynomial systems obtained from C in replacing g_s by g_s', g_s'' respectively, and C^+ be one obtained from C by adjoining h to it. Then it is clear that

$$\text{Zero }(C/JG) = \text{Zero }(C^+/JG) + \text{Zero }(C'/hJG) + \text{Zero }(C''/hJG).$$

Replace now Zero (C/JG) in the decomposition of Step 1 by the above union of 3 sets of zeros and proceed to the next step.

Step 3. Let us say that one polynomial system is of higher or lower rank than or equal rank to another according as their basic sets are so related. Then we see that each polynomial system S_i occurring in the decomposition of Step 1 is of lower rank than S. Moreover, each of the polynomial systems C^+, C', or C'', eventually occurring in Step 2, is clearly of lower rank than C, and hence of lower rank than S too.

Treat now in turn each set of zeros occurring in the decomposition of Step 2 in returning to Step 1, removing any empty set of zeros if it appears, and proceeding further as before.

As the polynomial systems occurring in the sets of zeros of the successive decompositions are of steadily decreasing ranks so we have to stop after a finite number of steps. Thus finally we arrive at either an empty set or a decomposition as described in the structure theorem.

As immediate applications of the above structure theorem we may cite the following ones:

(a) The usual unique decomposition of an algebraic variety into irreducible components can be deduced from our decomposition into sets of zeros with $G=1$. Moreover, this can be done in a constructive manner instead of a mere existential one.

(b) The Nullstellen-Satz of Hilbert can be deduced in a quite simple manner by our methods and again in a constructive manner, contrary to the usual mere existential proofs.

(c) Let the basic field K be either Q, R, or C, and only zeros in the extension field $\widetilde{K} = C$ are to be considered. It is clear from the decomposition formula that the set of zeros (i.e. C-zeros) is finite iff each of the irreducible ascending sets A_i occurring in the decomposition formula is composed of n polynomials where n is the number of variables x_i. These zeros may be then found by the usual method in solving successively the polynomial equations of each set A_i. The following example is taken from a paper of Buchberger[2] and may be used as an illustration of our method in comparison with several known methods as given in [2] and [3].

Example. Problem. Solve the following system of equations:

$$f_0 = 2x_3^2 - x_1^2 - x_2^2 = 0,$$
$$f_1 = x_1x_3 - 2x_3 + x_1x_2 = 0,$$
$$f_2 = x_1^2 - x_2 = 0.$$

For the method of Lazard one has to consider a matrix of 35 rows and 50 columns, with elements involving 4 auxiliary variables $U_1, \cdots, U_4$. One has then to decide whether the rank of this matrix is < 35 or $= 35$. In the latter case one has to decompose a certain polynomial, which is the determinant of a sub-matrix of the highest rank 35, into linear factors of U_i. The coefficients of U_i in these factors give then the solutions, including those at infinity, of the given system of equations.

The method of Buchberger consists of first determining for the ideal $a = (f_0, f_1, f_2)$ a Gröbner basis of 6 polynomials in all in the present example. Next a basis of the algebra $R[x_1, x_2, x_3]/a$ together with a multiplication table of the basis are determined. In case the basis of the algebra is finite as in the present example one proceeds then to determine successively a system of polynomials

$$p_1(x_1),\ p_2(x_1, x_2),\ p_3(x_1, x_2, x_3).$$

It is proved that all zeros, now finite in number, of the given system of equations are to be found among the system of equations $p_i = 0$ by the usual methods.

Our method runs briefly as follows.

First find the characteristic set C of the system $S = (f_0, f_1, f_2)$ which consists of the 3 polynomials g_i below:

$$g_0 = x_1^6 + 4x_1^5 - 5x_1^4 + 4x_1^3 - 4x_1^2,$$
$$g_1 = - x_2 + x_1^2,$$
$$g_2 = (x_1 - 2)x_3 + x_1x_2.$$

The non-constant initial is $J = x_1 - 2$. Denote the system with J adjoined to S by S', then we have

$$\text{Zero } (S) = \text{Zero } (C/J) + \text{Zero } (S').$$

Remark that in the case for which the number of polynomials in C is the same as that of variables x as in the present example, it is not necessary to factorize as in the general procedure indicated in the structure theorem. We just solve the equations $C = 0$ directly and successively to get all the solutions for which $J \neq 0$. The solutions are found by usual methods to be 6 in number: (0,0,0) counted twice, (1,1,1), and 3 others.

To determine the set Zero (S'), we find first the characteristic set C' of S' to be consisting of a non-zero constant or S' is contradictory. The set Zero (S') is thus empty and the totality of solutions is formed by the 6 zeros of the set Zero (S/J) as indicated above.

To find the solutions at infinity we have only to replace each of the polynomials f_i by $f_i^\wedge$ in keeping only the terms of the highest degree. Thus we have to consider the system $S^\wedge = (f_0^\wedge, f_1^\wedge, f_2^\wedge)$ given by

$$f_0^\wedge = 2x_3^2 - x_1^2 - x_2^2,$$
$$f_1^\wedge = x_1x_3 + x_1x_2,$$
$$f_2^\wedge = x_1^2.$$

The characteristic set $C^\wedge$ of $S^\wedge$ is seen to be consisting of 3 forms, viz.

$$g_0^\wedge = x_1^2,$$
$$g_1^\wedge = x_1x_2^2,$$
$$g_2^\wedge = x_1x_3 + x_1x_2.$$

The non-constant initials are $I_1 = x_1, I_2 = x_1$, with product $J = x_1^2$. Hence we have

$$\text{Zero } (S^\wedge) = \text{Zero } (C^\wedge/J) + \text{Zero } (S_1^\wedge) + \text{Zero } (S_2^\wedge),$$

in which $S_1^\wedge$ and $S_2^\wedge$ are enlarged systems of $S^\wedge$ adjoined by I_1 and I_2 respectively. The set Zero ($C^\wedge/J$) is clearly empty. Treating as before the sets Zero ($S_i^\wedge$) we find an infinity of solutions given by

$$x_1 = 0, \quad 2x_3^2 = x_2^2,$$

which represent two points at infinity

$$(x_1 : x_2 : x_3) = (0:1:1/\sqrt{2})$$

and $(0:1:-1/\sqrt{2})$.

Remark. In practice the equations to be solved usually have coefficients complicate real numbers, which, being arisen from measurements, are only approximate ones. We may thus assume that multivariate polynomials occurring in the procedure are irreducible ones so that the most difficult step of reducibility considerations may be entirely avoided. This will make the method particularly efficient for practical applications.

REFERENCES

[1] Wu Wen-tsün, Basic principles of mechanical theorem proving in elementary geometries, *J. Sys. Sci. & Math. Scis.*, **4** (1984), 207—235.

[2] Buchberger, B., Ein algorithmisches Kriterium für die Löxsbarkeit eines algebraischen Gleichungssystems, *Aeq. Math.*, **4** (1970), 374—383.

[3] Lazard, D., *Systems of Algebraic Equations*, EUROSAM, 1979, 88—94.

系 统 科 学 与 数 学
J. Sys. Sci. & Math. Scis.
5 (4) (1985), 290—302

ON THE PLANAR IMBEDDING OF LINEAR GRAPHS

WU WEN-JUN (WU WEN-TSÜN)
(*Institute of Systems Science, Academia Sinica, Beijing*)

1. INTRODUCTION

The present paper is a reproduction of the results already published in chinese from 1973 onwards. It is concerned with the problem of planar imbedding of linear graphs (supposed to be connected and possessing no loops henceforth). The problem may be separated into four parts:

P1. Decide whether a connected linear graph (or *graph* for short) G is imbeddable in the plane (or *imbeddable* for short).

P2. Decide, in the case of a non-imbeddable graph G, a minimal set of edges the removal of which will render the remaining part of G imbeddable.

P3. Give a method of imbedding G in the plane in the case G is imbeddable.

P4. Give a description of the totality of possible imbeddings of G in the plane in the case G is imbeddable.

The problem P1 was already solved in the early thirties. Thus, Kuratowski has given the following simple and elegant criterion [KU1]: Let $K1$ be the graph with five vertices and all edges connecting any two of them. Let $K2$ be the graph with two triads of vertices and all edges connecting pairs of vertices one from each triad. Then we have the following

Theorem of Kuratowski. *A graph G is imbeddable if and only if it does not contain any subgraph of type $K1$ or $K2$.*

Similar criteria have been given by Whitney and MacLane, also in the thirties. However, all these criteria are only *existential* in character, although they settle the problem P1 quite satisfactorily at least in a theoretical sense. In fact, these criteria give no means of a *constructive* manner for deciding whether a graph concretely given is planar or not. For example, for the Kuratowski criterion we have no means of detecting subgraphs of type $K1$ or $K2$ well hidden in a concretely given graph. This fact thus has deprived these criteria of any practical value.

After more than twenty years of silence the interest in the problem revived in the early sixties owing seemingly to practical needs. This time however, the interest lay no more on theoretical imbeddability of a linear graph, but rather, on practical decision of the imbeddability of any given graph in giving algorithmic procedures. Beginning from a paper by Auslander and Parter [AP1], the study culminated in a paper of Hopcroft and Tarjan [HT1] in giving an efficient planarity algorithm for a linear graph. Nevertheless their method gives

Received January 16, 1985.

merely an answer to problem P1 from the practical side and leaves problems P2—4 completely untouched. As mentioned by the authors themselves in their joint paper, their "planarity algorithm··· tests a graph G for planarity, but it does not actually construct a planar representation of G." We remark that it is just the latter part corresponding to problems P2—3 above that renders the study of planarity of graphs so important in applications.

On the other hand the present author discovered in 1967 a solution to problem P1 which is both of theoretical interest and of practical value in being algorithmic. The method was based on a theory of imbedding and immersion of complexes in a Euclidean space [WU1] and was applied this time to linear graphs, i. e. complexes of dimension 1. It leads to the criterion that a graph is imbeddable if and only if a certain system of linear equations on mod 2 coefficients is solvable in integers mod 2. These results, owing to circumstances, were not published until late 1973, cf. [WU2]. Now each equation in the linear system of our criterion has either two or four variables. In 1978 Liu Yan-pei made an important complement to our methd in reducing each such equation to one with only two variables [L1]. This enables the decision of planarity to be carried out actually without any computation and is extremely feasible. However, in either [HT1] or [L1] or [TU1] only criteria of imbeddability were given, with the important problem of actual imbedding in the case the graph is imbeddable entirely untouched. In the meantime the present author arrived at a complete solution of all problems P1—4 listed above and the proofs were purely algebraic with no more use of algebraic topology. These results were published as an appendix to the chinese version of the book [WU1], cf. [WU4].

The present paper has the aim of giving an English version of all these results, so far published only in Chinese, with due modifications.

To fix the ideas, throughout the paper the following notations will be adopted:

We will always work over integers mod 2 and the field of mod 2 integers will be denoted as usual by $Z2$.

The plane in which graphs are to be imbedded is denoted by $R2$.

The graph (connected without loops) is denoted by G, with numbers of vertices Nv and number of edges Ne.

The vertices of G are Vi, with i running over some index set I. The collection of all such vertices will be denoted by $V(G)$, or simply V.

The edges of G are Eq, with q running over some index set Q. The collection of all such edges will be denoted by $E(G)$, or simply E.

The letters $i, j, k, l, \cdots$ will be used for indices in I, and the letters $q, r, s, t, \cdots$ for those in E.

The set of all unordered pairs of edges (Er, Es), with Er, Es disjoint from each other, will be denoted by $D2(G)$, or simply $D2$.

The set of all pairs (Vi, Eq), with Vi not an end of Eq, will be denoted by $D1(G)$, or simply $D1$.

The collection of all functions

$$A: D1 \to Z2$$

forms naturally an additive group and will be denoted by $C1(G)$, or simply $C1$.

The collection of all functions

$$F: D2 \to Z2$$

forms naturally an additive group and will be denoted by $C2(G)$, or simply $C2$.

For any pair (Vi, Eq) in D1, the function A in $C1$ which takes the value 1 on (Vi, Eq) but 0 on any other pair in $D1$ will be denoted by $\langle Vi, Eq\rangle$.

Similarly, for any pair (Eq, Er) in $D2$, the function F in $C2$ which takes the value 1 on (Eq, Er) but 0 on any other pair in $D2$ will be denoted by $\langle Eq, Er\rangle$.

The morphism

$$d: C1(G) \to C2(G)$$

defined by

$$dA/(Eq, Er) = A/(Vi, Er) + A/(Vj, Er) + A/(Vk, Eq) + A/(Vl, Eq)$$

for

$$A \text{ in } C1, \ Eq = ViVj, \ Er = VkVl, \ (Eq, Er) \text{ in } D2,$$

will be called the *differential* in G.

For any two broken lines $L1, L2$ in $R2$ not both closed to become polygons for which the ends of $L1$ (resp. $L2$), which exist if not closed, are disjoint from $L2$ (resp. $L1$), there is a well-defined *intersection number* in $Z2$ which will be denoted by Int $(L1, L2)$.

For any closed polygon P with possibly self-intersections and a point A not on P there is the well defined *order* of A with respect to P in $Z2$ which will be denoted by Ord (A, P).

If B is another point in $R2$ not on P and L is a broken line joining A and B, we would have the following relation in $Z2$:

$$\text{Ord}(A, P) + \text{Ord}(B, P) = \text{Int}(L, P).$$

If A, B, P are as above with P a simple closed polygon and

$$\text{Ord}(A, P) = \text{Ord}(B, P)$$

in $Z2$, then by the theorem of Jordan A, B can be joined by a simple broken line in $R2$ disjoint from P.

2. A Criterion for Imbeddability

Without loss of generality we shall restrict maps of G in the plane $R2$ to piecewise linear ones which will always be so assumed in what follows. A (piecewise linear) map

$$f: G \to R2$$

is called an *imbedding* if it is topological or 1-1.

Let H be any subgraph of G. Then a map f will be called an *H-immersion* of G if the following conditions (a)—(e) are observed:

(a) The images of vertices are all different.

(b) The image of each edge is a simple broken line.

(c) The image of any vertex is not on the image of any edge except at the possible end.

(d) f is an imbedding when restricted on H, while for any edge Eq not in H, $f(Eq)$ will not meet $f(H)$ except possibly at vertices common to H and Eq.

(e) The images of any two edges will meet at most at a finite number of points besides the possible common ends.

The H-immersion will simply be called an *immersion* of G if H is the empty sub-graph.

Consider now any immersion

$$f: G \to R2.$$

Definition. The element $c(f($ in $C2(G) = C2$ defined by

$$c(f)/(Eq, Er) = \text{Int}\,(fEq, fEr), \text{ for } (Eq, Er) \text{ in } D2,$$

will be called the *immersion element* defined by f.

Theorem 1. *For any two immersions f and g of G in the Plane*

$$f, g: G \to R2,$$

the elements $c(f), c(g)$ belong to the same class of the quotient group $C2/dC1$.

Proof. Consider first the case f and g coincide on all vertices of G and all edges of G except a single one, say the edge Es.

Now fEs, gEs form a polygon P (with possibly self-intersections). For any vertex V_k of G disjoint from Es let us set

$$Ok = \text{Ord}(fVk, P), \text{ for } Vk \text{ disjoint from } Es.$$

Define now an element c in $C1$ by

$$c = \text{SUM } Ok.\langle Vk, Es\rangle,$$

the summation being over all vertices Vk disjoint from Es.

Now for any edge Eq disjoint from Es we have $fEq = gEq$. Therefore, with Vk, Vl as the two ends of Eq, we would have

$$c(f)/(Eq, Es) + c(g)/(Eq, Es) = \text{Int}(fEq, fEs) + \text{Int}(gEq, gEs)$$
$$= \text{Int}(fEq, P) = \text{Ord}(fVk, P) + \text{Ord}(fVl, P).$$

On the other hand if the ends of Es are Vi and Vj, then we would have

$$dc/(Eq, Es) = c/(Vi, Eq) + c/(Vj, Eq)$$
$$+ c/(Vk, Es) + c/(Vl, Es) = Ok + Ol.$$

Comparing, we have

$$c(f) + c(g) = dc/(Eq, Es).$$

For any pair (Eq, Er) in $D2$ with Eq, Er both different from Es, it is clear that

$$c(f) + c(g) = dc = 0/(Eq, Er).$$

So the assertion is proved in the above case.

Consider now the case that f and g coincide on all vertices of G and are arbitrary otherwise. We may always define a sequence of immersions $h0$, $h1, \cdots$, hs such that $h0$ coincides with f, hs with g, and each hr coincides with the preceding one with the exception of a single edge. By the preceding case each element $c(hr)$ will belong to the same class of $C2/dC1$ as the element c of the preceding immersion h in the sequence. It follows that the assertion still holds true in the present case.

Consider now the general case with f, g arbitrary, with however the images of all vertices under both f and g different from each other.

For any vertex Vi let us draw a simple broken line Li in the plane with ends fVi and gVi such that, what is clearly possible, these broken lines are mutually disjoint.

For each edge Eq let us join now the ends of fEq by a broken line Lq disjoint from all Li except possibly at the images of their common ends.

Define now an immersion h and an immersion h' of G by

$$h(Eq) = Lq,$$
$$h'(Eq) = Li + Lq + Lj,$$

where Vi, Vj are the two ends of Eq.

From the construction we see that the elements $c(h)$ and $c(h')$ belong to the same class of $C2/dC1$. On the other hand by the preceding cases already proved $c(h)$ and $c(h')$ are in the same classes of $C2/dC1$ as $c(f)$ and $c(g)$ respectively. Hence $c(f)$ and $c(g)$ belong to the same class of $C2/dC1$ in this case too.

Finally, for two arbitrary immersions f and g let us take an immersion h such that both f, h and g, h are pairs of immersions as in the preceding case. Then both $c(f)$ and $c(h)$, as well as both $c(g)$ and $c(h)$, will belong to the same class of $C2/dC1$. Hence $c(f)$ and $c(g)$ belong to the same class too.

The theorem is thus proved in all respects.

From the above theorem the following definition is legitimate:

Definition. The class in $C2/dC1$ of the elements $c(f)$ for any immersion f of G in the plane will be called the *imbedding class* of G and will be denoted by $I(G)$ in what follows.

From the very definition of imbedding it is clear that for G to be imbeddable in the plane it is necessary that

$$I(G) = 0.$$

We shall prove that this condition is not only necessary but also sufficient. For this purpose we shall prove first some preliminary lemmas as follows.

Lemma 1. *Let G' be a subgraph of G. Then the natural restriction will induce morphisms $r1$ and $r2$ so that the diagram below is commutative:*

$$\begin{array}{ccc} C1(G') & \xrightarrow[d]{} & C2(G') \\ {\scriptstyle r1}\uparrow & & \uparrow{\scriptstyle r2} \\ C1(G) & \xrightarrow[d]{} & C2(G) \end{array}$$

Moreover, for the morphism thus induced

$$r': C2(G)/dC1(G) \to C2(G')/dC1(G')$$

we have

$$r'(I(G)) = I(G').$$

The proof is evident and will be omitted.

Lemma 2. *Let G' be a subdivision of G. Then there are natural morphisms $s1$ and $s2$ so that the diagram below is commutative:*

$$\begin{array}{ccc} C1(G) & \xrightarrow[d]{} & C2(G) \\ {\scriptstyle s1}\uparrow & & \uparrow{\scriptstyle s2} \\ C1(G') & \xrightarrow[d]{} & C2(G') \end{array}$$

Moreover, for the morphism thus induced

$$s': C2(G')/dC1(G') \to C2(G)/dC1(G)$$

we have

$$s'(I(G')) = I(G).$$

Proof. Suppose G' is obtained from G by introducing a single new vertex V' on some edge Eq of G with ends Vi and Vj. Define now $s1$ and $s2$ in the following way. Let us denote the edges ViV' and VjV' in G' derived from Eq of G by Eq' and Eq'' respectively. Then for any elements $c1$ in $C1(G')$ and $c2$ in $C2(G')$ we define $s1c1$ in $C1(G)$ and $s2c2$ in $C2(G)$ by

$$\begin{aligned} s1c1/(Vk, Er) &= c1/(Vk, Er), \text{ for } Er \langle\ \rangle Eq, \\ s1c1/(Vk, Eq) &= c1/(Vk, Eq') + c1/(Vk, Eq''), \\ s2c2/(Er, Es) &= c2/(Er, Es), \text{ for } Er, Es \langle\ \rangle Eq, \\ s2c2/(Eq, Er) &= c2/(Eq', Er) + c2/(Eq'', Er). \end{aligned}$$

It is easy to verify that $ds1c1 = s2dc1$ and $s'(I(G')) = I(G)$. The lemma is thus true for this simple case. Since any subdivision of G is formed of a sequence of such elementary subdivisions of the above type, the lemma is proved.

Lemma 3. *For Kuratowski's graphs $K = K1$ or $K2$ we have*

$$I(K) \langle\ \rangle 0.$$

Proof. Let us consider e. g. the first Kuratowski's graph $K = K1$. Denote the 5 vertices of K by $V1, \cdots, V5$ and immerse K in the plane in the usual way with images Wi of Vi forming a regular pentagon and images of the edges the respective sides and diagonals of the pentagon. The element $c(f)$ of the corresponding immersion is then given by

$$c(f)/(V1V3, V2V4) = 1,$$

$$c(f)/(V1V3,\ V2V5) = 1,$$
$$c(f)/(V2V4,\ V3V5) = 1,$$
$$c(f)/(V1V4,\ V2V5) = 1,$$
$$c(f)/(V1V4,\ V3V5) = 1,$$
$$c(f)/(ViVj,\ VkVl) = 0,$$

for any other pair $(ViVj,\ VkVl)$ in $D2(k)$. In other words, we have

$$c(f) = \langle V1V3,\ V2V4\rangle + \langle V1V3,\ V2V5\rangle + \langle V2V4,\ V3V5\rangle$$
$$+ \langle V1V4,\ V2V5\rangle + \langle V1V4,\ V3V5\rangle.$$

The differential d is defined by

$$d\langle V1,\ V2V3\rangle = \langle V1V4,\ V2V3\rangle + \langle V1V5,\ V2V3\rangle, \text{ etc.}$$

Consider any element

$$c1 = \text{SUM}\, Xijk \cdot \langle Vi,\ VjVk\rangle$$

in $C1(K)$, in which $Xijk = Xikj$ are all mod 2 integers in $Z2$ and the summation is over all triples i, j, k chosen from $1, 2, \cdots, 5$ which are mutually distinct. If $c(f)$ is the d-image of $c1$, then the following set of equations should be true:

$$X124 + X324 + X213 + X413 = 1,$$
$$X125 + X325 + X213 + X513 = 1,$$
$$X235 + X435 + X324 + X524 = 1,$$
$$X125 + X425 + X214 + X514 = 1,$$
$$X135 + X435 + X314 + X514 = 1.$$
$$X134 + X234 + X312 + X412 = 0,$$
$$X135 + X235 + X312 + X512 = 0, \text{ etc.,}$$
$$X145 + X245 + X412 + X512 = 0.$$

In all there are 15 such equations. By adding all these equations we get $0 = 1$ since we are working in the domain Z2. This proves that $c(f)$ cannot be the d-image of any element in $C1(K)$ or $I(K)\langle\rangle 0$ for $K = K1$. The proof of the case $K = K2$ is similar and will be omitted.

Fundamental Theorem I. *For a graph G to be imbeddable in the plane it is necessary and sufficient that*

$$I(G) = 0.$$

Proof. It is enough to prove only its sufficiency. Suppose G is not imbeddable. By Theorem of Kuratowski G should contain a certain subgraph G' which is some subdivision of either $K1$ or $K2$. By Lemmas 2 and 3 we should have $I(G')\langle\ \rangle 0$. By Lemma 1 we have then a fortiori $I(G)\langle\ \rangle 0$. Hence $I(G) = 0$ would imply that G is imbeddable. This proves the theorem.

As a complement we have also the following

Theorem 2. *For any element $c2$ in $C2(G)$ belonging to the imbedding class $I(G)$ in $C2(G)/dC1(G)$ there is an immersion f of G such that*

$$c2 = c(f).$$

Proof. This is clearly a direct consequence of the following

Lemma 4. *Let* $c2$ *be any immersion element of* G *in the imbedding class* $I(G)$. *Then for any functon* $\langle Vk, Eq\rangle$ *of* $C1(G)$ *the element*

$$c2' = c2 + d\langle Vk, Eq\rangle$$

is also an immersion element of some immersion of G.

Proof. Let g be any immersion of G defining the element $c2$. On the interior of the image under g of Eq let us take a small segment L with ends Vi', Vj' which contains no image points of any other edges of G. Replace now L with a simple broken line L' joining Vi', Vj' such that $L + L'$ will form a loop with only image of Vk and with no image of any other vertices in its interior. Let g' be the map of G in the plane which coincides with g everywhere except that L is replaced now by L'. We may also clearly choose L' in such a way that g' is a well-defined immersion. It is now easy to verify that $c(g') = c2'$ and the lemma is thus proved.

Remark. All the theorems and proofs here are of an algebraico-topological character, but we have avoided the use of any such terminology as done in the original paper [WU2]. It is also clear that the concepts and results in this section may be naturally extended with no essential changes to graphs G not necessarily connected.

3. Reduction of Criterion to Solvability of Linear Equations

Consider the graph G and an arbitrary immersion f as before which will determine an immersion element

$$c(f) = \text{SUM } Iqr \cdot \langle Eq, Er\rangle \text{ in } C2, \tag{1}$$

in which the summation is to be extended over all pairs (Eq, Er) of $D2$ and

$$Iqr = \text{Int}(fEq,\ fEr). \tag{2}$$

According to the fundamental theorem in Sect. 2, the planar imbeddability of G depends then on the existence of a function $c1$ in $C1$ such that

$$dc1 = c(f). \tag{3}$$

As $C1$ has a basis consisting of functions $\langle Vi, Er\rangle$ with (Vi, Er) running over all pairs in $D1$, we may set such a $c1$ of $C1$ in the form

$$c1 = \text{SUM } Xir \cdot \langle Vi, Er\rangle, \tag{4}$$

with summation extended over all pairs (Vi, Er) in $D1$ and Xir unknowns in $Z2$ to be sought for. Form $dc1$ and compare both sides of (3). We then get a system of linear equations

$$(\text{EQN}f):\ Xir + Xjr + Xkq + Xlq = Iqr$$

with one equation corresponding to each pair (Eq, Er) in $D2$, the ends of Eq being supposed to be Vi, Vj, and the ends of Er to be Vk, Vl. Fundamental Theorem I in Sect. 2 can thus be reformulated in the following form:

Fundamental Theorem II. *For a graph G to be imbeddable, it is necessary and sufficient that for an arbitrary immersion f of G in the plane, the system of linear equations (EQNf) possess solutions of ths X's in mod 2 integers.*

We remark that the theorem remains true even if G is not connected.

Now the number of equations in (EQNf) is about $Ne \wedge 2$ and that of the unknowns X is about $Nv * Ne$ where Nv and Ne are the numbers of vertices and edges in G respectively. The determination of solvability of this system seems to be thus untractable in appearence. However this is not the case. In fact, owing to the particular form of these equations we can treat the system in a quite feasible algorithmic manner which will lead to a complete solution of both problems P1 and P2.

To see this let us arrange the edges of G in a definite order, say $E1, E2, \cdots, Eq, \cdots, En$, in which $n = Ne$. For each edge Er of G with $r > 1$ let us denote the set of equations in (EQNf) corresponding to pairs (Eq, Er) with $q = 1, 2, \cdots, r-1$ by (EQNr) which may eventually be non-existent. We denote also the subgraph (not necessarily connected) formed of $E1, \cdots, Er$ by Gr. Beginning from $r = 2$, let us arrange the sets of equations (EQNr) successively in an echelon form by the so-called Gaussian elimination with certain sets of equations, to be eventually forsaken. We remark in passing that the method of Gaussian elimination occured in fact already in the early Chinese classic *Nine Chapters of Arithmetic* together with introduction of negative numbers which appeared more than 2000 years ago. Now the set (EQN2) may be either empty or consisting of a single equation so that it is already in the echelon form. To start with we shall put (EQN2′) to be the same set as (EQN2), empty or not, and introduce a further empty set to be denoted by (DEL2). We set also $G2' = G2$.

Consider now $r > 2$ and put $s = r - 1$. Suppose that the sets of equations (EQNq) with $q = 1, 2, \cdots, s$ have already been treated with the result of a set of equations (EQNs') in echelon form as well as a set (DELs) of edges chosen from Gs such that the subgraph Gs' formed by edges in Gs but not in (DELs) is imbeddable in the plane. Remove now from the set, if non-empty, of equations (EQNr) those corresponding to the pairs (Eq, Er) with Eq in (DELs) and denote the set of remaining equations by (EQNr''). If the set (EQNr'') is non-empty, then adjoin this set to (EQNs') and arrange these in echelon form by Gaussian elimination. Two cases may then occur. In the first case the equations newly adjoined will render the whole set a contradictory one. The system of equations is then unsolvable so that the graph Gs' with Er adjoined will become non-imbeddable. We delete thus Er from G and adjoin Er to (DELs) to form (DELr). The subgraph Gr' will be set to be identical to Gs', and the system (EQNr') to (EQNs'). In the second case the reduction to echelon form can be caried out without arriving at contradiction. The system of equations arrived at consisting of (EQNs') and the newly adjoined (EQNr'') in reduced echelon form will then be denoted by (EQNr'). The set (DELr) will remain the same as (DELs) and Gr' will be Gs' adjoined by Er. We remark that as the set (EQNr'') is at most $r - 1$ in number and each equation in it has at most 4 unknowns with coefficients in $Z2$, the reduction to echelon form requires actually at most $8 * (r - 1)$ additions of mod 2 integers.

Finally, if the set (EQNr) or (EQNr'') is empty, then we shall proceed to the next step with (EQNr'), (DELr) the same as (EQNs'), (DELs), and Gr' as Gs' with Er

adjoined.

From the above we get thus the following

Theorem 3. *There is an algorithmic procedure which permits to determine in a finite number of steps whether a given graph G is imbeddable or not, and in the case it is not imbeddable, a set of edges should be deleted from the given graph so that the remaining graph is imbeddable.*

The above method settles thus both the problems P1 and P2 and can be easily programmed. To apply it we have to choose first an arbitrary immersion of the graph G, form successively the set of equarions (EQNr) and proceed as indicated above, As already remarked, the whole procedure requires at most

$$Na = \text{SUM } 4*(Ne-1)*Ne < 4*Ne\wedge 2$$

mod 2 additions and is thus quite feasible. The only defect is that a large amount of memory space may be required. We shall discuss this matter in later sections.

4. An Alternative Reduction of Criterion to Solvability of Linear Equations

In the original paper ⟨WU1⟩ the author has described a method of reducing the criterion of imbeddability to the solvability of a system of linear equations on $Z2$, which is a little different from that given in Sect. 3. Though the proof of this reduction is rather involved, it has however the advantage of being able to greatly reduce the number of unknowns in the equations. What is more important is that this method will lead to a complete solution of problems P1—4, in comparison with the one in Sect. 3 which permits to solve only problems P1—2. We repeat the remark already made in the introduction that it is problem P3 that is the decisive part in view of applications.

In order to explain this method we shall first introduce some notions as well as notations. Henceforth G will be supposed to be connected.

By a *tree* of the graph G supposed connected we shall mean a maximal one belonging to G, i.e. one passing through all vertices of G. Let a tree T be taken and fixed in what follows.

With respect to tree T of G the vertices will be divided into two classes: *internal* ones and *terminal* ones. The edges of G will also be divided into two classes: those belonging to the tree and those not. We shall call these *tree-edges* and *external edges* and denote them by $Eu, Ev, Ew, \cdots$ and $Ea, Eb, Ec, \cdots$ respectively.

Among the terminal vertices of T we shall choose one as the root of T which will be denoted by O henceforth.

Without loss of generality we shall make the assumption that no external edges issue from O. In fact, in the contrary case we may adjoin an extra edge to the graph G with one end at O and the other free. We may then take that free end as the new root of the new graph. The problems are actually the same for the new graph and the original one so far as imbeddability is concerned. Hence we shall suppose that the above device has been adopted in what follows so that the above assumption is always verified.

For any vertex Vi of G different from O there is a unique path belonging to T which leads from Vi to O and will be denoted by Pi. For any external edge Ea there is also a unique path belonging to T which joins the two ends of Ea and will be denoted by Pa. The cycle of G formed by Ea and Pa will then be denoted by Ca.

For any two vertices Vi, $Vj(\langle\rangle O)$ the two paths Pi, Pj will begin to meet first at some vertex in running toward O which will then be called the *V-meet* of Vi, Vj or of Pi, Pj.

For any tree-edge Eu with ends Vi, Vj, one of them, say Vi, will have the path Pi containing the other end Vj. We shall then call Vi the *head* and Vj and the *tail* of Eu.

Each tree-edge Eu will divide the tree T into two disconnected parts, say $T'u$ and $T''u$. One, say $T'u$, will contain the head of Eu and the subtree formed by $T'u$ and Eu will be denoted by Tu. The set consisting of Eu as well as all edges with two ends one in $T'u$ and the other in $T''u$ will be denoted by CSu. In the network theory a set of edges is called a *cut-set* of G if in removing it G will split into two or more disconnected parts. The set CSu is such a cut-set and in the network theory it is proved that the collection of sets CSu corresponding to all the tree-edges Eu form a basis of all cut-sets in an evident sense, cf. e. g. [SB1].

For any two sets of edges $S1$, $S2$ of G the function in $C2$ taking the value 1 on all pairs (Eq, Er) of $D2$ with Eq in $S1$ and Er in $S2$ and the value 0 otherwise will be denoted by $[S1, S2]$. In other words

$$\langle S1, S2\rangle = \text{SUM } \langle Eq, Er\rangle,$$

in which the summation is to be extended over all pairs (Eq, Er) as above. The following lemma is now readily proved (cf. [WU1]):

Lemma 5. *The subgroup $dC1$ of $C2$ has a set of generators (not necessarily a basis) consisting of elements*

$$\langle CSu, CSv\rangle \text{ and } \langle CSu, Ea\rangle,$$

corresponding to all pairs (Eu, Ev) and (Eu, Ea), disjoint or not, respectively.

Introduce now sets of variables or unknowns on $Z2$ as follows. To each unordered pair (Eu, Ev) of tree-edges disjoint or not is associated an unknown $Xuv(=Xvu)$.

To each pair (Eu, Ea) of a tree-edge Eu and an external edge Ea disjoint or not is associated an unknown Yua.

By the lemma above and Fundamental Theorem I in Sect. 2 it follows that for G to be imbeddable it is necessary and sufficient that for an arbitrary T-immersion f of G the following system of linear equations in $Z2$ be solvable in the unknowns X and Y:

$$\text{SUM } Xuv\cdot\langle CSu, CSv\rangle + \text{SUM } Yua\cdot\langle CSu, Ea\rangle = \text{SUM } Iqr\cdot\langle Eq, Er\rangle,$$

in which the various summations are to be extended over respective ranges. Compare the terms of both sides and note that by the very definition of a T-immersion $Iqr = 0$ when Eq or Er or both are tree edges, we get:

$$Xuv = 0, \text{ for } (Eu, Ev) \text{ in } D2,$$
$$Yua = \text{SUM1 } Xuv, \text{ for } (Eu, Ea) \text{ in } D2,$$
$$Iab = \text{SUM2 } Yua + \text{SUM3 } Yub + \text{SUM4 } Xuv.$$

The various summations are respectively extended over the ranges as follows:

SUM1 over Ev in Pa,

SUM2 over Eu in Pb,

SUM3 over Eu in Pa,

SUM4 over pairs (Eu, Ev), disjoint or not, with Eu in Pa and Ev in Pb.

Corresponding to each pair (Eu, Ea) of a tree-edge Eu and an external edge Ea with Eu, Ea disjoint or not let us introduce a new unknown Xua in $Z2$ by setting

$$Xua = \text{SUM1 } Xuv + Yua,$$

so that by equations about Yua above,

$$Xua = 0, \text{ for } (Eu, Ea) \text{ in } D2.$$

The equation about Iab will then become

$$Iab = \text{SUM2 } (\text{SUM1 } Xuv + Xua) + \text{SUM3 } (\text{SUM 5 } Xuv + Xub) + \text{SUM 4 } Xuv,$$

with SUM5 given by

SUM5 over Ev in Pb.

As the terms SUM2 SUM1 Xuv and SUM3 SUM5 Xuv are actually the same they cancel each other in $Z2$. Taking into account the equation $Xua = 0$ for (Eu, Ea) in $D2$, we get then

$$\text{SUM0 } Xuv + \text{SUM}' \, Xua + \text{SUM}'' \, Xub = Iab, \tag{If}$$

in which the various summations are to be extended over ranges as follows:

SUM0 over pairs (Eu, Ev) non-disjoint with Eu in Pa and Ev in Pb,

SUM′ over pairs Eu, Ea with Eu in Pb and (Eu, Ea) non-disjoint,

SUM″ over pairs Eu, Eb with Eu in Pa and (Eu, Eb) non-disjoint.

This leads to the following

Theorem 4. *For a graph G to be imbeddable it is necessary and sufficient that for an arbitrary tree T and an arbitrary T-immersion f of G the system of equations* (If) *corresponding to pairs (Ea, Eb) in $D2$ be solvable in the unknowns X in $Z2$.*

(To be continued)

线性图的平面嵌入问题

吴 文 俊

(中国科学院系统科学研究所)

摘　　要

本文是 1967 年以来作者只用中文发表的所得结果的一个英文综述，在文中证明了连通线性图可嵌入平面的一个充要条件是某一组 mod 2 系数的线性方程组有解. 在该方程组有解因而线性图可嵌入平面时，又可考虑另一组仍为 mod 2 系数的二次方程组，并根据这两组方程必然存在的共同解答来作出图的具体嵌入. 若图的顶点数与棱数各为 N_v 与 N_e，而顶点的最大次数为 m，则这些方程中的未知数个数最多为

$$(m-3)*N_e+N_v,$$

且在决定能否嵌入时只须用到不超过 $4*N_e\wedge 2$ 的 mod 2 加法即可. 因之这一方法容易编成程序且是切实可行的.

系 统 科 学 与 数 学
J. Sys. Sci. & Math. Scis.
6 (1) (1986), 23—35

ON THE PLANAR IMBEDDING OF LINEAR GRAPHS (CONTINUED)

WU WEN-JUN (WU WEN-TSÜN)
(*Institute of Systems Science, Academia Sinica*)

5. FURTHER REDUCTION OF FUNDAMENTAL SYSTEM OF LINEAR EQUATIONS

The fundamental system of linear equations (If) in preceding Sect 4 can actually be put in a much simpler form. For this purpose let us denote by NT a certain neighborhood of T in G sufficiently small so that the T-immersions considered will be some imbeddings when restricted on NT and that all possible intersections of images of disjoint external edges are not in NT. We shall call these T-immersions also *NT-immersions.*

Lemma 6. *For NT-immersions of G the immersion elements are already completely determined by the restricted imbeddings of the neighborhood NT of the respective immersions.*

Proof. For any external edge Ea with ends Vi, Vj let us take two points on Ea lying in the neighborhood NT and denote them by Via, Vja. These points will be taken so near to the tree that the edge Ea will split into three parts from Vi to Via, from Via to Vja, and from Vja to Vj, disjoint from each other except at possible common ends. These parts will be denoted by Eia, Ea', Eja respectively.

Consider now any two NT-immersions f and g which coincide on NT as imbeddings. For any pair of edges (Ea, Eb) in $D2$ with Vi, Vj the ends of Ea and Vk, Vl the ends of Eb let us denote the intersection number Int (fEa, fEb) still by Iab while $\mathrm{Int}(gEa, gEb)$ by Jab. We have then

$$Iab = \mathrm{Int}(fEa, fEb') = \mathrm{Ord}(fVkb, fCa) + \mathrm{Ord}(fVlb, fCa),$$

$$Jab = \mathrm{Int}(gEa, gEb') = \mathrm{Ord}(gVkb, gCa) + \mathrm{Ord}(gVlb, gCa).$$

As f and g coincide on NT, so on $Z2$,

$$fCa + gCa = fEa' + gEa', say = Ca',$$

and we have therefore

$$Iab + Jab = \mathrm{Ord}(fVkb, Ca') + \mathrm{Ord}(fVlb, Ca').$$

It follows that

$$Iab = Jab,$$

since $fVkb$ and $fVlb$ are the ends of the broken line $fEkb + fPb + fElb$ disjoint from the polygon Ca'. The immersion elements $c(f)$ and $c(g)$ of f and g are therefore the same. The proof is completed.

In order to avoid tedious verifications in the case that an external edge ends at some internal vertex of the tree, we shall adopt the following devices: If the external edge Ea has some internal end(s) of the tree as an end Vi (or both ends Vi and Vj), then we shall replace, with notations as in the proof of the above lemma, the edge Ea with two (or three)edges $VaVia$, and $ViaVj$ (or $VaVia$, $ViaVja$, and $VjaVj$). By the lemma above as well as Lemma 2 of Sect. 2 we see that it is immaterial to replace the original graph by this new one. We shall suppose in what follows that such a modification of the graph has already been done so that we may assume the following conventions for the graph G have been observed:

Convention 1. The root O of the tree T is a terminal vertex.

Convention 2. The ends of any external edge are both terminal vertices.

Let us call an unordered pair of tree-edges (Eu, Ev) a *redundant* one if the tail of say Eu is the head of Ev. The corresponding unknown Xuv is then also said to be *redundant*. Consider now any set of mod 2 integers $(Aqrs)$ corresponding to all unordered triples of tree-edges (Eq, Er, Es) with one end in common for which each $Aqrs$ is independent of the order of indices(q, r, s). Let us also put for each such triple

$$Xqrs = Xqr + Xqs + Xrs,$$

which is also independent of the order of the indices (q, r, s). We have then

Lemma 7. *If the set of equations*

$$Xqrs = Aqrs$$

corresponding to all unordered triples of edges (Eq, Er, Es) with a common end is solvable for $Xuv=Xvu$ in $Z2$, then the same set of equations is *also solvable with all redundant unknowns $Xuv=0$.*

Proof. Let $(Xuv)=(Auv)$ with $Auv=Avu$ corresponding to each unordered pair of edges (Eu, Ev) with common end be a solution of the above system of equations. For each such pair (Eu, Ev) with common end Vm let Eq be the edge on the path Pm with Vm as head. Then we see easily that

$$Xuv = 0 \text{ for } (Eu, Ev) \text{ redundant and}$$
$$Xuv = Auq + Avq + Auv \text{ otherwise}$$

is also a solution of the above system of equations and the lemma is thus proved.

We are now in a position to simplify the system of equations (If) of Theorem 4 in Sect. 4. We shall denote the equation in(If) corresponding to the pair (Ea, Eb) in $D2$by (EQab) and the ends of Ea, Eb by Vi, Vj and Vk, Vl respectively.

First let us remark that owing to conventions 1 and 2 the terms in SUM′ Xua and SUM″ Xub in the equation (EQab) are no more existent. For the terms in SUM0 Xuv we distinguish three cases.

Case 1. Pa, Pb are disjoint.

We see that SUM0 is nonexistent and fEa, fEb do not meet. So the equation (EQab) becomes 0=0 and is redundant.

Case 2. Pa, Pb meet at a single vertex Vm.

Let the tree-edges with end Vm on the paths from Vi, Vj, Vk, Vl to Vm be respectively Ep, Eq, Er, and Es. Then we see that SUM0 reduces to four terms

$$Xpr + Xps + Xqr + Xqs.$$

The equation (EQab) can thus be written in the form

$$Xpqr + Xpqs = Iab.$$

Case 3. *Pa, Pb* have a tree-path in common with end vertices *Vm, Vn*.

We may suppose that *Vm* is on the tree-path *ViVk* and *Vn* on *VjVl*. Let the edges with end *Vm* on the tree-paths *ViVm, VkVm* and *VmVn* be respectively *Ep, Er* and *Ew*. Similarly, let the edges with end *Vn* on the tree-paths *VjVn, VlVn* and *VnVm* be respectively *Eq, Es* and *Ez*. Then we see that the equation (EQab) may be written in the form

$$Xprw + Xqsz = Iab.$$

Suppose now *G* is imbeddable so that (If) is solvable in the unknowns *Xuv=Xvu*. In view of the above analysis of the form of the equations (EQab) in (If) it follows from Lemma 7 that the [illegible]m (If) will [illegible] solvable in unknowns *Xuv=Xvu* with all redundant ones =0. [illegible]eover, for any different terminal vertices *Vi, Vj*, both different from root *O* with *Vm* the *V*-meet of the paths *Pi* and *Pj*, different from *O* owing to our conventions, let *Er, Es* be the edges on *Pi, Pj* with tail *Vm* and set

$$Xij = Xrs(=Xji).$$

Then, with all redundant *Xuv*=0 it is easily verified that the left-hand side of the equations (EQab) either in Case 2 or in Case 3 can always be written in the form

$$Xik + Xjk + Xil + Xjl$$

which may eventually be reduced to only two terms.

It follows that the system of equations (If) may be replaced by a system (Xf) below which is much simpler in form and the Fundamental Theorem II may also be re-stated as:

Fundamental Theorem II'. *A graph G is imbeddable if and only if, given a tree T, a root O, and a T-immersion f, the system of equations*

$$Xik + Xil + Xjk + Xjl = Iab, \qquad \text{(Xf)}$$

corresponding to pairs (*Ea, Eb*) in *D*2 with *Vi, Vj* ends of *Ea* and *Vk, Vl* ends of *Eb*, is solvable in *Z*2.

We remark that, under the conventions above, each *Xij* occuring in equations (Xf) is some *Xrs* for a pair of non-disjoint tree-edges (*Er, Es*) having a common tail. We shall call all such pairs (*Er, Es*) *admissible pairs* in what follows.

We see that each equation of (Xf) involves actually at most 2 or 4 unknowns of *X* and eventually has the trivial form 0=0. Morever, the number of unknowns of *X* are readily estimated as in the following

Theorem 5. *If the maximum order of vertices in the graph G is m, then the number of unknowns of X occurring in the fundamental system of equations (Xf) is at most*

$$Nx = (m-3) * Ne + Nv,$$

in which Nv and Ne are respectively the original number of vertices and edges of G before modification.

Proof. Let Ok be the number of vertices of order k in G. Then we have

$$\text{SUM } k * Ok = 2 * Ne,$$

the summation being over $k >= 1$. Now to each vertex of order k the associated number of unknowns of X that may occur in the equations (Xf) is clearly at most $(k-1) * (k-2)/2$. Hence we have.

$$\begin{aligned} Nx &<= \text{SUM}((k-1) * (k-2)/2) * Ok \\ &= \text{SUM}k * (k-2) * Ok/2 - 1/2 * (\text{SUM}(k-2) * Ok) \\ &<= (m-2) * Ne - Ne + \text{SUM } Ok \\ &= (m-3) * Ne + Nv. \end{aligned}$$

That Nv, Ne may be taken to be the original numbers of the unmodified graph is also clear.

To determine the imbeddability of G we can now proceed as in Sect. 3 with the result of getting a set of edges that is to be deleted and a set of solutions of remaining equations in mod 2 integers of the unknowns X. The operations require only mod 2 additions at most $4 * Ne \Lambda 2$ in number as before. The number of unknowns has however been reduced so that much les memory space will be required. In particular, if the graph G has only vertices of order $< =3$, then the number of unknowns will be $< =Nv$, the number of vertices of G.

The great advantage of this method lies in reality in the fact that actual imbedding of an imbeddable graph G, or more generally the imbedding of the imbeddable graph which remains after removal of a certain set of edges, can be constructed from the set of solutions obtained from the fundamental system of equations. This will form the object of study in the next sections.

6. Geometrical Interpretation of Unkowns X and Rotation Numbers Associated to a T-Immersion

As stated at the end of last section, the solutions of the fundamental system of equations (Xf) for a graph G supposed to be imbeddable or become imbeddable after removal of certain edges will lead to a method of actual imbedding of such a graph. To see this we shall first give in this section some geometrical interpretation of the unknowns X involved in these equations. In fact, by Lemma 6 of Sect. 5 the nature of the T-immersion f will actually be determined by f on NT and this in turn will be determined by how the edges at a common end are mutually situated when immersed by f. This suggests thus the introduction of the following notions.

Let $L1$, $L2$, $L3$ be three simple broken lines in the plane disjoint from each other except that they have one end in common. We shall attach then to this ordered triple of lines a *rotation number* (in $Z2$).

$$R(L1,\ L2, L3) = 0 \text{ or } 1$$

according as in passing from $L1$ to $L3$ through $L2$ we have to turn around their common end in a counter-clockwise or in a clockwise sense.

Consider now an NT-immersion f of G with conventions 1, 2 observed. For any admissible pair of edges (Er, Es) with common tail Vm we set then by definition

$$Rrs(f) = R(fPm, fEr, fEs).$$

For each pair of vertices Vi, Vj different from O which lead to the admissible pair (Er, Es) with Er on Pi and Es on Pj having a common tail we shall set by definition

$$Rij(f) = Rrs(f).$$

Remark that the order of the indices are important in that

$$Rrs(f) = Rsr(f) + 1, \quad Rij(f) = Rji(f) + 1.$$

Theorem 6. *Let f be a T-immersion of G with conventions* 1, 2 *observed. Then for any pair of edges (Ea, Eb) in $D2$ with Vi, Vj ends of Ea and Vk, Vl ends of Eb, we have*

$$Iab(f) = Rik(f) + Rjk(f) + Ril(f) + Rjl(f).$$

Remark. *The equation is of the same form as the corresponding one in the fundamental system of equations (Xf). However the numbers $Rik(f)$ do not form a solution of the system (Xf) since $Rik(f) <\ > Rki(f)$ while we are seeking for solutions with $Xik = Xki$.*

Proof. Let Vm be the V-meet of the paths Pi, Pj. Consider first the vertex Vk. According as fO and fVk are interior or exterior to the cycle fCa, and according as the path Pk does not meet Ca or first meets Ca on Pi or Pj, there are in all 12 cases to consider. We verify easily that in all cases we shall have

$$Rik(f) + Rjk(f) = \mathrm{Ord}(fVk, fCa) + \mathrm{Ord}(fO, fCa).$$

Similarly we have

$$Ril(f) + Rjl(f) = \mathrm{Ord}(fVl, fCa) + \mathrm{Ord}(fO, fCa).$$

Hence we get

$$\begin{aligned} &Rik(f) + Rjk(f) + Ril(f) + Rjl(f) \\ &= \mathrm{Ord}(fVk, fCa) + \mathrm{Ord}(fVl, fCa) \\ &= Iab. \end{aligned}$$

Consider now two NT-immersions f and g of G. For any admissible pair of edges (Er, Es) let us set by definition

$$Wrs(f, g) = Rrs(f) + Rrs(g).$$

For any pair of vertices Vi, Vj different from O and leading to the admissible pair (Er, Es) we set then by definition

$$Wij(f, g) = Rij(f) + Rij(g),$$

or, what is the same,

$$Wij(f, g) = Wrs(f, g).$$

Remark that unlike the R's the numbers W are no more dependent on the order of the indices:

$$Wrs(f,g) = Wsr(f,g), \quad Wij(f,g) = Wji(f,g).$$

If the common tail of the admissible pair *Er, Es* is at *Vm,* then

$$Wrs(f,g) = 0 \text{ or } 1$$

according as the configurations (*fPm, fEr, fEs*) and (*gPm, gEr, gEs*) have the same sense of rotations or not. Hence the set of numbers $Wrs(f, g) = Wsr(f, g)$ corresponding to all admissible pairs (*Er, Es*) serves to compare the configurations of the two imbeddings f/NT and g/NT. More precisely we have the following

Theorem 7. *Let f, g be two NT-immersions of G. Suppose the fundamental system of equations (Xg) corresponding to g is solvable and has a solution*

$$(Xrs) = (Brs),$$

in which $Brs = Bsr$ are numbers in Z2 corresponding to all admissible pairs of edges (Er, Es). Then the fundamental system of equations (Xf) corresponding to f is also solvable and has a solution

$$(Xrs) = (Ars),$$

in which

$$Ars = Brs + Wrs,$$

where we have put for simplicity

$$Wrs = Wrs(f,g).$$

Proof. Let us set by definition

$$Ars = Brs + Wrs(= Asr).$$

Set also by definition

$$Bij = Brs, \quad Aij = Ars,$$

if the pair of vertices *Vi, Vj* different from *O* will lead to the admissible pair of edges (*Er, Es*). Then we shall have also

$$Aij = Bij + Wij(= Aji).$$

Consider now any pair of edges (*Ea, Eb*) in *D*2 with *Vi, Vj* ends of *Ea* and *Vk, Vl* ends of *Eb*. Write *Iab, Jab* for their respective intersection index under *f* and *g* as before. As (*Brs*) is a solution of the system (Xg), so we have

$$Jab = Bik + Bjk + Bil + Bjl.$$

By Theorem 6 we have also

$$Jab = Rik(g) + Rjk(g) + Ril(g) + Rjl(g),$$

$$Iab = Rik(f) + Rjk(f) + Ril(f) + Rjl(f).$$

Adding all these three equations together and taking into account the definition of *A* and *W*, we get

$$Iab = Aik + Ajk + Ail + Ajl.$$

This shows that the set of numbers (Ars) in $Z2$ with $Ars=Asr$ forms a solution of (Xf) and the theorem is proved.

Suppose now in particular that G is imbeddable with g an imbedding not only of NT but also of G as a whole. Then we have $Jab=0$ for all pairs in $D2$ so that $Xrs=0$ for all admissible pairs (Er, Es) forms a trivial solution of the system (Xg). By the above theorem

$$(Xrs) = (Wrs)$$

will form then a solution of the system (Xf). From the meaning of Wrs we have therefore the following

Geometrical Interpretation of the unknowns Xrs:

The set $Xrs=Xsr$ will serve as a set of indicators whether for each admissible pair of edges (Er, Es) with common tail Vm their images under f should be modified to change the sense of rotations of the triple (fPm, fEr, fEs) so that the modified immersion of T may be extendable to an actual imbedding of the whole graph G. See however the next section.

7. Quadratic Relations among the Unknowns X

By Fundamental Theorem II we know that for an arbitrary NT-immersion f of G, if the fundamental system of equations (Xf) possesses a solution

$$(Xrs) = (Ars),$$

with $Ars=Asr$ corresponding to all admissible pairs of edges (Er, Es) in $D2$, then G is imbeddable. From the preceding section it seems further that from this solution we would get an actual imbedding of G by modifying the NT-immersion f to another one g in changing the mutual rotational relationships of edges at the same vertices according to the formulae

$$Wrs(f, g) = Ars.$$

However, this is entirely not the case. In fact, though G is imbeddable if the system of equations (Xf) is solvable, not every solution of (Xf) will lead to an actual imbedding of G in the above manner. The reason may be seen as follows.

Let us consider any triple of edges Er, Es, Et with same tail Vm. For any immersion g we have then a set of 3 rotation numbers in $Z2$, *viz.*

$$\text{(Rg)} \qquad Rrs(g), Rst(g), Rtr(g).$$

As each R may take a value of 0 or 1, so apparently there would be 8 such sets of values to be taken for (Rg). However there are only 6 different types of orientational relationships of the edges Er, Es, Et and Pm under g. This shows that among the 8 sets of values of (Rg) only 6 will actually be geometrically realizable. In fact, (Rg) can never take up the sets of values (0, 0, 0) and (1, 1, 1). The problem thus arising is to pick out these 6 sets of values among the 8 sets. The solution of this problem will be furnished by the following device introduced in the original paper [WU4].

For an immersion g and 3 edges Er, Es, Et with same tail Vm let us set by definition

$$Qrst(g) = Rrs(g) * Rrt(g) + Rst(g) * Rsr(g) + Rtr(g) * Rts(g).$$

Remark that though the numbers $Rrs(g)$, etc. depend on the order of indices (r, s) etc., $Qrst(g)$ is independent of the order of the indices (r, s, t). We have now the following

Lemma 8. *The rotation numbers (Rg) satisfy always the relation*

$$Qrst(g) = 1.$$

Proof. Let us first remark that if g is such that in turning around the common end gVm, we shall get successively gPm, gEr, gEs, gEt in the counter-clockwise order, then we have $Rrs(g)=0$, $Rst(g)=0$, $Rtr(g)=1$ so that $Qrst(g)=1$.

Suppose next that $Qrst(g)=1$ for a certain configuration of gPm, gEr, gEs, gEt in the plane with e.g. gEr, gEs neighboring to each other in the arrangement. Let us interchange the orientational relationship between gEr, gEs but leave the others unchanged to get a new immersion g'. Then we have

$$Rrs(g') = Rrs(g) + 1,$$
$$Rst(g') = Rst(g), \quad Rtr(g') = Rtr(g).$$

It follows that

$$Qrst(g') = Qrst(g) + Rrt(g) + Rst(g) = Qrst(g),$$

since, with gEr, gEs neighboring to each other in the plane, $Rrt(g) = Rst(g)$. This proves the lemma since any other configuration of Pm, Er, Es, Et under any immersion may be got from the one under g by a number of such interchanges of immersed neighoring edges.

For any immersion f of G let us now introduce by definition a system of quadratic forms

$$\begin{aligned} Qrst(f, X) = &(Xrs + Rrs(f)) * (Xrt + Rrt(f)) \\ &+ (Xst + Rst(f)) * (Xsr + Rsr(f)) \\ &+ (Xtr + Rtr(f)) * (Xts + Rts(f)) \end{aligned}$$

corresponding to each triple (Er, Es, Et) with a common tail. Introduce also the system of quadratic equations

$$Qrst(f, X) = 1 \tag{Qf}$$

corresponding to all such triples. Note that in the equations $Xrs=Xsr$ while $Rrs(f)=Rsr(f)+1$, etc. However $Qrst(f, x)$ is independent of the order of indices (r, s, t). We have now the following

Fundamental Theorem III. *If corresponding to a T-immersion of G the fundamental system of equations (Xf) is solvable, then the systems of equations (Xf) and (Qf) taken together are also solvable.*

Proof. As the system (Xf) is solvable, by Fundamental Theorem II′ G is imbeddable with a certain g as an imbedding of G as a whole. Corresponding to each admissible pair of edges (Er, Es) let us put

$$Wrs = Wrs(f, g)$$

for simplicity. Then as in the proof of Theorem 7 of the last section, the system (Xf) will have a solution

$$(Xrs) = (Wrs).$$

By Lemma 8 above the set of numbers $(Rrs(g))$ will satisfy the relations

$$Qrst(g) = 1$$

corresponding to all triples of edges (Er, Es, Et) with common tails. As

$$Wrs = Rrs(f) + Rrs(g),$$

we see that $Qrst(f, X)$ will become $Qrst(g)$ when Xrs, etc. are substituted by Wrs, etc. This shows that $(Xrs)=(Wrs)$ will satisfy both systems of (Xf) and (Qf).

8. Actual Imbedding of Imbeddable Graphs

We are now ready to settle problem P3 of actually imbedding an imbeddable graph G in the plane, assuming that certain edges have already been removed to make G the remaining imbeddable part if necessary. For this purpose we shall first prove a converse of Lemma 8 of the preceding section, *viz.*

Lemma 9. *Let* (Nrs) *be a set of numbers in Z2 with* $Nrs=Nsr+1$ *corresponding to all admissible pairs of edges (Er, Es) with same tail which satisfies the relations*

$$Nrs * Nrt + Nst * Nsr + Ntr * Nts = 1 \qquad \text{(Nrst)}$$

corresponding to all triples of edges (Er, Es, Et) with common tails. Then there is an immersion g *of* G *such that the rotation numbers under* g *coincide with the corresponding numbers* N, *i.e.*

$$Rrs(g) = Nrs$$

for all admissible pairs of edges (Er, Es) *of* G.

Proof. Let us consider the simple case that G consists of an edge OVm and a finite set of edges

$$Er, Es, \cdots, Et, \cdots, \qquad \text{(E)}$$

all having Vm as common end. The tree T is then the same as G and O will be chosen as the root. If the number n of the edges in (E) is $n=3$, then it is clear by the preceding section that such immersion (in fact an imbedding) g of $G=T$ in the plane exists. We shall now proceed to prove this in general by induction on n.

Suppose thus the number of edges in (E) is $n>3$. By induction there is an immersion g' of $G=T$ such that for all pairs of edges (Ep, Eq) chosen from the set

$$Es, \cdots, Et, \cdots \qquad \text{(E')}$$

we have

$$Rpq(g') = Npq.$$

Suppose that in turning around the common end $g'Vm$ in a counter-clockwise sense on starting from $g'Pm$ ($Pm = OVm$) we shall pass in succession

$$\cdots, g'Es, \cdots, g'Ep, g'Eq, \cdots, g'Et, \cdots.$$

Suppose that among the numbers

$$\cdots, Nsr, \cdots, Npr, Nqr, \cdots, Ntr, \cdots$$

in this order the first non-zero number is Nqr so that

$$\cdots = Nsr = \cdots = Npr = 0, \text{ while } Nqr <> 0.$$

The equation in N corresponding to a triple of indices (r, q, t) with Et in the partial set of edges after Eq in the above order is given by

$$Nrq * Nrt + Nqt * Nqr + Ntr * Ntq = 1.$$

As

$$Nqt = Rqt(g') = 0, \quad Nrq = Nqr + 1 = 0, \quad Ntq = Rtq(g') = 1$$

we get

$$Ntr = 1, \text{ or } Nrt = 0.$$

Modify now the immersion g' to an immersion g such that g will be the same as g' on Pm and on all edges in (E') while gEr will be brought to a position between $gEp = g'Ep$ and $gEq = g'Eq$. Then we see that for any edge Es in the partial set of edges before Ep in the above order and any edge Et in the partial set of edges after Eq in that order,

$$Rrs(g) = 1 = Nrs, \quad Rrt(g) = 0 = Nrt,$$

$$Rrp(g) = 1 = Nrp, \quad Rrq(g) = 0 = Nrq.$$

For the other number R's we have say $Rst(g) = Rst(g') = Nst$. Hence g will have its rotation numbers all equal to the corresponding numbers N. The induction is thus completed and the lemma is proved for the above special graph G.

For the general graph we shall proceed in just the same manner with the modification that each time we bring a certain edge Er to a new position, we shall bring the whole sub-tree Tr to such new position at the same time. Arrange now the vertices different from O in a definite order and treat each vertex in turn as for the special graph above, with the above modification taken into due account. The rotation numbers of the new immersion for admissible pairs of edges with common tail at that vertex will be identical with the corresponding numbers N. Remark that the interchanges at one vertex will not affect the results of interchanges at other vertices. Hence in proceeding successively we shall finally arrive at a T-immersion with the desired property. The lemma is thus completely proved.

We have now the following

Fundamental Theorem IV. *If corresponding to a T-immersion f of G the* *fundamental systems of equation (Xf) and (Qf) taken together possess in Z2 a solution*

$$(Xrs) = (Ars)$$

with Ars=Asr corresponding to all admissible pairs of edges (Er, Es) with common tails, then there is an imbedding g of G as a whole in the plane with

$$Rrs(g) = Ars + Rrs(f)$$

for all such pairs (Er, Es).

Proof. Set for each admissible pair of edges (Er, Es)

$$Nrs = Ars + Rrs(f) \ (=Nsr + 1).$$

Then by the hypothesis of the theorem the set of numbers Nrs will satisfy all relations of the form (Nrst) corresponding to triple of edges (Er, Es, Et) with common tails. By Lemma 9 above there will be some T-immersion g of G with

$$Rrs(g) = Nrs$$

for all admissible pairs (Er, Es).

By Theorem 7 of Sect. 6, the fundamental system of equations (Xg) corresponding to g will have now a solution given by

$$Xrs = Ars + Wrs,$$

where

$$Wrs = Wrs(f, g) = Rrs(f) + Rrs(g).$$

Consequently (Xg) will have a solution identical to 0:

$$Xrs = 0$$

for all admissible pairs of edges (Er, Es). It follows from Fundamental Theorem II′ in Sect. 5 that for any pair of external edges (Ea, Eb) in $D2$ we should have

$$Iab = 0$$

or

$$\mathrm{Int}(gEa, gEb) = 0.$$

Arrange now all the external edges of G in a definite order, say

$$Ea, \cdots, Eb, \cdots, Ec, Ed, \cdots. \qquad (\mathrm{E})$$

Our aim is to extend the part of the T-immersion g restricted to T successively to the external edges of (E) to get each time an imbedding of T with successively adjoined edges as a whole. The final imbedding achieved in this way will then be a required imbedding of G in the plane as a whole.

Such an extension to Ea is trivial. Suppose that

$$Ea, \cdots, Eb, \cdots, Ec \qquad (\mathrm{E}')$$

in the ordered set (E) have been extended so that we have an imbedding g' of $G'=$

$T+(E')$ in the plane with g'/T identical to g. Let us try to extend g' to an imbedding including the next new edge Ed in the set (E). Consider any external edge disjoint from Ed in (E″), say Eb. By Lemma 6 of Sect. 5, we have for the pair (Eb, Ed) in $D2$

$$\text{Int}\,(g'Eb, g'Ed) = \text{Int}\,(gEb, gEd) = 0.$$

This means that if the ends of Ed are Vi and Vj, then we should have

$$\text{Ord}(g'Vi, g'Cb) = \text{Ord}(g'Vj, g'Cb).$$

Consequently $g'Vi$ and $g'Vj$ will lie in the same region in the plane separated by $g'T$, $g'Ea$, .. , $g'Ec$ of $g'G'$. We may thus join $g'Vi$ and $g'Vj$ by a simple broken line not meeting $g'G'$ except at the two ends. We extend than g'' to Ed by taking this broken line to be the image $g''Ed$. This achieves the induction and proves the thorem.

9. Procedure of Solving Problems P1---3 for a Graph

From the developments of the last sections it is now clear how to solve problems P1—3 for a given graph G. The procedure will be as follows.

Step 1. *Choose an arbitrary tree T of G as well as a root O. Modifications may be made according to Sect. 5 if required.*

Step 2. *Take an arbitrary T-immersion f of G.*

Step. 3. *Form the fundamental system of equations (Xf) successively and solve in the way as shown in Sect. 5. We get then a set DEL of edges to be removed from G to render the remaining graph G' imbeddable. Denote the restriction of f to G' by f'. As no ambiguity can occur we shall denote G' and f' again by G and f. The set of solutions of corresponding fundamental equations (Xf) will be denoted by (S).*

Step 4. *Form the system of quadratic equations (Qf) for G (i.e. (Qf') for G') and verify whether each solution in the set (S) is also a solution of (Qf) or not. By Fundamental Theorem III of Sect. 6, such solutions necessarily exist.*

Step 5. *For any solution of (Xf) and (Qf) taken together, modify f/T to a T-immersion g of G as in Sect. 8. Such a T-immersion g may then be extended to get an imbedding of G in the plane as a whole as shown in the Fundamental Theorem IV of Sect. 8.*

Remark. By introducing new unknowns and new system of equations it can be shown that the totality of all possible imbeddings of the imbeddable graph essentially different from each other will be obtained in correspondence with the solutions of the three systems of equations taken altogether. This gives the solution of problem P4 as stated in Sect. 1. We shall not however enter into this and will leave the details to the original paper [WU4].

References

[AP1] Auslander, L. & Parter, V., On imbedding graphs in the sphere, *J. Math. Mech.*, **10** (1961), 517—523.

[HT1] Hopcroft, J. & Tarjan, R., Efficient planarity testing, *JACM*, **21** (1974), 549—568.

[KU1] Kuratowski, C., Sur le problème des courbes gauches en topologie, *Fund. Math.*, **15** (1930), 271—283.

[L1] Liu Yan-pei, Modulo-2 programming and planar imbedding, *Acta Math. Appl. Sinica*, **1** (1978), 321—329 (in Chinese).

[L2] ———, On the linearity of testing planarity of graphs, to be published in *Annals of Chinese Math.*

[R1] Rosenthiel, P., Preuve algebrique du critere de planarite de Wu-Liu, *Annals of Discrete Math.*, **9** (1980), 67—78.

[SB1] Seshu, S. & Balabanian, N., Linear Network Analysis, 1959.

[TU1] Tutte, W. T., Toward a theory of crossing numbers, *J. Comb. Theory*, **8** (1970), 45—53.

[WU1] Wu Wen-tsün, A theory of Imbedding, Immersion, and Isotopy of Polytopes in a Euclidean Space, Science Press, Beijing, 1965.

[WU2] ———, A mathematical problem in the design of integrated circuits, *Mathematics in Practice and Theory*, 1 (1973), 20—40 (in Chinese).

[WU3] ———, Planar imbedding of linear graphs, *Kexue Tongbao*, 2(1974), 226—228 (in Chinese).

[WU4] ———, Layout problem in printed circuits and integrated circuits, Appendix to Chinese Edition of [WU1], 1978, 213—261.

系统科学与数学
J. Sys. Sci. & Math. Scis.
6 (3) (1986), 204—216

A MECHANIZATION METHOD OF GEOMETRY AND ITS APPLICATIONS

I. DISTANCES, AREAS AND VOLUMES

WU WEN-JUN (WU WEN-TSÜN)
(*Institute of Systems Science, Academia Sinica, Beijing*)

1. INTRODUCTION

The present paper is the first of a series of papers dealing with a mechanization method of geometry and related domains, with emphasis on its application aspects. It bears the same title as a short paper [WU6] presented to *Kuxue Tongbao* last year but with more details and also more materials. It is mainly the content of part of lectures given by the author for a course on mechanical theorem proving in the Graduate School of Academia Sinica during the September-December semester, 1984.

To begin with, let us first recall briefly some fundamental notions and facts that originated from works of J. F. Ritt, cf. [R1, 2]. Let a finite set of polynomials (abbr. pols and polset) PS in variables $X1, \cdots, Xn$ with coefficients in a basic field K of characteristic 0 be given. The method of Ritt then permits to deduce from it by rational operations alone a polset CHS of special type, called the CHARACTERISTIC SET (abbr. CHAR-SET) of PS. In the case that PS is not CONTRADICTORY or CHS does not consist of a single pol which is a non-zero constant of K, then we can divide the variables Xi into two parts $U1, \cdots, Ud, Y1, \cdots, Ye$ with $d + e = n$ such that CHS will be of the forms

$$C1 = I1 * Y1 \wedge M1 + \text{lower degree terms in } Y1,$$

$$C2 = I2 * Y2 \wedge M2 + \text{lower degree terms in } Y2,$$

$$\cdots,$$

$$Ce = Ie * Ye \wedge Me + \text{lower degree terms in } Ye.$$

In the expressions the coefficients of powers of Yj in Cj are all pols in $U1, \cdots, Ud, Y1, \cdots, Yj'$ alone with $j' = j - 1$. Moreover, all the coefficients are REDUCED with respect to the subset of preceding $C1, \cdots, Cj'$ in the sense that for each Yi with $i < j$ there are only terms of degree $< Mi$. In particular the coefficients Ii of the leading terms of the Ci's are called INITIALS and play a particularly important role in the theory of Ritt. For any polset PS and pol G let us denote the totality of zeros of PS in any arbitrary extension field of K for which $G \langle \rangle 0$ by Zero (PS/G). For $G = 1$ we write simply Zero (PS) for Zero $(PS/1)$. Then the following two formulae constitute what we have called

Received February 4, 1986.

the Ritt Well-Ordering Principle:

$$\text{Zero } (CHS/J) \prec \text{Zero } (PS) \prec \text{Zero } (CHS), \tag{1}$$

$$\text{Zero } (PS) = \text{Zero } (CHS/J) + \text{SUM}i \text{ Zero } (PSi). \tag{2}$$

In the formulae CHS is the char-set of PS, J is the product of all initials Ii of pols in CHS and PSi is the enlarged set of PS with Ii adjoined to it for each i. We use here and here after the notations $\prec$ and $\succ$ to mean "is contained in" and "contains" respectively. We remark in passing that for practical purposes we may understand CHS to be in some broader sense with REDUCED in less stringent conditions asdefined above. The formulae (1), (2) will, however, remain to be true. Furthermore, we have a more general formula describing the structure of Zero (PS/G) which we have called the ZERO DECOMPOSITION THEOREM and can be put in the following form:

$$\text{Zero}(PS/G) = \text{SUM}j \text{ Zero } (IRRj/Gj). \tag{3}$$

Note that the right-hand side may eventually be empty, which means in this crse that G vanishes for all zeros of PS or Zero$(PS) \prec$ Zero (G). In the formula each $IRRj$ is a polset of similar forms as CHS (the U's are naturally different for different j's) and is moreover IRREDUCIBLE in some sense which we shall not enter. Each Gj is also some pol with non-zero REMAINDER with respect to $IRRj$ which we shall not enter either. What is important to us is that PS (and G) being given, the right-hand side of (1)—(3) will be completely determined in a mechanical manner so that it can be accordingly programmed on computers. For more details we refer to [WU3, 4, 5, 7].

The formulae (1)—(3) and their natural generalizations to the differential case constitute the basis of our mechanization method of (elementary and differential) geometry and related domains. The method has diverse applications in a variety of directions, besides solution of arbitrary systems of algebraic equations and mechanical theorem-proving and theorem-discovering of geometries as we have explained in various occasions. As one of further applications, our method permits to determine automatically unknown relations between various geometry entities in a quite simple manner. To serve as examples of illustration, the present paper will deal with relations involving distances, areas and volumes in either euclidean or non-euclidean spaces. The method is however a general one and may be explained as follows.

To fix the ideas, let a, b, c be, say three known magnitudes and x be a magnitude already known to be completely determined by a, b, c without knowing however the exact relation which is to be found. Suppose that by given hypothesis x is connected by polynomial relations with a, b, c through certain other magnitudes d, e, etc. Denote now the known magnitudes a,b,c by $X1$,$X2$,$X3$ and the unknown magnitude x by $X4$. Denote the other magnitudes d, e, etc. by $X5$,$X6$, etc. The given relations will then form a polset PS consisting of pols in the variables Xi. Let us form the char-set CHS of PS. By (1) any zero of PS is necessarily a zero of CHS. The first pol $C1$ of the pols CHS is therefore one in $X1$, $\cdots$, $X4$ alone whose vanishing will give the exact relation to be sought for.

2. The Qin-Heron Formula of Area of a Triangle

The simplest example is perhaps the determination of the area of a triangle in terms of the lengths of its three sides a, b, c. Let $s = (a + b + c)/2$. Then it is well-known that the area will be given by

$$\text{Area} \wedge 2 = s * (s - a) * (s - b) * (s - c). \tag{4}$$

In present-day geometry this formula (4) is usually attributed to Heron, whose life ranged, according to historians of mathematics, from 2c B. C. to 10c A. D. In the famous books of Heath there is a proof of this Heron's formula, see [H], pp. 87—88. The proof is however so intricate and unnatural that the present author cannot refrain from suspecting that it was only supplemented, in following the euclidean pattern of proofs, by someone much later than the actual discovery of the formula in whatever manner and in whatever time.

On the other hand, in the Chinese classic *Shu Shu Jiu Zhang*, or *Mathematical Treatise in Nine Sections*, 1247, of Qin Jiushao in the Sung Dynasty, there appeared a formula which may be described as follows.

Let the three sides of a triangle be given by gr(=great), mid(=middle) and sm (= small), then the area of the triangle will be given by

$$\text{Area} \wedge 2 = [\text{sm} \wedge 2 * \text{gr} \wedge 2 - (\text{gr} \wedge 2 + \text{sm} \wedge 2 - \text{mid} \wedge 2) \wedge 2]/4. \tag{5}$$

Clearly this is equivalent to (4), but is expressed in a form quite involved and is somewhat mysterious at first appearance. The author of the above classic gave no indication of its source or any idea of proof. However, based on the tradition of Chinese geometry entirely different from the tradition of Euclid, the author has constructed a proof of (5) which has the peculier character of arriving quite naturally at this peculier formula. It is therefore not unreasonable to guess that this is just the same proof which was known to our ancestors. Now it is a simple matter to transform (5) into the neat elegant form (4). But once (4) is known, it would be completely insensible to turn it into a form so ugly in looking like (5). For this reason the present author has drawn the conclusion that Qin (or someone in earlier dates) discovered formula (5) at least independently of Heron. For more details cf. [WU8].

Return now to our mechanization method in dealing with the above.

PROB 1. Determine the area of a triangle in terms of its three sides.

For the determination let us choose for the sake of simplicity of computations the coordinate system so that the three vertices of the triangle will be given by

$$a0 = (0, 0), \quad a1 = (x5, 0), \quad a2 = (x6, x7).$$

Let the lengths of the three sides be

$$a0a1 = x1, \quad a0a2 = x2, \quad a1a2 = x3,$$

and the area be

$$\text{area} = x4.$$

Note that the order of the variables Xi is chosen in accordance with the principle stated in Sect. 1. The hypothesis polset PS consists of then 4 pols given by

$$p1 = +2 * x4 - 1 * x5 * x7,$$

$$p2 = +1 * x5 - 1 * x1,$$

$p3 = +1*x2\wedge 2 - 1*x6\wedge 2 - 1*x7\wedge 2,$

$p4 = +1*x3\wedge 2 - 1*x6\wedge 2 - 1*x5\wedge 2 + 2*x6*x5 - 1*x7\wedge 2.$

The char-set *CHS* is readily found to consist of 4 pols of which the first one is given by

$$c1 = +2*x1\wedge 2*x2\wedge 2 - 1*x3\wedge 4 + 2*x2\wedge 2*x3\wedge 2 + 2*x1\wedge 2*x3\wedge 2 - 1*x2\wedge 4 - 1*x1\wedge 4 - 16*x4\wedge 2.$$

The relation $c1 = 0$ is just the Qin-Heron formula to be sought for.

3. Further Problems in Euclidean Space

We give some further examples about distances, areas, volumes, ect. in a euclidean plane or space in what follows. First of all, the problem about the area of a triangle as given in Sect. 2 can be naturally extended to the case of the volume of a tetrahedron in a euclidean space. We lay down thus the following

PROB 2. Determine the volume of a tetrahedron T in a euclidean space in terms of its 6 edges.

To solve this problem let us take coordinates so that the 4 vertices of the tetrahedron T are given by

$$a0 = (0, 0, 0),\ a1 = (x8, 0, 0),$$
$$a2 = (x9, x10, 0), a3 = (x11, x12, x13). \qquad (6)$$

Set the lengths of the edges to be

$$a0a1 = x1,\ a0a2 = x2,\ a0a3 = x3,\ a1a2 = x4,\ a1a3 = x5,\ a2a3 = x6. \qquad (7)$$

Denote the volume of T by $x7$. Then the problem is equivalent to finding a relation between $x7$ and $x1, \cdots, x6$.

We remark in passing that, following the general principle as described in Sect. 1, we have chosen the first 6 variables x to be those supposed known and the next one $x7$ the volume to be determined. This is indeed the crucial point in applying our mechanization method.

The conditions implied by the problem are now $p1 = 0, \cdots, p7 = 0$ with the p's given below:

$p1 = +6*x7 + 1*x8*x10*x13,$

$p2 = +1*x8 - 1*x1,$

$p3 = +1*x2\wedge 2 - 1*x9\wedge 2 - 1*x10\wedge 2,$

$p4 = +1*x3\wedge 2 - 1*x11\wedge 2 - 1*x12\wedge 2 - 1*x13\wedge 2,$

$p5 = +1*x4\wedge 2 - 1*x9\wedge 2 - 1*x8\wedge 2 + 2*x9*x8 - 1*x10\wedge 2,$

$p6 = +1*x5\wedge 2 - 1*x11\wedge 2 - 1*x8\wedge 2 + 2*x11*x8 - 1*x12\wedge 2 - 1*x13\wedge 2,$

$p7 = +1*x6\wedge 2 - 1*x11\wedge 2 - 1*x9\wedge 2 + 2*x11*x9 - 1*x12\wedge 2 - 1*x10\wedge 2 + 2*x12*x10 - 1*x13\wedge 2.$

The corresponding char-set is readily found by our program to be consisting of pols $c1, \cdots,$, $c7$ as shown below:

$$\begin{aligned} c1 = & -1 * x4 \wedge 2 * x3 \wedge 4 + 1 * x2 \wedge 2 * x4 \wedge 2 * x5 \wedge 2 - 1 * x2 \wedge 4 * x5 \wedge 2 \\ & + 1 * x1 \wedge 2 * x2 \wedge 2 * x5 \wedge 2 - 1 * x2 \wedge 2 * x5 \wedge 4 \\ & - 1 * x1 \wedge 2 * x2 \wedge 2 * x4 \wedge 2 + 1 * x2 \wedge 2 * x3 \wedge 2 * x5 \wedge 2 \\ & + 1 * x1 \wedge 2 * x4 \wedge 2 * x3 \wedge 2 - 1 * x4 \wedge 4 * x3 \wedge 2 \\ & + 1 * x2 \wedge 2 * x4 \wedge 2 * x3 \wedge 2 - 1 * x1 \wedge 2 * x3 \wedge 2 * x5 \wedge 2 \\ & - 1 * x1 \wedge 4 * x6 \wedge 2 - 1 * x1 \wedge 2 * x6 \wedge 4 \\ & + 1 * x1 \wedge 2 * x3 \wedge 2 * x6 \wedge 2 + 1 * x1 \wedge 2 * x2 \wedge 2 * x6 \wedge 2 \\ & - 1 * x4 \wedge 2 * x5 \wedge 2 * x6 \wedge 2 + 1 * x2 \wedge 2 * x5 \wedge 2 * x6 \wedge 2 \\ & + 1 * x1 \wedge 2 * x5 \wedge 2 * x6 \wedge 2 + 1 * x4 \wedge 2 * x3 \wedge 2 * x6 \wedge 2 \\ & - 1 * x2 \wedge 2 * x3 \wedge 2 * x6 \wedge 2 + 1 * x1 \wedge 2 * x4 \wedge 2 * x6 \wedge 2 \\ & + 1 * x4 \wedge 2 * x3 \wedge 2 * x5 \wedge 2 - 144 * x7 \wedge 2, \\ c2 = & +1 * x8 - 1 * x1, \\ c3 = & -1 * x4 \wedge 2 + 1 * x2 \wedge 2 + 1 * x1 \wedge 2 - 2 * x1 * x9, \\ c4 = & +1 * x2 \wedge 2 - 1 * x9 \wedge 2 - 1 * x10 \wedge 2, \\ c5 = & -1 * x5 \wedge 2 + 1 * x3 \wedge 2 + 1 * x1 \wedge 2 - 2 * x1 * x11, \\ c6 = & -1 * x6 \wedge 2 + 1 * x3 \wedge 2 + 1 * x9 \wedge 2 - 2 * x11 * x9 - 2 * x12 * x10 \\ & + 1 * x10 \wedge 2, \\ c7 = & +6 * x7 + 1 * x1 * x10 * 13. \end{aligned}$$

Remark that the char-set given here is in some broader sense as indicated in Sect. 1. The equation $c1 = 0$ may now be put in the form

$$\begin{aligned} 144 * x7 \wedge 2 = & -\mathrm{SUM1}(x1 \wedge 2 * x6 \wedge 4 + x6 \wedge 2 * x1 \wedge 4) \\ & + \mathrm{SUM2}(x2 \wedge 2 * x1 \wedge 2 * x5 \wedge 2) \\ & - \mathrm{SUM3}(x1 \wedge 2 * x2 \wedge 2 * x4 \wedge 2). \end{aligned} \tag{8}$$

In the formula the SUM's are summations to be extended over respective ranges as below:

SUM1 over 3 pairs of opposite edges $(x1, x6)$, $(x2, x5)$, $(x3, x4)$,

SUM2 over 12 triples of non-closed edges like $(x2, x1, x5)$,

SUM3 over 4 triples of edges forming a triangle like $(x1, x2, x4)$.

Equation (8) gives now the expression of the volume $x7$ in terms of the edges $x1, \cdots, x6$ as required.

PROB 3. Find the relation between the 6 distances $a0a1, \cdots, a2a3$ of the 4 points $a0, a1, a2, a3$ in a euclidean plane.

In fact if the three distances $a0a1$, $a0a2$, $a1a2$ are known, then the triangle $a0a1a2$ is already rigid in form. With know distances $a0a3$ and $a1a3$ there are then just two positions for $a3$ to take and then the distance $a2a3$ will be determined. Form this we see that the relation to be sought for should be quadratic in $a2a3 \wedge 2$ or quartic in $a2a3$ and hence also quartic in each of the 6 distances involved. This can be readily verified by the computer as before.

Let us take thus coordinates so that

$$a0 = (0, 0),\ a1 = (x8, 0),\ a2 = (x9, x10),\ a3 = (x11, x12).$$

Set the 6 distances as in (7). Then the hypothesis polset will be the same as $p2, \cdots, p7$ in PROB 2 in setting $x13 = 0$. The first pol $c1$ of the char-set should be a pol in $x1, \cdots, x6$ alone which is readily seen to be

$$c1 = -\text{SUM1} + \text{SUM2} - \text{SUM3}, \tag{9}$$

in which the SUM's are just the same expressions as occured on the right-hand side of (8). The relation to be found is then given by $c1 = 0$ which corresponds to the fact that the volume of the tetrahedron $a0a1a2a3$ is this time 0.

We may ask the same question as PROBs 2, 3 about the volumes of a 4-simplex in a 4-dimensional euclidean space $R4$, of a 5-simplex in $R5$, and the relation between the distances of any 5 points in $R3$, or that of any 6 points in $R4$, and even those for a euclidean space of arbitrary but fixed dimension. One can solve these problems in just the same manner as above so far the dimension of the euclidean space is fixed in advance and so far the limitations of the memory space and running time of a computer permit to carry out such a computation. On the other hand, the determination of such a formula for a general dimension n is entirely out of the reach of our mechanization method. Remark that such formulae have already been discovered by Cayley et al by ingenious manipulations of determinants. This shows that, although the use of computers furnishes a powerful tool in mathematical studies, we cannot solely rely upon it without resort to usual methods of mathematics.

As a further example let us consider the following interesting problem raised by Gauss in a short paper (cf. [G]), viz.

PROB 4. Determine the area of a planar pentagon $a0a1a2a3a4$ in terms of the areas of 5 triangles with vertices taken from $a0, \cdots, a4$.

Gauss pre-supposed in fact that the pentagon is a convex one and the 5 triangles considered are

$$a0a1a2,\ a1a2a3,\ a2a3a4,\ a3a4a0,\ a4a0a1. \tag{10}$$

To deal with this problem by means of our method let us choose oblique coordinates such that

$$a0 = (0, 0),\ a1 = (x21, 0),\ a2 = (x22, x23),\ a3 = (x24, x25),$$
$$a4 = (0, x26).$$

There are 10 triangles in all which can be formed form the 5 vertices of the pentagon. Denote the areas of the above 5 triangles by $x1, \cdots, x5$ in that order. Denote the areas of the other 5, viz.

$$a0a2a3,\ a1a3a4,\ a2a4a0,\ a3a0a1,\ a4a1a2$$

by $x11, \cdots, x15$ in that order. We remark that by the area of a triangle the order of the vertices or the orientation of the triangle has been taken into account here. Similarly the area of the oriented pentagon $a0a1a2a3a4$ with vertices given in that order will be denoted by $x10$. For the various areas we have then relations $p1 = 0, \cdots, p15 = 0$ with the pols p given below:

$$p1 = +2*x1 - 1*x21*x23 \quad (\text{area}(a0a1a2) = x1),$$

$$p2 = +2*x2 - 1*x21*x23 - 1*x22*x25 + 1*x23*x24 + 1*x21*x25 \quad (\text{area}(a1a2a3) = x2),$$

$$p3 = +2*x3 - 1*x22*x25 - 1*x24*x26 + 1*x22*x26 + 1*x23*x24 \quad (\text{area}(a2a3a4) = x3),$$

$$p4 = +2*x4 - 1*x24*x26 \quad (\text{area}(a3a4a0) = x4),$$

$$p5 = +2*x5 - 1*x21*x26 \quad (\text{area}(a4a0a1) = x5),$$

$$p6 = +1*x10 - 1*x1 - 1*x11 - 1*x4 \quad (\text{area}(a0a1a2a3a4) = x10 = \text{area}(a0a1a2) + \text{area}(a0a2a3) + \text{area}(a0a3a4)),$$

$$p7 = +1*x10 - 1*x2 - 1*x12 - 1*x5 \quad (\text{area}(a0a1a2a3a4) = \text{area}(a1a2a3) + \text{area}(a1a3a4) + \text{area}(a1a4a0)),$$

$$p8 = +1*x10 - 1*x3 - 1*x13 - 1*x1 \quad (\text{area}(a0a1a2a3a4) = \text{area}(a2a3a4) + \text{area}(a2a4a0) + \text{area}(a2a0a1)),$$

$$p9 = +1*x10 - 1*x4 - 1*x14 - 1*x2 \quad (\text{area}(a0a1a2a3a4) = \text{area}(a3a0a1) + \text{area}(a3a4a0) + \text{area}(a3a1a2)),$$

$$p10 = +1*x10 - 1*x5 - 1*x15 - 1*x3 \quad (\text{area}(a0a1a2a3a4) = \text{area}(a4a0a1) + \text{area}(a4a1a2) + \text{area}(a4a2a3)),$$

$$p11 = +2*x11 - 1*x22*x25 + 1*x23*x24 \quad (\text{area}(a0a2a3) = x11),$$

$$p12 = +2*x12 - 1*x21*x5 - 1*x24*x26 + 1*x21*x26 \quad (\text{area}(a1a3a4) = x12),$$

$$p13 = +2*x13 - 1*x22*x26 \quad (\text{area}(a2a4a0) = x13),$$

$$p14 = +2*x14 - 1*x21*x25 \quad (\text{area}(a3a0a1) = x14),$$

$$p15 = +2*x15 - 1*x21*x23 - 1*x22*x26 + 1*x21*x26 \quad (\text{area}(a4a1a2) = x15).$$

Suppose that we are interested as in the original work of Gauss, in the determination of the area $x10$ of the pentagon in terms of the areas of the 5 triangles in the set (10). Then we may take our polset PS to be consisting of the 7 pols $p1, \cdots, p5, p6$ and $p11$. We readily find that the first pol $c1$ of the char-set is given by

$$\begin{aligned}
&- 1*x10*x1*x4*x3 + 1*x1*x4*x2*x3 + 1*x1*x4\wedge 2*x3 \\
&- 1*x10*x5*x2*x3 - 1*x1*x5*x2*x3 \\
&+ 1*x4*x5*x2*x3 + 1*x5*x2\wedge 2*x3 + 1*x10*x1*x5*x3 \\
&- 1*x1*x4*x5*x3 + 1*x10\wedge 2*x5*x2 \\
&- 1*x10*x5*x2\wedge 2 - 1*x10*x4*x5*x2 - 1*x1\wedge 2*x5*x2
\end{aligned}$$

$$+ 1*x1*x5*2\wedge 2 + 1*x10\wedge 2*x1*x4 - 1*x10*x1\wedge 2*x4$$
$$-1*x10*x1*x4\wedge 2 + 1*x1*x4\wedge 2*x5 + 1*x1\wedge 2*x4*x2$$
$$+ 1*x1\wedge 2*x4*x5-1*x10*x1*x4*x2 - 1*x10\wedge 2*x1*x5$$
$$+ 1*x10*x1\wedge 2*x5 - 1*x10*x5\wedge 2*x2 + 1*x1*x5\wedge 2*x2$$
$$+ 1*x4*x5\wedge 2*x2 + 1*x10*x1*x5\wedge 2 - 1*x1\wedge 2*x5\wedge 2$$
$$-1*x1*x4*x5\wedge 2.$$

Now this pol $c1$, considered as one in $x10$, has its initial

$$I1 = +1*x2*x5 + 1*x1*x4 - 1*x1*x5$$

as non-trivial content. Removing this content from $c1 = 0$ which is equivalent to disregarding certain degeneracy cases, we get the required relation

$$x10\wedge 2 - (x1 + x2 + x3 + x4 + x5)*x10$$
$$+ (x1*x2 + x2*x3 + x3x4 + x4*x5 + x5*x1) = 0. \qquad (11)$$

This is just the formula found by Gauss by an ingenious method. We remark however that, while Gauss had to suppose the pentagon to be a convex one, our method has no such restrictions and works for all cases, even for degenerate pentagons for which some of the vertices may be collinear. Moreover, if we are interested in the expression of the area of the pentagon in terms of the areas of any other 5 of the 10 triangles, we may just choose the due polset from the 15 pols given above and treat in the same manner as above.

Let us consider now a problem of slightly different character, viz.

PROB 5. Determine the ex-radius of the circumscribing sphere S of a tetrahedron in terms of its 6 edges.

Let us take the coordinates so that the 4 vertices and the 6 edges of the tetrahedron will be given as in (6), (7). Let the ex-center of S be $(x14, x15, x16)$ and the ex-radius be $x7$. Then we have to consider the polset PS consisting of the following 10 pols with the first 6 ones the same as $p2, \cdots, p7$ in the polset of PROB 2, while the remaining 4 are given by

$$p7 = +1*x7\wedge 2 - 1*x14\wedge 2 - 1*x15\wedge 2 - 1*x16\wedge 2,$$
$$p8 = +1*x7\wedge 2 - 1*x14\wedge 2 - 1*x8\wedge 2 + 2*x14*x8 - 1*x15\wedge 2 - 1*x16\wedge 2,$$
$$p9 = +1*x7\wedge 2 - 1*x14\wedge 2 - 1*x9\wedge 2 + 2*x14*x9 - 1*x15\wedge 2 - 1*x10\wedge 2 + 2*x15*x10 - 1*x16\wedge 2,$$
$$p10 = +1*x7\wedge 2 - 1*x14\wedge 2 - 1*x11\wedge 2 + 2*x14*x11 - 1*x15\wedge 2 - 1*x12\wedge 2 + 2*x15*x12 - 1*x16\wedge 2 - 1*x13\wedge 2 + 2*x16*x13.$$

The first pol $c1$ of the char-set which furnishes the solution is found to be given by

$$-4*x4\wedge 2*x3\wedge 4*x7\wedge 2 + 4*x2\wedge 2*x4\wedge 2*x5\wedge 2*x7\wedge 2$$
$$- 4*x2\wedge 4*x5\wedge 2*x7\wedge 2 + 4*x1\wedge 2*x2\wedge 2*x5\wedge 2*x7\wedge 2$$
$$- 4*x2\wedge 2*x5\wedge 4*x7\wedge 2 - 4*x1\wedge 2*x2\wedge 2*x4\wedge 2*x7\wedge 2$$

$$+ 4 * x2 \wedge 2 * x3 \wedge 2 * x5 \wedge 2 * x7 \wedge 2$$
$$+ 4 * x1 \wedge 2 * x4 \wedge 2 * x3 \wedge 2 * x7 \wedge 2$$
$$- 4 * x4 \wedge 4 * x3 \wedge 2 * x7 \wedge 2 + 4 * x2 \wedge 2 * x4 \wedge 2 * x3 \wedge 2 * x7 \wedge 2$$
$$- 4 * x1 \wedge 2 * x3 \wedge 2 * x5 \wedge 2 * x7 \wedge 2$$
$$- 4 * x1 \wedge 4 * x6 \wedge 2 * x7 \wedge 2 - 4 * x1 \wedge 2 * x6 \wedge 4 * x7 \wedge 2$$
$$+ 4 * x1 \wedge 2 * x3 \wedge 2 * x6 \wedge 2 * x7 \wedge 2$$
$$+ 4 * x1 \wedge 2 * x2 \wedge 2 * x6 \wedge 2 * x7 \wedge 2$$
$$- 4 * x4 \wedge 2 * x5 \wedge 2 * x6 \wedge 2 * x7 \wedge 2$$
$$+ 4 * x2 \wedge 2 * x5 \wedge 2 * x6 \wedge 2 * x7 \wedge 2$$
$$+ 4 * x1 \wedge 2 * x5 \wedge 2 * x6 \wedge 2 * x7 \wedge 2$$
$$+ 4 * x4 \wedge 2 * x3 \wedge 2 * x6 \wedge 2 * x7 \wedge 2$$
$$- 4 * x2 \wedge 2 * x3 \wedge 2 * x6 \wedge 2 * x7 \wedge 2$$
$$+ 4 * x1 \wedge 2 * x4 \wedge 2 * x6 \wedge 2 * x7 \wedge 2$$
$$+ 4 * x4 \wedge 2 * x3 \wedge 2 * x5 \wedge 2 * x7 \wedge 2$$
$$- 2 * x1 \wedge 2 * x2 \wedge 2 * x5 \wedge 2 * x6 \wedge 2 + 1 * x4 \wedge 4 * x3 \wedge 4$$
$$+ 1 * x1 \wedge 4 * x6 \wedge 4 - 2 * x2 \wedge 2 * x4 \wedge 2 * x3 \wedge 2 * x5 \wedge 2$$
$$- 2 * x1 \wedge 2 * x4 \wedge 2 * x3 \wedge 2 * x6 \wedge 2 + 1 * x2 \wedge 4 * x5 \wedge 4.$$

In comparison with the pol $c1$ of PROB 2, we see readily that the coefficient of $x7 \wedge 2$ in the present $c1$ is equal to $4 * 144 * \mathrm{VOL} \wedge 2$, where VOL is the volume of the tetrahedron. Denote the diameter of the circumscribing sphere by DIAM and let us set

$$S = (A + B + C)/2,$$

with

$$A = x1 * x6, \quad B = x2 * x5, \quad C = x3 * x4 \tag{12}$$

corresponding to the 3 pairs of opposite edges $x1$, $x6$; $x2$, $x5$; and $x3$, $x4$. Then the present relation can be put in the following neat form, viz.

$$9 * \mathrm{VOL} \wedge 2 * \mathrm{DIAM} \wedge 2 = S * (S - A) * (S - B) * (S - C). \tag{13}$$

The author is at a loss where to find this formula in the literature of pasttimes. Note the connection of the right-hand side of (13) with the Ptolemy formula about 4 points on the same circle.

4. Problems in Hyperbolic Plane or Space

Our mechanization method may also be applied to same problems in other kinds of geometries. To fix the ideas, let us consider the case of plane hyperbolic geometry for which Beltrami coordinates will be used in what follows. We remark that in studies of non-euclidean geometries transcendental functions are intensively used. On the other hand our method has to deal solely with polynomials of pure ALGEBRAIC character. This is however not an unsurmountable barrier to the applications of our method since we are dealing actually with ALGEBRAIC relations between the TRANSCENDENTAL functions. It has been explained in the book [WU3], Chap 6 and in the earliest paper about mechanical theorem proving [WU1] that there are already some examples illustrating how we can deal with such problems involving transcendental functions. In what follows one can see again how this is done.

PROB 6. Find the relation between the 6 distances of 4 points $a0, \cdots, a3$ in a hyperbolic plane.

Let us take the coodinate system such that the 4 points are given in Beltrami coordinates by

$$a0 = (0, 0), \quad a1 = (x7, 0), \quad a2 = (x8, x9), \quad a3 = (x10, x11).$$

Set also for the 6 distances:

$$x1 = \cosh a0a1, \quad x2 = \cosh a0a2, \quad x3 = \cosh a0a3,$$
$$x4 = \cosh a1a2, \quad x5 = \cosh a1a3, \quad x6 = \cosh a2a3.$$

We have then a polset as given below:

$$p1 = +1 * x1 \wedge 2 - 1 * x1 \wedge 2 * x7 \wedge 2 - 1,$$
$$p2 = +1 * x2 \wedge 2 - 1 * x2 \wedge 2 * x8 \wedge 2 - 1 * x2 \wedge 2 * x9 \wedge 2 - 1,$$
$$p3 = +1 * x3 \wedge 2 - 1 * x3 \wedge 2 * x10 \wedge 2 - 1 * x3 \wedge 2 * x11 \wedge 2 - 1,$$
$$p4 = +1 * x4 - 1 * x1 * x2 + 1 * x1 * x2 * x7 * x8,$$
$$p5 = +1 * x5 - 1 * x1 * x3 + 1 * x1 * x3 * x7 * x10,$$
$$p6 = +1 * x6 - 1 * x2 * x3 + 1 * x2 * x3 * x8 * x10$$
$$+ 1 * x2 * x3 * x9 * x11.$$

The first pol $c1$ of the char-set is given by

$$\begin{aligned} &+2 * x1 * x4 * x3 * x6 + 1 * x6 \wedge 2 - 1 * x4 \wedge 2 * x3 \wedge 2 - 1 * x2 \wedge 2 * x5 \wedge 2 \\ &\quad + 2 * x1 * x2 * x5 * x6 + 1 * x3 \wedge 2 + 1 * x5 \wedge 2 + 2 * x4 * x2 * x5 * x3 \\ &\quad - 2 * x1 * x5 * x3 - 2 * x4 * x5 * x6 - 1 * x1 \wedge 2 * x6 \wedge 2 - 2 * x2 * x3 * x6 \\ &\quad - 2 * x1 * x4 * x2 + 1 * x2 \wedge 2 + 1 * x4 \wedge 2 + 1 * x1 \wedge 2 - 1. \end{aligned} \tag{14}$$

The equation $c1 = 0$ is then the relation to be sought for.

PROB 7. Determine the area of a triangle in the hyperbolic plane in terms of its 3 sides.

To solve the problem let us take a coordinate system with the 3 vertices of the triangle given in Beltrami coordinates as

$$a0 = (0, 0), \quad a1 = (x15, 0), \quad a2 = (x16, x17).$$

Let A be the area of the triangle so that we have

$$pi - A = a0 + a1 + a2, \tag{15}$$

in which the a's denote also the 3 internal angles of the triangle. Set now

$$x1 = \cosh a0a1, \quad x2 = \cosh a0a2, \quad x3 = \cosh a1a2, \quad x4 = \cos A,$$

$$x21 = \cos a0, \quad x22 = \sin a0, \quad x23 = \cos a1, \quad x24 = \sin a1,$$

$$x25 = \cos a2, \quad x26 = \sin a2.$$

Then we have a polset $p1, \cdots, p12$ below:

$$p1 = +1*x1\wedge 2 - 1*x1\wedge 2*x15\wedge 2 - 1,$$

$$p2 = +1*x2\wedge 2 - 1*x2\wedge 2*x16\wedge 2 - 1*x2\wedge 2*x17\wedge 2 - 1,$$

$$p3 = +1*x1*x2 - 1*x1*x2*x15*x16 - 1*x3,$$

$$p4 = +1*x18\wedge 2 - 1*x16\wedge 2 - 1*x17\wedge 2,$$

$$p5 = +1*x19\wedge 2 - 1*x17\wedge 2 - 1*x1\wedge 2*x15\wedge 2 - 1*x1\wedge 2*x16\wedge 2 \\ +2*x1\wedge 2*x15*x16,$$

$$p6 = +1*x18*x21 - 1*x16,$$

$$p7 = +1*x19*x23 - 1*x1*x15 + 1*x1*x16,$$

$$p8 = +1*x18*x22 - 1*x17,$$

$$p9 = +1*x19*x24 - 1*x17,$$

$$p10 = +1*x18*x19*x25 - 1*x1*x17\wedge 2 + 1*x1*x15*x16 - 1*x1*x16\wedge 2,$$

$$p11 = +1*x2*x18*x26 - 1*x1*x15*x24,$$

$$p12 = +1*x21*x24*x26 + 1*x22*x23*x26 + 1*x22*x24*x25 \\ - 1*x21*x23*25 - 1*x4.$$

The first pol $c1$ of the char-set is found to be one of 43 terms which may be considered as a pol in $x4$ of degree 1 with coefficients themselves pols in $x1, x2, x3$, as shown below:

$$-1*x2\wedge 2*x3 + 1*x1*x2\wedge 3 - 2*x1\wedge 2*x2\wedge 2*x3 + 3*x1*x2*x3\wedge 2 + \cdots$$

$$-1*x3\wedge 3 + 1*x1\wedge 3*x2 - 1*x1*x2 - 1*x1\wedge 2*x3 + 1*x3 - 1*x2\wedge 3 + \cdots$$

$$-1*x1\wedge 2*x2 + 1*x2 + 3*x1*x2\wedge 2*x3 - 2*x1\wedge 2*x2*x3\wedge 2 \\ - 1*x2*x3\wedge 2 + 1*x1*x3\wedge 3 + 1*x1\wedge 3*x3 - 1*x1*x3 - 1*x1\wedge 3 \\ + 1*x1 + 1*x2\wedge 3*x3 + \cdots$$

$$-2*x1*x2\wedge 2*x3\wedge 2 - 1*x2*x3 - 1*x1*x2\wedge 2 + 3*x1\wedge 2*x2*x3 + \cdots,$$

$$-1*x1*x3\wedge 2 + 1*x2*x3\wedge 3 + 1*x1\wedge 2*x2\wedge 2*x3\wedge 2 + 1*x1\wedge 2*x2\wedge 2 \\ + \cdots - 1*x1\wedge 3*x2*x3 - 1*x1*x2*x3 - 1*x1*x2*x3\wedge 3 \\ + 1*x2\wedge 2*x3\wedge 2 + \cdots$$

$$-1*x1*x2\wedge 3*x3 + 1*x1\wedge 2*x3\wedge 2 + 1*x1\wedge 2*x4 \\ - 1*x1\wedge 2*x3\wedge 2*x4 + \cdots$$

$$-1*x2\wedge 2*x3\wedge 2*x4 - 1*x1\wedge 2*x2\wedge 2*x4 + 1*x2\wedge 2*x4 - 1*x4 \\ + 1*x3\wedge 2*x4 + 1*x1\wedge 2*x2\wedge 2*x3\wedge 2*x4.$$

After the removal of the content $(x1 - 1)*(x2 - 1)*(x3 - 1)$ we get a pol $c1$ of 18 terms only, viz.

$$c1' = +3*x2*x1*x4 + 3*x2*x4 + 3*x1*x4 + 3*x4 + 2*x1*x4 \\ + 2*x4 + 1*x4 + 4 - 1*x3\wedge 2 + 3*x2*x1 - 1*x3*x2 \\ -1*x3*x1 - 1*x3 - 1*x2\wedge 2 - 1*x2*x1 - 1*x2 - 1*x1\wedge 2 \\ - 1*x1.$$

We remark that this equation can be put in the form

$$(x1+1)*(x2+1)*(x3+1)*(x4-1)$$
$$=-2*x1*x2*x3+x1\wedge 2+x2\wedge 2+x3\wedge 2-1. \quad (16)$$

Replacing $x1$, $x2$, $x3$, $x4$ by cosh $a0a1$, cosh $a0a2$, cosh $a1a2$ and cos A respectively we see that this equation is also equivalent to a form as given in Ex. 27 of [GR], p. 361.

PROB 8. Determine the volume V of a tetrahedron in the hyperbolic space in terms of its 6 edges.

This is a difficult problem which seems to be not yet completley settled. In fact, partial results known involve already such transcendental functions called Lobachevsky functions whose properties are yet not quite clear, cf. e. g. [MIL]. Moreover, Dehn has pointed out that no expressions like (15) involving dihedral angles, etc. of the tetrahedron can exist, cf. e. g. [KL], p. 203. Now a comparison of (8) and (9) shows that it would be legitimate, taking into account PROB 6, to conjecture that the final relation to be found is of the form

$$A*\mathrm{Tr}(V)=B,$$

in which Tr ia a certain transcendental function, A and B are certain pols in $x1,\cdots,x6$, with $x1=\cosh a0a1$, etc. (the a's being the vertices of the tetrahedron), and B is given by the pol in (14). We hope that some day we may return to this problem again.

References

[BL] Bledsoe, W. W. & Loveland, D. W., Ed., Automated theorem proving: After 25 years, AMS, 1983.

[CH], Chou, S. C., Proving elementary geometrical theorems using Wu's algorithm, in [BL], 243—286.

[G] Gauss, K. F., Aufloesung einer geometrischer Aufgabe, Gessam, Werke, Bd 4 (1880), 406-7.

[GR] Greenberg, M. J., Euclidean and non-euclidean geometries, development and history, Freeman & Co., 1980.

[H] Heath, T. L., The 13 books of Euclid's Elements, vol. 2, Dover, 1956.

[KL] Klein, F., Vorlesungen uber nicht-euklidische Geometrie, Goettingen, 1927.

[MIL] Milnor, J., Hyperbolic geometry: the first 150 years, *Bull. Amer. Math. Soc.*, 6 (1982), 9—24.

[R1] Ritt, J. F., Differential equations from the algebraic standpoint, *Amer. Math. Soc.*, 1932.

[R2] Ritt, J. F. Differential algebra, *Amer. Math. Soc.*, 1950.

[WU1] Wu Wen-tsün, On the decision problem and the mechanization of theorem-proving in elementary geometry, *Scientia Sinica*, **21:** 2(1978), 159—172; re-published in [BL], 213—234.

[WU2] ——, Some recent advance in mechanical theorem-proving of geometries, in [BL], 235—242.

[WU3] ——, Basic Principles of Mechanical Theorem Proving in Geometries (Part on

Elementary Geometries) (in Chinese), Science Press, 1984.

[WU4] ——, Basic principles of mechanical theorem-proving in elementary geometries, *J. Sys. Sci. & Math. Scis.*, **4** (1984), 207—235.

[WU5] ——, On zeros of algebraic equations——An application of Ritt principle, *Kuxue Tongbao*, **31** (1986), 1—5.

[WU6] ——, A mechanization method of geometry and its applications, I. Distances, areas, and volumes in euclidean and non-euclidean geometry, *Kuxue Tongbao* (to appar).

[WU7] ——, A mechanization method of geometry, I. Elementary geometry, *Chinese Quarterly J. of Math.*, **1** (1986).

[WU8] ——, The out-in complementary principle, in Ancient China's Technology and Science, Foreign Lang. Press, 1983, 66—89.

几何学机械化方法及其应用

I. 距离、面积与体积

吴 文 俊

(中国科学院系统科学研究所)

摘 要

本文是一系列文章的第一篇，这些文章阐述一种几何学及其有关领域的机械化方法，而着重在方法的应用方面．本文专门讨论欧氏或非欧平面或空间中有关距离、面积与体积间关系的自动推导．特别是，我们用这一方法发现了四面体体积 VOL 与外接球直径 DIAM 间的一个关系式

$$9*\mathrm{VOL}\wedge 2*\mathrm{DIAM}\wedge 2=S*(S-A)*(S-B)*(S-C),$$

其中，$S=(A+B+C)/2$，而 A, B, C 则是三组对棱的乘积．

Proceedings of the International Congress of Mathematicians
Berkeley, California, USA, 1986

Recent Studies of the History of Chinese Mathematics

WU WEN-TSUN

1. Introduction. We shall restrict ourselves to the study of Chinese mathematics in ancient times, viz., from remote ancient times up to the fourteenth century. In recent years such studies were vigorously pursued both in China and in foreign countries. Much deeper understandings have since been gained about what Chinese ancient mathematics really was. The author will freely use their results but will be solely responsible for all points of view expressed in what follows.

Two basic principles of such studies will be strictly observed, viz.:

P1. All conclusions drawn should be based on original texts fortunately preserved up to the present time.

P2. All conclusions drawn should be based on reasonings in the manner of our ancestors in making use of knowledge and in utilizing auxiliary tools and methods available only at that ancient time.

For P1 we shall mention only [**AR**, **AN**, **SI**, **MA**], which will be referred to repeatedly in what follows.

For P2 we shall emphasize that the use of algebraic symbolic manipulations or parallel-line drawings should be strictly forbidden in any deductions of algebra or geometry since they were seemingly nonexistent in ancient Chinese classics. In fact, Chinese ancient mathematics had its own line of development, its own method of thinking, and even its own style of presentation. It is not only independent of, but even quite different from the western mathematics as descendents of Greeks. Before going into more details of concrete achievements, we shall first point out some peculiarities of Chinese ancient mathematics.

First, instead of calculations of pencil-paper type, the ancient Chinese made all computations in manipulating rods on counting boards. This was possible because the Chinese already possessed, in very remote times, the most perfect place-valued decimal system; it allowed them to represent the integers by properly arranged rods placed in due positions on the board. In particular, the number 0 in, or as, a decimal integer was just represented by leaving some empty place in the right position. In fact the word "arithmetic," the usual terminology for "mathematics," was just a literal translation of Chinese characters "Suan Shui" meaning "counting methods."

Secondly, results were usually presented in the form of separate problems, each of which was divided into several items, as follows. 1. Statement of the problem with numerical data. 2. Numerical answer to the problem. 3. "Shui," or the method of arriving at the result. It was most often just what we call today the "algorithm," sometimes also just a formula or a theorem. Note that the numerical values in Item 1 play no role at all in the method, which was so general that any other numerical values could be substituted equally well. Item 1 thus served just as an illustrative example. 4. Sometimes "Zhu," or demonstrations which explained the reason underlying the method in Item 3. In Song Dynasty and later, there was often added a further item: 5. "Cao," or drafts which contained details of the calculations for arriving at the final result.

2. Theoretical studies involving integers. In this section, by an integer we shall always mean a positive one.

In ancient Chinese mathematics there were no notions of prime number and factorization or its likeness. However, there was a *Mutual-Subtraction Algorithm*, for finding the GCD of two integers; its name literally meant *equal*. The algorithm ran as follows:

"Subract the less from the more, mutually subtract to diminue, in order to get the *equal*."

As a trivial example, the *equal* (:= GCD) of 24 and 15 is found to be 3 in the following manner:

$$(24,15) \dashrightarrow (9,15) \dashrightarrow (9,6) \dashrightarrow (3,6) \dashrightarrow (3,3). \tag{2.1}$$

The underlying principle is, as pointed out by Liu Hui in [**AN**], that during the procedure the integers are steadily diminished in magnitudes while the *equal* duplicates remain the same.

In spite of the fact that the prime number concept was never introduced in our ancient times, there were some theoretical studies involving integers which were not at all trivial. We shall cite two of these mainly based on works of S. K. Mo at Nanking University and J. M. Li at Northwestern University, China.

The GouGu form (:= right-angled triangle) was a favorite object of study throughout the lengthy period of development of mathematics in ancient China. In particular, the triples of integers which can be attributed to 3 sides Gou, Gu, and Xuan (:= shorter arm, longer arm, and hypothenuse) of a GouGu form had been completely determined early in the classic [**AR**]. Thus, in the GouGu Chapter 9 of [**AR**] there appeared eight such triples, viz.,

$$\begin{array}{llll} (3,4,5), & (5,12,13), & (7,24,25), & (8,15,17), \\ (20,21,29), & (20,99,101), & (48,55,73), & (60,91,109). \end{array}$$

The occurence of such triples was not merely an accidental one. In fact, in Problem 14 of that chapter a method of general formation of such integer triples was implied. We record this problem.

"Two persons start from same position. A has a speed-rate 7 while B rate 3. B goes eastward while A goes first southward 10 units and then meets B in going northeasternwise. Find the units traversed by A and B."

The Shui (:= method or algorithm) for the solution was:

"Squaring 7, also 3, taking half of the sum, this will be the slantwise unit-ratio of A. Subtract this unit-ratio from square of 7, rest is the southern unit-ratio. Multiply 7 by 3 is eastern unit-ratio of B."

As already mentioned in §1, the particular numbers 7 and 3 in the problem serve merely as illustrations and we may equally well substitute these numbers by any pair of integers say m, n with $m > n > 0$. The Shui then says that the 3 sides are in the ratio

$$\text{Gou}: \text{Gu}: \text{Xuan} = [m^2 - (m^2 + n^2)/2]: m * n: (m^2 + n^2)/2.$$

The eight triples given above may then be determined by the pairs

$$(m, n) = (2, 1),\ (3, 2),\ (4, 3),\ (4, 1),\ (5, 2),\ (10, 1),\ (8, 3),\ (10, 3).$$

In Liu Hui's [**AN**] a demonstration or a proof of geometrical character was given which was based on some general *Out-In Complementary Principle*, and it will be explained in more detail in §3. We note here that Liu's proof showed also that $m : n$ is in reality the ratio of Gou + Xuan to Gu which will be a ratio in integers if and only if the three magnitudes Gou, Gu, Xuan are in ratio of integers. The Shui had thus given an exhaustive list of integer triples for the three sides of the GouGu form.

As a second example let us cite the *Seeking*-1 *Algorithm* which is now well known as the Chinese Remainder Theorem. Recent studies have shown that the algorithm originated in calendar-making since Hans Dynasty, and there was a sufficiently clear line of development until the appearance of the classic [**MA**] of Qin in 1247 A.D. In Qin's preface to his work he stated that the method was not contained in [**AR**] and no one knows how it was deduced, but it was widely applied by calendarists. The method was well-explained for the first time in the first part of [**MA**] and contained nine problems, ranging from calendar-making, dyke-erection, treasure-computing, tax-distribution, rice-selling, military-expedition, brick-architecture, up to even a case of stealing. All the problems were reduced to one which, in modern writings, would be of the form (:=: stands for "congruent to")

$$U :=: Uj \mod Mj, \qquad 1 \le j \le r, \tag{2.2}$$

with integers Uj, Mj known and U to be found. The integers Mj were called by Qin Ting-Mu (:= moduli), literally meaning fixed-denominators which were not necessarily prime to each other. Qin first gave an algorithm for reducing the problem to one with the moduli prime to each other two by two in applying successively the *Mutual-Subtraction Algorithm*. We shall therefore restrict ourselves, in what follows, to the case of Uj pairwise prime.

To a modern mathematician a solution to (2.2) would be found in the following manner (cf., e.g., [**AP**, p. 250]).

Let $\phi(N)$ be the Euler function of the integer N which can be determined from a factorization of N into prime numbers. Set

$$\begin{aligned} M &= M1 * \cdots * Mr, \\ Nj &= (M/Mj)^{\phi(Mj)}, \qquad 1 \le j \le r. \end{aligned} \tag{2.3}$$

Then the solution of (2.2) will be given by

$$U :=: \sum_j Uj * Nj \mod M.$$

Both the method and the result are really simple and elegant. However, in view of the difficulty of factorization and the amount of computation involved in (2.3), it would be rather difficult to get final answers to the nine problems in Qin's classic, even with the aid of modern computers.

On the other hand, the method of Qin ran as follows.

As a preliminary step let us take the remainder Rj of M/Mj mod Mj which was called Qi-Shu, literally meaning odd-number, but just some technical term. Now determine numbers Kj such that

$$Kj * Rj :=: 1 \mod Mj. \tag{2.4}$$

The final answer to be found is then given by

$$U :=: \sum_j Uj * Kj * (M/Mj) \mod M. \tag{2.5}$$

The integers Kj were called, by Qin Cheng-Lui, also a technical term literally meaning multiplication-rate (multiplier below). The algorithm for determining Kj to satisfy (2.4) was called, by Qin, *da-yan qiu-yi shui*, for which *qiu-yi* literally means *seeking*-1, while *da-yan* is some philosophical term of little interest to us. The first step of the Seeking-1 Algorithm consisted then in placing four known numbers 1, 0 (i.e., *empty*), Rj, Mj in the left-upper (LU), left-lower (LL), right-upper (RU), and right-lower (RL)

$$\text{corners of a square:} \begin{array}{|cc|} \text{LU} & \text{RU} \\ \text{LL} & \text{RL} \end{array} = \begin{array}{|cc|} 1 & Rj \\ & Mj \end{array}$$

We remark that these four numbers verify the trivial congruences

$$\text{LU} * Rj :=: \text{RU} \mod Mj, \qquad \text{LL} * Rj :=: -\text{RL} \mod Mj. \tag{2.6}$$

The next steps of the algorithm consisted then of manipulating the four numbers in the square by steadily reducing their magnitudes while keeping the validity of congruences (2.6). After a finite number of steps the number, say RU, will be reduced to 1, and according to (2.6) the number LU is then the multiplier Kj to be found. The underlying principle of this Seeking-1 Algorithm, as listed below in details, is thus essentially the same as the Mutual-Subtraction Algorithm in finding the *equal* (:= GCD) of two integers, only much more complicated. The algorithm was:

"Put Qi at RU, and Ting at RL, and put Tian-Yuan-1 at LU. First divide RL by RU. Multiply quotient with 1 of LU and put it to LL. Next take the numbers RU and RL, mutually divide the more by the less. Then mutually multiply quotients to numbers in LU and LL. Stop until odd-1 in RU. Verify then the number in LU and take it as multiplier."

As a concrete example let us consider Problem 9 of Chin's classic which dealt with a stealing case. The judge in charge of the case was able to determine the amount of rice stolen by each of the three thiefs by means of the algorithm. For one of the thiefs the determination of the corresponding multiplier ran as follows:

$$\begin{bmatrix} 1 & 14 \\ & 19 \end{bmatrix} \dashrightarrow \begin{bmatrix} 1 & 14 \\ & 5 \end{bmatrix} 1 \dashrightarrow \begin{bmatrix} 1 & 14 \\ 1 & 5 \end{bmatrix} \dashrightarrow \begin{bmatrix} 1 & 4 \\ 1 & 5 \end{bmatrix} 2 \dashrightarrow \begin{bmatrix} 3 & 4 \\ 1 & 5 \end{bmatrix} \dashrightarrow$$

$$\begin{bmatrix} 3 & 4 \\ 1 & 1 \end{bmatrix} 1 \dashrightarrow \begin{bmatrix} 3 & 4 \\ 4 & 1 \end{bmatrix} \dashrightarrow \begin{bmatrix} 3 & 1 \\ 4 & 1 \end{bmatrix} 3 \dashrightarrow \begin{bmatrix} 15 & 1 \\ 4 & 1 \end{bmatrix} \dashrightarrow \text{stop: } k = 15.$$

One may compare this sequence of computations with the trivial one (2.1).

The numerical data in the above example is the simplest one among the nine problems of Chin's classic, but already not an easy one in using the mentioned method with Euler functions. The other eight problems will eventually involve astronomically large numbers which may be eventually out of reach of the Euler-function method, but were still done with ease by Qin in using the Seeking-1 Algorithm.

3. Geometry. In contrast to what one usually believes, geometry was intensely studied, in addition to being well-developed, in ancient China. The misunderstanding is likely due to the fact that Chinese ancient geometry was of a type quite different from that of Euclid, both in content and presentation. Thus, there were no deductive systems of euclidean fashion in the form of definition-axiom-theorem-proof. On the contrary, the ancient Chinese formulated, instead of a lot of axioms, a few general plausible principles on which various geometrical results were then discovered and proved in a deductive manner, as shown by Liu Hui [**AN**].

The points of emphasis in Chinese ancient geometry and in the geometry of Euclid were also quite different. Thus, the Chinese ancestors paid no attention at all to the parallelism but, on the contrary, showed great interest in orthogonality of lines. In fact, the GouGu form, or the right-angled triangle, had incessantly occupied a central position among the geometrical objects to be studied throughout thousands of years of development. Secondly, the Chinese ancestors showed little interest in angles but heavily emphasized distances. Thirdly, geometrical studies were always closely connected with applications so that measurements, determination of areas and volumes were among the central themes of study. Finally, geometry was always developed in step with algebra, which culminated in the algebrization of geometry in Song-Yuan Dynasties. This later discovery was rightly pointed out, e.g., by Needham to be the first important step (and indeed, the decisive step) toward the creation of analytical geometry.

We shall illustrate these points with a few examples.

EXAMPLE 1. The Sun-Height Formula. On the earth-level plane erect two gnomons G1, G2 of equal height with a certain distance apart. The sun-shadows of the gnomons are then measured and the sun's height over the level plane is given by

$$\text{Sun-hgt} = \text{Gnomon-hgt} * \text{Gnomon-dist/Shadow-difference} + \text{Gnomon-hgt}.$$

This formula, already depicted in some classic of early Hans Dynasty and cited very often in later calendarical works, was clearly too rough an estimate to rely on. Liu Hui had, however, translated the formula into earth measurements by replacing the sun by some sea-mountain, thus turning the Sun-Height Formula into a realistic Sea-Island Formula. His classic [**SI**] contained all nine such formulae beginning with the above one as the simplest. There were proofs as well as diagrams accompanying this classic; they are still mentioned in some classics of Song Dynasty but have since been lost. Based on fragments and incomplete colored diagrams of some classic by Zhao Shuang in 3c A.D., the author has reconstructed a proof of the above Sun-Height or Sea-Island Formula by rearranging the arguments in that classic as follows (Y =yellow, B =blue):

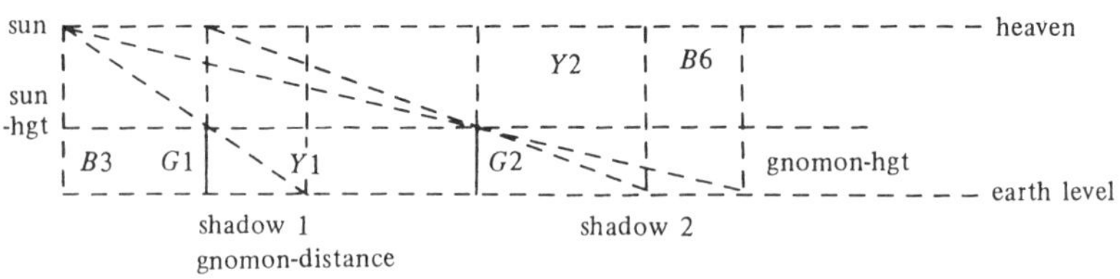

"$Y1$ and $Y2$ are equal in areas. $Y1$ connected with $B3$ and $Y2$ with $B6$ are also equal in areas. $B3$ and $B6$ are also equal in area. Multiply gnomon-distance by gnomon-height to be the area of $Y1$. Take shadow-difference as breadth of $Y2$ and divide, one gets height of $Y2$. The height rises up to same level as sun. From diagram gnomon-height is to be added."

With the accompanying diagram the proof of the formula is evident.

EXAMPLE 2. The Out-In Complementary Principle (OICP). In Example 1, various area-equalities were all consequences of a certain general Out-In Complementary Principle which was clearly formulated in the classic [**AN**] in very concise terms. It means simply that whenever a figure, planar or solid, is cut into pieces and moved to other places, then the sum of areas or volumes will remain unchanged. This seemingly most common-place principle had been applied successfully to problems of extreme diversity, sometimes unexpected, besides that of Example 1. As further examples consider the GouGu form with three sides: Gou, Gu, and Xuan. One may form various sums and differences from them

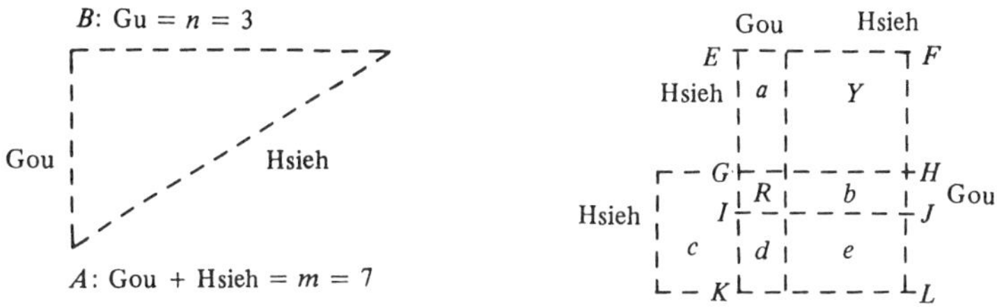

FIGURE 1

$c + d = \text{Hsieh}^2 - \text{Gou}^2 = d + e = \text{Gu}^2 = n^2,$
$2 * EFGH = EFKL = m^2 + n^2 = (\text{Gou+Hsieh})^2 + \text{Gu}^2,$
$a + Y = \text{Hsieh} * m = \text{Hsieh} * (\text{Gou+Hsieh}),$
$b + R = EFIJ - EFGH = \text{Gou} * m = \text{Gou} * (\text{Gou+Hsieh}) = m^2 - (m^2 + n^2)/2.$

like Gou-Gu sum, Gou-Xuan difference, etc. In the GouGu Chapter 9 of [**AR**], there were a number of problems for determining Gou, Gu, and Xuan from two of these nine entities, and all were solved by means of this principle. In particular, the general formula of Gou-Gu integers as described in §2 was obtained by applying the principle to Problem 14 by considering as known the ratio of Gou-Xuan sum to Gu. Liu Hui then demonstrated the result by OICP as shown in Figure 1 (R =red, Y =yellow).

In [**MA**] there was formula for determining the AREA of a triangle with three sides: the GReatest one, the SMallest one, and the MIDdle one in the form

$$4 * \text{AREA}^2 = \text{SM}^2 * \text{GR}^2 - [(\text{GR}^2 + \text{SM}^2 - \text{MID}^2)/2]^2.$$

This formula is clearly equivalent to the Heron one. It cannot be deduced from the latter since it is so ugly, in form, in comparison to the elegant latter formula. By applying some formula given in [**AN**] about Problem 14, based on OICP, the author has reconstructed a proof which is in accordance with Chinese tradition and leads naturally to Qin's formula.

We note that the Chinese ancient methods of (square and cubic) root-extraction and quadratic-equation solving were in fact all based on OICP geometrical in character. We also note that all the formulae in [**SI**], in quite intricate form, will be arrived at in a natural manner by applying OICP. On the other hand it seems difficult, or at least a roundabout, unnatural manner, to get these formulae if the euclidean method is to be used.

EXAMPLE 3. Volume of solids. With the OICP alone the areas of any polygonal form can be determined. This will not be the case for volumes of polyhedral solids, and Liu Hui was well aware of it. Liu Hui had, however, completely solved the problem in reasoning as follows. Let us cut a rectangular parallelopiped slantwise into two equal parts called Qiandu, and then cut the Qiandu slantwise into two parts called Yangma (a pyramid) and Bienao (a tetrahedron on special type). Using an ingenious reasoning corresponding to a certain limiting process,

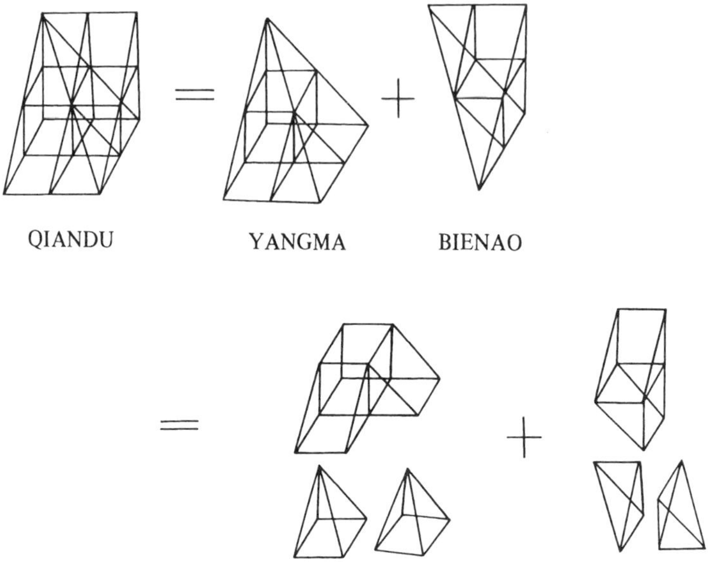

FIGURE 2

he made some assertion which the author has baptized as the *Liu Hui Principle*, viz.,

"Yangma occupies two and Bienao one, that's an invariable ratio."

Together with the OICP the volume of any polyhedral solid can then be determined, and a lot of beautiful formulae for various kinds of solids were determined in this way in the Sang-Gong Chapter 5 of [**AR**]. Liu Hui's demonstration of his principle, which was both elegant and rigorous, consisted of cutting a big QIANDU into smaller yangma's, etc., as in Figure 2.

From Figure 2 it is now clear that

$$1 \text{ YANGMA} - 2 \text{ BIENAO} = 2(1 \text{ yangma} - 2 \text{ bienao}).$$

Continuing, the right-hand side will become smaller and smaller and can be ultimately neglected, as argued by Liu Hui:

"The more they are cut into smaller halves, the smaller will be the remains. The ultimate smallness is infinitesimal, and infinitesimal is formless. Accordingly it is no need to take into account the remain."

For more details see [**WA**], a remarkable paper by Wagner.

Liu Hui had also considered the determination of curvilinear solids, notably that of a sphere. He showed that the solution will depend on the determination of the volume of a curious solid defined as the intersection of two inscribed cylinders in a cube. Liu Hui himself cannot solve this problem and left it, being rigorous in thinking and strict in attitude, to later generations, saying that

"Fearing loss of rightness, I dare to leave the doubts to gifted ones."

The keen observation of Liu Hui had been closely followed and ripened finally to a complete solution of the problem in 5c A.D. by Zu Geng, son of great mathematician, astronomer, and engineer Zu Congtze. In fact, Zu Geng had formulated a general principle which was equivalent to the later rediscovered Cavalieri Principle, viz.,

"Since areas in equal height are equal the volumes cannot be unequal."

We shall leave Zu Geng's beautiful proof about the formula of volume of sphere to other known works. On the other hand, this principle was, in reality, already used by Liu Hui himself in deriving formulae of volumes of various simple curvilinear solids treated in [**AR**], though without an explicit statement. For this reason the author has proposed to use the name *Liu-Zu Principle* instead of the name *Zu-Geng Principle* which is usually used by our Chinese colleagues.

In a word, the OICP, the Liu Hui Principle, and the Liu-Zu Principle were sufficient to edify the whole theory of solids, curvilinear or not, in a satisfactory manner as done by the Chinese ancestors.

4. Algebra. Algebra was no doubt the most developed part of mathematics in ancient China. It should be pointed out that algebra at that time was actually a synonym for method of equation-solving. The problems of equation-solving seem to come from two different sources. One of the sources was rudimentary commerce or goods-exchange which led to the *Excess-Deficiency Shui* in very remote times up to *Fang-Cheng Shui* as depicted in Chapter 8 of [**AR**]. This Chapter 8 dealt with methods of solving simultaneous linear equations along with the introduction of negative numbers. The title "Fang Cheng," the same terminology for "equations" used in Chinese texts nowadays, could be better interpreted as "square matrices." In fact, "Fang" literally means square or rectangle while "Cheng," as explained in Liu Hui's [**AN**], was just data arranged on the counting board in the form of a matrix, viz.

"Arranged as arrays in rows, so it is called Fang Cheng."

Furthermore, the method of solution was just manipulations of rows and columns as in elimination nowadays. Details of such stepwise reduction of arrays to normal forms in some examples can also be found in [**AN**].

A second source of equations was from measurements or geometrical problems. Thus, in the study of sun-heights there were formulae for both sun-height and sun's level distance from the observer. The sun-observer distance was then determined by means of the Gou-Gu Theorem, well known in quite remote times, which then required extraction of square roots. Both the proof of the Gou-Gu Theorem and the method of square root extraction were seemingly based on the OICP—so, also, for the cubic root extraction. Now in Gou-Gu Chapter 9 of [**AR**] there was also a problem which led naturally, by OICP, to a quadratic equation. There was some technical terminology for solving such an equation literally meaning "square-root extraction with an extra term Cong," which clearly implied the origin as well as the method of solving such equations.

The second line was developed further to solving cubic equations in early Tang Dynasty, at the latest, and culminated in the method of numerical solution of higher degree equations in Song Dynasty, identical, actually, to the later rediscovered Horner's method in 1819.

A discovery of utmost importance during Song-Yuan Dynasties (10–14c) was the introduction of the notion "Tian-Yuan," literally meaning "Heaven-Element," which was nothing but what we call an *unknown* nowadays. Though equation-solving occupied a central position in the development of mathematics for thousands of years already, this was perhaps the first time that precise notion and systematic use of *unknowns* were thereby introduced. The Chinese mathematicians at that time recognized very well the power of this method of Tian-Yuan as expressed in some classic of Zhu Szejze:

"To solve by Tian-Yuan not only is clear the underlying reasons and is versatile the method but also saved large amounts of efforts."

The method of Tian-Yuan was further developed in Song-Yuan Dynasties up to the solving of simultaneous high-degree equations involving four unknowns. Along with it, algebrization of geometry, manipulations of polynomials, and the method of elimination were also developed. The two lines of development of equation-solving thus merged into one which was closer to algebra in the modern sense. The limitation to four unknowns was largely due to the fact that all manipulations had to be carried out on counting boards with coefficients of different-type terms of a polynomial to be arranged in definite positions on the board. If one was to get rid of the counting board in adapting another system, as was fairly probable since communications with the outside arabic world were more influential than ever, then mathematics would face an exceedingly fertile era of flourishment. However, all further developments stopped and mathematics actually came to death since the end of Yuan Dynasty. When Matteo Ricci came to China at the end of Ming Dynasty, almost no Chinese high intellectuals knew about "Nine Chapters"!

5. Conclusion. We shall leave other achievements about limit concept, high-difference formulae, series summation, etc. owing to space limitation. In short, Chinese ancient mathematics was mainly constructive, algorithmic, and mechanical in character so that most of the *Shuis* can be readily turned into computerized programs. Moreover, it used to draw intrinsic conclusions from objective facts, then sum up the conclusions into succinct principles. These principles, plain in reasoning and extensive in application, form a unique character of ancient Chinese mathematics. The emphasis has always been on the tackling of concrete problems and on simple, seemingly plausible principles and general methods. The same spirit permeates even such outstanding achievements as the algebrization and the place-value decimal system of numbers. In a word, Chinese ancient mathematics had its own merits, and, of course, also its inherited deficiencies. It is surely inadmissible to neglect the brilliant achievements of our ancestors, as was the case in the Ming Dynasty. It would also be absurd not to absorb the superior techniques of the foreign world, as was the case of early Tang Dynasty. At that time the writing system of Indian numerals was imported, but its use as an alternative for the counting board system was rejected. In fully recognizing the powerfulness of our traditional method of thinking, and in absorbing at the same time the highly developed foreign techniques, we foresee a novel new era of achievements in Chinese mathematics.

BIBLIOGRAPHY

[**AN**] Liu Hui, *Annotations to "Nine chapters of arithmetic"*, 263 A.D.

[**AP**] Donald E. Knuth, *The art of computer programming*, Vol 2, Addison-Wesley, Reading, Mass., 1969.

[**AR**] *Nine chapters of arithmetic*, completed in definite form in 1c A.D.

[**MA**] Qin Jiushao, *Mathematical treatise in nine chapters*, 1247 A.D.

[**SI**] Liu Hui, *Sea island mathematical manual*, 263 A.D.

[**WA**] Donald B. Wagner, *An early Chinese derivation of the volume of a pyramid: Liu Hui, third century A.D.*, Historia Math. **6** (1979), 169–188.

INSTITUTE OF SYSTEMS SCIENCE, ACADEMIA SINICA, CHINA

Acta Mathematica Sinica, New Series
1987, Vol. 3, No. 3 pp.227—236

On Chern Numbers of Algebraic Varieties with Arbitrary Singularities

Wu Wentsün (吴文俊)
Institute of Systems Science, Academia Sinica

Received December 5, 1986

Abstract. In 1965 the author introduced the notion of Chern classes for an algebraic variety with arbitrary singularities. Based on this definition the well-known Miyaoka-Yao inequalities have been proved and extended by quite simple direct computations.

In 1977 Miyaoka and Yau (see [M] and [Y]) have proved independently a remarkable inequality about Chern numbers of a SMOOTH algebraic surface S, viz.

$$c_1^2(S) <= 3 * c_2(S). \qquad \text{(MY)}$$

Some results and conjectures of similar nature have also been anounced for high dimensional algebraic varieties (see e.g. [T], [Y]). Their considerations are all restricted to algebraic varieties without any singularities since tools for complex manifolds were used throughout. Now in 1977 MacPherson [MP] has introduced the notion of Chern classes for any algebraic variety with arbitrary singularities. It is natural to ask whether the above inequality remains true for this general case for which the present author is quite ignorant of the present status. On the other hand early in 1965 the present author has already generalized the notion of Chern classes to arbitrary algebraic varieties in an entirely different way from that of MacPherson et al, cf. [WU1—3]. It turns out that the formula (MY) and its alike can be easily dealt with by our treatment for varieties with singularities. This will be the main theme of the present paper. Other applications of our method will be dealt with later.

We use in this paper notations which, being readily done by computer-printing, are somewhat different from the usually adopted ones. For the convenience of the reader a comparison between these two types of some of these notations used are tabulated below:

new notations	usual notations	explanations
$-<$	$\subset$, $\in$	"is contained in" or "belongs to"
$>-$	$\supset$	"contains"
$<=$	$\leqq$	"less than or equal to"

> =	$\geqq$	"greater than or equal to"
< >	$\neq$	"not equal to"
*		"multiply by"
^		"to the power of" or "intersects with"
$(n//m)$	$\binom{n}{m}$	binomial coefficient

Sect 1. The Composite Grassmann Variety

Let us recall first some fundamental facts about composite Grassmann variety due to Ehresmann et al (see e.g. [EH], [HP], [CHOW], and [WU2]), which is at the basis of our treatment. Note that we are working in the complex domain so the modifier "complex" will often be omitted.

Consider thus a projective space CPn of dimension n. The linear subspaces of dimension k will be denoted by $[k]$, $[k]'$, Sk, $S'k$, etc. For fixed integers p,q with $0 < p < q < n$ the totality of pairs $([p], [q])$ with $[p] - < [q] - < [n] = CPn$ will be denoted henceforth as $GR(n;p,q)$. It is a special kind of composite Grassmann variety and is an irreducible algebraic variety without singularities so that intersection can be well defined in it, see e.g. [HP], Chap.XI.

Following Ehresmann let us consider a fixed sequence of linear subspaces

$$S0 \ - < \ S1 \ - < \ \cdots \ - < \ Sn = CPn. \tag{1.1}$$

Let Ai, Bj be integers verifying

$$0 \ < = \ A0 \ < \ A1 \ < \ \cdots \ < \ Ap \ < = \ n, \tag{1.2}$$

$$0 \ < = \ B0 \ < \ B1 \ < \ \cdots \ < \ Bq \ < = \ n. \tag{1.3}$$

With respect to (1.1) we shall denote by the Ehresmann symbol of the form

$$[A0, \ A1, \ \cdots, \ Ap/B0, B1, \ \cdots \ Bq] \tag{1.4}$$

the totality of pairs $([p], [q])$ such that

($E1$) dim $([p] \wedge [SAi]) \ > = \ i$, for $0 \ < = \ i \ < = \ p$;

($E2$) dim $([q] \wedge [SBj]) \ > = \ j$, for $0 \ < = \ j \ < = \ q$;

($E3$) each Ai is some Bj.

The variety (1.4), usually called a Schubert variety, has a dimension

$$\dim \ [A0, \ A1, \ \cdots, \ Ap/B0, B1, \ \cdots, \ Bq] = \mathrm{SUMi} \ (Ai - i) + \mathrm{SUM'j} \ \ (Bj - j). \tag{1.5}$$

In (1.5) SUMi is to be extended over i from 0 to p while SUM'j is over only such j from 0 to q for which Bj is not equal to any Ai. In particular, $GR(n;p,q)$ is itself such a Schubert variety with symbol and dimension given by

$$GR(n;p,q) = [(n-p,\cdots,n)/(n-q,\cdots,n)], \tag{1.6}$$

$$\dim\ GR(n;p,q) = (n-p)*(p+1)+(n-q)*(q-p). \tag{1.7}$$

We remark that the bracket ( ) in (1.6) means that the integers therein are consecutive ones.

Take now a second fixed sequence of linear subspaces

$$S'0 -< S'1 -< \cdots -< S'n = CPn \tag{1.1$'$}$$

for which all $S'i$ in (1.1)′ and Sj in (1.1) are in general position. Denote the Schubert variety corresponding to (1.4) defined however with respect to (1.1)′ by

$$[A0, A1, \cdots, Ap/B0, B1, \cdots, Bq]'. \tag{1.4$'$}$$

We shall set
$[A'p, \cdots, A'1, A'0/B'q, \cdots, B'1, B'0]' =$ Dual $[A0, A1, \cdots, Ap/B0, B1, \cdots, Bq]$, in which $A'i = n - Ai$ and $B'j = n - Bj$. We see that any Schubert variety will intersect its dual in a single point.

According to Chow (see [CHOW]), the variety $GR(n;p,q)$ has a rational dissection formed of all the above Schubert varieties defined with respect to (1.1) with boundaries removed and the totality of such Schubert varieties will represent a basis of the group of rational equivalence classes of $GR(n;,p,q)$. The rational dissection defined with respect to (1.1)′ is then said to be *DUAL* to the rational dissection above defined with respect to (1.1) in the sense of [WU2]. It easily follows that the totality of Schubert varieties (1.4) (or (1.4)′) form also a basis of the group of algebraic equivalence classes of $GR(n;p,q)$. For an algebraic variety with arbitrary singularities V let us denote by $RATr(V)$ respectively $ALGr(V)$ the group of rational respectively algebraic equivalence classes in dimension r of V. Denote also for any subvariety W of dimension r of V, its rational respectively algebraic equivalence class, by $R\text{–}Cls(W)$ respectively $A\text{–}Cls(W)$. If V is devoid of any singularity, then the sum of $ALGr(V)$ for all r will possess an intersection

$$ALGr(V) * ALGs(V) -< ALGt(V)$$

with $t = r + s - \dim V$ which turns the sum into an intersection ring or CHOW RING of the nonsingular variety V. In particular for the nonsingular $GR(n;p,q)$ we have for any Schubert varieties E, E' the formulae of intersection

$$A\text{–}Cls(E) * A\text{–}Cls\ \text{Dual}\ E = 1, \text{ while } A\text{–}Cls(E) * A\text{–}Cls\ (E') = 0, \text{ for}$$

$$E' <> \text{Dual E, and } \dim\ E' + \dim\ E = \dim\ GR(n;p,q).$$

Furthermore, the association of any Schubert variety to its dual will induce a natural morphism

$$\text{Dual: } ALGs(GR(n;p,q)) \longrightarrow ALGt(GR(n;p,q)),$$

in which $t = \dim\ GR(n;p,q) - s$.

Sect 2. The Intersection Ring of $GR(n;0,d)$

For the purpose of the present paper we shall restrict ourselves henceforth to the particular case of the composite grassmann $GR(n;p,q)$ with $p = 0$, $q = d$. The dimension of our grassmannian is then given by

$$\dim GR(n;0,d) = (n-d)*(d+1)+d.$$

For reasons to be explained later we are particularly interested in algebraic equivalence classes below:

$$GAMst = A\text{-}Cls\ [s\text{-}t/(0,\cdots,d-t),(d-t+2,\cdots,d+1)]\ -<\ ALGs\ GR(n;0,d)$$

for $0 <= t <= s <= d$, and

$$CHs = SUMt\ (\mathrm{sgn}(t) * (d-t+1//d-s+1) * GAMst)\ -<\ ALGs\ GR(n;0,d)$$

for $0 <= s <= d$, in which SUMt means summation extended over t from 0 to s.

Remark. In [WU1–3] there are some misprints in sign in the binomial coefficients.

For $s = 1, 2$ or 3 we have in particular

$$CH1 = (d+1) * GAM10 - GAM11$$
$$= (d+1) * A\text{-}Cls\ [1/(0,\cdots d)] - A\text{-}Cls\ [0/(0,\cdots,d-1),d+1], \tag{2.1}$$

$$CH2 = d*(d+1)/2*GAM20 - d*GAM21 + GAM22$$
$$= d*(d+1)/2*A\text{-}Cls\ [2/(0,\cdots,d)] - d*A\text{-}Cls\ [1/(0,\cdots,d-1),d+1]$$
$$+ A\text{-}Cls\ [0/(0,\cdots,d-2),d,d+1]. \tag{2.2}$$

$$CH3 = (d+1//3)*A\text{-}Cls\ [3/(0,...,d)] - (d//2)*A\text{-}Cls[2/(0,...,d-1),d+1]$$
$$+ (d-1)*A\text{-}Cls[1/(0,\cdots,d-2),d,d+1]$$
$$- A\text{-}Cls\ [0/(0,\cdots,d-3),d-1,d,d+1]. \tag{2.3}$$

The intersection structure or Chow ring of $GR(n;0,d)$ will only be partially determined but will be sufficient for our purposes. For this we shall first prove the following lemmas.

Lemma 1. *For the Schubert variety* $A = [n\text{-}Ai/n\text{-}Ad,\cdots,n\text{-}Ai,\cdots,n\text{-}A0] = \mathrm{Dual}\ [Ai/A0,\cdots,Ai,\cdots,Ad]$ *to have a dimension* $>=\ \dim\ GR(n;0,d)-d$ *it is necessary that*

$$A0 = 0, A1 = 1,\cdots, Ai = i$$

so that

$$A = [n-i/n\text{-}Ad,\cdots,n\text{-}Aj,(n-i,\cdots,n)]$$

in which we have put $i+1=j$.

Proof. The hypothesis implies that

$$\dim\ [Ai/A0,\cdots,Ai,\cdots,Ad] = SUMk\ (Ak-k)+i <= d.$$

Now the integers Ak should verify the conditions

$$0 <= A0 < A1 < \cdots < Ad <= n, \text{ or}$$

$$0 <= A0 <= A1-1 <= \cdots <= Ai-i <= Aj-j <= \cdots <= Ad-d.$$

It follows that $Ai > i$ would imply $Ak > k$ for all $k >= i$ so that $SUMk\ (Ak-k)+i > d$ contractory to the inequality given above. Consequently $Ai = i$ and whence $A0 = 0$, $A1 = 1$, etc. up to $Ai = i$ as to be proved.

Lemma 2. *Let* $j = i+1$. *Then*

$$A\text{-}Cls\ [n-i/n\text{-}Ad,\cdots,n\text{-}Aj,(n\text{-}i,\cdots,n)]$$
$$= A\text{-}Cls\ [n/n\text{-}Ad,\cdots,n\text{-}Aj,(n\text{-}i,\cdots,n)] * A\text{-}Cls\ [n-i/(n-d,\cdots,n)]. \quad (2.4)$$

Proof. Denote the Schubert varieties involved in the above equality by A, B, C respectively defined with respect to sequences of linear subspaces like (1.1) as follows.

For B: $[0]-<[1]-<\cdots-<[n]=CPn$,

For C: $[0]'-<[1]'-<\cdots-<[n]'=CPn$

with $[n-i-1]-<[n-i]'$ but otherwise the $[r]$ and $[s]'$ are in general position. The variety A is then defined with respect to the sequence

$$[0]-<[1]-<\cdots-<[n-i-1]-<[n-i]'-<[n-i+1]'-<\ \cdots-<[n]=CPn.$$

Clearly an element $(S0,Sd)$ of $GR(n;,0,d)$ will belong to the intersection of B and C if and only if

$$\dim(Sd\ \wedge\ [n\text{-}Ak])>=d\text{-}k,$$

for $k=i+1,\cdots,d$ and $S0\ -<\ [n\text{-}i]'$, i.e.

$$(S0,Sd)\ -<\ A.$$

Now

$$\dim\ B=\dim GR(n;0,d)-SUMk(Ak-k),$$

$$\dim\ C=\dim GR(n;0,d)-i,\ \text{and}\ \dim\ A=\dim GR(n;0,d)-SUMk\ (Ak-k)-i,$$

in which $SUMk$ is to be extended over k from $i+1$ to d. Now by Ehresmann the variety A is an irreducible one. It follows then from dimensionality considerations that the right-hand side should be equal to an integral multiple of the class of A. This integer is the intersection multiplicity of B and C and is easily seen to be 1. This proves the formula (2.4) of the Lemma.

It is clear that

$$A\text{-}Cls\ [n\text{-}i/(n\text{-}d,\cdots,n)] * A\text{-}Cls\ [n\text{-}j/(n\text{-}d,\cdots,n)] = A\text{-}Cls\ [n\text{-}i\text{-}j/(n\text{-}d,\cdots,n)]. \quad (2.5)$$

Furthermore we have also in the right dimensions

$$A\text{-}Cls\ [n/n\text{-}Ad,\cdots,n\text{-}A0] * A\text{-}Cls\ [n/n\text{-}Bd,\cdots,n\text{-}B0]$$
$$= SUMc\ A\text{-}Cls\,[n/n\text{-}Cd,\cdots,n\text{-}C0], \quad (2.6)$$

in which

$$A\text{-}Cls\,[n\text{-}Ad,\cdots,n\text{-}A0] * A\text{-}Cls\,[n\text{-}Bd,\cdots,n\text{-}B0]$$
$$= SUMc\ A\text{-}Cls\,[n\text{-}Cd,\cdots,n\text{-}C0] \quad (2.7)$$

is the intersection formula in the ordinary grassmannian as shown in [HP], Chap. XIV, which can in turn be explicitely determined by means of the well-known formulae of Pieri and Giambelli.

Let us now introduce some classes as follows:

$$P=A\text{-}Cls\ [1/(0,\cdots,d)] \quad (2.8)$$

$$Qh=A\text{-}Cls\ [0/(0,\cdots,d-1),d+h] \quad (2.9)$$

$$P'=\text{Dual}\ P=A-cls\,[n-1/(n-d,...,n)] \quad (2.8)'$$

$$Q'h=\text{Dual}\ Qh=A\text{-}Cls\ [n/n-d-h,(n-d+1,\cdots,n)] \quad (2.9)'$$

for $0 <= h <= n-d$. From the above we get easily the following.

Theorem. *The Chow ring of algebraic equivalence classes of the composite grassmannian $GR(n;0,d)$ is generated by the classes P' and $Q'h$ in the dimension* $>=$ dim $GR(n;O,d)-d$. *The multiplicative structure in that range is completely determined by the formulae* (2.4)–(2.9)$'$.

By the theorem we deduce from (2.1)–(2.3):

$$\text{Dual } CH1 = (d+1)*P' - Q'1, \tag{2.10}$$

$$\text{Dual } CH2 = d*(d+1)/2*P'\wedge 2 - d*P'*Q'1 + (Q'1\wedge 2 - Q'2), \tag{2.11}$$

$$\begin{aligned}\text{Dual } CH3 = {} & (d+1)*d*(d-1)/6*P'\wedge 3 - d*(d-1)/2*P'\wedge 2*Q'1 \\ & + (d-1)*P'*(Q'1\wedge 2 - Q'2) - (Q'1\wedge 3 - 2*Q'1*Q'2 + Q'3).\end{aligned} \tag{2.12}$$

Sect 3. Ehresmann Classes of an Algebraic Variety with Arbitrary Singularities

Let Vd be an irreducible algebraic variety of dimension d and V' a subvariety containing all singularities of Vd. By considering subvarieties of a fixed dimension s, the author has introduced in [WU1] the notion of group of UNNEGLIGIBLE algebraic equivalence classes modulo V' for each dimension s, with methods as described in Chap. XI of [HP], which will be denoted by $ALGs(Vd/V')$ in what follows. There is also a natural morphism for each dimension s, viz.

$$Js\colon\ ALGs(Vd/V') \longrightarrow ALGs(V').$$

Let Wd be also some irreducible algebraic variety of same dimension d with W' a subvariety containing all singularities of Wd. Let T be a birational transformation of Wd to Vd verifying the following properties:

P1. T is everywhere defined on Wd.

P2. $T(x) -< V'$ if and only if $x -< W'$.

P3. T is biunivoque on $Wd - W'$.

It is proved in [WU1] that under these conditions the birational transformation T will induce in each dimension s a natural morphism

$$Ts\colon\ ALGs(Wd/W') \longrightarrow ALGs(Vd/V').$$

Note that for these groups of unnegligible algebraic equivalence classes no mulplicative structure will be introduced in their sum.

Let Ge be now an irreducible algebraic variety of dimension e in a complex projective space with no singularities so that intersection may be defined in Ge in the usual manner. Let Wd be an irreducible subvariety of dimension d in Ge and W' a subvariety of Wd containing all singularities of Wd if exist. As Ge is in a complex projective space any subvariety of Ge is algebraically equivalent to some one which will intersect simply with both Wd and W'. From this we easily deduce that, by considering intersections with Wd in Ge, there will be natural morphisms

$$Is\colon\ ALGs(Ge) \longrightarrow ALGt(Wd/W'),$$

in which $t = s + d - e$.

Consider now an irreducible algebraic variety Vd of dimension d in a projective space CPn of dimension n. Take an arbitrary generic point $P0$ of Vd and let Pd be the tangent space of Vd at $P0$. With the pair $(P0, Pd)$ as a generic point there will be a determined irreducible subvariety Wd

of dimension d in the composite grassmannian $GR(n;0,d)$ which may be considered as a subvariety of a projective space of sufficiently high dimension. Now any pair $(P0', Pd')$ of Wd is a specialization of $(P0, Pd)$ which implies that $P0'$ is a specialization of $P0$ and is thus a well-defined point of Vd. Clearly, if the singular subvariety of Vd is V', then the subvariety W' of Wd consisting of all points $(P0', Pd')$ with $P0'$ in V' will contain all the singular points of Wd if there are any. The correspondence

$$T\colon\ (P0', Pd') \longrightarrow P0'$$

is thus a birational one verifying the properties P1–3 with $V' = T(W')$. For the pair $G = GR(n;0,d)$ and Vd we have then a sequence of morphisms

$$ALGs(GR(n;O,d)) \xrightarrow{Is} ALGt(Wd/W') \xrightarrow{T} ALGt(Vd/V') \xrightarrow{Jt} ALGt(Vd),$$

in which $t = s + d - e$, with $e = \dim\, GR(n;0,d)$. Besides we have also the dual morphism

$$\text{Dual: } ALGs(GR(n;0,d)) \longrightarrow ALGs'(GR(n;0,d))$$

in which

$$s' = \dim GR(n;0,d)) - s.$$

Consider now any Ehresmann symbol

$$EH = [A0/B0, B1, \cdots, Bd] \text{ with } s = SUM'k\ (Bk - k) + A0$$

and

$$r = s' + d - \dim GR(n;0,d) = d - s,$$

in which $SUM'k$ is to be extended over k from 0 to d for which $Bk <> A0$. We shall lay down the following

Definition. The algebraic equivalence class

$$Jr\ T\ Is'\ \text{Dual}\ EH\ -<\ ALGr(Vd)$$

will be called the EHRESMANN CLASS of Vd corresponding to the symbol EH and will be denoted by

$$EH(Vd) = [A0/B0, B1, \cdots, Bd](Vd).$$

More generally, for any algebraic equivalence class $ACLS -< ALGs(GR(n;O,d))$, we shall set by definition

$$ACLS(Vd) = Jr\ T\ Is'\ \text{Dual}\ ACLS\ -<\ ALGr(Vd).$$

As particular Ehresmann classes we have also GAMKRELIDZE CLASSES and CHERN CLASSES defined respectively by $(r = d - s)$

$$GAMst(Vd) = [s - t/(0, \cdots, d - t), (d - t + 2, \cdots, d + 1)](Vd)\ -<\ ALGr(Vd),$$

$$\text{CHs}(Vd) = SUMt\ (\text{sgn}(t) * (d - t + 1)//(d - s + 1) * GAMst(Vd))\ -<\ ALGr(Vd).$$

We note that in case that Vd is devoid of any singularities so that Vd may be considered as a complex manifold in a complex projective space, then according to Gamkrelidze the homology classes defined by the algebraic equivalence classes $CHs(Vd)$ are just the dual of the usual Chern classes. This justifies the terminologies introduced above, cf. [G].

Remark that in the notations $EH(Vd)$, $CHs(Vd)$, etc., integer n, the dimension of the ambiant projective space in which lies the variety Vd, does not enter into play, as is natural and easy to see.

Sect 4. Chern Numbers of an Algebraic Variety with Arbitrary Singularities

Let Vd be an irreducible algebraic variety of dimension d with arbitrary singularities. Consider any Ehresmann symbol

$$EH = [A0/B0, B1, \cdots, Bd] \text{ with } SUM'k\ (Bk - k) + A0 = d,$$

in which $SUM'k$ is to be extended over k from 0 to d for which $Bk <> A0$. The Ehresmann class $EH(Vd)$ of $ALG0(Vd)$ is in the image of $ALG0(Vd/V')$ under the morphism $J0$ and can thus be identified to an integer, to be called the EHRESMANN CHARACTER of Vd corresponding to the symbol EH in what follows. From the definition it is clear that all such characters are of projective nature and were known as PROJECTIVE CHARACTERs of the variety in the sense of Severi, cf. e.g. [SR]. Among these Ehresmann characters we have in particular CHERN CHARACTERs to be defined as follows.

A sequence of integers $p = (a,\ b,\ \cdots,\ c)$ will be said to be a partition of d if

$$0 < a <= b <= \cdots <= c, \text{ and } a + b + \cdots + c = d.$$

Define now $CHp\ -<\ ALGd(GR(n; 0, d))$ by

$$\text{Dual } CHp = \text{Dual } CHa * \text{Cual } CHb * \cdots * \text{Dual } CHc.$$

The integer identified to the algebraic equivalence class $CHp(Vd)\ -<\ ALG0(Vd)$ will then be called the CHERN CHARACTES of Vd corresponding to the partition p. By the intersection formulae in $GR(n; 0, d)$ as developed in the preceding sections it is clear that any such CHp can be expressed by means of algebraic equivalence classes P and Qh.

Let us consider as an example the case $d = 2$, i.e. the case of an algebraic surface $V2$ with arbitrary singularities. For such a $V2$ we have 4 Ehresmann characters and 2 Chern classes besides the trivial one $CH0$, viz.

$$[2/0, 1, 2](V2) = \text{Classical } Mu0(V2) = \text{Order of } V2,$$

$$[1/0, 1, 3](V2) = \text{Classical } Mu1(V2) = \text{Rank of } V2,$$

$$[0/0, 2, 3](V2) = \text{Classical } Mu2(V2) = \text{Class of } V2,$$

$$[0/0, 1, 4](V2) = \text{Classical } Nu2(V2) = \text{Type of } V2.$$

The last terminology is for $V2$ in CPn with $n > 3$ alone, but we shall keep this term for $V2$ in $CP3$ too. Cf. [SR], Chap. IX.

$$CH1(V2) = 3 * [1/0, 1, 2](V2) - [0/0, 1, 3](V2) = (\text{Dual}(3 * P' - Q'1))(V2),$$

$$CH2(V2) = 3 * [2/0,\ 1,\ 2](V2) - 2 * [1/0,\ 1,\ 3](V2) + [0/0,\ 2,\ 3](V2)$$
$$= (\text{Dual}(3 * P' \wedge 2 - 2 * P' * Q'1 + Q'1 \wedge 2 - Q'2))(V2).$$

There are 2 Chern characters $CH11(V2)$ and $CH2(V2)$ for which we have for the former

$$CH11(V2) = (\text{Dual}(9 * P' \wedge 2 - 6 * P' * Q'1 + Q'1 \wedge 2))(V2).$$

It follows that

$$3 * CH2(V2) - CH11(V2) = 2 * (\text{Dual } Q'1 \wedge 2)(V2) - 3 * (\text{Dual } Q'2)(V2).$$

Now in the grassmannian $GR(n; 0, d)$ we have the multiplication formula

$$A\text{–}Cls[n/n-3,\ n\text{–}2, n] = A\text{–}Cls\ [n/n-3, n-1, n] \wedge 2 - A\text{–}Cls\ [n/n-4, n-1, n].$$

Taking the dual of both sides we get

$$[0/0, 2, 3](V2) = (\text{Dual } Q'1 \wedge 2)(V2) - (\text{Dual } Q'2)(V2).$$

As the left side is $Mu2(V2)$ and the last term is $Nu2(V2)$ we get

$$2 * (\text{Dual } Q'1 \wedge 2)(V2) - 3 * (\text{Dual } Q'2)(V2) = 2 * Mu2(V2) - Nu2(V2).$$

If $V2$ is in $CP3$ then $Nu2(V2)$ is clearly 0 and hence we get the following

Theorem. *For a surface $V2$ in $CP3$ with arbitrary singularities we have for the Chern characters the inequality*

$$3 * CH2(V2) >= CH11(V2).$$

Suppose that the surface $V2$ has no singularities so that it is a SMOOTH complex surface. Then $CH2(V2)$ is just the usual Chern number $c_2(V2)$ and $CH11(V2)$ the Chern number $c_1^2(V2)$. The above inequality becomes then the Miyaoka–Yau inequality stated in the beginning of the paper. The above theorem can therefore be considered as a generalization of the Miyaoka–Yau inequality to the case of algebraic surfaces with arbitrary singularities lying in $CP3$.

On the other hand suppose that the variety $V2$ is not in $CP3$. Then there are known examples for which

$$2 * Mu2(V2) < Nu2(V2).$$

Cf. formulae (10) and (11) on [SR], p.221. It follows that the Miyaoka–Yau inequality is not true in general for surfaces in CPn with singularities present. We leave open the question of the truth of the inequality in case of NON–SINGULAR $V2$ in CPn with $n > 3$.

Consider now any hypersurface Vd of dimension d in CPn with $n = d + 1$. We have then from the very definition

$$Q'h = 0 \text{ for } h >= 2.$$

For $d = 3$ in particular we would have then from (2.10)—(2.12):

$$\text{Dual } CH1 = 4 * P' - Q'1,$$

$$\text{Dual } CH2 = 6 * P' \wedge 2 - 3 * P' * Q'1 + Q'1 \wedge 2,$$

$$\text{Dual } CH3 = 4 * P' \wedge 3 - 3 * P' \wedge 2 * Q'1 + 2 * P' * Q'1 \wedge 2 - Q'1 \wedge 3.$$

There are 3 partitions (1, 1, 1), (1, 2) and (3) of the integer $d = 3$ for which we have

$$\text{Dual } CH111 = (4 * P' - Q'1) \wedge 3 = 64 * P' \wedge 3 - 48 * P' \wedge 2 * Q'1 + 12 * P' * Q'1 \wedge 2 - Q'1 \wedge 3,$$

$$\text{Dual } CH12 = (4 * P' - Q'1) * (6 * P' \wedge 2 - 3 * P' * Q'1 + Q'1 \wedge 2)$$

$$= 24 * P' \wedge 3 - 18 * P' \wedge 2 * Q'1 + 7 * P' * Q'1 \wedge 2 - Q'1 \wedge 3.$$

whence

$$4 * CH12(V3) - 8 * CH3(V3) - CH111(V3) = 5 * (\text{Dual } Q'1 \wedge 3)(V3).$$

As (Dual $Q'1 \wedge 3)(V3)$ is necessarily non–negative we get the following generalization of a theorem due to Tai (cf. [T]), viz.

Theorem. *For a hypersurface V3 of dimension 3 in CP4 with arbitrary singularities we have for the Chern characters the inequality*

$$4 * CH12(V3) - 8 * CH3(V3) - CH111(V3) >= 0.$$

Clearly the method is entirely general which will permit us to get generalizations of other theorems of Tai to case of algebraic hypersurfaces with arbitrary singularities. We can also investigate possible generalizations of inequalities of Miyaoka–Yau type in the case of higher dimensions. We shall however not enter into these problems since the method of treatment is quite clear.

References

[CH] Chern, S.S., On the characteristic classes of complex sphere bundles and algebraic varieties, *Amer.Math.Soc.*, **75** (1953), 565–597.

[CHOW] Chow W.L., Algebraic varieties with rational dissections, *Proc. Nat. Acad. Sci.*, **42** (1956), 116–9.

[EH] Ehresmann, C., Sur la topologie de certaines espaces homogenes, *Ann. of Math.*, **35** (1934), 396–443.

[G] Gamkrelidze, P.B., Chern cycles of complex algebraic manifolds, *IZV. ACAD. SCIS, CCCP, math. ser.*, **20** (1956), 685–706 (in Russian).

[HP] Hodge, W. V. D. and Pedoe, D., Methods of algebraic geometry, Cambridge, **1** (1947), **2** (1952).

[MP] MacPherson, R.D., Chern classes for singular algebraic varieties, *Ann. of Math.*, **100** (1974), 423–432.

[M] Miyaoka, Y., On the Chern numbers of surfaces of general type, *Invent. Math.*, **42**(1977), 225–237.

[SR] Semple, J.G. and Roth, L., Introduction to algberaic geometry, Oxford, 1949.

[T] Tai, S., A class of symmetric functions and Chern numbers of algebraic varieties, Preprint, 1985.

[VdV] Van de Ven, On the Chern numbers of surfaces of general type, *Invent. Math.*, **36**(1976), 285–293.

[VdW] Van der Waerden, Einfuerung in die algebraischen Geometrie, Berlin, 1945.

[WU1] Wu Wen-tsun, On Chern characteristic systems of an algebraic variety (in Chinese), *Shuxue Jinzhan*, **8** (1965), 395–401.

[WU2] ——, On algebraic varieties with dual rational dissections (in Chinese), *Shuxue Jinzhan*, **8** (1965) 402–409.

[WU3] ——, Chern classes on algberaic varieties with arbitrary singularities, in Several Complex Variables, Proc. 1981 Hangzhou Conf. Eds. Kohn et al, Birkhauser, (1984) 247–249.

[Y] Yau, S.T., Calabi's conjecture and some new results in algberaic geometry, *Proc. Nat. Acad. Sci.*, 74 (1977), 1798–1799.

MM Research Preprints, 53–59
KLMM, AMSS, Academia Sinica
Vol. 1, Feb 1987

Mechanical Derivation of Newton's Gravitational Laws from Kepler's Laws

Wu Wen-tsun
Institute of Systems Science, Academia Sinica

It is an important historical event that Newton derived his laws from Kepler's laws. However, how the former ones can be deduced from latter ones is rarely touched upon in current texts on calculus or mechanics, though the deduction of the latter ones from the former ones is treated rather often in such texts, e.g. [1]. The present preliminary report aims at such a deduction, and, what is perhaps more important for our purposes, a deduction in a MECHANICAL manner. The author owes for this report much to Professor Gabriel of Argonne National Laboratory. In fact, during a visit to Argonne in 1986 Prof. Gabriel told the author such a problem for which he was already quite successful in applying his own automated reasoning method based on works of Ritt can be applied as well to deal with such kind of problems.

To begin with, let us first recall some fundamental notions and the basic principles underlying such method for which we refer for more details to [3, 4] and [5, 6].

Let F be a DIFFERENTIAL FIELD (abbr. d-FIELD) which for the present paper may be understood to be simply the d-field of all rational functions of some parameter t considered as the time. To any DIFFERENTIAL POLYNOMIAL (abbr. d-POL) P ($\neq 0$) in some indeterminates $X_1, \ldots, X_n$ over the basic d-field F we shall associate a 4-tuple of integers $[t\ c\ r\ d]$, called the INDEX-SET of P, in notation ind(P), as follows.

t = number of actual terms in P,

c = the greatest subscript c for which X_c occurs actually in P, to be called the CLASS of P, and be denoted as $cls(P)$.

r = the highest order r for which the r-th derivative D_rX_c of the above X_c occurs actually in P, to be called the ORDER of P and to be denoted as $ord(P)$.

d = the highest degree d of the above D_rX_c which occurs actually in P, to be called the DEGREE of P and to be denoted by $deg(P)$.

For a d-pol P with $cls(P) = c$, $ord(P) = r$, and $deg(P) = d$, we shall call the derivative D_rX_c the LEAD of P, to be denoted by $lead(P)$. Let L be this lead. Then the coefficient of L^d, which is itself a d-pol, is called the INITIAL of P, to be denoted as $init(P)$. The formal partial derivative of P w.r.t. L is then called the SEPARANT of P, to be denoted by $sep(P)$. Naturally, all these terminologies come from works of Ritt.

For d-pols in indeterminates $X_1, \ldots, X_n$ over the d-field F we shall consider a partial ordering $\ll$ defined in the following way. Let P_1, P_2 be d-pols with index sets $[t_1\, c_1\, r_1\, d_1]$ and $[t_2\, c_2\, r_2\, d_2]$ resp. We say then $P_1 \ll P_2$ if one of the following cases occurs:

(a) $c_1 < c_2$,
(b) $c_1 = c_2$, but $r_1 < r_2$,

(c) $c_1 = c_2$, $r_1 = r_2$, but $d_1 < d_2$.

With respect to such a partial ordering of d-pols we can then introduce the notions of DIFFERENTIAL ASCENDING SET, DIFFERENTIAL BASIC SET, and DIFFERENTIAL CHARACTERISTIC SET (abbr. d-ASC-SET, d-BAS-SET, and d-CHAR-SET resp.)just as in the case of ordinary polynomial. We define also the notion of d-REDUCED as that of REDUCED in the ordinary case.

Consider now a d-asc-set d-ASC consisting of d-pols

$$P_1, P_2, \ldots, P_s \tag{d-ASC}$$

with

$$0 < cls(P_1) < cls(P_2) < \cdots < cls(P_s).$$

For any d-pol G we have then the following REMAINDER FORMULA:

$$\prod_i (I_i^{L_i}) \prod_j (S_j^{M_j}) G = \sum_k Q_k P_k + R.$$

in which I_i, S_j are the respective initials and separants of d-pols in d-ASC, L_i and M_j are certain non-negative integers which will be taken to be as small as possible, and Q_k, R d-pols with R d-reduced w.r.t. d-ASC. The d-pol R is accordingly called the d-REMAINDER (abbr. d- REMDER) of G w.r.t. d-ASC, to be denoted as d-remdr(G/d-ASC).

A finite set of non-zero d-pols is called a DIFFERENTIAL POLSET (abbr. d-POLSET). Let such a d-polset DPS be given. A d-pol in the same indeterminates X_i but over an arbitrary DIFFERENTIAL EXTENSION FIELD (abbr. d-EXT-FIELD), F' of F will be said to be a SOLUTION (abbr. SOL) or d-ZERO of the set DPS if it satisfies all the equations $P = 0$ for P in DPS. The totality of all such solutions or d-zeros will be denoted by d-zero (DPS) and the totality of only those which are not d-zero of a given d-pol G will be denoted by d-zero (DPS/G).

Given a d-polset DPS we can deduce, just as in the ordinary case, a d-char-set DCHR in a mechanical way. We have then, also as in the ordinary case, the formulas below:

$$\text{d-zero (DPS)} \subset \text{d-zero (DCHR)}, \tag{I}$$

$$\text{d-zero (DCHR}/K) \subset \text{d-zero (DPS)}, \tag{II}$$

$$\text{d-zero (DPS)} = \text{d-zero (DCHR}/K) + \sum_k \text{d-zero (DPS}_k). \tag{III}$$

in which K is the product of all initials and separants of d-pols in DCHR, and DPS$_k$ are d-polsets which are the enlarged DPS with one of the initials or the separants adjoined to it.

The formulas (I) – (III) are at the basis of all our considerations about mechanization of mathematics in the case involving differentiation.

Come now to the problem proper as cited in the title of the present paper. Let us first formulate the Kepler's laws (K) and the Newton's laws (N) in the manner as given below:

(K_1) The planets move in elliptic orbits around the sun as focus.
(K_2) The vector from the sun to the planet sweeps equal areas in equal times.

(K_3) The squares of periods of planet is motions are proportional to the cube of the semi major axis of the elliptic orbits.

(N_1) The acceleration of a planet is inversely proportional to the square of the distance from the sun to the planet.

(N_2) The acceleration vectors of planets are directed toward the sun.

In order to deduce mechanically the Newton's laws $(N_1), (N_2)$ from Kepler's laws (K_1) – (K_3) (actually only (K_1) – (K_2) will be sufficient) let us take first coordinates and transform the various laws into equation forms as follows.

Take polar coordinates with the sun at the pole and the major axis of the elliptic orbit as the polar axis. Then the orbit will have an equation of the form

$$r = p/(1 - e * \cos w) \tag{1}$$

in which w is the angle between the polar axis and the vector from the sun to the planet. The Kepler's law (K_1) corresponds to the equation (1) and also (2) – (3) below taken together:

$$p = \text{const} \text{ , or } p' = 0, \tag{2}$$

$$e = \text{const} \text{ , or } e' = 0, \tag{3}$$

in which the prime means derivative w. r. t. the time t. Similarly Kepler's law (K_2) will correspond to the equations (4), (5) below:

$$r^2 w' = h, \tag{4}$$

$$h' = 0. \tag{5}$$

Let us take also rectangular coordinates (x, y) associated to the above polar coordinates (r, w). Then the Newton's laws N_1, N_2 will correspond to the following set of equations:

$$r^4[(x'')^2 + (y'')^2] = k, \tag{6}$$

$$k' = 0, \qquad (N_1, 7)$$

$$xy'' = yx''. \qquad (N_2, 8)$$

Now between the polar and the rectangular coordinates we have also the equations (9) – (13) below:

$$x = r \cos w, \tag{9}$$

$$y = r \sin w, \tag{10}$$

$$\cos^2 w + \sin^2 w = 1, \tag{11}$$

$$(\cos w)' = -(\sin w)w', \tag{12}$$

$$(\sin w)' = +(\cos w)w'. \tag{13}$$

To proceed further let us first remark that it is immaterial whether the equations (9) – (13) are dependent or not. What is important for us is that the computer can not recognize any irrational or transcendental entities like $\sin w$ or $\cos w$. This can however be remedied simply by treating $\cos w$ and $\sin w$ just like indeterminates connected by relations (11) - (13). To apply our implemented programs let us now introduce indeterminates in replacing the various functions by x's as given below:

$$(p\,,\,e\,,\,r\,,\,x\,,\,y\,,\,w\,,\,\cos w\,,\,\sin w\,,\,h\,,\,k) = (x_{21}, x_{22}, x_{31}, x_{32}, x_{33}, x_{41}, x_{42}, x_{43}, x_{51}, x_{52}).$$

With this change of notations the equations (1) - (13) will turn to be the equations $P_i = 0$ with P_i given by (1′) - (13′) as shown below:

$$+1 * x_{31} - 1 * x_{31} * x_{22} * x_{42} - 1 * x_{21}, \tag{1'}$$

$$+1 * x_{21}', \tag{2'}$$

$$+1 * x_{22}', \tag{3'}$$

$$+1 * x_{31}^2 * x_{41}' - 1 * x_{51}, \tag{4'}$$

$$+1 * x_{51}', \tag{5'}$$

$$+1 * x_{31}^4 * (x_{32}'')^2 + 1 * x_{31}^4 * (x_{33}'')^2 - 1 * x_{52}, \tag{6'}$$

$$+1 * x_{52}', \tag{7'}$$

$$+1 * x_{32} * x_{33}'' - 1 * x_{33} * x_{32}'', \tag{8'}$$

$$+1 * x_{31} * x_{42} - 1 * x_{32}, \tag{9'}$$

$$+1 * x_{31} * x_{43} - 1 * x_{33}, \tag{10'}$$

$$+1 * x_{42}^2 + 1 * x_{43}^2 - 1, \tag{11'}$$

$$+1 * x_{42}' + 1 * x_{43} * x_{41}', \tag{12'}$$

$$+1 * x_{43}' - 1 * x_{42} * x_{41}'. \tag{13'}$$

Take now the d-polset DPS to consisting of the 11 d-pols (1′) - (6′), (9′) - (13′) of the above set. Remark that the planets move in true non-degenerate elliptic orbits so that we have

$$\begin{aligned} x_{21} = p \neq 0\,,\; x_{22} = e \neq 0, \\ x_{31} = r \neq 0\,,\; x_{33} = y \neq 0\,. \end{aligned} \tag{14}$$

In applying our algorithm for the finding of d-char-set DCHR of DPS we can then remove any factors x_{21}, x_{22}, x_{31} and x_{33} during the procedure. The DCHR is found to be the 3-th d-bas-set consisting of the 10 d-pols C_i given below:

$$+1 * x_{21}',$$

$$+1 * x_{22}',$$

$$\begin{aligned} -1 * x_{31} * x_{21}^2 * x_{31}'' + 1 * x_{31} * x_{21} * (x_{31}')^2 + \cdots \\ -1 * x_{21}^2 * (x_{31}')^2 + 2 * x_{31}^2 * x_{21} * x_{31}'' + \cdots \\ +1 * x_{31}^3 * x_{22}^2 * x_{31}'' - 1 * x_{31}^3 * x_{31}'', \end{aligned}$$

$$-1 * x_{31} + 1 * x_{21} + 1 * x_{22} * x_{32},$$
$$+1 * x_{31}^2 - 2 * x_{31} * x_{21} + 1 * x_{21}^2 + 1 * x_{22}^2 * x_{33}^2 + \cdots$$
$$-1 * x_{21} * x_{31}' + 1 * x_{31} * x_{22} * x_{33} * x_{41}',$$
$$+1 * x_{31} - 1 * x_{31} * x_{22} * x_{42} - 1 * x_{21},$$
$$+1 * x_{31} * x_{43} - 1 * x_{33},$$
$$+1 * x_{31}^2 * x_{41}' - 1 * x_{51},$$
$$+1 * x_{31}^4 * (x_{32}'')^2 + 1 * x_{31}^4 * (x_{33}'')^2 - 1 * x_{52}.$$

The CPU-time for bringing up this d-char-set is 146 sec.. The non-trivial initials are:

$I_3 = -1 * x_{31} * x_{21}^2 + 2 * x_{31}^2 * x_{21} + 1 * x_{31}^3 * x_{22}^2 - 1 * x_{31}^3 = +1 * x_{31} * x_{22}^2 * x_{33}^2,$
$I_4 = +1 * x_{22},$ ect..

The separants are essentially the same as the initials, with at most a further factor of x_{33}.

The proof of the Newton's laws is now readily done. In fact, we find the d-remdrs of the d-pols ($7'$) and ($8'$) to be both 0 w.r.t. the above dpolset DCHR. By the equation (I) and the remainder formula we see then the Newton's laws are true at least in the non-degenerate case (14). The degenerate case for which one of $x_{21}, x_{22}, x_{31}, x_{33}$ is zero can be dealt with in a similar but much easier way.

The Newton's laws have thus been derived in a mechanical way from the Kepler's laws as required. However, in proving that the remainders are zero it requires, somewhat unexpected, a quite long time, viz, a CPU-time of 10875.6 sec.. This defect comes seemingly from two sources. One is due to inadequacy of programming in the procedure of reductions so that improvement of the implementation of program is needed. A second one is due to inadequate choice of coordinate systems. Thus, instead of a mixed use of polar and rectangular coordinate systems we have tried to use the rectangular system alone. In this way the Kepler's law (K_1) will correspond to following equations

$$r = p + ex, \tag{15}$$
$$r^2 = x^2 + y^2 \tag{16}$$

together with equations (2) and (3). Similarly, the Kepler's law (K_2) becomes the equation

$$xy' - yx' = h \tag{17}$$

with h satisfying (5). Replacing now the various functions by the x's as before we have then to consider a d-polset DPS' consisting of 7 d-pols ($2'$), ($3'$), ($5'$), ($6'$) and those corresponding to (15) - (17), viz.

$$+1 * x_{31} - 1 * x_{22} * x_{32} - 1 * x_{21},$$
$$+1 * x_{21}',$$

$$+1 * x'_{22},$$
$$+1 * x_{32}^2 + 1 * x_{33}^2 - 1 * x_{31}^2,$$
$$+1 * x_{32} * x'_{33} - 1 * x_{33} * x'_{32} - 1 * x_{51},$$
$$+1 * x'_{51},$$
$$+1 * x_{31}^4 * (x''_{32})^2 + 1 * x_{31}^4 * (x''_{33})^2 + \cdots$$
$$-1 * x_{52}.$$

The d-char-set DCHR' is readily found in a CPU-time of 106.2 sec to be consisting of the following 7 d-pols C'_i as the 2-th d– bas-set, viz.

$$+1 * x'_{21},$$
$$+1 * x'_{22},$$
$$+1 * x_{21}^2 * (x'_{31})^2 + 1 * x_{31}^3 * x''_{31} + \cdots$$
$$-2 * x_{31}^2 * x_{21} * x''_{31} + 1 * x_{31} * x_{21}^2 * x''_{31} + \cdots$$
$$-1 * x_{31}^3 * x_{22}^2 * x''_{31} - 1 * x_{31} * x_{21} * (x'_{31})^2,$$
$$+1 * x_{31} - 1 * x_{22} * x_{32} - 1 * x_{21},$$
$$+1 * x_{32}^2 + 1 * x_{33}^2 - 1 * x_{31}^2,$$
$$+1 * x_{32} * x'_{33} - 1 * x_{33} * x'_{32} - 1 * x_{51},$$
$$+1 * x_{31}^4 * (x''_{32})^2 + 1 * x_{31}^4 * (x''_{33})^2 + \cdots,$$
$$-1 * x_{52}.$$

The remainders of the d-pols (7′) and (8′) w. r. t. DCHR' are again found to be zero in a shorter CPU-time of 5949.7 sec. The Newton's laws are thus again deduced from the Kepler's laws in a mechanical way a little simpler than the way before. It seems that improvement of the programming will further simplify the proofs in shortening the CPU-time to probably less than half an hour. We remark that times are naturally all referred to the computer which we are in use.

The proof presuppose that the Newton's Laws are already known and require merely a verification. Now suppose that we are in the stage of knowing the Kepler's experimental Laws alone, but entirely ignorant of what will be the form of the underlying Laws of Motion. The Principle in the form of (I) – (III) now furnishes us a method of automatically discover such unknown governing Laws. For this purpose let us introduce the acceleration a by $a^2 = (x'')^2 + (y'')^2$ arranging the order of the various entities involved in setting

$$(p, e, x, y, r, h, a^2) = (x_{21}, x_{22}, x_{31}, x_{32}, x_{12}, x_{51}, x_{11}).$$

Remark that we have deliberately arranged a and r to be the first two indeterminates in expecting to find some relation between them as few first d-pols in the d-char-set which would give us the Laws of Motion to be found. The hypothesis d-polset is now consisting of 7 pols below:

$$\begin{aligned} H_1 &= +1 * x_{12} - 1 * x_{22} * x_{31} - 1 * x_{21}, \\ H_2 &= +1 * x_{21}', \\ H_3 &= +1 * x_{22}', \\ H_4 &= +1 * x_{31}^2 + 1 * x_{32}^2 - 1 * x_{12}^2, \\ H_5 &= +1 * x_{31} * x_{32}' - 1 * x_{32} * x_{31}' - 1 * x_{51}, \\ H_6 &= +1 * x_{51}', \\ H_7 &= +1 * (x_{31}'')^2 + 1 * (x_{32}'')^2 - 1 * x_{11}^2. \end{aligned}$$

In a CPU-time of about 21 min., we find the final d-char-set to be consisting of 7 d-pols of which the first two d-pols are one in $x_{11} = a^2$ alone and the other in $x_{12} = r$ and $x_{11} = a^2$. The first one gives us thus a differential equation observed by the acceleration. This equation and the second one between a and r are both too complicate to be of any interest. However, during the process there appears a d-pol in the 4-th d-polset given by:

$$B = +4 * x_{12}' * x_{11} + 1 * x_{12} * x_{11}'.$$

By our general principle of MTD $B = 0$ should be a consequence of the original d-polset, i.e., a consequence of Kepler's Laws. The equation $B = 0$ is however nothing else but the Newton's inverse square law $r^2 * a = \text{const.}$. We have thus discovered in an automatic manner the Newton's Law (N_1) from the Kepler's Laws by means of our general Principle. Moreover, the d-pol

$$H_8 = H_5 + H_6 = +1 * x_{31} * x_{32}'' - 1 * x_{32} * x_{31}''$$

has its d-remainder already 0 w.r.t. the first d-bas-set BS_1 consisting of the successive d-pols H_2, H_3, H_1, H_4, H_5. Hence we have also automatically discovered during the procedure the theorem $H_8 = 0$, i.e., Newton's Law (N_2).

References

[1] Courant, R., Differential and integral calculus, vol 2 (1936).

[2] Gabriel, J.R., SARA - A small autometed reasoning assistant. Preprint, Argonne National Laboratory (1986).

[3] Ritt,J.F., Differential equations from the algebraic standpoint, Amer. Math. Spc., (1950).

[4] Ritt,J.F., Differential algebra, Amer. Math. Soc., (1950).

[5] Wu Wen-tsun, Basic principles of mechanical theorem-proving in elementary geometries, J.Sys.Sci. & Math. Scis., 4 (1984) 207-235. Re-published in J. of Automated Reasoning, 2 (1986) 221-252.

[6] Wu Wen-tsun, A constructive theory of differential algebraic geometry. Preprint, to be published in Proc. DD6– Symposium in 1985 at Shanghai.

A MECHANIZATION METHOD OF GEOMETRY AND ITS APPLICATIONS

——II. CURVE PAIRS OF BERTRAND TYPE

WU WENJUN (WU WEN-TSUN 吴文俊)

(*Institute of Systems Science, Academia Sinica, Beijing*)

Received November 25, 1985.

Let K be a differential field (abbr. d-field) of characteristic 0 and DPS be a set of differential polynomials (abbr. d-pols) in indeterminates $X1, X2, \cdots, Xn$ with coefficients in K. Let G be any other d-pol in the same indeterminates Xi. Then we shall denote by Zero (DPS/G) the collection of all zeros of DPS (or solutions of equations $DPS = 0$) in any extension d-field of K which are, however, net zeros of G (or solutions of $G = 0$). According to theory of Ritt, one determines from DPS a set of d-pols CHS to be called the characteristic set (abbr. char-set) of DPS. Any such char-set (not unique) will enjoy the following properties: (i) CHS is an ascending set (abbr. asc-set) in the sense of Ritt. (ii) Each zero of DPS is also a zero of CHS. (iii) Let the initials of the d-pols $C1$, $C2, \cdots, Cs$ in CHS, say s in number, be $I1, I2, \cdots Is$, then each zero of CHS, which is not a zero of the product $J = I1 * I2 * \cdots * Is$, is also a zero of DPS.

More precisely, we have the following Ritt Well-Ordering Principle:

$$\text{Zero}(DPS) = \text{Zero}(CHS/J) + SUMi\,\text{Zero}(DPSi), \tag{1}$$

in which each $DPSi$ is the set of d-pols DPS with Ii adjoined to it. Furthermore, we have also the following Zero Decomposition Theorem:

$$\text{Zero}(DPS/G) = SUMj\ \text{Zero}(ASCj/Rj), \tag{2}$$

in which each $ASCj$ is some irreducible asc-set and Rj some d-pol with non-zero remainder with respect to $ASCj$. The $ASCj$ and Rj can all be determined in a mechanical manner from the given DPS and the d-pol G. The determination of CHS from DPS is also a mechanical one and we have accordingly programmed on some small computer. In fact, it is on the formulae (1) and (2) that relies our method of mechanical theorem proving and discovering of differential geometries. For more details, See [1]. We remark that as usual all theorems are to be understood in some generic sense.

In this note, we shall consider curve pairs of Bertrand type in metric or affine space as an illustration of how our method can be applied to discovering theorems connected with such curve pairs.

Consider thus a pair of curves C and C' in one-to-one correspondence with arc lengths s, s' as parameters in the ordinary metric space. Let us attach moving frames (P, $E1$, $E2$, $E3$) and (P', $E'1$, $E'2$, $E'3$) to C and C' at corresponding points P and P'. The curvature and torsion of C, C' will be denoted by K, T, and K', T' respectively. Let

$$P' = P + SUMjAj \cdot Ej,$$
$$E'i = SUMjUij \cdot Ej, i = 1, 2, 3.$$

Introduce now indeterminates $X1$, $X2$, etc. and change the notations as follows:

$$ds'/ds\ A1\ A2\ A3\ U11\ U12\ U13\ U21\ U22\ U23\ U31\ U32\ U33\ K\ T\ K'\ T'$$
$$= X5\ X6\ X7\ X8\ X9\ X10\ X11\ X12\ X13\ X14\ X15\ X16\ X17\ X25\ X30\ X35\ X40.$$

From the Frenet formulae of C, C' we easily deduce the following set of d-pols in which $d1Xi$, $d2Xi$, $\cdots$ mean successive derivatives of indeterminate Xi with respect to s:

$$P1 = +1*X5*X9 - 1 - 1*d1X6 + 1*X25*X7,$$
$$P2 = +1*X5*X10 - 1*d1X7 - 1*X25*X6 + 1*X30*X8,$$
$$P3 = +1*X5*X11 - 1*d1X8 - 1*X30*X7,$$
$$P4 = +1*X35*X5*X12 - 1*d1X9 + 1*X25*X10,$$
$$P5 = +1*X35*X5*X13 - 1*d1X10 - 1*X25*X9 + 1*X30*X11,$$
$$P6 = +1*X35*X5*X14 - 1*d1X11 - 1*X30*X10,$$
$$P7 = +1*X35*X5*X9 - 1*X40*X5*X15 + 1*d1X12 - 1*X25*X13,$$
$$P8 = +1*X35*X5*X10 - 1*X40*X5*X16 + 1*d1X13 + 1*X25*X12 - 1*X30*X14,$$
$$P9 = +1*X35*X5*X11 - 1*X40*X5*X17 + 1*d1X14 + 1*X30*X13,$$
$$P10 = -1*X40*X5*X12 - 1*d1X15 + 1*X25*X16,$$
$$P11 = -1*X40*X5*X13 - 1*d1X16 - 1*X25*X15 + 1*X30*X17,$$
$$P12 = -1*X40*X5*X14 - 1*d1X17 - 1*X30*X16,$$
$$P13 = +1*X9\hat{}2 + 1*X10\hat{}2 + 1*X11\hat{}2 - 1,$$
$$P14 = +1*X12\hat{}2 + 1*X13\hat{}2 + 1*X14\hat{}2 - 1,$$
$$P15 = +1*X15\hat{}2 + 1*X16\hat{}2 + 1*X17\hat{}2 - 1,$$
$$P16 = +1*X9*X12 + 1*X10*X13 + 1*X11*X14,$$
$$P17 = +1*X9*X15 + 1*X10*X16 + 1*X11*X17,$$
$$P18 = +1*X12*X15 + 1*X13*X16 + 1*X14*X17.$$

Consider now the cases (ij) for which the line of Ei coincides with that of $E'j$ at corresponding points. For example, the case (22) is the classical one of Bertrand curve pairs for which the principal normals of C, C' at corresponding points coincide. For this case we should add to (DPS) some further d-pols $Q1, \cdots Q8$ as shown below to form an enlargedset (DPS22) such that the 26 equations DPS22 = 0 will constitute the hypothesis set for such curve pairs:

$$Q1 = +1*X6,$$
$$Q2 = +1*X8,$$
$$Q3 = +1*X12,$$
$$Q4 = +1*X14,$$

$$Q5 = +1*X10,$$
$$Q6 = +1*X16,$$
$$Q7 = +1*X9 - 1*X17,$$
$$Q8 = +1*X11 + 1*X15.$$

Suppose that we are interested in finding the yet unknown relations between the curvature K and torsion T of C. For this purpose let us change the notations further in replacing $X25$ by $X1$ and $X30$ by $X2$ and denote the set of d-pols thus got from $(DPS22)$ by $(KT22)$. The char-set of $(KT22)$ is readily found to consist of 16 d-pols, of which the first one is

$$C1 = +1*d1X1*d2X2 - 1*d1X2*d2X1.$$

We have thus discovered the theorem $C1 = 0$ which is equivalent to the classical theorem of Bertrand saying that the curvature $X1 = K$ and the torsion $X2 = T$ of C are connected by a linear relation. Moreover, during the procedure we encounter various d-pols whose vanishing shows that the distance between corresponding points and the angle between the corresponding tangents are both constants. These classical theorems are thus rediscovered in an automatic manner, too.

The above example can be extended in various manners as shown below:

(A) Instead of relation between K and T for the case (22) we may also ask e. g. relations between T, T' of the curves C, C'. For this purpose we may first change $X30$, $X40$ in (DPS) to $X2$, $X4$. It turns out that the first d-pol of the char-set of the corresponding set of d-pols is

$$C1 = +1*X2*d1X4 + 1*X4*d1X2.$$

We rediscover thus automatically the theorem due to Schell that the product of torsions at corresponding points of a Bertrand curve pair is a constant. We may try also other pairs of geometrical entities to see whether they are connected by any interesting geometrical relations.

(B) We may also treat the cases (23), (32), and (33) in the same manner. These are the cases for which at corresponding points of C and C' we have respectively:

principal normal of C = binormal of C',

binormal of C = binormal of C'.

We find the first d-pol $C1$ of the respective char-set with $X_1 = K$, $X2 = T$, $X4 = T'$ to be as follows:

$$(KT23): C1 = -1*X1\hat{}2*d1X1 + 1*X2\hat{}2*d1X1 - 2*X1*X2*d1X2.$$

$$(TT'23): C1 = +1*X2\hat{}2*X4\hat{}2*d1X4\hat{}2 + 2*X2*X4\hat{}3*d1X2*d1X4$$
$$+ 1*X4\hat{}4*d1X2\hat{}2 - 4*X2\hat{}4*d1X4\hat{}2.$$

$(KT32)$ or $(K'T'23)$: $C1$

$$= +1*X1\hat{}2*d2X2\hat{}2*X2 - 2*X1*d1X2*d2X2*d1X1*X2$$
$$+ 1*d1X2\hat{}2*d1X1\hat{}2*X2 - 2*X1\hat{}2*d1X2\hat{}2*d2X2$$
$$+ 2*X1*d1X2\hat{}3*d1X1 + 4*X1\hat{}4*d1X2\hat{}2*X2.$$

$(KT33): C1 = +1*X2.$

The formulae $(KT23)$ and $(TT'23)$ show that for a Bertrand curve pair of type (23) we should have $K\wedge 2 + T\wedge 2 = \text{const}*K$, a theorem due to Mannheim, and also $T*T'\wedge 2 = \text{const}*(T\pm T')$. On the other hand, the formula $(KT33)$ shows that only planar curves can form Bertrand pairs of type (33). All these theorems are discovered in an automatic manner.

(C) We may also consider curve pairs C, C' of Bertrand type in an affine space. In fact, let ds be the affine arc element of C, then we may attach Fre ıet-Darboux frames $(P, E1, E2, E3)$ to C such that

$$dP/ds = E1,$$
$$dE1/ds = E2,$$
$$dE2/ds = E3,$$
$$dE3/ds = -T.E1 - K.E2,$$

in which K, T are the affine curvature and affine winding of C. Similarly for C'. If the affine principal normals of C coincide with those of C' at corresponding points of C, C', then treating the pairs as before in the case (23), we rediscover various theorems due to Ogiwara, cf. [2]. For the relation between K and T of such a pair we get, however, a d-pol of 30 terms involving $d3T$ and $d2K$ both to the power 2 which seems to be too complicated to have any geometrical interest. Other cases can be treated in the same manner.

(D) We may consider also curve pairs connected by some relations of geometrical interest, e. g. with tangents, principal normals, or binormals parallel to each other at corresponding points, etc. We may also consider curve pairs in a projective space, a conformal space, etc., with certain significant lines of these curves at corresponding points connected by certain geometrical relations, etc. Clearly our method applies equally well to dealing with all these cases to discover possible new theorems of geometrical interest whatsoever.

References

[1] Wu Wen-tsun, A constructive theory of differential algebaric geometry based on works of J. F. Ritt, with particular applications to mechanical theorem-proving of differential geometries, Preprint for Symp. DD6 at Shanghai, 1985.

[2] Ogiwara, S., Über affine Bertrand'sche Kurven, *Jap. J. Math.*, **4**(1927), 93—99.

Vol. 1 No. 1 SYSTEMS SCIENCE AND MATHEMATICAL SCIENCES Aug., 1988

A MECHANIZATION METHOD OF GEOMETRY AND ITS APPLICATIONS

III. MECHANICAL PROVING OF POLYNOMIAL INEQUALITIES AND EQUATION-SOLVING*

Wu Wenjun (Wu Wen-tsün 吴文俊)

(*Institute of Systems Science, Academia Sinica, Beijing, China*)

Abstract

This is the third paper of the series entitled "A mechanization method of geometry and its applications", cf. [WU3—5]. In the present paper it is shown how inequalitie can be proved by means of the author's Zero Decomposition Theorem of equations solving. Numerous examples are given which deal with definiteness of polynomials, inequalities between symmetric polynomials, trigonometrical inequalities, and geometric inequalities.

1. Some Generalities

Let *PS* be a polset in variables $X1, X2, \cdots, Xn$ on the basic coefficient field K of characteristic 0. Let us form the charset *CS* of *PS* with initials Ii of pols in *CS*. During the procedure we may remove certain factors Fj for the sake of lessening the computational work. Let us denote by J the product of all such initials and removed factors. Then we have the following formulas for the sets of zeros ($-\!\!<$ stands for "is contained in"):

$$\text{Zero } (CS/J) -\!\!< \text{Zero } (PS) -\!\!< \text{Zero } (CS), \tag{I}$$

$$\text{Zero } (PS) = \text{Zero } (CS/J) + \text{SUM}\, k \text{ Zero } (PSk), \tag{II}$$

in which each PSk is the polset *PS* enlarged by djoining to it either an initial Ii or a removed factor Fj. By treating each of PSk in the same manner and proceeding onwards, we shall finally arrive at the following ZERO DECOMPOSITION FORMUIA (in the weak form):

$$\text{Zero } (PS) = \text{SUM}\, k \text{ Zero } (ASCk/Jk), \tag{III}$$

in which each $ASCk$ is an ascending set with Jk the product of all the initials in $ASCk$ and the zeros will be understood to be in a definite extension field of K and in a certain pre-determined open domain O of the $(X1, \cdots, Xn)$-space S. Formulas (I)—(III) offer us a general method of solving an arbitrary system of polynomial ẻquations. Moreover, these formulas and their extensions to the differential case are also on the basis of our general mechanization method of geometry. Some applications of this method have been described in the previous onesones of this series, cf. [WU3—5]. As a further application we shall take K to be the field and show in the present paper how to apply the method to the mechanical proving and hence also mechanical discovering, of polynomial inequalities. In fact, for the proving of inequalities a general method is already furnished by elemenfary calculus. We have only to proceed a little further in solving the respective equations by applying formulas (I)—(III).

* Received December 21, 1987.

This work is supported by NSFC Grant JI S5312.

In more betails it may be described as follows.

In the simplest case a polynomial inequality may be put into the form for some constant a,

$$G(X) >= a (\text{or} \quad G(X) > a) \tag{1.1}$$

in which G is a pol on real (actually rational or even integral) coefficients and X stands for $(X1, \cdots, Xn)$ restricted to the given open domain O. (1. 1) is equivalent to saying that the greatest lower bound glb of $G(X)$ for X in O is $>= a$ (or $> a$). In problems like linear programming O will be the interior of a convex polyhedron and the glb will be attained as the least value of $G(X)$ at some vertex on the boundary of this polyhedron. On the other hand, for a large number of inequalities (1.1) which one encounters in mathematics it could be verified by direct computations, geometrical considerations, or any other means, that the following condition would hold true for R = closure of O:

(C) The values of pol $G(X)$ for X in R will attain its least value in the interior of R.

We shall now restrict ourselves only to inequalities verifying (C). For such inequalities the problem is now reduced to proving that

$$\min G(X) >= a \ (\text{or} \quad \min G(X) > a) \tag{1.2}$$

for (X) in O. Now the points (X) in O which render G to be extremal, i.e. either a local minimum or a local maximum, should satisfy the following equations as NECESSARY conditions:

$$Di\ G = 0, \ i = 1, \ \cdots, \ n, \tag{1.3}$$

in which Di means derivative w. r. t. Xi. The proving of inequalities (1.1) is thus reduced as a first step, under condition (C), to solving equations(1.3), which can be done by means of (I)—(III). Among the solutions found we shall choose those whic will render the value of G to be the smallest, say $G0$. We have then to verify that $G0 >= a$ (or $> a$) and to test whether this is really a local minimum. The last test can be done by forming the second variation, viz.

$$V = G(Z+E) - G(Z), \tag{1.4}$$

in which $Z = (Z1, \cdots, Zn)$ is the zero of G in question and $E = (E1, \cdots, En)$ is some small variation of Z. It is then enough for the proof of inequalities (1.1) if we can prove, as a SUFFICIENT condition, that the form V is positive definite for Ei sufficiently small. The positive definiteness of such a form V is again a problem of the same type as above and may be reduced to equations-solving. Moreover, let Q be the quadratic part of the second variation V, viz. the quadratic form

$$Q = \text{SUM}ij\ DijG * Ei * Ej, \tag{1.5}$$

in which Dij means the second derivative w.r.t. Xi, Xj with values taken at the corresponding point. Then it will be sufficient to show the definiteness of Q which is easily done.

More generally we have to prove an inequality of the form (1.1) under some restricted conditions (Hi again real pols)

$$Hi(X) = 0, \ i = 1, \ 2, \ \cdots, \ m. \tag{1.6}$$

We shall restrict ourselves again only for the case with condition (C) verified, with however R to be understood as the closure of the intersection of O and the algebraic variety defined by (1.6). In this case, again as in the elementary calculus we may form the IAGRANGIAN pol

$$L=G+\mathrm{SUM}i\, Mi*Hi, \tag{1.7}$$

in which Mi are the lagrangian multipliers. Then the points rendering G minimum with (1.6) satisfied can usually be sought among the solutions of the system of equations (1.6) and (1.8) below:

$$Di\ L=0. \tag{1.8}$$

With zeros found by means of (I)—(III) we may then proceed as before. In general equations (1.6) may however be dependent. In such cases the method of lagrangian multipliers is usually non-applicable and some preliminary treatment, e. g. formation of char-sets, may be necessary. Let us suppose that this has already been done in case of necessity. Our general procedure of proving inequalities under condition (C) can thus be described as follows.

Step 1. Denote the system of pols Hi in (1.6) by PS and decompose Zero (PS) as in (III).

Step 2. For each ascending set $ASCk$ with successive pols Akj and corresponding initials Ikj let us form the lagrangian pol

$$Lk=G+\mathrm{SUM}j\ Mkj * Akj \tag{1.9}$$

with lagrangian multipliers Mkj.

Step 3. Form for Lk the equations

$$Di\ Lk=0, \quad i=1, \cdots, \quad n. \tag{1.10}$$

Solve for the system (1.6) and (1.10) together by our general method of equations-solving and determine all such zeros lying in the open domain O and rendering $Jk<>0$.

Step 4. Test whether each such zero in Step 3 is a real minimal point by forming the second variation or its quadratic part as in (1.4) or (1.5). Note that not only the zero Z, but also the variation $Z+E$, should be so chosen to verify equations (1.6). Suppose that this has been done. Choose then among the minimal values the smallest G' with corresponding zero $Z=(Z1, \cdots, Zn)$.

Step 5. Test the validity of the lagrangian multiplier method in verifying that the rank of the matrix $[Di\ Hj]$ at the corresponding zero Z is equal to m. We suppose that this is really the case.

Step 6. Determine the zero at which rank of the matrices $[D_i\ H_j]$ is less than m and verify whether they do not furnish the smallest value of G.

If the tests in Steps 4—6 are all done in the affirmative, then the procedure has succeeded and (1.1) is proved as required under the restricted conditions (1.6), so far (C) is assumed to be true.

The above procedure may be modified in various manners. We may, for example, either omit Step 1 or, instead of using the complete decomposition (III) in Step 1, use only (II) and proceed further with Zero (PSk) if it is necessary. We may also omit Step 4 or 5 in case the definiteness of the quadratic part Q can be ascertained by direct geometrical or other considerations.

The above method of proving inequalities is a mechanical one but not at all a complete one. In fact, even in such a simple case of a polynomial G there are no existant conditions, which are both necessary and sufficient, for G to be a local maximum or minimum at a certain point. We have therefore to satisfy ourselves with methods which are sufficiently efficient to prove non-trivial inequalities, leaving aside the question of completeness. We remark also that the procedure described above is just the same as in any textbook on elementary calculus. The only difference is perhaps the use of the method of polynomial equations solving, as embodied in the formulas (I)—(III), which has greatly enlarged the field of inequalities proving. Moreover, the method described, though far from being complete, can however be programmed and worked out on a computer and is already found to be quite efficient as may be seen from examples in the following sections. It furnishes us with a means of discovering "new" inequalities without knowing a priori their possible forms as seen from these examples too. These examples are somewhat of a typical character and hundreds of quite non-trivial inequalities can be in fact proved according to their pattern.

2. Definiteness of Polynomials

Let O be an open domain in the $(X1, \cdots Xn)$-space S. Then a real pol G in $X=(X1, \cdots, Xn)$ is said to be POSITIVE DEFINITE or SEMI-DEFINITE in O if for all X in O, we have $G(X)>0$ or $G(X)>=0$. Similarly for NEGATIVE DEFINITENESS or SEMI-DEFINITENESS. The determination of the defiitneeiss of such a real pol can be reduced to a problem of equations-solving which, quite often, can be achieved in the manner described in Sect. 1, if $G(X)$ attains its least value >0 or $>=0$ at points in the interior of O. The following is a concrete example which serves as a simple illustration of our general method.

Example 1. The Motzkin polynomial

$$G=1+X1^2 * X2^4+X1^4 * X2^2-3 * X1^2 * X2^2 \tag{2.1}$$

is semi-definite positive in the whole $(X1, X2)$-plane.

Proof. Let X move along a line through the origin to the infinity. The values of G will clearly become plus infinity if the line is different from the $x1$-or $x2$-axis on which G has the constant value $+1$. It follows that G will attain its least value in some finite part of the plane or that condition (C) is observed in this case. We may thus apply the procedure as described in Sect. 1.

Form thus the derivatives

$$D1G=2 * X1 * X2^2 * (X2^2+2 * X1^2-3).$$

$$D2G=2 * X1^2 * X2 * (2 * X1^2+X2^2-3).$$

The set Zero $(D1\,G,\ D2\,G)$ consists of 3 parts, viz.

$$\{(0,\ X2)\},\ \{(X1,\ 0)\},\ \{(1,\ 1),\ (1,\ -1),\ (-1,\ 1),\ (-1,\ -1)\}.$$

We have $G=1$ for $(0,\ X2)$ or $(X1,\ 0)$, but $G=0$ for $(1,\ 1)$, etc. Consider e. g. the point $(1,\ 1)$. Set

$$X1=1+E1, \quad X2=1+E2$$

with $E1$, $E2$ small. The corresponding quadratic part Q of the second variation is found to be

$$Q=4 * E1^2+4 * E2^2-4 * E1 * E2>=0$$

with $Q=0$ only for $E1=E2=0$. It follows that G takes its minimums at the point (1, 1), and also at (1, −1), etc. with value 0. Hence $G>=0$ as asserted.

It may happen that a real pol is definite or semidefinite positive in some smal domains but not so in a certain larger open domain O. For example, let a, b, c be the three sides of a triangle so that besides

$$a>0, \quad b>0, \quad c>0 \tag{2.2}$$

we have also

$$b+c-a>0, \quad c+a-b>0, \quad a+b-c>0. \tag{2.3}$$

Now the area A of the triangle is given by the formula

$$16 * A^2=-a^4-b^4-c^4+2 * b^2 * c^2+2 * c^2 * a^2+2 * a^2 * b^2$$
$$=(b+c-a) * (c+a-b) * (a+b-c) * (a+b+c).$$

The pol in the right side of the first line is thus a positive definite one in the open domain O defined by inequalities (2.2) and (2.3). It is clearly not so in the larger open domain defined by (2.2) alone. The following is another example.

Example 2. The pol

$$G=-a^2-b^2-c^2+2 * b * c+2 * c * a+2 * a * b \tag{2.4}$$

is definite positive in the open domain O defined by (2.2) and (2.3).

Proof. In view of (2.3) we can set

$$\begin{aligned} b+c-a&=2 * x^2,\\ c+a-b&=2 * y^2,\\ a+b-c&=2 * z^2 \end{aligned} \tag{2.5}$$

with $x>0$, $y>0$, $z>0$ so that

$$\begin{aligned} a&=y^2+z^2,\\ b&=z^2+x^2,\\ c&=x^2+y^2. \end{aligned} \tag{2.6}$$

By direct compution we find G of (2.4) is given by

$$G=4 * (y^2 * z^2+z^2 * x^2+x^2 * y^2)>0$$

as to be proved.

Remark that G of (2.4) is again non-definite positive in the larger domain defined by (2.2) alone.

The method of proof here in introducing (2.5) and (2.6) may be quite useful in proving inequalities involving sides of a triangle.

3. Inequalities Involving Symmetric Polynomials

There are numerous inequalities involving symmetric pols. We shall prove a few below to illustrate our general method.

Example 3. For $a, b, c>0$ we have

$$(b+c)*(c+a)*(a+b)>=8*a*b*c, \tag{3.1}$$

and the equality occurs only for $a=b=c$.

Proof. In order to take into account $a, b, c>0$, let us introduce new variables $x21$, $x22$, $x23$ according to Seidenberg in setting

$$x11=a>0,\ x12=b>0,\ x13=c>0, \tag{3.2}$$

$$x21^2*x11=1,\ x22^2*x12=1,\ x23^2*x13=1. \tag{3.3}$$

We have then to show that

$$\min G=8$$

under conditions (3.3) in which

$$G=(x12+x13)*(x13+x11)*(x11+x12)*x21^2*x22^2*x23^2. \tag{3.4}$$

Form thus the lagrangian pol

$$L=G+x101*(x21^2*x11-1)+x102*(x22^2*x12-1)$$
$$+x103*(x23^2*x13-1), \tag{3.5}$$

in which $x101$, $x102$, and $x103$ are the respective lagrangian multipliers. We shall denote also by $x10$ the extremal value of G to be found.

Denote by PS the polset consisting of the pols corresponding to (3.3), the 6 derivatives DiL with $i=11$, 12, 13, 21, 22, 23 and the pol $G-x10^2$. Then we have to find extremal values of G by determining the set Zero(PS). Now the char-set CS (in the weak sense) of PS is readily found to be consisting of 9 pols with the first three given by

$$C1=x10^2-8,$$
$$C2=3*x12-x11*(x10^2-5),$$
$$C3=-6*x13+x11*(x10^2-2).$$

During the procedure we have however removed the following factors

$$x11,\ x12,\ x13,\ x10,\ x11+x12,$$
$$x10^2+1,\ 4*x10^2+1,\ x10^4+1,$$

which are all necessarily non-zero. The only non-trivial initials of CS is $x11$ and is also non-zero. It follows that (II) applied to PS becomes simply

$$\text{Zero}(PS)=\text{Zero}(CS).$$

We find thus the only possible solutions of our problem:

$$x10^2=8,$$
$$x11=x12=x13>0,$$
$$x21=x22=x23=+\text{or}-\text{sqr}(x11).$$

To see whether these values furnish the true maximum or minimum, let us conside e. g. the point for which $x11$, $x12$, $x13$, $x21$, $x22$, $x23$ are all equal to 1. Take a nearby point by setting

$$x11=1+x11',\ x12=1+x12',\ x13=1+x13',$$

$$x21=1+x21',\ x22=1+x22',\ x23=1+x23'$$

with small x' such that $(x11, \cdots, x23)$ satisfy all the above equations too. It is readily verified that for the second variation the quadratic part of $[G(1+x')-G(1)]/2$ is given by

$$x11'^2+x12'^2+x13'^2-x12' * x13'-x13' * x11'-x11' * x12'$$

which is >0 for $x11'$, $x12'$, $x13'$ not all equal and $=0$ otherwise. In the latter case we have however always $G(1+x')=G(1)$. It follows that $x10>0$ with $x10^2=8$ is a true minimum and (3.1) is proved.

In the above proof we have introduced new variables $x21$, $x22$, $x23$ to take into account conditions (3.2). This would cause unnecessary complications in computations. In fact, we can avoid this in removing simply any such factors $x11$, $x12$, $x13>0$ during the procedure. Instead of treating Ex. 3 in this way let us consider another example below as an illustration.

Example 4. For a, b, $c>0$ we have

$$3 * (b+c) * (c+a) * (a+b)<=8 * (a^3+b^3+c^3). \tag{3.6}$$

Proof. Let us introduce $x11$, $x12$, $x13$ as in (3.2) and set also

$$x21=x12+x13,\ x22=x11+x13,\ x23=x11+x12, \tag{3.7}$$

$$x30=x11^3+x12^3+x13^3, \tag{3.8}$$

$$x21 * x22 * x23=x10 * x30. \tag{3.9}$$

The polynomial to be extremized is then

$$G=x10$$

under the above conditions (3.7)—(3.9). Denote the pols corresponding to equations in (3.7)—(3.9) by $P1$, $\cdots$, $P5$ and form the lagrangian pol

$$L=x10+x101 * P1+\cdots+x105 * P5.$$

We have to find the set Zero(PS) where PS consists of the 5 pols Pi and the 8 pols DiL with $i=30$, 23, 22, 21, 13, 12, 11 and 10. The char-set of PS is readily found to be consisting of 12 pols of which the first three are

$$C1=3 * x10-8,$$

$$C2=x12-x11,$$

$$C3=-18 * x13+15 * x11 * x10-22 * x11.$$

The factors removed during the procedure are $x101$, $x10$, $x21$, $x11$.

The non-trivial initial occuring in the final char-set is $x11$. It is readily seen that there is no necessity to proceed further to study Zero(PS') where PS' is PS enlarged by adjoining any one of the above factors or initials. Hence the only solutions to our problem are given by

$$x10=8/3,\ x11=x12=x13 \text{ or } a=b=c.$$

We omit the verification that they furnish actually maximums and (3.6) is thus proved.

Let us write any pol symmetric in n variables $V1, \cdots, Vn$ with a typical term

$$T=V1^{E1} * \cdots * Vn^{En}$$

as SYM n T. Between the various typical terms of fixed total degree there may be introduced a partial ordering by majoration, in notation: $T1 \ll T2$ or $T2 \gg T1$ if $T1$ is majorized by $T2$. For exact definition we refer to [HLP], p. 45. A theorem of Muirhead says that between two symmetric pols SYM n $T1$ and SYM n $T2$ in the same variables $Vi>0$ and of the same total degree there is some inequality connecting them if and only if $T1$, $T2$ are comparable in the above partial ordering. More precisely, for $T1 \ll T2$ we have

$$N2 * \text{SYM}\, n\ T1 <= N1 * \text{SYM}\, n\ T2,$$

in which $N1, N2$ are the number of terms in the two symmetric pols. In particular, for $n=3$ we have

$$V1^3 \gg V1^2 * V2 \gg V1 * V2 * V3.$$

By the Muirhead theorem we have therefore

$$2 * \text{SYM3}\ V1^3 >= \text{SYM3}\ V1^2 * V2,$$
$$\text{SYM3}\ V1^2 * V2 >= 6 * \text{SYM3}\ V1 * V2 * V3.$$

Examples 3 and 4 follow immediately from these two formulas and are thus only very special cases of the general Muirhead theorem. In view of this it is therefore of interest to consider the following example which is not covered by the Muirhead theorem.

Example 5. For a triangle with sides a, b, c and perimeter

$$a+b+c=2*s \tag{3.10}$$

we have the following inequality due to Santalo:

$$\text{sqr}(s-a)+\text{sqr}(s-b)+\text{sqr}(s-c) <= \text{sqr}(3*s). \tag{3.11}$$

Proof. Let us set, besides (3.2),

$$x10=s, \quad x15=\text{sqr}(s), \tag{3.12}$$

$$x21=\text{sqr}(s-a), \quad x22=\text{sqr}(s-b), \quad x23=\text{sqr}(s-c). \tag{3.13}$$

Then the pol to be extremized is $x22$ give by

$$x15 * x20 = x21+x22+x23. \tag{3.14}$$

Let Pi, $i=1, \cdots, 6$, be the pols in x corresponding to equations in (3.12)—(3.14) in taking account of (3.10). Form the lagrangian pol

$$L=x20+x101 * P1+\cdots+x106 * P6.$$

Let PS be the polset consisting of Pi and DjL for $j=10$, 11, 12, 13, 15, 20, 21, 22 and 23. The char-set is readily found to be consisting of 14 pols of which the first five are

$$\begin{aligned} &C1=3*x11-2*x10,\\ &C2=2*x12-2*x10+x11,\\ &C3=x13-2*x10+x11+x12,\\ &C4=x15^2-x10,\\ &C5=x20^2*x10-3*x15^2. \end{aligned} \tag{3.15}$$

The factors removed during the procedure and non-trivial initials in the char-set are

$$x\,101,\ x\,23,\ x\,20,\ x\,15,\ x\,11,\ x\,10,\ x\,11-x\,10.$$

All these are >0 except possibly $x101$ and $x11-x10$ which occur both as removed factors. The latter one causes collapse of the triangle and may be discarded. We have therefore

$$\text{Zero}(PS)=\text{Zero}(CS)+\text{Zero}(PS1), \tag{3.16}$$

in which $PS1$ is PS enlarged by adjoining the pol $x\,101$. It is readily seen that $PS1$ is contradictory so that (3.16) reduces to

$$\text{Zero}(PS)=\text{Zero}(CS). \tag{3.17}$$

From the expressions of Ci in CS as given by (3.15) we see that the only solutions are given by

$$x\,11=x\,12=x\,13=2*x\,10/3, \text{ etc.}$$

with the corresponding extremized value given by

$$x\,20=\text{sqr}(3).$$

It may be verified as usual that this value of $x\,20$ is a maximum. Now the open domain O of the problem is defined by (3.2) and

$$x\,12+x\,13>x\,11,\ x\,11+x\,13>x\,12,\ x\,11+x\,12>x\,13.$$

For a point on the boundary of O with $x\,12+x\,13=x\,11$, we have $s=b+c, s-a=0$, $s-b=c$, $s-c=b$ so that

$$x\,20=(\text{sqr}(b)+\text{sqr}(c))/\text{sqr}(b+c)<=\text{sqr}(2)<\text{sqr}(3).$$

Similarly for a point on the boundary of O with $x\,11=0$ we have $a=0$, $s=b=c$, $s-a=s$, $s-b=0$, $s-c=0$, so that

$$x\,20=1<\text{sqr}(3).$$

Condition (C) is thus seen to be observed and the inequality (3.11) is proved.

4. Trigonometrical Inequalities

Inequalities involving trigonometrical functions occur quite often in examinations or problems-solving of elementary mathematics, as may be seen from the columns on Problems and Solutions of Amer. Math. Monthly. We shall show that our method works also for such kind of problems, in spite of the fact transcendental functions are involved in the inequalities. The point is to replace such transcendental functions by their interrelated algebraic relations which, but not the functions themselves, are the only factors playing the essential role in the inequalities. We remark that this principle was already pointed out in the first paper on mechanical theorem proving of the author ([WU 1]) and has been applied to various kinds of problems. The following example is now one of the simplest to apply the principle this time to inequalities.

Example 6. For a triangle ABC with angles

$$A+B+C=pi, \tag{4.1}$$

we have

$$\sin A+\sin B+\sin C<=3*\text{sqr}(3)/2, \tag{4.2}$$

$$\sin A * \sin B * \sin C <= 3 * \mathrm{sqr}(3)/8, \tag{4.3}$$

$$\cos A + \cos B + \cos C <= 3/2, \tag{4.4}$$

$$\cos A * \cos B * \cos C <= 1/8. \tag{4.5}$$

Moreover, in each of the inequalities the equality occurs only if the triangle is an equilateral one.

Remark. Let us form as in elementary calculus, for e.g. (4.2), the lagrangian pol

$$L = \sin A + \sin B + \sin C + M * (A + B + C - pi)$$

with M the lagrangian multiplier. From

$$DL/DA = DL/DB = DL/DC = 0$$

we find

$$\cos A = \cos B = \cos C$$

and (4.2) is readily proved. This is the same for (4.3)—(4.5) and many other trigonometrical inequalities of similar type. However, such a method may lead in more general cases to transcendental equations which it is almost impossible to deal with. For this reason we prefer to apply our general method as indicated above in the beginning of this section.

Proof. Let us set

$$\sin A = x11,\ \sin B = x12,\ \sin C = x13, \tag{4.6}$$

$$\cos A = x21,\ \cos B = x22,\ \cos C = x23. \tag{4.7}$$

Then we have between the x's the following algebraic relations:

$$x11^2 + x21^2 = 1,\ x12^2 + x22^2 = 1,\ x13^2 + x23^2 = 1. \tag{4.8}$$

Moreover, from (4.1) we have $\sin A = \sin(B+C)$, $\cos A = -\cos(B+C)$, etc. Hence we have also

$$x11 = x12 * x23 + x13 * x22,\ x21 = -x22 * x23 + x13 * x12, \tag{4.9}$$

$$x12 = x13 * x21 + x11 * x23,\ x22 = -x23 * x21 + x11 * x13, \tag{4.10}$$

$$x13 = x11 * x22 + x12 * x21,\ x23 = -x21 * x22 + x12 * x11. \tag{4.11}$$

For the case (4.2) the pol to be extremized under the restricted conditions (4.6)—(4.11) is $x11 + x12 + x13$. The open domain O in which the extremal points are to be found is defined by

$$0 < x11 < 1,\ 0 < x12 < 1,\ 0 < x13 < 1, \tag{4.12}$$

$$0 < x21 < 1,\ 0 < x22 < 1,\ 0 < x23 < 1. \tag{4.13}$$

In the present case the lagrangian multiplier method is inapplicable since equations (4.8)—(4.11) are not independent. We have first to find an independent set of conditions equivalent to (4.8)—(4.11) by determining the char-set of the polset corresponding to these equations. This char-set is readily found to be consisting of 4 pols of which the first one is given by

$$\begin{aligned} P1 = {} & x13^4 + x12^4 + x11^4 + 4 * x11^2 * x12^2 * x13^2 \\ & -2 * x12^2 * x13^2 - 2 * x11^2 * x13^2 - 2 * x11^2 * x12^2. \end{aligned} \tag{4.14}$$

We remark that $P1=0$ corresponds to the equality of two different expressions for the area of the triangle. We remark also that the derivation of $P1$ follows our general method of determining unknown relations, in the present time between $\sin A$, $\sin B$ and $\sin C$ of a triangle, as described in [WU 4, 5]. It follows that we may replace the polset corresponding to (4.8)—(4.11) by the pol $P1$ alone with the corresponding open domain O defined by (4.12).

We are thus in a position to form the lagrangian pol

$$L = x11 + x12 + x13 + x101 * P1.$$

Let PS be the polset consisting of pol $P1$, the derivatives DiL with $i=13$, 12, 11 and also the pol

$$x11 + x12 + x13 - x10$$

with $x10$ as the extremal value to be found. By our general method we find that the char-set of PS is a contradictory one if we remove during the procedure the factors

$$x12,\ x11,\ x13 - x12,\ x12 - x11.$$

As $x12>0$ and $x11>0$ we have by our general formulas

$$\text{Zero}(PS) = \text{Zero}(PS1) + \text{Zero}(PS2), \tag{4.15}$$

in which $PS1$ and $PS2$ are the polsets PS enlarged by adjoining to it $x13-x12$ and $x12-x11$ respectively.

Consider first Zero $(PS1)$. The char-set of $PS1$ is readily found with first pol

$$C1 = (4 * x10^2 - 27) * (x10^2 - 4)^2.$$

The factors removed during the procedure are $x11$, $x10$ and $4 * x10^2 + 9$ which are all >0. Hence we may separate Zero $(PS1)$ into two parts with $PS1$ enlarged by adjoining $4 * x10^2 - 27$ and $x10^2 - 4$ respectively. For the first part the char-set is given by

$$\begin{aligned} C1 &= 4 * x10^2 - 27, \\ C2 &= -3 * x11 + x10, \\ C3 &= 2 * x12 + x11 - x10, \\ C4 &= x13 - x12, \text{ etc.} \end{aligned}$$

No factors have been removed and no non-trivial initials appear. The only zeros for this part are thus given by

$$x10 = 3 * \text{sqr}(3)/2,\ x11 = x12 = x13 = x10/3. \tag{4.16}$$

For the second part we find that the char-set is contradictory with $x11>0$ as the only factor removed. Hence it contributes no new zeros and the set Zero $(PS1)$ is thus solely given by (4.16).

Consider next the set Zero $(PS2)$. The first pol of the char-set is found to be $-(4 * x10^2 - 27) * (x10^2 - 4)^3$ with the following factors removed during the procedure:

$$x10,\ 2 * x11 - x10,\ x10^2 - 54,\ 5 * x10^2 + 216.$$

The set Zero $(PS2)$ is thus formed by 4 parts: the zero-sets of $PS2$ enlarged by

$4*x10^2-27$, $x10^2-4$, $2*x11-x10$ and $x10^2-54$ respectively. For the first part the char-set is given by

$$C1=4*x10^2-27,$$
$$C2=3*x11-x10,$$
$$C3=x12-x11,$$
$$C4=-x13-x12-x11+x10, \text{ etc.}$$

The zeros are thus again given by (4.16). On the other hand the other 3 parts are all contradictory and furnish no zeros at all.

In summary we see therefore from (4.15) that Zero(PS) is composed of the set given by (4.16).

It remains to verify that the value of $x10$ taken on the boundary of O is $<=$ the one given in (4.16), which is easily done. It remains also to verify that $x10$ in (4.15) is really a maximum, which is also easily done by direct computation. This completes the proof of (4.2).

We have carried out the proofs of (4.3)—(4.5) in the same manner. We remark only that in the case of (4.4) or (4.5) we have set, in order to simplify the computations, instead of (4.6),

$$\sin A=x31, \ \sin B=x32, \ \sin C=x33. \qquad (4.6)'$$

With relations as in (4.8)—(4.11) the char-aet of corresponding pols is then given by

$$P1'=x23^2+x22^2+x21^2+2*x21*x22*x23-1, \text{ etc.}$$

Remark that $P1'=0$ gives the relation between $\cos A$, $\cos B$ and $\cos C$ under condition (4.1). The remaining proofs are similar to the one above for (4.2) and are a little involved. We can however apply a much simpler method of proof as described below.

Example 7. An alternative method of proving (4.2)—(4.5).

We remark that both $\sin A$ and $\cos A$ can be expressed rationally in terms of $\tan A/2$, and similarly for the others. Let us set therefore

$$\tan A/2=x11, \ \tan B/2=x12, \ \tan C/2=x13, \qquad (4.17)$$
$$\sin A=x21, \quad \sin B=x22, \quad \sin C=x23, \qquad (4.18)$$
$$\cos A=x31, \quad \cos B=x32, \quad \cos C=x33. \qquad (4.19)$$

We have then

$$x33*(1+x13^2)=2*x13, \ x23*(1+x13^2)=1-x13^2, \qquad (4.20)$$
$$x32*(1+x12^2)=2*x12, \ x22*(1+x12^2)=1-x12^2, \qquad (4.21)$$
$$x31*(1+x11^2)=2*x11, \ x21*(1+x11^2)=1-x11^2, \qquad (4.22)$$
$$x11*x12+x11*x13+x12*x13=1. \qquad (4.23)$$

Remark that (4.23) is the relation between $\tan A/2$, $\tan B/2$, $\tan C/2$ because of condition (4.1).

Let us consider e. g. the case (4.2). Let $P1, \cdots, P4$ be the pols corresponding to the equations in (4.23) and those in $x21$, $x22$, $x23$ of (4.20)—(4.22). Form now the lagrangian pol

$$x21+x22+x23+x101*P1+\cdots+x104*P4$$

and its derivatives w. r. t. $x11$, $x12$, $x13$, $x21$, $x22$, $x23$. Consider now the pol-set PS consisting of these derivatives, the pols Pi, and the pol

$$x21+x22+x23-x10$$

with $x10$ the extremal value to be found. The char-set CS is readily found to be consisting of the following pols:

$$C1=4*x10^2-27,$$
$$C2=-9*x11+2*x10,$$
$$C3=-x12+x11,$$
$$C4=x13*x12+x13+x12*x11-1, \text{ etc.}$$

The factors removed during the procedure as well as the non-trivial initials of the char-set are

$$x13,\ x12+x11,\ x11^2+1,\ x12^2+1,\ x13^2+1$$

which are all>0. We thus find again Zero(PS)=Zero(CS) is composed of

$$x10=3*\mathrm{sqr}(3)/2,$$
$$x11=x12=x13=\mathrm{sqr}(3)/3,$$
$$x21=x22=x23=\mathrm{sqr}(3)/2.$$

Inequality (4.2) is thus again proved in a manner much simpler than the one given in Example 5. Similarly for (4.3)—(4.5).

The inequalities (4.2)—(4.5) are symmetric in the angles A, B, C and the final result is easy to guess. For a non-symmetric inequality of which the final answer is not easy to foresee let us consider the following

Example 8. For $x1$, $x2$, $x3>0$ we have for a triangle ABC

$$x1*\cos A+x2*\cos B+x3*\cos C$$
$$<=(x2*x3/x1+x3*x1/x2+x1*x2/x3)/2. \qquad (4.24)$$

Proof. Let PS be the polset consisting of pols Pi corresponding to equations in (4.20) — (4.23) not involving $x21$, $x22$, $x23$ and the derivatives of the lagrangian pol

$$L=x1*x31+x2*x32+x3*x33+x101*P1+\cdots+x104*P4$$

w. r. t. $x11$, $x12$, $x13$, $x31$, $x32$, $x33$ as well as the pol

$$x1*x31+x2*x32+x3*x33-x10$$

where $x10$ is the extremal value to be found. The char-set of PS is readily found with the first pol given by

$$C1=-2*x10*x3*x2*x1+x3^2*x2^2+x3^2*x1^2+x2^2*x1^2. \qquad (4.25)$$

The factors removed during the procedure and the non-trivial initials of the char-set are all>0. (4.25) gives thus the extremal value $x10$ and (4.24) follows now easily.

The inequalities in Examples 6—8 can all be proved in a quite simple manner as indicated in the Remark. The following example is, however, one which cannot be treated in this way while our general method will furnish equally well the required solution.

Example 9. For a triangle ABC we have

$$\cos A * \cos B+\cos A * \cos C+\cos B * \cos C<=3/4.$$

Proof. Let us set $\tan A/2=x11$, etc. as in (4.17)—(4.19) and set also

$$x31 * x32+x31 * x33+x32 * x33=x20. \tag{4.26}$$

with $x20$ the extremal value to be determined. We have now 5 pols Pi corresponding to the equations in (4.20) — (4.23) not involving $x21$, $x22$, $x23$ and (4.26). Form the lagrangian pol

$$x20+x101 * P1+\cdots+x105 * P5.$$

The char-set of the polset PS formed by the 5 pols Pi and the derivatives of L w. r. t. $x33$, $x32$, $x31$, $x20$, $x13$, $x12$, $x11$ is given by

$$C1=x11^2-2,$$
$$C2=x12-x11 * (x11^2-1),$$
$$C3=x13 * (x12+x11)+x11 * x12-1, \text{ etc.}$$

The factors removed during the procedure are

$$x13-x12,\ x13-x11,$$
$$x12+x11,\ x11^2+1,\ x12^2+1$$

which are all>0 except the first two. Now froms

$$C2=C2=C3=0$$

with $x11=\tan A/2>0$ we get successively

$$x11=+\mathrm{sqr}(2),$$
$$x12=x11 * (x11^2-1)=+\mathrm{sqr}(2),$$
$$x13=(1-x11 * x12)/(x11+x12)=-\mathrm{sqr}(2)/4<0.$$

The last equation shows that the char-set is a contradictory set in the open domain defined by $x11>0$, $x12>0$, $x13>0$, etc. It follows from our general formula (II) that

$$\mathrm{Zero}(PS)=\mathrm{Zero}(PS1)+\mathrm{Zero}(PS2),$$

in which the polsets $PS1$ and $PS2$ are both PS enlarged by adjoining to it the removed factors $x13-x12$ and $x13-x11$ respectively. We may now treat each PSi in the same manner as before and arrive finally at the conclusion as to be proved.

5. Geometrical Inequalities

We shall give in this section only a few examples of inequalities arising in geometry as mere illustrations of our general method.

Example 10. A triangle with given perimeter has a greatest area when it is an equilateral one.

Proof. Let the sides of the triangle be a, b, c with perimeter p and the area be A. Set

$$a=x11,\ b=x12,\ c=x13,\ p=x0,\ 4 * A=x20. \tag{5.1}$$

Then we have

$$x20^2 = -x11^4 - x12^4 - x13^4 + 2*x12^2*x13^2 + \\ +2*x13^2*x11^2 + 2*x11^2*x12^2. \tag{5.2}$$

The problem is to find the maximum of $x20$ under the restricted condition (5.2). We may proceed by our general method which is actually the same as in the ordinary elementary calculus. For a problem of the same type but not a trivial one let us consider the following.

Example 11. For a triangle of given perimeter in a hyperbolic plane the area is the greatest when it is an equilateral one.

Proof. This problem seemingly similar to Example 10, is clearly much less simple to settle. First, we have to find an expression of the area A of the triangle in question in terms of the three side-lengths, say a, b and c. Such an expression has already been found in an automatic manner by our general mechanization method of geometry, cf. [WU4, 5]. In fact, let

$$\cosh a = x21,\ \cosh b = x22,\ \cosh c = x23, \tag{5.3a}$$

$$\cos A - 1 = x20. \tag{5.3b}$$

Then we have

$$x20*(x21+1)*(x22+1)*(x23+1) \\ = -2*x21*x22*x23 + 1 - x21^2 - x22^2 - x23^2. \tag{5.4}$$

Next, because the perimeter $2*s$ of the triangle is given, $\cosh a$, $\cosh b$ and $\cosh c$ are no more independent. Therefore we have to determine the relations between them, which can also be done by our general mechanization method. To this end let us set as in preceding Examples 7—9

$$\tanh a/2 = x11,\ \tanh b/2 = x12,\ \tanh c/2 = x13, \tag{5.5}$$

$$\tanh s = x0. \tag{5.6}$$

Then from $\tanh(a+b+c)/2 = \tanh s$ we get

$$x11*x12*x13 + x11 + x12 + x13 \\ = (x12*x13 + x11*x13 + x11*x12 + 1)*x0. \tag{5.7}$$

For $\cosh a$, etc. we have also

$$x23*(1-x13^2) = 1 + x13^2, \tag{5.8}$$

$$x22*(1-x12^2) = 1 + x12^2, \tag{5.9}$$

$$x21*(1-x11^2) = 1 + x11^2. \tag{5.10}$$

The problem is now to determine the extremal value of area A, or rather $x20$, under the restricted conditions (5.4), (5.7) — (5.10) with corresponding pols $P1, \cdots, P5$. The lagrangian pol will be thus

$$L = x20 + x101*P1 + \cdots + x105*P5.$$

The char-set CS of the polset PS consisting of the pols Pi and the various derivatives of L, is found to be contradictory if we remove during the procedure the following factors:

$$x13 - x12,\ x13 - x11,\ x12 - x11, \tag{5.11}$$

$$x\,13-1,\ x\,12-1,\ x\,11-1,\ x\,0-1, \tag{5.12}$$

$$x\,13+1,\ x\,12+1,\ x\,11+1,\ x\,0+1, \tag{5.13}$$

$$x\,11 * x\,12-1,\ x\,20+2, \tag{5.14}$$

$$x\,11^2 * x\,0-2 * x\,11+x\,0, \tag{5.15}$$

$$x\,12^2 * x\,0-2 * x\,12+x\,0. \tag{5.16}$$

It is clear that $x\,20+2=\cos A+1>0$ and all the other factors in (5.12)—(5.14) are non-zero since $0<\tanh x<1$ for any $x>0$. Moreover, for the triangle we have $b+c>a$ so that $s>a$. Whence $\tanh s>\tanh a$ or $x\,11^2 * x\,0-2 * x\,11+x\,0>0$. Similarly the factor in (5.16) is also>0. It follows therefore from our general formula (II) that

$$\mathrm{Zero}(PS)=\mathrm{Zero}(PS\,1)+\mathrm{Zero}(PS\,2)+\mathrm{Zero}(PS\,3),$$

in which the polsets PSi are PS enlarged by adjoining to it the three factors in (5.11) respectively. We may treat each of the Zero(PSi) in turn and proceed in the same way as in the preceding examples to arrive at the final conclusion.

Example 12 (Pedoe Inequality, cf. [P]). Let ABC and $A'B'C'$ be two triangles in the same plane with sides a, b, c; a', b', c'; and areas A, A' respectively. Then we have always

$$a'^2 * (b^2+c^2-a^2)+b'^2 * (c^2+a^2-b^2)+c'^2 * (a^2+b^2-c^2)$$
$$>=16 * A' * A. \tag{5.17}$$

Moreover, the equality occurs only when the two triangles are similar.

Proof. Let us set

$$a=x\,11,\quad b=x\,12,\quad c=x\,13, \tag{5.18}$$

$$a'=x\,21,\ b'=x\,22,\ c'=x\,23. \tag{5.18$'$}$$

We are naturally restricted to the open domain defined by $x\,11>0$, etc. Set

$$4 * A'=x\,30. \tag{5.19}$$

Then we have

$$x\,30^2=-x\,21^4-x\,22^4-x\,23^4+2 * x\,21^2 * x\,22^2$$
$$+2 * x\,21^2 * x\,23^2+2 * x\,22^2 * x\,23^2. \tag{5.20}$$

Introduce also $x\,25$ by setting

$$x\,25 * x\,30=x\,21^2 * (x\,12^2+x\,13^2-x\,11^2)$$
$$+x\,22^2 * (x\,13^2+x\,11^2-x\,12^2)$$
$$+x\,23^2 * (x\,11^2+x\,12^2-x\,13^2). \tag{5.21}$$

Let us consider the triangle ABC as already given while $A'B'C'$ is a variable one. Then the problem reduces to the determination of the minimum value of $x\,25$ in terms of known values a, b, c under the restricted conditions (5.20) and (5.21). Let PS be the pol-set consisting of pols corresponding to (5.20) and (5.21) as well as the derivatives of the lagrangian pol

$$x\,25+x\,101 * P1+x\,102 * P2$$

w. r. t. $x\,30$, $x\,25$, $x\,23$, $x\,22$ and $x\,21$. Then we have to determine Zero(PS) for the extremal value of $x\,25$. Now the char-set of PS is readily found to be consisting of the pols below:

$$C1=-x22*x11+x21*x12,$$

$$C2=-x23^2*x11^2+x22^2*x11^2+x21^2*x13^2-x21^2*x12^2,$$

$$C3=-x25^2-x11^4-x12^4-x13^4$$
$$+2*x11^2*x12^2+2*x11^2*x13^2+2*x12^2*x13^2,$$

$$C4=x30*(x13^2-x12^2-x11^2)$$
$$-x25*(x23^2-x22^2-x21^2).$$

The factors removed during the procedure and the nono-trivials of the char-set are

$$x101,\ x25,\ x11,\ x21,\ x11*x22+x12*x21,$$
$$x11*x23+x13*x21 \text{ and } x11^2+x12^2-x13^2.$$

All these pols are non-zero except perhaps the last one which means that the given triangle is a right-angled one. Leaving this case aside we see from the expressions of the char-set that $x25$ will reach its extreme value $4*A'$ in case $x21$, $x22$, $x23$ are proportional to $x11$, $x12$, $x13$, or that the triangles ABC, $A'B'C'$ are similar to each other, as asserted. The case of ABC being right-angled can also be treated in the same manner by adding at the outset the restricted condition $x11^2+x12^2=x13^2$.

References

[HLP] Hardy, G. H., et al, Inequalities, Cambridge, 1934.

[P] Pedoe, D., An inequality for two triangles, *Proc. Camb. Phil. Soc.*, **38** (1942), 397—398.

[WU1] Wu, W. T., On the decision problem and the mechanization of theorem-proving in elementary geometry, *Scientia Sinica*, **21** (1978), 159—173; re-published in Automated Theorem Proving: after 25 Years, Ed. W. W. Bledsoe & D. W. Loveland, 1984, 235—242.

[WU2] Wu, W. T., On zeros of algebraic equations — an application of Ritt principle, *Kexue Tongbao*, **31** (1986), 1—5.

[WU3] Wu, W. T., A mechanization method of geometry, I Elementary geometry, *Chinese Quarterly J. of Math.*, 1 (1986), 1—14.

[WU4] Wu, W. T., A mechanization method of geometry and its applications, I. Distances, areas, and volumes, *J. Sys. Sci. & Math. Scis.*, **6** (1986), 204—216.

[WU5] Wu. W. T., A mechanization method of geometry and its applications, II. Distances, areas, and volumes in euclidean and non-euclidean geometries, *Kuxue Tongbao*, **32** (1987), 436—440.

[WU6] Wu, W. T., A zero structure theorem for polynomial equations-solving and its applications, MM-Preprints, Institute of Systems Science, 1987.

ON THE FOUNDATION OF ALGEBRAIC DIFFERENTIAL GEOMETRY

Wu Wen-tsun

(*Institute of Systems Science, Academia Sinica, Beijing* 100080, *China*)

1. Introduction

By algebraic differential geometry we shall mean one which is so related to the ordinary algebraic geometry just as what the metric, the affine, or the projective differential geometry is related to the metric, the affine, or the projective geometry. As in the ordinary case, the first step in laying down a foundation of algebraic differential geometry is to define the notion of algebraic differential variety and to prove an irreducible decomposition of such varieties. Such a foundation may be traced back to the works of Riquier, Janet, Cartan, Thomas, and particularly Ritt, cf. the references at the end of the paper. We remark that, while the exposition of Ritt et al was highly analytical in character, we have removed all traces of analytical reasonings to render the theory a purely algebraic one. Furthermore, while Ritt et al were aimed at a study of differential equations, we are also interested in its geometrical aspects as well as its applications, particularly for the mechanical or automatic theorem proving of differential geometries, cf. e.g. [WU3–8].

The main topic consists in the study of the structure of an algebraic differential variety defined as the zero set of a finite set differential polynomials. Various structure or decomposition formulas are given for such zero sets which correspond to the ordinary ones for ordinary polynomials and can be carried out by mere computations, cf.e.g.[WU1, 2]. Such decompositions can then be applied to differential geometries and other related subjects which render the proving of differential geometrical theorems to mere computations. The applications are however not limited to theorem proving as seen from [WU7] and the example given in the last section of the paper.

2. Ordering Tuples

Let m be a positive integer fixed throughout the present paper.

DEF. An ordered sequence of m non-negative integers

$$t = (I_1, I_2, \cdots, I_m)$$

is called an ORDERING m-TUPLE or simply a TUPLE. I_i is then called the i-th COORDINATE of t, to be denoted by $COOR_i(t) = I_i$. The sum of all these coordinates is called the ORDER of t, to be denoted by

$$Ord(t) = SUM_i COOR_i(t).$$

DEF. For any two tuples u and v, we say u is a MULTIPLE of v or v is a DIVISOR of

$$COOR_i(u) \geq COOR_i(v), \quad i = 1, 2, \cdots, m.$$

We write then $u >> v$ or $v << u$.

Notation. The totality of ordering m-tuples will be denoted by Tot. For any finite set of tuples $T- < Tot$, we shall set $Tot(T)$ = Totality of multiples of some t in T.

1

DEF. For any two tuples u and v, their PRODUCT $uv = vu$ is the tuple with

$$COOR_i(uv) = COOR_i(u) + COOR_i(v), \quad i = 1, 2, \cdots, m.$$

We introduce now an ordering among all the tuples according to the following

DEF. For any two tuples u and v we say that u is HIGHER THAN v or v is LOWER THAN u if either (1) or (2) below holds true:

(1) $Ord(u) > Ord(v)$.

(2) $Ord(u) = Ord(v)$ and there is some $k > 0$ and $\leq m$ such that

$$COOR_i(u) = COOR_i(v),\ i > k, \quad COOR_k(u) > COOR_k(v).$$

We write then: $u > v$ or $v > u$.

DEF. A finite set of tuples T is said to be PRIME if no t in T is a multiple of another t' in T.

DEF. For any finite set of tuples T, the MAXIM of T, to be denoted by $Max(T)$, is the tuple defined by

$$Max(T) = n - tuple(MAX_1(T), \cdots, MAX_m(T)), \quad \text{with}$$

$$MAX_i(T) = Max\{COOR_i(t)/t \quad \text{in} \quad T\}.$$

DEF. For any finite set of tuples T, the COMPLETION of T, to be denoted by $Comp(T)$, is the set of tuples defined by

$$Comp(T) = \{u/u << Max(T)u >> t \quad \text{for some} \quad t \quad \text{in} \quad T\}.$$

DEF. For any finite set of tuples T and any tuple $t << Max(T)$, the integer i ($1 \leq i \leq m$) is called a MULTIPLIER of t w.r.t T if

$$COOR_i(t) = MAX_i(T).$$

Otherwise it is called NON-MULTIPLIER of t w.r.t T. In that case we have

$$COOR_i(t) < MAX_i(T).$$

Notation. For any finite set of tuples T and any tuple t, we shall set

$$Mult(t/T) = \text{set of all multipliers of} \quad t \quad \text{w.r.t} \quad T,$$

$$Nult(t/T) = \text{set of all non} - \text{multipliers of} \quad t \quad \text{w.r.t} \quad T.$$

DEF. For $t << Max(T)$, the set of all multiples tu of t with $COOR_i(u) = 0$ for i in $Nult(t/T)$ is called the TOTAL MULTIPLE SET of t w.r.t. T, to be denoted by

$$TMU(t/T) = \{tu/COOR_i(u) = 0 \quad \text{for} \quad i \quad \text{in} \quad Nult(t/T)\}.$$

THEOREM. Let T be a finite set of tuples. For any tuple v there is a unique tuple $t << Max(T)$ such that $v = tu$ for some tuple u with

$$v- < TMU(t/T) \quad \text{or} \quad COOR_i(u) = 0 \quad \text{for} \quad i- < Nult(t/T).$$

Moreover, if v is in $Tot(T)$, then t is in $Comp(T)$.

Proof. t is determined as $COOR_i(t) = Min(COOR_i(v), MAX_i(T))$.

TUPLE–DECOMPOSITION THEOREM.

$$Tot(T) = SUM_t TMU(t/T), \qquad Tot = SUM'_t TMU(t/T),$$

in which SUM_t runs over t in $Comp(T)$, while SUM'_t runs over $t << Max(T)$. Moreover, the sets $TMU(t/T)$ in the sums are disjoint from each other.

Proof. This follows directly from the preceding theorem.

2

3. Differential Field and Derivatives

DEF. An ordinary field of characteristic 0 will be called an ELEMENTARY FIELD (abbr. e-field). An ordinary integral domain containing the ring of integers as subring will be called an ELEMENTARY RING(abbr. e-ring).

DEF. A DIFFERENTIAFIEL (abbr. d-field) DF resp. a DIFFERENTIAL RING (abbr. d-ring) DR in INDEPENDENTs $X_1, \cdots, X_m$ is an e-field resp. an e-ring with m further DIFFERENTIAL OPERATIONs D_i, $i = 1, 2, \cdots, m$ verifying the following relations:

$$\begin{aligned}
&D_i X_i = 1, \quad D_i X_j = 0 \quad \text{for} \quad j <> i,\\
&D_i(A1 + A2) = D_i A1 + D_i A2,\\
&D_i(A1 * A2) = D_i A1 * A2 + A1 * D_i A2,\\
&D_i(D_j A) = D_j(D_i A),
\end{aligned}$$

for $A, A1, A2$ in DF resp. DR.

Notation. $D_i D_j A = D_i(D_j A), D_0 A = A$, for A in DF on DR.

DEF. DERIVATIVEs of A in a d-field DF are elements in DF of the form

$$DERtA = D_m \cdots D_m \cdots D_1 \cdots D_1 A,$$

in which $t = (I_1, \cdots, I_m)$ is an ordering m-tuple, and each D_i occurs I_i times, $i = 1, 2, \cdots, m$.

DEF. A d-field DF' is called a d-SUB-FIELD of another d-field DF if DF' is a sub-field of DF in the ordinary sense and for any element A of DF', all derivatives of A are the same whether they are considered as elements of DF' or DF. The d-field DF is then called a d-EXTENSION-FIELD (abbr. d-ext-field) of DF.

DEF. The e-field resp. the e-ring consisting of same elements as a d-field DF resp. a d-ring DR with relations of differentiation neglected is said to ASSOCIATED to the d-field DF resp. the d-ring DR and will be denoted by $Elem(DF)$ resp. $Elem(DR)$.

With a d-field DF given let $Y_1, Y_2, \cdots, Y_n$ be in some d-ext-field of DF which will be called INDETERMINATEs and will be fixed throughout the whole paper.

DEF. For any derivative $DERuY_i$ with u a tuple we call $Ord(u)$ the ORDER of $DERuY_i$.

We now introduce among certain derivatives in two different types as follows.

DEF. For any tuples u and v we say that $DERuY_i$ is HIGHER THAN $DERvY_j$ or $DERvY_j$ is LOWER THAN $DERuY_i$ if the following holds true:

For type 1: Either $u > v$, or $u = v$, and $i > j$.

For type 2: Either $i > j$, or $i = j$, and $u > v$.

We write in either type 1 or type 2

$$DERuY_i > DERvY_j \quad \text{or} \quad DERvY_j < DERuY_i.$$

4. Differential Polynomials and Their Ordering

Throughout the whole paper we shall suppose fixed a d-field $d - BF$ which will be referred to as the d-BASIC FIELD.

DEF. An ordinary polynomial (abbr. pol) in certain indeterminate with coefficient is an e-field will be called a ELEMENTAR POL (abbr. e-pol).

DEF. A DIFFERENTIAL POLYNOMIAL(abbr. d-pol) is an ordinary pol is X_i ($i = 1, 2, \cdots, m$), $Y_j (j = 1, 2, \cdots, n)$ and their derivatives with coefficients in $d - BF$. The DERIVATIVEs $DER_i DP$ and $DERtDP$ of DP for any $i = 1, 2, \cdots, m$ or tuple t are then defined in the usual manner.

DEF. A X-POL is a d-pol in which no Y_j and their derivatives occur.

DEF. Given a d-pol DP let us consider it as an e-pol in all $X_i (i = 1, 2, \cdots, m)$, Y_j $(j = 1, 2, \cdots, n)$ and derivatives $DERtY_i$ appearing in DP as independent indeterminates, then this e-pol will be said to be ASSOCIATED to DP and will be denoted by $Elem(DP)$.

DEF. Let DP be d-pol which is not a X-pol. Then the highest derivative occuring in DP is called the LEADING DERIVATIVE or simply the LEAD of DP. The order of the lead is called the ORDER of DP and the subscript of Y in the lead of DP is called the CLASS of DP. Consider DP as an ordinary pol in this lead the highest degree is called the DEGREE of DP. The coefficient, as a d-pol, of the highest degree term in the lead of DP is called the INITIAL of DP and the form a partial derivative of D w.r.t. the lead of DP the SEPARANT of DP.

Notation. For a d-pol DP which is not a X-pol we write

$$Ld(DP) = \text{Lead of} DP, \quad Ord(DP) = \text{Order of} DP,$$

$$Cls(DP) = \text{Class of} DP, \quad Deg(DP) = \text{Degree of} DP,$$

$$Init(DP) = \text{Initial of} DP, Sep(DP) = \text{Separant of} DP.$$

Thus, for such a DP with $LD = Ld(DP)$, $d = Deg(DP)$, $I = Init(DP)$, we may write DP in the form

$$DP = I * LD \wedge d + \text{lower degree terms in} LD,$$

with coefficient of each term a d-pol, which, if not a constant or a X-pol, will have its lead lower than LD.

We now introduce a partial ordering among the d-pols in the following way.

DEF. Let DP, DQ be non-zero d-pols. Then we say that DQ is HIGHER THAN DQ or DQ is LOWER THAN DP and we write $DP > DQ$ or $DQ < DP$ if one of the following cases (1)–(2) takes place:

(1) DQ is a X-pol while DP is not.

(2) Both DP, DQ are not X-pols and either

$$Ld(DP) > Ld(DQ) \quad \text{or} \quad Ld(DP) = Ld(DQ) \quad \& \quad Deg(DP) > Deg(DQ).$$

DEF. If DP, DQ are non-zero d-pols for which neither one is higher than the other, then we say that DP, DQ are INCOMPARABLE in ORDER and we write in this case

$$DP <=> DQ.$$

DEF. A non-zero d-pol DQ is daid to be REDUCED w.r.t. a non-zero d-pol DP if DP is not a X-pol and no proper derivative of lead DL of DP occurs in DQ. Furthermore, either DL does not occur in DQ, or DL occurs in DQ with a degree $< Deg(DP)$.

The following proposition is clear from the very definitions:

PROP. Any sequence of d-pls steadily decreasing in order

$$DP1 > DP2 > \cdots$$

is necessarily finite.

Remark. As in the case of derivatives, the partial ordering of d-pols and others in later sections may be either of type 1 or type 2.

5. d-Polset, d-Zero, and Algebraic Differential Variety

DEF. A finite collection of non-zero e-pols resp. d-pols is called a e-POLSET resp. a d-POLSET.

DEF. For a d-polset DPS the e-polset EPS consisting of epols associated to d-pols in DPS is said to be ASSOCIATED to DPS and we write then $EPS = Elem(DPS)$.

DEF. For any e-polset EPS we say that two e-pols EF_1 and EF_2 are e-CONGRUENT w.r.t. EPS and we write then $EF_1 = EF_2, \quad e-\mathrm{mod}(EPS)$, if there exist a finite number of e-pols EA_i such that

$$EF_1 - EF_2 = SUM_i EA_i * EP_i,$$

in which EP_i are e-pols in EPS.

DEF. For any d-polset DPS we say that two e-pols DF_1 and DF_2 are d-CONGRUENT w.r.t. DPS and we write then $DF_1 = DF_2, \quad d-\mathrm{mod}(DPS)$, if there are a finite number of d-pols DAt_i such that

$$DF_1 - DF_2 = SUM_i[SUMt_i DAt_i * DERtDP_i],$$

in which SUM_i runs over a finite set of indices i, $SUMt_i$ runs over a finite number of tuples t corresponding to rach i and DP_i are all d-pols in DPS.

Notation. We write for simplicity $DF_1 = DF_2 \quad e-\mathrm{mod}(DPS)$ if

$$Elem(DF_1) = Elem(DF_2)\, e - \mathrm{mod}(Elem(DPS)).$$

Below $d-BF'$, $d-BF''$ will denote some d-ext-fields of $d-BF$.

DEF. $Z' = (Z'_1, \cdots, Z'_n)$ in $(d-BF')\wedge n$ is a $d-BF'-ZERO$ of a d-pol DP if $DP(Z') = 0$ or $DP = 0$ for $(Y_1, \cdots, Y_n) = (Z'_1, \cdots, Z'_n)$.

DEF. $Z' = (Z'_1 \cdots, Z'_n)$ in $(d-BF')\wedge n$ is a $d-BF'-ZERO$ of a d-polset DPS if Z' is a $d-BF'-zero$ of all d-pols in DPS.

Notation. Let $IDPS$ be any set of d-pols which may be either finite or infinite, and DG be any d-pol, we shall write

$d-BF'-Zero(IDPS/DG)$ = Totality of $d-BF'-zeros$ of $IDPS$ which are not $d-BF'-zeros$ of DG.

$d-Zero(IDPS/DG)$ = Totality of $d-BF'-zeros$ of $IDPS$ which are not $d-BF'-zeros$ of DG for all d-ext-fields $d-BF'$ of $d-BF$.

$d-BF'-Zero(IDPS) = d-BF'-Zero(IDPS/1)$,

$d-Zero(IDPS) = d-Zero(IDPS/1)$,

$d-Zero(DP) = d-Zero(\{DP\})$ for a d-polset$\{DP\}$ consisting of a single non-zero pol DP.

Remark. When $d-BF'$ is evident from the context or unnecessary to specify, we write also simply, if no confusion can arise,

$d-BF'-Zero(IDPS/DG) = d-Zero(IDPS/DG)$,

$d-BF'-Zero(IDPS) = d-Zero(IDPS)$.

DEF. An ALGEBRAIC DIFFERENTIAL VARIETY (abbr. alg-d-var) over $d-BF$ as d-BASIC FIELD is the set $d-Zero(DPS)$ for some d-polset DPS.

DEF. An alg-d-var is said to be d-IRREDUCIBLE if it is not the union of two different alg-d-vars different both from the given one.

DEF. For Z' in $(d-BF')\wedge n$ and Z'' in $(d-BF'')\wedge n$, we say that Z'' is a SPECIALIZATION of Z' if for any d-pol DP with Z' in $d-Zero(DP)$ or $DP(Z') = 0$, one has also Z'' in $d-Zero(DP)$ or $DP(Z'') = 0$. Notation: The totality of all specializations of Z' will be denoted by $Spec(Z')$.

DEF. For any infinite set $IDPS$ of d-pols we say that a d-polset FBS is a FINITE BASIS of $IDPS$ if for any d-pol DP in $IDPS$, there is some positive integer p such that

$$DP \wedge p = 0, \quad d-\mathrm{mod}(FBS).$$

FINITE BASIS THEOREM. For any infinite set of d-pols $IDPS$ there is a d-polset FBS such that

$$d-Zero(FBS) = d-Zero(IDPS).$$

Proof. By the theorem of Ritt and Raudenbush (cf. [R1, 2]) there is a finite FBS of $IDPS$ which may be served as the FBS in the assertion.

DEF. Any FBS in the theorem is called a FINITE BASIS of the set IDPS.

THEOREM. For any Z' in $(d-BF')\wedge n$, $Spec(Z')$ is a d-irreducible alg-d-var with

$$Spec(Z') = d - Zero(FBS),$$

in which FBS is a finite basis of the infinite set $IDPS$ of d-pols having Z' as a d-zero.

Proof. That $Spec(Z')$ is an alg-d-var follows from the evident equalities

$$Spec(Z') = d - Zero(IDPS) \quad \text{and} \quad d - Zero(IDPS) = d - Zero(FBS).$$

If $d - Zero(DPS)$ is some alg-d-var contained in $Sepc(Z')$ and contains the point Z', then any point Z'' in $Sepc(Z')$ will be a d-Zero of any d-pol in DPS so that $Spec(Z')$ coincides with $d - Zero(DPS)$.

This proves the d-irreducibility of $Spec(Z')$.

In later sections we shall prove the converse of the above theorem, viz.

THEOREM. For any d-irreducible alg-d-var $d - Zero(DPS)$ there is a Z' in $d - BF' \wedge n$ for some d-ext-field $d - BF'$ of $d - BF$ such that

$$d - Zero(DPS) = Spec(Z')$$

and any alg-d-var is a finite union of such d-irreducible ones.

6. d-Ascending-Set and d-Remainder

DEF. A d-ASCENDING-SET (abbr. d-asc-set) is either a single non-zero X-pol and is then said to be TRIVIAL or a finite sequence of non-zero d-pols none of which are X-pols

$$(d-ASC) DP_1, DP_2, \cdots, DP_r$$

such that

$$DP_1 < DP_2 < \cdots < DP_r$$

with each DP_i reduced w.r.t. any preceding $DP_j, j < i$.

DEF. An IS-POWER-PRODUCT of a non-trivial d-asc-set $(d-ASC)$ is any power product of all these initials and separants of d-pols in $(d-ASC)$. In particular, the product of all these initials and separants is called simply the IS-PRODUCT of $(d-ASC)$. The IS-PRODUCT of a trivial d-asc-set is defined to be 1.

DEF. A d-pol DR is said to be REDUCED w.r.t. a d-asc-set $(d-ASC)$ if $(d-ASC)$ is non-trivial and DR is reduced w.r.t. each d-pol in $(d-ASC)$.

d-REMAINDER THEOREM. For any d-pol G and a non-trivial d asc-set $(d-ASC)$, there is a unique IS-power-product J of $(d-ASC)$ such that

$$\text{(R)} \qquad J * DG = DR \qquad d - \text{mod}(d - ASC)$$

with DR reduced w.r.t. $(d-ASC)$.

Proof. Cf. [R1, 2].

DEF. DR in the above theorem is called the d-REMAINDER of DG w.r.t. $(d-ASC)$ and the formula (R) is called the d-REMAINDER FORMULA of DG w.r.t. $(d-ASC)$.

Notation. $DR = d - Remdr(DG/(d-ASC))$.

We now introduce a partial ordering of d-ascending-sets in the following way.

Given two non-trivial d-asc-sets

$$(d-ASC-P) DP_1, DP_2, \cdots DP_r, \qquad (d-ASC-Q) DQ_1, DQ_2, \cdots, DQ_s,$$

we shall say that $(d-ASC-P)$ is HIGHER THAN $(d-ASC-Q)$ or $(d-ASC-Q)$ is LOWER THAN $(d-ASC-P)$ if either (a) or (b) below holds true:

(a) There is some k such that $DP_i <=> DQ_i$ for $i < k$ while $DP_k > DQ_k$.

(b) $r < s$ and $DP_i <=> DQ_i$ for all $i \leq r$.

DEF. A trivial d-asc-set is said to be LOWER THAN any non-trivial one or a non-trivial d-asc-set is HIGHER THAN any trivial one.

DEF. Two d-asc-sets are said to be INCOMPARABLE in ORDER if neither one is higher thaan the other.

Notation given two d-asc-sets $(d - ASC_1)$ and $(d - ASC_2)$, trivial or not, we shall write

$$(d - ASC_1) > (d - ASC_2), (d - ASC_1) < (d - ASC_2) \quad \text{or} \quad (d - ASC_1) <=> (d - ASC_2),$$

according as whether $(d - ASC_1)$ is higher than, lower than, or incomparable to $(d - ASC_2)$.

LEMMA 1. Any sequence of d-asc-sets steadily decreasing in order

$$(d - ASC_1) > (d - ASC_2) > \cdots$$

is necessarily finite.

Proof. Similar to the ordinary case in [WU1].

DEF. A d-BASIC-SET (abbr. d-bas-set) of a d-polset DPS is any lowest d-asc-set contained in DPS.

From the very definition we have the following

PROP. Any two d-bas-sets of a d-polset are incomparable in order.

From this proposition we see that the following definition is legitimate:

DEF. For two d-polsets DPS_1 and DPS_2 we say that DPS_1 is HIGHER THAN, LOWER THAN, or INCOMPARABLE in ORDER to DPS_1 accor-to a d-bas-set of DPS_1 is higher than, lower than, or incomparable in order to a d-bas-set DPS_2 or not.

Notation. We write $DPS_1 > DPS_2$ or $DPS_2 < DPS_1$, $DPS_1 < DPS_2$ or $DPS_2 > DPS_1$, and $DPS_1 <=> DPS_2$ resp. according as DPS_1 is higher than, lower than, and incomparable to DPS_2 resp.

From Lemma 1 we have also

LEMMA 2. Any sequence of d-polsets steadily decreasing in order

$$DPS_1 > DPS_2 > \cdots$$

is necessarily finite.

The condition (b) in the definition of ordering of d-asc-sets furnishes us a means of lowering the order if a d-polset, viz.

LEMMA 3. Let DBS be a d-bas-set of a d-polset DPS and DR be a d-pol reduced w.r.t. DBS. Then the d-polset enlarged by adjoining DR to DPS is lower than DPS.

7. Completion of a d-Asc-Set

Given a non-trivial d-asc-set

$$(d - ASC) \quad DF_1, \cdots, DF_r,$$

we shall separate derivatives of Y_p for each p in various classes and then define the COMPLETION of $(d - ASC)$ in the following way.

DEF. Derivatives of leads of DF_i in $(d - ASC)$ are called the PRINCIPAL DERIVATIVEs of $(d - ASC)$.

DEF. Leads of DF_i in $(d - ASC)$ are called LEADING DERIVATIVEs or PROPER PRINCIPAL DERIVATIVEs of $(d - ASC)$, while other principal derivatives are called IMPROPER PRINCIPAL DERIVATIVEs of $(d - ASC)$.

DEF. Derivatives not principal ones are called PARAMETRIC DERIVATIVEs of $(d - ASC)$.

Notation. For $1 \leq p \leq n$,

$LTUP_p(d-ASC)$=Totality of tuples t such that $DERtY_p$ is the leading derivative of some DF_i in $(d-ASC)$.

$$MAX_p(d-ASC) = Max(LTUP_p(d-ASC), \text{i.e. the tuple}(Mp_1, Mp_2, \cdots, Mp_m)$$

with $Mp_i = Max\{COOD_i(t)/t \text{ in } LTUP_p(d-ASC)\}$.

$CTUP_p(d-ASC) = CompLTUP_p(d-ASC)$, i.e. totality of tuples u which are divisors of $MAX_p(d-ASC)$ and at the same time mulltiples of some tuples t in $CTUP_p(d-ASC)$.

DEF. A principal derivative of the form $DERtY_p$ with tuple t in $CTUP_p(d-ASC)$ is called a C-PRINCIPAL DERIVATIVE or C-DERIVATIVE of $(d-ASC)$.

The condition that each d-pol in $(d-ASC)$ reduced is w.r.t. any preceding ones implies readily the following

PROP. The tuple set $LTUP_p(ASC)$ of a d-asc-set is a prime set.

DEF. For tuple v in $CTUP_p(d-ASC)\backslash LTUP_p(d-ASC)$, let us form the d-remainder $d-Remdr(DERvY_p)/(d-ASC)) = DRv_p$ w.r.t. $(d-ASC)$ so that

$$J * DERvY_p = DRv_p \qquad d-\text{mod}(d-ASC)$$

in which J is an IS-power-product of $(d-ASC)$. Then:

DEF. The d-pol $J * DERvY_p - DRv_p$ is called a DERIVED d-POL of $(d-ASC)$ rel. v and p.

Let the set of all derived d-pols of $(d-ASC)$ be arranged in increasing orders:

$$DG_1, DG_2, \cdots, DG_s.$$

DEF. The sequence consisting of all DF_i and DG_j arranged in increasing orders $(g = r+s)$: $(d-ASC+)DH_1, \cdots, DH_g$ is called the COMPLETION of $(d-ASC)$.

PROP. For the derived d-pol $DH_k = J * DERvY_p - DRv_p$ rel. v, p as above we have

$$Ld(DH_k) = DERvY_p, Init(DH_k) = Sep(DH_k) = J, \text{and} Deg(DH_k) = 1.$$

Remark. The initials and separants of DH_k in $(d-ASC+)$ are all IS-power-products of $(d-ASC)$.

8. Integrability Pols of a d-Asc-Set

Let a non-trivial d-asc-set $(d-ASC)DF_1, DF_2, \cdots, DF_r$ be given with its completion $(d-ASC+)DH_1, DH_2, \cdots, DH_g$. to simplify the notation, we shall write simply $LTUP_p = LTUP_p(d-ASC)$, etc. with $d-ASC$ omitted.

DEF. An M-DERIVATIVE DM of $(d-ASC)$ is a d-pol of the form $DM = DERuDH_h$, with $Ld(DH_h) = DERtY_p, t- < CTUP_p$, $COOR_i(u) = 0$ for $i- < Nult(t/LTUP_p)$ or $u- < TMU(t/LTUP_p)$.

DEF. An M-PRODUCT of $(d-ASC)$ is a product of at least one M-derivatives.

DEF. An M-POL of $(d-ASC)$ is a linear sum of M-products with coefficients d-pols in leading and parametric derivatives alone.

Consider any d-pol DP in some M-derivatives DMh and other derivatives, parametric or principal, of $(d-ASC)$. Suppose that among these principal derivatives there are improper ones of which the highest one is $DERvY_p$. By tuple decomposition theorem we have then a unique product representation

$$v = ut \,\text{with}\, t- < CTUP_p\backslash LTUP_p, \text{and} COOR_i(u) = 0 \text{for} i- < Nult(t/LTUP_p),$$

or

$$u- < TMU(t/LTUP_p).$$

8

Let $DERtY_p$ be the lead of DH_k. Then we have

$$J1 * DERvY_p = DERuDH_k + DU,$$

in which J_1 is an IS-power-product of $(d - ASC)$, and DU a d-pol in parametric derivatives and principal derivatives lower than $DERvY_p$. Replacing $DERvY_p$ in DP by $(DERuDH_k + DU)/J_1$ and clearing of fractions, we get a d-pol $DP_1 = J1' * DP$, J_1' being a power of J_1, in such a form which involves besides the M-derivatives DMh and the new one $DM = DERuDH_k$, eventually also parametric and principal derivatives. The latter ones are however all lower than $DERvY_p$.

DEF. The above procedure

$$DP \longrightarrow DP_1 = J_1' * DP_1.$$

is called an M-REDUCTION of DP.

In the d-pol DP as before suppose that there is, besides the parametric derivatives, some leading derivatives not in the M-derivatives DMh already present. Suppose that the highest such leading derivatives $DERtY_p$ for which the corresponding d-pol in $(d - ASC)$ is DF_i has a degree $d > Deg(DF_i)$. In multiplying DP by some power J_2 of the initial of DF we can replace $J2 * (DERtY_p \wedge d$ in DP by some linear sum of d-pols DF_j in $(d - ASC)$ preceding DF_i and some d-pol DP_2 in which $DERtY_p$ will appear with a degree $< Deg(DF_i)$.

DEF. The above procedure

$$DP \longrightarrow DP_2 = J_2 * DP$$

is called an I-REDUCTION of the d-pol DP.

It is clear that in applying successive M-and I-reductions we will arrive finally at a d-pol $J * DP$ such that we can put it in the form

$$J * DP = M(DP) + N(DP)$$

possessing the following properties:

(1) J is a certain IS-power-product of $(d - ASC)$.

(2) $M(DP)$ is an M-pol.

(3) $N(DP)$ is a d-pol containing parametric and leading derivatives alone.

(4) The leading derivatives in $M(DP)$ and $N(DP)$ not appearing already in M-derivatives have each a degree less than the degree of that derivative in the corresponding d-pol DF_i of $(d - ASC)$.

DEF. In the above formula the d-pols $M(DP)$ and $N(DP)$ are called resp. the M-PART and the N-PART or the NULL-PART of DP.

Consider now any DH_h of $(d - ASC+)$ with lead $DERtY_p$ such that the tuple t has a non-multiplier i w.r.t. $LTUP_p$, or $i- < Nult(t/LTUP_p)$. We have then $DH_h = I * (DERtY_p) \wedge d+$ lower degree terms, in which $I = Init(DH_h)$. Hence we get

$$DER_i DH_h = S * DERuY_p + DU,$$

in which $S = Sep(DH_h)$, $DERu = DER_i DERt$, and DU is a d-pol lower than $DER_i DH$ and its lead $DERuY_p$. As $LTUP_p$ is a prime set we have

$$u- < CTUP_p \backslash LTUP_p$$

and $DERuY_p$ is the lead of some DH_k in $(d - ASC+)$. We have then

$$DH_k = I' * DERuY_p + DV,$$

in which $I' = Init(DH_k)$ and DV is a d-pol lower than DH_k and its lead $DERuY_p$. It follows that we have an identity of the form

$$J_1 * DER_i DH_h \underset{9}{=} J_2 * DH_k + DW,$$

in which J_1, J_2 are IS-power-products of $(d-ASC)$ and DW is a d-pol lower than the lead $DERuY_p$ of both DH_k and DER_iDH_h.

DEF. The null-part of the above d-pol $DW = J_1 * DER_iDH_h - J_2 * DH_k$, or what is the same, the null-part of $J_1 * DER_iDH_h$, is called the INTEGRABILITY POL of $(d-ASC)$ corresponding to DH_h and the non-multiplier i.

DEF. A non-trivial d-asc-set is said to be PASSIVE if all its integrability pols are zero.

9. Passivity Theorem

Let a non-trivial d-asc-set

$$(d-ASC)DF_1, DF_2, \cdots, DF_r$$

be given as in Section 8.

Below we shall denote by J_1, J_2, etc. any IS-power-product of d-pols in $(d-ASC)$, by DM_1, DM_2, etc. any M-derivative, and by MP_1, MP_2, etc. any M-pol.

PASSIVITY THEOREM. If $(d-ASC)$ is passive, then all derivatives of d-pol in $(d-ASC+)$ when multiplied be some IS-power-product of $(d-ASC)$, have their null-parts zero. Moreover, any such derivative DE can be written in the form

$$J * DE = J' * DM + MP,$$

in which DM is an M-derivative having same lead as that of the given derivative DE, while MP is an M-pol in which all M-derivative are lower than DE.

Proof. Consider any DH_h of $(d-ASC+)$ with lead $DERtY_p$:

$$Ld(DH_h) = DERtY_p, t- < CTUP_p, \tag{1}$$

and any derivative DER_iDH_h where $1 \leq i \leq m$. If

$$i- < Nult(t/LTUP_p),$$

then the null-part of DER_iDH_h is an integrability-pol corresponding to DH_h and i so that it is zero by passivity of $(d-ASC)$. If

$$i- < Mult(t/LTUP_p),$$

then DER_iDH_h is itself an M-derivative so that its null-part is trivially zero. Hence the theorem is true for DH_h and any DER_iDH_h with $1 \leq i \leq m$.

Consider now any derivative $DERwDH_h$ with $Ord(w) > 1$ and suppose that the theorem has been proved for all derivatives of d-pols in $(d-ASC+)$ which are lower than $DERwDH_h$.

Let us write the lead of DH_h as in (1). If $w- < TMU(t/LTUP_p)$, then $DERwDH_h$ is itself an M-derivative and there is nothing to prove. Suppose therefore the contrary. Then we can write $DERw$ as $DERvDER_i$ with $1 \leq i \leq m$, and $i- < Nult(t/LTUP_p)$. As $(d-ASC)$ is passive, we will have an identity of the form

$$J_1 * DER_iDH_h = J_1' * DH_k + MP_1, \tag{2}$$

in which DH_k has the same lead as DER_iDH_h, while MP_1 is an M-pol with all its M-derivatives lower than DER_iDH_h.

Form now $DERv$ of both sides of (2), we get then an identity of the form

$$\begin{aligned} J_1 * DERwDH_h = {} & J_1' * DERvDH_k + DERvMP_1 \\ & -SUMu(DAu * DERuDH_h) - SUMx(DBx * DERxDH_k), \end{aligned} \tag{3}$$

10

in which

$$Ord(u) < Ord(w), \quad Ord(x) < Ord(v) < Ord(w).$$

By induction hypothesis we have then

$$J_2 * DERvDH_k = J_2' * DM_2 + MP_2, \tag{4}$$

$$Ju * DERuDH_h = Ju' * DMu + MPu, \tag{5}$$

$$Jx * DERxDH_k = Jx' * DMx + MPx. \tag{6}$$

Now each term in $DERvMP_1$ has at least one factors of the form $DERzDH_j$ with $0 \leq Ord(z) \leq Ord(v) < Ord(w)$. Again by induction hypothesis we may write for $Ord(z) > 0$

$$Jz * DERzDHj = Jz' * DMz + MPz, \tag{7}$$

in which DMz has the same lead as $DERzDH_j$, while all M-derivatives in MP_z are lower than DER_zDH_j

Substituting (4)–(7) into (3), we get an identity of the form

$$J * DERwDH_h = J' * DM + MP,$$

in which DM has same lead as $DERwDH_h$ while all M-derivatives in MP are lower than $DERwDH_h$. This proves the theorem for $DERwDH_h$.

As the theorem is already seen to be true for DH_h and any DER_iDH_h for $1 \leq i \leq m$, the theorem is proved by induction.

10. Irreducibility of d-Asc-Set

Let it be given a non-trivial d-asc-set

$$(d - ASC)DF_1, DF_2, \cdots, DF_r$$

and an e-asc-set or an ordinary asc-set $(e - ASC)EF_1, EF_2, \cdots, EF_s$.

DEF. A PARTIAL d-ASC-SET of $(d - ASC)$ at STAGE i is the d-asc-set consisting of the first i d-pols in $(d - ASC)$, viz.

$$(d - ASC_i)DF_1, DF_2, \cdots, DF_i.$$

DEF. A PARTIAL e-ASC-SET if $(e - ASC)$ at STAGE i is the e-asc-set consisting of the first i e-pols in $(e - ASC)$, viz.

$$(e - ASC_i)EF_1, EF_2, \cdots, EF_i.$$

DEF. $(d - ASC)$ resp. $(e - ASC)$ is d-IRREDUCIBLE resp. e-IRREDUCIBLE if for any $1 \leq i \leq r$ resp. $\leq s$ there cannot exist any relation of the form

$$DH_i * DF_i = DF_i' * DF_i'' \qquad d - mod(d - ASC_j) \tag{1d}$$

resp.

$$EF_i * EF_i = EF_i' * EF_i'' \qquad e - mod(e - ASC_j), \tag{1e}$$

in which $j = i - 1$, and for each i, DF_i' and DF_i'', resp. EF_i', EF_i'' are d-pols resp. e-pols having same lead as DF_i resp. EF_i, while DH_i resp. EH_i is some d-pol resp. e-pol with lower lead and reduced w.r.t. the partial d-asc-set $d - ASC_j$ resp. the partial e-asc-set $e - ASC_j$. In the contrary case we say that $(d - ASC)$ resp. $(e - ASC)$ is d-REDUCIBLE resp. e-REDUCIBLE.

To a d-asc-set $(d-ASC)$ over a d-basic field $d-BF$ let us associate an e-asc-set $(e-ASC)$ over the e-field $e-BF = Elem(d-BF)$ with $s=r$ and $EF_i = Elem(DF_i)$ considered as e-pols in the indeterminates $Et_p = Elem(DERtY_p)$ corresponding to all derivatives $DERtY_p$ for which those corresponding to the parametric derivatives of $(d-ASC)$ are to be considered as independent ones. Furthermore, the indeterminates are to be arranged in such an order that those corresponding the parametric derivatives come before those corresponding to the principal ones in arbitrary way, while the latter ones are in the same order as the original derivatives.

DEF. The e-asc-set determine in the above manner is said to be ASSOCIATED to the d-asc-set $(d-ASC)$ and will be denoted by $Elem(d-ASC)$.

IRREDUCIBILITY THEOREM. For a d-asc-set $(d-ASC)$ over some d-field $d-BF$ to be d-irreducible over $d-BF$, it is necessary and sufficient that its associated e-asc-set $(e-ASC)$ be e-irreducible over the associated e-field $e-BF = Elem(d-BF)$.

Proof. $(\Longrightarrow)$ Suppose that $(e-ASC)$ is e-reducible so that for some i there is an identity of the form (1e) or more explicitely,

$$EH_i * EF_i = EF_i' * EF_i'' + SUM_k EQ_k * EF_k, \tag{1e}'$$

in which the summation is over $k=1,2,\cdots,j$ where $j=i-1$. This is a pure algebraic identity in the indeterminates X and all $Et_p = Elem(DERtY_p)$ considered as independent ones. A fortiori we have therefore necessarily an identity of the form

$$DH_i * DF_i = DF_i' * DF_i'' + SUM_k DQ_k * DF_k, \tag{1d}'$$

in which DH_i, DF_i', DF_i'', and DQ_i are the d-pols with $Elem\ (DH_i) = EH_i$, etc., by recovering the various Et_p to the original derivatives $DERtY_p$. This proves that $(d-ASC)$ is d-reducible if the associated $(e-ASC)$ is e-reducible, or what is the same, $(e-ASC)$ is e-irrecducible if $(d-ASC)$ is d-irreducible.

$(\Longleftarrow)$ Suppose that $(d-ASC)$ is d-reducible is that for some i we have an identity of the form (1d) as described above. We may rewrite this identity in a more explicit form:

$$DH_i * DF_i = DF_i' * DF_i'' + SUM_k SUMt_k (DQt_k * DERtDF_k), \tag{2d}$$

in which $SUMt_k$ is to be taken over some tuple-set T_k and SUM_k over $k=1,2,\cdots,j$ with $j=i-1$. Taking *Elem* at both sides we get then an identity of the form

$$EH_i * EF_i = EF_i' * EF_i'' + SUM_k SUMt_k (EQt_k * EFt_k), \tag{2e}$$

in which $EH_i = Elem(DH_i)$, etc., while $EFt_k = Elem(DERtDF_k)$.

Suppose that in the sums of (2d) there are $DERtDF_k$ with $Ord(t) > 0$. Let $DERvDF_h$ be the highest such one with lead $DERuYe$. There is then some identity of the form

$$DS * DERuYe = DA + DERvDF_h. \tag{3d}$$

Here $DS = Sep(DF_h)$ and DA is some d-pol lower than $DERuYe$. As (2e) is a pure algebraic identity in all X and $Eue = Elem(DERuYe)$, etc. which are to be considered as independent indeterminates, we can substitute in it Eue by EA/ES with $EA = Elem(DA)$ and $ES = Elem(DS)$. Clearing of fractions and recovering to original form in derivatives we get then an identity still of the form (2d) but the term in the sum involving $DERvDF_h$ has been removed. Proceeding in the same manner we will finally remove all such terms and arrive at an identity of the form (1d)'. Taking *Elem* at both sides we get then an identity of the form (1e) with $EH_i = Elem(DH_i)$, etc. which shows that $(e-ASC)$ is e-reducible. Hence $(e-ASC)$ is e-reducible if $(d-ASC)$ is d-reducible or $(d-ASC)$ is d-irreducible if $(e-ASC)$ is e-irreducible.

11. Formal Taylor Series and Generic Zero Theorem

DEF. A CONSTANT w.r.t. the d-basic field $d - BF$ is an element in a d-ext-field of $d - BF$ for which all derivatives are zero.

PROP 1. The totality of constants w.r.t. $d - BF$ form an elementary ring under the natural arithmetical operations.

Let $O_1, \cdots, O_m$ be a set of constants taken in some d-ext-field of $d - BF$ which are independent altogether.

Notation. For a tuple $t = (I_1, \cdots, I_m)$, we write $(X - O) \wedge t = PROD_i((X_i - O_i) \wedge I_i)$. $t! = PROD_i(I_i!)$.

DEF. A FORMAL TAYLOR SERIES at ORIGIN $O = (O_1, \cdots, O_m)$ is a formal power series of the form

$$FTS = SUMt[(Ct/t!) * (X - O) \wedge t],$$

in which $SUMt$ is taken over all tuples t and Ct are all constants in some d-ext-field of $d - BF$.

DEF. The constant C_0 corresponding to the tuple $0 = (0, \cdots, 0)$ is called the VALUE of FTS at origin $O = (O_1, \cdots, O_m)$.

Notation. $C_0 = Val(FTS/O)$ or simply $Val(FTS)$.

PROP 2. Under natural arithmetical operations and differentiations the totality of formal taylor series at given origin O form a d-ring $RFTS$.

PROP 3. Val is an oridinary ring morphism of the elementary ring $Elem(RFTS)$ into the elementary ring of constants.

PROP 4. The value of $DERt(FTS)$ at the origin O is Ct.

Suppose now we are given a passive d-irreducible d-asc-set

$$(d - ASC) DF_1, DF_2, \cdots, DF_r$$

with its completion

$$(d - ASC+) DH_1, DH_2, \cdots, DH_g.$$

Below we shall construct for each Y_p a formal taylor series

$$FTS_p = SUMt[(Ct_p/t!) * (X - O) \wedge t)]$$

such that $FTS = (FTS_1, \cdots, FTS_n)$ is a d-zero of $(d - ASC)$. For this purpose let us introduce first for simplifications the following

Notation. For any d-pol DP we shall denote by DP' the d-pol get from DP by substituting O_i for X_i, and Ct_p for $DERtY_p$ whenever Ct_p has already been determined.

Construction of $FTS_p, p = 1, \cdots, n$:

Step 1. Take O_i and all Ct_p corresponding to parametric derivatives $DERtY_p$ as INDEPENDENT constants.

Step 2. For leading derivatives $Ld(DF_i) = DERuY_h$.

Let KO be the e-ext-field of $e - BF = Elem(d - BF)$ get by adjoining to it all the constants in Step 1 as independent transcendental elements.

Determine now inductively on i the constant Cuh as a zero of the e-pol DF_i' in some e-ext-field of the e-field $K_j, j = i - 1$. By the irreducibility theorem DF_i' is e-irreducible as an ordinary pol in the lrad of DF_i on coefficients in K_j. The e-ext-field of K_j which is an algebraic extension by adjoining Cuh by means of the equation

$$DF_i' = 0 \tag{1}$$

is defined inductively as K_i.

A consequence of this determination is the following

PROP 5. For any non-zero d-pol DR reduced w.r.t. $(d - ASC)$ we have $(DR)' <> 0$. In particular, for any IS-power-product J, we have $J' <> 0$.

Step 3. For C-derivative $DERvY_p$ not any leading derivative.

We have in this case a derived d-pol DH_k in $(d - ASC+)$ such that

$$DH_k = J * DERvY_p - DRv_p,$$

in which J is an IS-power-product of $(d - ASC)$ and DRv_p is the d-remainder of $DERvY_p$ w.r.t. $(d - ASC)$. In DRv_p there appear only either parametric or leading derivatives so that $(DRv_p)'$ is well-defined by Step 1 and Step 2. Since $J' <> 0$ by PROP 5 we can determine Cv_p by

$$J' * Cv_p = (DRv_p)', \quad \text{or} \quad (DH_k)' = 0. \tag{2}$$

Step 4. Consider now any improper principal derivative $DERvY_p$ which is not a C-derivative. By tuple-decomposition theorem there is a unique C-derivative $DERtY_p$ such that $v = ut$ with $t- < CTUP_p$ and $v- < TMU(t/LTUP_p)$.

Let $DERtY_p$ be the lead of DH_k. Then we have an identity of the form

$$J_1 * DERvY_p = DERuDH_k + DU.$$

In the identity J_1 is an IS-power-product of $(d - ASC)$ and DU a d-pol in which all derivatives are either parametric ones or principal ones lower than $DERvY_p$.

Suppose that values have already been attributed to such principal derivative lower than $DERvY_p$ so that DU' and J_1' have already been well-determined. As the value J_1' cannot be zero in view of the PROP 5, we can determine a value Cv_p by

$$J_1' * Cv_p = DU' \quad \text{or} \quad (DERuDH_k)' = 0. \tag{3}$$

This gives an inductive procedure of determining the values of improper principal derivatives which are not C-derivatives in terms of those of lower ones. The induction has to end at C-derivatives whose values have already been determined in Steps 1, 2 and 3. Consequently the values of any improper principal derivatives which are not C-derivatives will be determined.

Notation. For any d-pol DP, the formal taylor series get from DP by substituting FTS_p for Y_p will be denoted by DP''.

Now for any derivative $DERtY_p$ we have from PROP 4

$$Val(DERtY_p)'' = Val(DERtFTS_p) = Ct_p = (DERtY_p)'.$$

From PROP 3 we get therefore

PROP 6. For any d-pol we have $Val(DP'') = DP'$.

DEF. A point $GZ = (Z_1, \cdots, Z_n)$ with Z_i in some d-ext-field of $d - BF$ is called a GENERIC d-ZERO of $(d - ASC)$ if it is a d-Zero of all DH_k in $(d - ASC+)$ while it is not a d-zero of any non-zero d-pol reduced w.r.t. $(d - ASC)$.

GENERIC ZERO THEOREM. If $(d - ASC)$ is passive and d-irreducible then the point $GZ = (FTS_1, \cdots, FTS_n)$ with FTS_p constructed as above is generic d-zero of $(d - ASC)$.

Proof. That GZ is not d-zero of any non-zero d-pol reduced w.r.t. $(d - ASC)$ is shown already by PROP 5. It is sufficient therefore to prove for any DH_h of $(d - ASC+)$

$$DH_h'' = 0. \tag{4}$$

In turn we have to prove for any tuple u,

$$Val(DERu(DH_h)'') \quad \text{or} \quad Val(DERuDH_h)'' = 0.$$

By PROP 6, this is equivalent to

$$(DERuDH_h)' = 0. \tag{5}$$

Now according to the construction FTS_p we see from (1)–(3) that (5) is true for any $DERuDH_h$ which is an M-derivative DM :

$$(DM)' = 0. \tag{6}$$

It follows that for any M-pol MP we should have

$$(MP)' = 0. \tag{7}$$

By the passivity theorem we see that for any derivative DE of some DH_h in $(d-ASC+)$, we have an identity of the form

$$J_1 * DE = J_2 * DM + MP, \tag{8}$$

in which J_1, J_2, are IS-power-products of $(d-ASC)$, DM is an M-derivative, and MP an M-pos. As $(J_1') <> 0$ by PROP 5 we get from (6)–(8) $(DE)' = 0$, or (5) for any derivative of DH_h a theorem is thus proved.

The following theorem shows the significance of generic zero and will play an important role in the whole theory.

GENERIC-ZERO-REMAINDER THEOREM. Let $(d-ASC)$ be a d-asc-set which is passive and d-irreducible and GZ be a generic d-zero. Then a d-pol DP will have GZ as its d-zero or $DP(GZ) = 0$ if and only if

$$d - Remdr(DP/(d-ASC)) = 0.$$

Proof By the d-remainder theorem we have

$$J * DP = d - Remdr(DP/(d-ASC)) \qquad d - mod(d-ASC)$$

in which J is some IS-power-product. As the d-remainder, if nonzero, is reduced w.r.t.$(d-ASC)$, the theorem follows directly from PROPs 5 and 6.

12. d-Char-Set of a d-Polset

For any d-polset DPS we form a certain particular d-asc-set DCS according to the following scheme:

$$\begin{aligned} DPS = DPS_0 - &< DPS_1 - < \cdots - < DPS_s \\ DBS_0 > DBS_1 &> \cdots DBS_s = DCS \\ RIS_0 \quad RIS_1 &\cdots RIS_s = Empty \end{aligned} \tag{S}$$

In the above scheme (S) each DBS_i is a d-bas-set of DPS_i, each RIS_i is the d-polset consisting of all non-zero integrability pols of DBS_i as well as all non-zero d-remainders of d-pols in $DPS_i - DBS_i$ w.r.t. DBS_i. Furthermore, each d-polset DPS_i is the union of DPS_j and RIS_j with $j = i-1$. It is clear that the d-bas-sets $DBS_0 > DBS_1 > \cdots$ are steadily decreasing in order so that the construction should end in a finite number of steps and in certain stage s we should have $RIS_s = Empty$.

DEF. The corresponding d-bas-set $DBS_s = DCS$ in the above scheme is called a d-CHARACTERISTIC SET (abbr. d-char-set) of the given d-polset DPS.

THEOREM(Well-Ordering Principle). Let I_i and S_i be the initials and separants of d-pols in DCS and J be the IS-product of DCS. Then

$$d - Zero(DCS/J) - < d - Zero(DPS) - < d - Zero(DCS) \tag{I}$$

$$\begin{aligned} d - Zero(DPS) = d - Zero(DCS/J) &+ SUM_i d - Zero(DPS_i') \\ &+ SUM_i d - Zero(DPS_i''). \end{aligned} \tag{II}$$

In these formulas each DPS_i' resp. DPS_i'' is the enlarged d-polset obtained from DPS by adjoining to it I_i resp. S_i.

Proof. From the construction we see that

$$d - Zero(DPS) = d - Zero(DPS_0) = \cdots = d - Zero(DPS_s). \tag{1}$$

From the d-remainder formula and the emptiness of RIS_s we have readily

$$\begin{aligned} d - Zero(DPS_s) = d - Zero(DBS_s/J) &+ SUM_i d - Zero(DPSs_i') \\ &+ SUM_i d - Zero(DPSs_i''), \end{aligned} \tag{2}$$

in which each $DPSs'_i$ resp. $DPSs''_i$ is the enlarged d-polset get from DPS_s by adjoining to it I_i resp. S_i. It is also clear from (1) that for each i we have

$$d - Zero(DPSs'_i) = d - Zero(DPS'_i), \tag{3$'$}$$

$$d - Zero(DPSs''_i) = d - Zero(DPS''_i). \tag{3$''$}$$

From (1), (2), (3)$'$ and (3)$''$ we get then (II). The formula (I) is also immediate from the construction.

PROP. Each d-polset DPS'_i or DPS''_i in (II) is lower than the given d-polset DPS in the partial ordering of d-polsets.

Proof. As both I_i and S_i are reduced w.r.t. DBS_s we see that $DBSs'_i$ and $DBSs''_i$ are both lower than DBS_s. As $DBS_0 > DBS_s$ we have $DBS_0 > DBSs'_i$ and $DBSs''_i$. The implie $DPS > DPS_i$ and $> DPS''_i$ as to be proved.

13. Zero Decomposition or Zero-Structure Theorems

Given a d-polset DPS let us form a d-char-set DCS according to the scheme (S). In the formula (II) the d-polset DCS is a certain passive d-asc-set. For the other d-polsets DPS'_i or DPS''_i we may treat in the same manner as for DPS so that each $d - Zero(DPS'_i)$ or $d - Zero(DPS''_i)$ will be further splitted into a sum of d-zero-set a (II). The same procedure can be carried on further so far some d-polset not already in the form of passive d-asc-set still appear in the sum. Since all DPS'_i and DPS''_i are lower than DPS the procedure has to end in a finite number of steps and so we get finally the following

ZERO DECOMPOSITION THEOREM (Weak Form). There is an algorithmic procedure which permits to give for any d-polset DPS a decomposition of the following form:

$$d - Zero(DPS) = SUM_k d - Zero(DCS_k/J_k). \tag{III}$$

In the formula each DCS_k is a passive d-asc-set and J_k is the IS-product of DCS_k.

Suppose that a d-asc-set

$$(ASCF) \qquad DF_1, \cdots, DF_r$$

is d-reducible at a certain stage i so that we have a relation of the form

$$DH_i * DF_i = DF'_i * DF''_i \qquad d - mod(ASCF),$$

in which DF'_i, DF''_i are d-pols having same lead as DF_i while DH_i is a d-pol with lower lead and is reduced w.r.t. the partial d-asc-set $(j = i - 1)$

$$(ASCF_j) \qquad DF_1, \cdots, DF_j.$$

Let DFS' and DFS'' be the d-polsets obtained from $(ASCF)$ in replacing DF_i by DF'_i and DF''_i respectively and DFS the enlarged d-pols obtained from $(ASCF)$ by adjoining to it the d-pol DH_i. Then have for any d-pol G

$$\begin{aligned} d - Zero(ASCF/G) = d &- Zero(DFS'/DH_i * G) \\ &+ d - Zero(DFS''/DH_i * G) + d - Zero(DFS/G). \end{aligned}$$

We remark that the d-polsets DFS', DFS'', and DFS are all lower than the original d-polset $(ASCF)$.

Applying now to each of DFS', DFS'', and DFS the Zero Decomposition Theorem we would get then a decomposition of the form

$$d - Zero(ASCF/G) = SUM_j d - Zero(ASC_j/G_j * J_j).$$

In the formula each ASC_j is a passive d-asc-set lower than $ASCF$, each J_j is the IS-product of ASC_j, and each G_j is some d-pol determined in the procedure.

Consider now any DCS_k in (III). If it is d-reducible then we can decompose $d-Zero(DCS_k/J_k)$ into the sum of d-zero-sets for which the d-asc-sets are lower than DCS_k. We can continue this procedure so far there are still d-zero-sets of d-reducible d-asc-sets appearing in the sum. The procedure has to end in a finite number of steps so that we get the following

ZERO DECOMPOSITION THEOREM (Strong Form). There is an algorithmic procedure which permits to give for any d-polset DPS a decomposition of the form

$$d - Zero(DPS) = SUM_k d - Zero(IRR_k/J_k * G_k), \tag{IV}$$

in which each IRR_k is a d-irreducible passive d-asc-set and J_k is the ID-product of IRR_k.

Let the generic d-zero of IRR_k be GZ_k. As $Spec(GZ_k)$ is a d-irreducible alg-d-var which is in turn determined by IRR_k, we shall denote it by $Var(IRR_k)$:

$$Spec(GZ_k) = Var(IRR_k).$$

Remark that in (IV) we may suppose without loss of generality that GZ_k is not a d-zero of G_k, since otherwise the set $d - Zero(IRR_k/G_k * J_k)$ will be empty and can be removed from the sum. With this understood we have then the further

VARIETY DECOMPOSITION THEOREM. For any d-polset DPS we have a decomposition of the form

$$d - Zero(DPS) = SUM_k Var(IRR_k), \tag{V}$$

in which $Var(IRR_k)$ is the alg-d-var determined as $Spec(GZ_k)$ of certain d-irreducible passive d-asc-set IRR_k.

Proof. Let us consider the decomposition (IV). As each d-zero of IRR_k for which $J_k <> 0$ is in $Var(IRR_k)$ it is clear that

$$d - Zero(DPS) - < SUM_k Var(IRR_k).$$

Consider now any d-zero Z in some $Var(IRR_k)$ so that Z is a specialization of GZ_k. As GZ_k is not a d-zero of G_k and J_k we see from (IV) that GZ_k is a d-zero of DPS, or a d-zero of all d-pols in DPS. The point Z, being a specialization of GZ_k, is then also a d-zero of all d-pols in DPS, or a d-zero of DPS. It follows that

$$SUM_k Var(IRR_k) - < d - Zero(DPS)$$

too and hence (V) is proved.

Combining the last two theorems we see that (IV) can be improved to the following form:

$$d - Zero(DPS) = SUM_k d - Zero(IRR_k/J_k). \tag{IV$'$}$$

Remark. For each k let FBS_k be a finite basis of the d-polset $IDPS_k$ consisting of d-pols having GZ_k as a d-zero. Then $Var(IRR_h) - < Var(IRR_k)$ if and only if GZ_h is in $Var(IRR_k)$ or $Remdr(DF/IRR_h) = 0$ for any d-pol DF in FBS_k. This permits to remove any redundant components in (V) to make the decomposition uncontractable and hence also unique. In particular, this implies as a colorrary the theorem stated in the end of Section 5.

14. Basic Principles of Mechanical Theorem Proving

DEF. A THEOREM is consisting of a d-polset called HYPOTHESIS SET (abbr. hyp-set) and a d-pol called CONCLUSION d-pol (abbr. conc-pol).

DEF. Let the hyp-set and the conc-pol of a theorem T be resp. HYP and $CONC$. Then we say:

(1) T is TRUE if $d - Zero(HYP) - < d - Zero(CONC)$.

(2) T is GENERICALLY TRUE under NON-DEGENERACY CONDITIONs $DEG_i <> 0$ for DEGENERACY d-POLs (abbr. deg-pol) DEG_i if

$$d - Zero(HYP/PROD_i DEG_i) - < d - Zero(CONC).$$

(3) T is TRUE on a part $PADV$ of an alg-d-var $d - Zero(HYP)$ if

$$PADV - < d - Zero(CONC).$$

Remark. A d-zero in $d - Zero(HYP)$ is nothing but a geometrical configuration verifying the hypothesis of the theorem T and $d - Zero(HYP)$ is just the alg-d-var of all such geometrical configurations.

For the hyp-set HYP of a theorem T let us form a d-char-set DCS according to scheme (S). Let the initials and separants of d-pols in DCS be I_i and S_i resp. and J the IS-product of DCS. Let HYP_i' and HYP_i'' be the enlarged d-polsets obtained from HYP by adjoining to it I_i and S_i resp. Then by the Well-ordering Principle we have:

$$d - Zero(HYP) = d - Zero(DCS/J) + SUM_i d - Zero(HYP_i') \\ + SUM_i d - Zero(HYP_i'').$$

From now the d-remainder DR of the conc-pol $CONC$ of theorem T w.r.t. $DCS : DR = d - Remdr(CONC/DCS)$. The d-remainder formula gives then

$$J' * CONC = DR \quad d - mod(DCS), \tag{1}$$

in which J' is an IS-power-product of DCS. Moreover, we have

$$d - Zero(DCS/J) - < d - Zero(HYP) - < d - Zero(DCS). \tag{2}$$

Suppose that $DR = 0$ as a d-pol. Then we see from (1–2) that $HYP = 0$ would imply that $CONC = 0$ so far $J' <> 0$ or no I_i or S_i is 0 i.e.

$$d - Zero(HYP/PROD_i(I_i * S_i) - < d - Zero(CONC). \tag{3}$$

The theorem T is thus seen to be generically true under the non-degeneracy conditions

$$I_i <> 0, \quad S_i <> 0. \tag{4}$$

If the passive d-asc-set DCS is furthermore d-irreducible, then the d-generic zero GZ of DCS is a d-zero of DCS for which no I_i or S_i is 0 and so by (2) it is also a d-zero of HYP. If the theorem T is generically true under the non-degeneracy condition (4) so that in particular GZ is a d-zero of $CONC$ by (3), then it follows from (1) that we have necessarily $DR = 0$ as a d-pol. Hence we have:

PRINCIPLE of MECHANICAL THEOREM PROVING or MTP-PRINCIPLE I (Weak Form). If the d-remainder DR of $CONC$ w.r.t. d-char-set DCS of HYP is 0, then the theorem T with hyp-set HYP and conc-pol $CONC$ is generically true under the non-degeneracy conditions (4) for which I_i and S_i are the initials and separants of d-pols in DCS respectively. Moreover, if the d-char-set DCS is d-irreducible the above condition $DR = i$ also necessary one for the theorem T to be generically true under the above non-degeneracy conditions.

Let us now apply the zero-decomposition theorem (weak form) (III) to the hyp-set HYP so that

$$d - Zero(HYP) = SUM_k d - Zero(DCS_k/J_k),$$

in which DCS_k are passive d-asc-sets and J_k is the IS-product of DCS_k. Let DR_k be the d-remainder of $CONC$ w.r.t. DCS_k. Then we have the following

MTP-PRINCIPLE I′(General Weak Form). If $DR_k = 0$, then the theorem T is true on the part $d - Zero(DCS_k/J_k)$ of the whole variety of geometrical configurations verifying the hyp-set HYP. The condition $DR_k = 0$ is also necessary if DCS_k is d-irreducible.

More generally, let us apply the variety-decomposition theorem to HYP so that we have

$$d - Zero(HYP) = SUM_k Var(IRR_k),$$

in which IRR_k are all d-irreducible passive d-asc-sets. Then we have the following:

MTP-PRINCIPLE II(Strong Form). For a theorem T with hypset HYP and conc-pol $CONC$ to be true on the d-irreducible component $Var(IRR_k)$ of the variety $d - Zero(HYP)$, it is necessary and sufficient that the d-remainder of $CONC$ w.r.t. IRR_k is 0:

$$d - Remdr(CONC/IRR_k) = 0.$$

15. An Example

The following example is taken from the works of Pommaret (cf. [P1, 2]) which will serve as an illustration of our general method. For simplicity of notations we shall denote e.g. by $D_{ij}Y$ the derivative $DERt$ for the tuple $t = (i, j)$ with $m = 2$.

Ex. Let Y, Z be functions of independent variables X_1, X_2 connectes by the relations $DP_1 = 0, DP_2 = 0$ where

$$DP_1 = 2 * Y + D10Z + Z \wedge 2,$$
$$DP_2 = 2 * D20Y + 4 * Y * Z \wedge 2 + 8 * Y \wedge 2 - 4 * Z * D10Y - D01Z.$$

Then we have $DQ_1 = 0$ and $DQ_2 = 0$ where

$$DQ_1 = D30Y + D01Y + 12 * Y * D10Y,$$
$$DQ_2 = D30Z + D01Z - 6 * Z \wedge 2 * D10Z.$$

These two equations correspond to the usual as well as a modified KdV equation respectively.

In order to prove $DQ_1 = 0$ let us set $Y = Y_1, Z = Y_2$ so that we have a d-polset $DPS = \{DP_1, DP_2\}$ with

$$DP_1 = D10Y2 + 2 * Y_1 + Y_2 \wedge 2,$$
$$DP_2 = -D01Y_2 + 4 * Y_1 * Y_2 \wedge 2 - 4 * D10Y_1 * Y_2 + 2 * D20Y_1 + 8 * Y_1 \wedge 2.$$

Let us take the second type of ordering so that

$$Ld(DP_1) = D10Y_2, \quad Ld(DP_2) = D01Y_2.$$

The d-bas-set DBS_0 of $DPS_0 = DPS$ is then $\{DB_1, DB_2\}$ with $DB_1 = DP_1$ and $DB_2 = DP_2$. The completion of DBS_0 is consisting of 3 d-pols, viz. $DH_1 = DB_1$, $DH_2 = DB_2$ and a further one which one readily finds to be

$$DH_3 = D11Y_2 + 8 * Y_1 * Y_2 \wedge 3 - 8 * D10Y_1 * Y_2 \wedge 2 + 16 * Y_1 \wedge 2 * Y_2$$
$$+4 * D20Y_1 * Y_2 - 24 * Y_1 * D10Y_1 - 2 * D30Y_1.$$

The integrability d-pol corresponding to DH_1 and non-multiplier X_2 is found to be

$$IP_1 = D01Y_1 + D30Y_1 + 12 * Y_1 * D10Y_1.$$

On the other hand the integrability d-pol corresponding to DH_2 and non-multiplier X_1 is readily found to be 0. The d-polset DPS_1 is thus consisting of the three d-pols DP_1, DP_2 and IP_1. The d-bas-set DBS_1 is clearly the sequence IP_1, DP_1, DP_2 which is both passive and d-irreducible and is

19

thus the final d-char-set. As the initials and separants are all non-zero, the alg-d-var $d-Zero(DPS)$ is d-irreducible, being consisting of a single component of which a generic d-zero may be found from DBS_1 by computations. In particular IP_1 is the same as DQ_1 so that $DQ_1 = 0$ is automatically proved.

In order to prove $DQ_2 = 0$ let us set $Y = Y_2, Z = Y_1$ so that

$$DP_1 = 2 * Y_2 + D10Y_1 + Y_1 \wedge 2,$$
$$DP_2 = 2 * D20Y_2 - 4 * Y_1 * D10Y_2 + 8 * Y_2 \wedge 2 + 4 * Y_1 \wedge 2 * Y_2 - D01Y_1.$$

The d-bas-set DBS_0 of $DPS_0 = DPS = \{DP_1, DP_2\}$ is then consisting of the single d-pol $DB_1 = DP_1$ with lead Y_2, the ordering being still of type 2. The d-remainder of DP_2 is readily found to be $DR = D - Remdr(DP_2/DBS_0) = -D30Y_1 - D01Y_1 + 6 * Y_1 \wedge 2 * D10Y_1$ which coincides with $-DQ_2$. This proves $DQ_2 = 0$. Morover, the final d-char-set is consisting of the two d-pols DR and DP_1 which shows again that the alg-d-var $d - Zero(DPS)$ is d-irreducible with a single component whose generic d-zero may be computed by means of the above d-char-set.

References

[C] E. Cartan, Les systemes differentiels exterieurs et leurs applications geometriques, Hermann (1946).

[J] M. Janet, Lecons sur les systemes d'equations aux derivees partielles, Gauthier Villars(1920).

[P1] J. F. Pommaret, Geometrie differentialle algebrique et theorie du controle, C. R. Acad. Sc. Paris, 302, Ser. I, (1986), 547–550.

[P2] ——, Lie pseudogroups and mechanics, Gordon and Breach, (1988).

[R1] J. F. Ritt, Differential equations from the algebraic stand-point, Amer. Math. Soc., (1932).

[R2] ——, Differential algebra, Amer. Math. Soc., (1950).

[RQ] C. H. Riquier, La methode des fonctions majorantes et les systemes d'equations aux derivees partielles, Gauthier–Villars(1928).

[T1] J. M. Thomas, Differential systems, Amer. Math. Soc. (1937).

[T2] ——, Riquier's existence theorems, Annals of Math. 30(1929), 286–310; 35(1934), 306–311.

[WU1] Wu Wen-tsun, Basic principles of mechanical theorem-proving in elementary geometries, J. Sys. & Sci. Math. Scis., 4(1984), 207–235. Republished in J. of Automated Reasoning, 2(1986), 221–252.

[WU2] ——, On zeros of algebraic equations— an application of Ritt principle, Kexue Tongbao 31(1986), 1-5.

[WU3] ——, On the mechanization of theorem proving in elementary and differential geometry, Scientia Sinica, Math. Supp. (I), 94-102, (1979)(in Chinese).

[WU4] ——, Mechanical theorem proving in elementary geometry and differential geometry, in Proc. 1980, Beijing DD-Symposium, Beijing, V. 2, 1073-1092(1982).

[WU5] ——, A constructive theory of differential algebraic geometry, Proc. DD6-Symposium, 1985 Shanghai(1987), 173–189.

[WU6] ——, A mechanization method of geometry and its applications, 2. Curve pairs of Bertrand type, Kexue Tongbao, 32(1987), 585–588.

[WU7] ——, Mechanical derivation of Newton's Gravitational Laws from Kepler's Laws, MM-Res. Preprints, Institute of Systems Sciences, No. 1(1987), 53–61.

[WU8] ——, Automatic derivation of Newton's Gravitational Laws from Kepler's Laws, to appear in New Trends in Automated Mathematical Reasoning, Eds. A. Ferro et al.

MM Research Preprints

No. 4, 1989, 1-21

On the Generic Zero and Chow Basis of an Irreducible Ascending Set

WU WEN-TSUN
(Institute of Systems Science, Academia Sinica)

Introduction.

For both equations-solving and theorem-proving, the two main topics of study of our mathematics mechanization, there are two methods of approach to follow, viz. the zero-set approach and the ideal-theoretical approach. While the latter one dominates the present-day mathematics, we have proposed instead the former one and has shown its advantages in certain instances. One of such instances is furnished by the theory of elimination, as shown in [WU6] and some other papers in this Preprints. More evident is its role in mechanical geometry theorem proving, to be abbreviated as MTP in what follows. Along with the zero-set approach and the ideal-theoretical approach (mainly through the use of Groebner Basis, cf. e.g. [BU], [KP], and [K-S]), Chou in [CH] has given a comparison through a large scale experiment. In the last section of this paper we show that even in the intricate reducible case of MTP, the zero-set approach is still very effective while other approaches may become difficult to deal with.

We remark that, for the algebraic geometry as an example, the more intuitive and more direct zero-set approach occured much earlier than the ideal-theoretical one. In fact, the classical treatise of Van der Waerden on algebraic geometry, viz. [VdW], was written in a manner with no mention of ideals at all. The whole theory was based on the two central concepts: GENERIC POINT and SPECIALIZATION. Only in later times algebraic geometry becomes more and more ideal-theoretical in character. The notions of generic points and specialization gradually disappeared. The only trace of zero-set approach which remains seems to be the very definition of an algebraic variety as the zero-set of a set of polynomials. It goes without saying that the ideal-theoretical approach is a method which has been proved to be very powerful and very fruitful. However, there are still interests in renewing the rather old-fashioned zero-set approach of algebraic geometry. For example, the author has introduced a simple and computational definition of Chern Classes of an irreducible algebraic variety with arbitrary singularities via its generic point, cf. [WU8,9]. As a consequence one may prove and even discover in a simple computational way a lot of inequalities of such varieties, including the Miyaoka-Yau inequalities in particular. Cf. [WU9] and a series of papers of Shi beginning with [SH1].

In the present paper we shall base our method of mathematics mechanization on the rather naive notion of zero-set. We gather thereby results which have been scattered in the diverse papers of the author. Briefly, an irreducible algebraic variety (affine or projective) is completely determined by its generic point and such a generic point may be describled in the explicit form of what we have called an IRREDUCIBLE ASC-SET. From such an explicit form we can form, by means of the so-called Chow Form, a basis of the prime ideal associated to the corresponding algebraic variety in question. Such a basis will then be called a CHOW BASIS of the irreducible asc-set. The first 3 sections of this paper give a summary of the interrelations of the concepts involved. The 4-th section gathers together the various Decomposition Theorems on which are based our methods of equations-solving and also the thereby deduced various Principles on which are based our methods of MTP. The last two sections give some examples of MTP to illustrate our methods in view of jetting some lights on our zero-set approach in comparison with other ideal-theoretical approaches. Cf. in particular [WU4,7].

§ 1. Generic Point of an Irreducible Ascending Set.

Let K be a basic field of characteristic 0 and $X_1, ..., X_n$ be a set of indeterminates fixed throughout the whole paper. Moreover, EK, K', and K'' will denote throughout some extension fields of K. Let us introduce a further indeterminate X_0. For any point $X' = (X'_1, ..., X'_n)$ in the affine n-space $EK(n)$ with X'_i in some extension field EK of K the point $(1 : X'_1 : ... : X'_n)$ in the corresponding projective n-space $PEK(n)$ of homogeneous coordinates $(X_0 : X_1 : ... : X_n)$ will be denoted by $Pr(X')$. Conversely, a point $X' = (X'_0 : X'_1 : ... : X'_n)$ in $PEK(n)$ will be said to be AT INFINITY if $X'_0 = 0$, and if $X'_0 \neq 0$, then the point in $EK(n)$ represented by $(X'_1/X'_0, ..., X'_n/X'_0)$ will be denoted by $Af(X')$.

By a POL we shall mean a polynomial in $K[X_1, ..., X_n]$ unless otherwise stated. Then a homogeneous polynomial in $K[X_0, X_1, ..., X_n]$ will be simply called an H-POL. The *H-pol* in $X_0, X_1, ..., X_n$ obtained from a pol F in making it a homogeneous one by means of adjoining X_0 in a natural way will then be denoted by $Pr(F)$. Conversely, the pol obtained from an *H-pol* F by setting X_0 to 0 will be denoted by $Af(F)$.

Consider now the case for which EK is a $FINITE$ extension field of K. For the point X' of $EK(n)$ as above let $U_1, ..., U_d$ be those X'_i which are altogether transcendental over K and arranged in the same order as in $(X_1, ..., X_n)$. Denote the extension field of K in adjoining $U = (U_1, ..., U_d)$ to K by K_0. The X'_i other than U_j are all algebraic over K_0, which, in the same order as in $(X'_1 ..., X'_n)$, will be renamed as $(Y'_1, ..., Y'_e)$, with $e + d = n$.

The element Y'_1, being algebraic over K_0, will satisfy an equation of the form $F_1 = 0$, where

$$F_1 = I_1 * Y_1^{D_1} + \text{ lower degree terms in } Y_1,$$

is a polynomial in $K[U, Y_1]$, irreducible over K_0. Let K_1 be the algebraic extension field of K_0 by adjoining Y'_1. Then Y'_2 is algebraic over K_1 and satisfies an equation of the form $F'_2 = 0$, where F'_2 is obtained from a polynomial

$$F_2 = I_2 * Y_2^{D_2} + \text{ lower degree terms in } Y_2$$

of $K[U, Y_1, Y_2]$ in replacing Y_1 by Y'_1 and F'_2 is irreducible over K_1. We adjoin now Y'_2 to K_1 to get a field K_2. The procedure can be continued and finally we get an equation $F'_e = 0$ where F'_e is obtained from a polynomial

$$F_e = I_e * Y_e^{D_e} + \text{ lower degree terms in } Y_e$$

of $K[U, Y'_1, ..., Y'_{e-1}]$ in replacing $Y_1, ..., Y_{e-1}$ by $Y'_1, ..., Y'_{e-1}$ resp., and F'_e is irreducible over $K_{e-1} = K_0(Y'_1, ..., Y'_{e-1})$. The final algebraic extension field obtained from K_{e-1} by adjoining Y'_e will be denoted by K_e. It is a finite extension field of K and is intrinsically determined by X' up to a field isomorphism preserving K_0.

Now the sequence of pols

$$(IRR) \qquad F_1, F_2, ..., F_e$$

forms what we have called an IRREDUCIBLE ASCENDING SET or IRREDUCIBLE ASC-SET of pols in $X_1, ..., X_n$ over K. The above thus shows how an irreducible asc-set can be deduced from a point in an affine n-space over a finite extension field of K.

Conversely, let an irreducible asc-set of the form (IRR) as above be given. Rename the principal variables of the pols P_i by Y_i and the other X as $U_1, ..., U_d$ in the same order as in $X_1, ..., Xn$. Then F_1 is a polynomial of $K[U, Y_1]$ which is irreducible in $K_0 = K(U)$. Let Y'_1 be a zero of P_1 in a certain extension field of K_0 and set $K_1 = K_0(Y'_1)$ as the algebraic extension field by adjoining Y'_1 to K_0. Then F_2 is a polynomial of $K[U, Y_1, Y_2]$ for which F'_2 obtained from F_2 in replacing Y_1 by Y'_1 is irreducible in K_1. Take any root Y'_2 of F'_2=0 in a certain extension field of K_1. Adjoin Y_2' to K_1 we get then a field K_2. The procedure can be continued and we get a point $X' = Perm(U_1, ..., U_d, Y_1', ..., Y_e')$ with Y_i' in some extension field EK of K where $Perm$ denotes the permutation which brings $(U_1, ..., U_d, Y_1, ..., Y_e)$ back to $(X_1, ..., X_n)$. It is clear that the irreducible asc-set which one would get from X' is no other but the given (IRR) except some unimportant factors for each Fi. The relation between an irreducible asc-set (IRR) and the point X' as above is a reciprocal one and we shall call the point X' thus determined from (IRR) a GENERIC POINT of (IRR).

Let (IRR) be now an irreducible ascending set as given above and

$$GZ = (Z_1', Z_2', ..., Z_n')$$

be a generic point of (IRR). The importance of the notion of a generic point may then be seen from the following theorems.

Theorem 1.1. The generic zero GZ of an irreducible asc-set (IRR) is not a zero of any pol P reduced w.r.t. (IRR).

Theorem 1.2. The generic zero GZ of an irreducible asc-set (IRR) will be a zero of a pol P if and only if the remainder R of P w.r.t. (IRR) is 0.

The Theorem 1.1 follows directly from the construction of GZ. The Theorem 1.2 follows then from the remainder formula of P, viz.

$$IP * P = SUM_i\, A_i * F_i + R,$$

in which R is the remainder and IP is a certain power product of initials of pols F_i in (IRR). In fact, as all such initials are reduced w.r.t. (IRR), GZ is not a zero of IP. As GZ is a zero of all F_i, so from the remainder formula we see that GZ will be a zero of P if and only if it is a zero of R, as to be proved.

The set of all pols having GZ as a zero forms clearly a prime ideal in the ring $K[X_1, ..., X_n]$. This ideal will englobe all the pols F_i and hence all the linear combinations of F_i with arbitrary pols as coefficients. However, in general this ideal will englobe much more pols than these mere combinations, as we shall see in later sections. To avoid confusion with the ideal generated by F_i which is usually denoted as $(F_1, ..., F_e)$ we shall denote the ideal of pols having GZ as zero by $Ideal[GZ]$ or $Ideal[IRR]$. On the other hand the usual ideal $(F_1, ..., F_e)$ will be denoted by $Ideal(IRR)$.

The ideal $Ideal[IRR]$ is consisting of an infinity of pols which has, however, by the finite-basis theorem of Hilbert, necessarily a finite basis. Simple examples show that in general such a finite basis will necessarily contain pols not any linear combination of F_i, i.e. pols not in $Ideal(IRR)$. To determine explicitely such a finite basis of $Ideal[IRR]$ is however not an easy matter. One of such determinations is furnished by the notion of Chow Forms which will be explained in the latter sections.

§ 2. Generic Point of an Algebraic Variety.

Let HPS be a finite set of *H-pols* in $X_0, X_1, ..., X_n$ over K. A zero $X' = (X'_0 : X'_1 : ... : X'_n)$ of all *H-pols* in HPS with X'_i not all zero and all X'_i in some extension field of K will be called an *H-ZERO* of HPS. The set of all such *H-zeros* will be denoted by $HZero(HPS)$. A PROJECTIVE ALGEBRAIC VARIETY in the projective space $PK(n)$ in the ordinary sense is just the set $HZero(HPS)$ for some finite set of *H-pols* HPS. Any zero in $HZero(HPS)$, considered as a point in $PEK(n)$ for some extension field EK of K, will also be called a *POINT* of the variety $HZero(HPS)$. Such a projective algebraic variety is said to be IRREDUCIBLE if it is not the union of two distinct projective algebraic varieties both distinct from the given one.

Let $Z' = (Z'_0 : ... : Z'_n)$ be a point in the projective n-space $PK'(n)$ with Z'_i not all 0 and all Z'_i in some extension field K' of K. Similarly let $Z" = (Z_0" : ... : Z_n")$ be a point in $PK"(n)$ for some extension field $K"$ of K. Then the point $Z"$ is called an H-SPECIALIZATION of Z' if any *H-pol* having Z' as a zero will have $Z"$ as an *H-zero* too. A point Z in a projective space $PEK(n)$ for some extension field EK of K is said to be an H-GENERIC POINT of the projective algebraic variety $HZero(HPS)$ if any point of $HZero(HPS)$ is an H-specialization of Z. In particular the generic point Z itself is then a point of that variety. A fundamental theorem of algebraic geometry says now:

Theorem 2.1P. A projective algebraic variety is irreducible if and only if it has an H-generic point.

Let $Z = (Z_0 : ... : Z_n)$ be a point in the projective n-space $PEK(n)$ for some extension field EK of K. The set of all H-specializations of Z will be denoted by $HSpec(Z)$. On the other hand the set of all H-pols with Z as a zero forms a prime homogeneous ideal which will be denoted by $HIdeal(Z)$. It is clear that

$$HSpec(Z) = HZero(HIdeal(Z)).$$

Now by Hilbert Finite Basis Theorem the ideal $HIdeal(Z)$ has a finite set of H-pols, say FB, as its basis so that

$$HZero(HIdeal(Z)) = HZero(FB).$$

The last one is by definition a projective algebraic variety in $PK(n)$ and is irreducible since the ideal $HIdeal(Z)$ is prime. Moreover it is easily seen from the very definitions that Z is an H-generic point of this variety. Hence we have

Theorem 2.2P. For any point Z in a projective n-space $PEK(n)$ with EK some extension field of K the set $HSpec(Z)$ is a projective irreducible algebraic variety in $PK(n)$ with Z as an H-generic point.

The above notions can be naturally extended to the affine case. Thus, for extension fields K', K'' of K, a point X'' in the affine n-space $K''(n)$ will be called a SPECIALIZATION of a point X' in the affine n-space $K'(n)$ if any pol having X' as a zero will have X'' as a zero too. The set of all specializations of X' will be denoted by $Spec(X')$. For any polset, i.e. a finite set of pols over K, say PS, the set $Zero(PS)$ will be called an AFFINE ALGEBRAIC VARIETY in the affine space $K(n)$ and any zero in $Zero(PS)$ is called a *POINT* of this variety. The variety is said to be IRREDUCIBLE if it is not the union of two distinct affine algebraic varieties both distinct from the given one. Furthermore, a point in $K'(n)$ for some extension field K' of K is called a *GENERIC POINT* of an affine algebraic variety if any point of that variety is a specialization of that point. Given any point Z' of $K'(n)$ the set of all pols having Z' as zero which forms a prime ideal will be denoted by $Ideal(Z')$. Analogous to the previous theorems we have then the following ones.

Theorem 2.1A. An affine algebraic variety is irreducible if and only if it has a generic point.

Theorem 2.2A. For any point Z in an affine n-space $EK(n)$ with EK some extension field of K the set $Spec(Z)$ is an affine irreducible algebraic variety in $K(n)$ with Z as a generic point.

Now the notions of projective resp. affine algebraic varieties have some close relations as described below.

Let $V = HZero(HPS)$ be a projective irreducible algebraic variety in $PK(n)$ with HPS a finite set of *H-pols*. Let $GZ = (Z_0' : Z_1' : \ldots : Z_n')$ be an H-generic point of V. Suppose that V is not wholly at infinity. Then Z_0' is unequal to 0 so that $Af(GZ)$ is well-defined. The affine irreducible algebraic variety with $Af(GZ)$ as generic point will then be denoted by $Af(V)$. It is clear that $Af(V) = Zero(PS)$ where PS is the polset consisting of pols $Af(F)$ with F an H-pol in HPS and any point in $Af(V)$ is of the form $Af(X)$ with X a point of V. Conversely, given an affine irreducible algebraic variety $V' = Zero(PS)$ with PS some polset let GZ' be a generic point of V'. Then there will be determined a projective irreducible algebraic variety $V = HZero(HPS)$ with $Pr(GZ')$ as H-generic point and HPS the set of H-pols of the form $Pr(F)$ for F in PS. The points of V which are not at infinity are just those of the form $Pr(X')$ with X' points of V'. The relations between $Af(V)$ and $Pr(V')$ for V a projective irreducible algebraic variety not wholly at infinity and V' an affine irreducible algebraic variety are clearly a reciprocal one in the sense that

$$Pr(Af(V)) = V, \text{ and } Af(Pr(V')) = V'.$$

In conclusion, we see that the notions of (H-)generic point, irreducible ascending set, and irreducible algebraic variety (projective or affine) are in essence equivalent ones in that one may be determined from the other.

§ 3, Chow Form of an Irreducible Algebraic Variety or an Irreducible Ascending Set.

In [C-VdW] Chow introduced the notion of ZUGEORDNETE FORM of a projective irreducible algebraic variety and then extended to that of an arbitrary projective algebraic variety via its irreducible components. This notion was later called the CAYLEY FORM by Hodge in [H-P] and was called CHOW FORM by the French school of algebraic geometers. We shall adopt the terminology of CHOW FORM owing to its originator which is quite current in the literature.

The Chow form of a projective irreducible algebraic variety is in fact determined by its generic point as follows. Let $V = HZero(HPS)$ be a projective irreducible algebraic variety with HPS a finite set of H-pols. Suppose that V is not wholly at infinity. Then V will have an H-generic point of the form

$$GZ = (1 : Z'_1 : \ldots : Z'_n).$$

Let the degree of transcendency of Z over K be d which is in fact the DIMENSION of the variety V. Introduce now a set of independent indeterminates U_{ij} with (i,j) running over the range

$$R : i = 0, 1, \ldots, d; \quad j = 1, \ldots, n.$$

Adjoin these U_{ij} to K to form the transcendental extension field UK over K. Set now $Z'_0 = 1$ and introduce $d + 1$ elements U_{i0} by

$$SUM_j\ U_{ij} * Z'_j = 0, \quad i = 0, 1, \ldots, d$$

in which SUM_j runs over j from 0 to n. As the $d + 1$ elements

$$U_{i0} = -SUM_{j'}\ U_{ij} * Z_{j'}$$

7

in which $SUM_{j'}$ runs over j from 1 to n are all algebraic over K and altogether have clearly a transcendence degree d over UK, there will be an algebraic relation among these U_{i0}. This relation, after clearing of fractions, will be of a form $CF = 0$, with CF a polynomial in $K[U_{ij}]$, in which (i,j) is over the range

$$UR: \quad i = 0, 1, ..., d; \quad j = 0, 1, ..., n.$$

This polynomial CF, which is determined up to certain non-zero constant in K, is then called the CHOW FORM of the projective irreducible algebraic variety in question.

In [C-VdW] it was described how a finite basis of H-pols CB can be determined from the Chow Form of a projective (irreducible) algebraic variety V such that $V = HZero(CB)$. In [H-P] a simpler method of such a determination is also given. This method depends on the following two propositions.

Prop.H1. ([H-P] X7,Th.IV) For any $(d+1)$-tuple of distinct integers $t = (I_0, I_1, ..., I_d)$ chosen from (0,1,...,n) let Pt be the determinant of a $(d+1)*(d+1)$ matrix of which the element in i-th row and j-th column is U_{ik} with $k = I_j$, i, j being both from 0 to d. Then the Chow form of a projective irreducible algebraic variety of dimension d is a polynomial G in Pt with t running over all $(d+1)$-tuples t.

Let G be now the Chow Form of a projective irreducible algebraic variety V of dimension d expressed in terms of Pt as in Prop. H1. Take a set of indeterminates A_{ij} with (i,j) running over the range

$$R': \quad i = 0, 1, ..., n-d-2; \quad j = 0, 1, ..., n.$$

Set $n - d - 2 = c$. For any $(c+1)$-tuple of distinct integers $s = (J_0, J_1, ..., J_c)$ chosen from (0,1,...,n) let Qs be the determinant of a $(c+1)*(c+1)$ matrix of which the element in the i-th row and j-th column is A_{ik} with $k = J_j$, i, j being both from 0 to c. In G let us replace each Pt by

$$Pt = Et * SUM_k\, (-1)^k * X_{rk} * Q_{rk}.$$

In the formula:

(1) SUM_k runs over k from 0 to $c1$;

(2) r is the $(n-d)$-tuple which one obtains from $(0, 1, \ldots, n)$ in deleting the integers appearing in t;

(3) Et means $+1$ or -1 according as the permutation from the ordered concatenated set of t and then r to the ordered set $(0, 1, \ldots, n)$ is an even or an odd one;

(4) Q_{rk} is Q_s with s the $(c+1)$-tuple in deleting the $(k+1)$-th integer in the $(n-d)$-tuple r, k being from 0 to c;

(5) X_{rk} is the variable X_h with h the k-th integer in r, k being from 0 to c.

Denote now by G' the polynomial in terms of Q_s obtained from G as above and write it in the form

$$G' = SUM_p\ H_p * K_p,$$

in which SUM_p runs over p indexing all distinct power products H_p of A_{ij}, and K_p are the corresponding coefficients which are H-pols in $X = (X_0, X_1, ..., X_n)$. Then we have ([H-P] X7,p45-46):

Prop.H2. The H-pols K_p form a finite basis CB, to be called a CHOW BASIS, such that

$$V = HZero(CB).$$

The above introduction of Chow forms and Chow basis and their determination for a projective algebraic variety can be naturally extended to affine irreducible algebraic varieties or irreducible asc-sets in the following way.

Consider an affine irreducible algebraic variety $V = Zero(PS)$ with a generic point GZ. Let $V' = Pr(V)$ be the projective irreducible algebraic variety with $Pr(GZ)$ as H-generic point. Let us form the Chow Form CF of this variety from $Pr(GZ)$ as above described. Then CF will be defined also as the CHOW FORM of the affine variety V.

Consider next any irreducible asc-set

$$(IRR) \qquad F_1,\ F_2, ...,\ F_e.$$

Let GZ be the generic point of (IRR) which is also a generic point of the affine irreducible algebraic variety $Spec(GZ) = Var[IRR]$. The Chow Form of Var[IRR] which may be determined from GZ will then be defined as the CHOW FORM of the irreducible asc-set (IRR).

The method described above for a projective irreducible algebraic variety gives also the means of determining the Chow Form of an affine irreducible algebraic variety or an irreducible asc-set In the same way the propositions H1 and H2 will give a Chow Basis CB such that the corresponding affine irreducible algebraic variety V or the variety $Var[IRR]$ of an irreducible ascending set (IRR) will be determined as $Zero(CB)$. Such a basis will also be called a CHOW BASIS of the ideal $Ideal[IRR]$ or simply a CHOW BASIS of the irreducible asc-set (IRR). As shown by examples in later sections, this basis may contain pols which differ from each by mere signs. The basis obtained by deleting the redundant ones will then be called a CONDENSED CHOW BASIS of the corresponding ideal $Ideal[IRR]$ or the irreducible asc-set (IRR). For such a basis CB we have then

$$Var[IRR] = Spec(GZ) = Zero(CB).$$

From the above and the last sections we see that the notions of an irreducible ascending set, an affine irreducible algebraic variety, the generic point, the Chow Form or Chow Basis, etc. are intimately interrelated and in essence they are equivalent to each other in that they may be determined one from the other. Among these concepts the irreducible asc-set may be considered as the central one since it is more explicit and from it others are relatively easier to be determined. The following diagram illustrates such interrelations with the irreducible asc-set IRR enjoying the central role. Cf. also the Sect 5 of the paper [WU7].

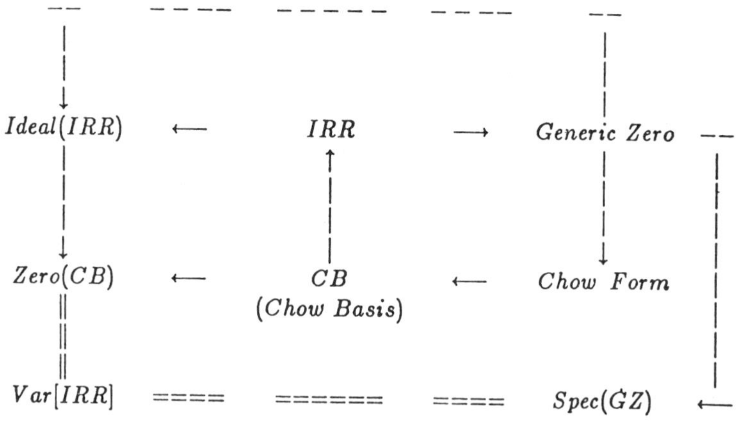

§ 4. Decomposition Theorems and Principles of Mechanical Geometry Theorem-Proving.

We recapitulate in this section the fundamental theorems in the form of DECOMPOSITION THEOREMs of zero-sets which are at the basis of our mechanization method of equations-solving and theorem-proving. For more details we refer to the relavant papers of the author.

DECOMPOSITION THEOREM D1. For any polset PS we have

$$Zero(CS/J) \subset Zero(PS) \subset Zero(CS), \qquad (I)$$

$$Zero(PS) = Zero(CS/J) + SUM_i Zero(PSi), \qquad (II)$$

in which CS is a char-set of PS, PS_i is the polset PS enlarged by adjoining to it the initial I_i of the i-th pol in CS, and J is the product of all these initials.

DECOMPOSITION THEOREM D2. For any polset PS we have a decomposition of the form

$$Zero(PS) = SUM_j \ Zero(ASC_j/J_j), \qquad (III)$$

in which each ASC_j is an asc-set and each J_j is the product of all initials of pols in ASC_j.

DECOMPOSITION THEOREM D3. For any polset PS we have a decomposition of the form

$$Zero(PS) = SUM_k \ Zero(IRR_k/J_k), \qquad (IV)$$

in which each IRR_k is an irreducible asc-set and J_k is the product of all initials of pols in IRR_k.

DECOMPOSITION THEOREM D4. For any polset PS we have a decomposition into irreducible components of the algebraic variety $Zero(PS)$ in the form

$$Zero(PS) = SUM_k \ Var[IRR_k]. \qquad (V)$$

In the decomposition (V) it may happen that one irreducible component is contained in the other. Whether this is so for two such components may be verified by mere computations via Chow Basis based on the following

Theorem 4.1. For any two irreducible asc-sets IRR and IRR' let CB be a Chow Basis of IRR. Then

$$Var[IRR'] \subset Var[IRR]$$

if and only if for any pol B in CB we have

$$Remdr(B/IRR') = 0.$$

Proof. By Sect.3, we have

$$Var[IRR] = Zero(CB).$$

Let GZ' be a generic point of IRR'. By Theorem 2.2A we have

$$Var[IRR'] \subset Zero(CB)$$

if and only if GZ' is in $Zero(B)$ for all B of CB. By Theorem 1.2, the latter is true if and only if $Remdr(B/IRR') = 0$. This proves the theorem.

It follows from this theorem that we can remove in (V) any redundant component $Var[IRR_i]$ by mere computations of remainders to render the decomposition uncontractible. That such an uncontractible decomposition is unique is clear. So we can strengthen the DEC.TH. D4 to the following

DECOMPOSITION THEOREM D5. The decomposition of $Zero(PS)$ in irreducible components as in (V) can be made uncontractible and unique by mere computations. Moreover, if CB_k is a Chow Basis of IRR_k in the above decomposition, then we may also write (V) in the form

$$Zero(PS) = SUM_k\ Zero(CB_k). \qquad (V')$$

The applications of the above decomposition theorems to equations-solving are self-evident. Consider now the problem of theorem-proving. The basic field K will then be understood to be the rational field Q.

Consider thus a theorem $T = \{HYP, CONC\}$ with HYP the hypothesis polset and $CONC$ the conclusion pol in same variables $X_1, ..., X_n$. Use MTP to stand for MECHANICAL THEOREM-PROVING. Then by DEC.TH.D1 we deduce the following

MTP PRINCIPLE P1. For a theorem $T = \{HYP, CONC\}$ let

$$Zero(HYP) = Zero(CS/J) + SUM_i\ Zero(HYP_i),$$

be the decomposition of $Zero(HYP)$ by means of DEC.TH.D1, in which CS is the char-set of HYP, J is the product of all initials I_i of pols in CS and HYP_i is the polset HYP enlarged by adjoining to it the i-th initial I_i of CS. Suppose that the remainder of $CONC$ w.r.t. the char-set CS of HYP is 0, i.e.

$$Remdr(CONC/CS) = 0, \qquad (VI)$$

then the theorem T is GENERICALLY TRUE under the NON-DEGENERACY CONDITION

$$J \neq 0,$$

or the NON-DEGENERACY CONDITIONNs (r = number of pols in CS)

$$I_i \neq 0, \quad i = 1, ..., r.$$

Moreover, if the char-set CS is an irreducible one, then the condition (VI) is also a necessary one for T to be generically true under the above non-degeneracy conditions.

By the various decomposition theorems about $Zero(HYP)$ we deduce also:

MTP PRINCIPLE P2. For a theorem $T = \{HYP, CONC\}$ let us form the decomposition

$$Zero(HYP) = SUM_k\ Zero(ASC_j/J_j)$$

in which each ASC_j is an asc-set and J_j is the product of all initials of pols in ASC_j. Then T will be generically true on the part $Zero(ASC_j/J_j)$ if

$$Remdr(CONC/ASC_j) = 0. \qquad (VII)$$

Conversely, if the asc-set ASC_j is an irreducible one, then the condition (VII) is also necessary for T to be generically true on the part $Zero(ASC_j/J_j)$.

MTP PRINCIPLE P3. For a theorem $T = \{HYP, CONC\}$ let us form the decomposition

$$Zero(HYP) = SUM_k\ Var[IRR_k]$$

in which each IRR_k is an irreducible asc-set. Then T will be true on the whole irreducible component $Var[IRR_k]$ if and only if

$$Remdr(CONC/IRR_k) = 0. \qquad (VIII)$$

We shall give in next sections some examples to illustrate how the above Principles are applied in concrete cases. The method used here will also throw some light on our zero-set approach vis-a-vis the usual ideal-theoretical approach at least in the case of mechanical theorem proving.

§ 5. Example: The Desargues Theorem.

We take again, as in [WU6,7], the Desargues Theorem as example to illustrate our general theory and method explained in previous sections.

Example 1. Desargues Theorem. Let ABC, $A'B'C'$ be triangles with corresponding sides parallel to each other. If two joining lines of pairs of corresponding vertices, say AA' and BB', meet in a point O, then the joining line of the third pair CC' will pass through O too.

To prove this let us take AA', BB' as coordinate axis. Let the points in question have resp. the coordinates:

$$A = (X_1, 0),\ \ B = (0, X_3),\ \ C = (X_4, X_5),$$
$$A' = (X_2, 0), B' = (0, X_6), C' = (X_7, X_8).$$

Then the hypothesis-set will be $HYP = \{H_1, H_2, H_3\}$ with

$$\begin{aligned} H_1 &= X_1 * X_6 - X_2 * X_3, & [AB \| A'B'] \\ H_2 &= X_4 * (X_8 - X_6) - X_7 * (X_5 - X_3), & [BC \| B'C'] \\ H_3 &= X_8 * (X_4 - X_1) - X_5 * (X_7 - X_2). & [CA \| C'A'] \end{aligned}$$

The conclusion pol is given by

$$CONC = X_4 * X_8 - X_5 * X_7. \qquad [O \text{ on } CC']$$

As pointed out in [WU6,7], no power of $CONC$ can be contained in the ideal (H_1, H_2, H_3) so that $CONC = 0$ is not a consequence of the equations $H_i = 0$, $i = 1, 2, 3$. This will be true only under certain non-degeneracy conditions in the form of inequalities which are not easily foreseen from the very hypothesis set HYP. This will cause inherent difficulties in the ideal-theoretical approach to the mechanical geometry theorem proving. However, in our zero-set approach the difficulties will be automatically resolved as explained in what follows.

Method I.

Based on Decomposition Theorem D1 we form first the char-set CS of HYP with

$$Zero(HYP) = Zero(CS/J) + SUM_i\ Zero(HYP_i).$$

In the formula $CS = C_1, C_2, C_3$ with

$$C_1 = H_1, \quad C_3 = H_2, \quad while$$
$$C_2 = (X_1 * X_3 - X_1 * X_5 - X_3 * X_4) * X_7 + (X_4 - X_1) * X_4 * X_6 + X_2 * X_4 * X_5$$

The initials are

$$I_1 = X_1, \quad I_2 = X_1 * X_3 - X_1 * X_5 - X_3 * X_4, \quad I_3 = X_4,$$

and J is the product $I_1 * I_2 * I_3$. Each polset HYP_i is HYP enlarged by adjoining to it the initial I_i.

It is readily verified that

$$Remdr(CONC/CS) = 0,$$

It follows that

$$Zero(HYP/J) \subset Zero(CONC),$$

or the Desargues Theorem is true so far

$$J \neq 0, \quad or$$
$$I_1 \neq 0, \quad I_2 \neq 0, \quad I_3 \neq 0.$$

Each of the conditions $I_i \neq 0$ has some evident geometrical meaning. The more troublesome one is the condition $I_2 \neq 0$ which says that the triangle ABC (and hence also the triangle $A'B'C'$) should not be degenerate into a line. If we are in doubt whether $I_2 \neq 0$ is really an unavoidable condition for the truth of Desargues Theorem we may just add I_2 to HYP to form a new hypothesis set $HYP2$ and apply our method to $HYP2$ as before.

Method II.

In order not to unnecessarilly complicate the computations let us consider X_1, X_2, X_3 to be constants assumed to be non-zero and $X_1 \neq X_2$. We will thus be working in an affine 5-space of coordinates $(X_4, X_5, X_6, X_7, X_8)$. A geometrical configuration verifying the hypothesis $HYP = 0$ is just a point on the algebraic variety $Zero(HYP)$ in that affine space.

Based on Decomposition Theorem D2 we find now

$$Zero(HYP) = SUM_i\ Zero(ASC_i/J_i),$$

in which SUM_i runs over $i = 1,2,3$ and each J_i is the product of initials of pols in corresponding asc-set ASC_i. The asc-sets ASC_i are resp. consisting of pols C_{ij} as follows.

$$\begin{aligned} &For\ \ ASC1: && C_{1j} = C_j,\ \ j = 1,2,3. \\ &For\ \ ASC2: && C_{21} = I_2,\ \ C_{22} = H_1,\ \ C_{23} = H_2. \\ &For\ \ ASC3: && C_{31} = I_3, C_{32} = H_1, C_{33} = X_7, \\ & && C_{34} = X_1 * X_8 + X_5 * (X_7 - X_2). \end{aligned}$$

It turns out that all these asc-sets ASC_i are irreducible ones so we shall rename them as IRR_i. The DEC.TH.D4 gives then:

$$Zero(HYP) = SUM_i\ Var[IRR_i].$$

It is readily found that

$$\begin{aligned} &Remdr(CONC/IRR_1) = 0, \quad Remdr(CONC/IRR_3) = 0, while \\ &Remdr(CONC/IRR_2) \neq 0. \end{aligned}$$

It follows that the Desargues Theorem is true on the whole irreducible components $Var[IRR_1]$ and $Var[IRR_3]$, but not so on the component $Var[IRR_2]$. The reason may be seen directly from geometrical considerations. In fact, A, A', B being fixed as X_1, X_2, X_3 are given constants, B' or X_6 is already well-determined. The variety $Var[IRR_1]$ is therefore consisting of the configuration of triangle pairs ABC, $A'B'C'$ with C in GENERIC position and of all other configurations obtained therefrom by specializations. The variety $Var[IRR_2]$ is consisting of configurations of such triangle pairs with C and C' taken arbitrarily on the lines AB and $A'B'$ resp. The variety $Var[IRR_3]$ is consisting of such configurations with C arbitrarily chosen on the axis OBB' and thereby C' is well-determined. We see also that these three components will have their dimensions equal to 2,2,1 resp., being the degrees of freedom of the configurations in question. Directly from geometrical considerations it is also readily seen that on the component $Var[IRR_2]$ the theorem can not be true. The above computations by means of remainder-formation serve to give systematic verifications of such geometrical evidences. Furthermore, we see from mere geometrical considerations that all configurations in $Var[IRR_3]$ are in fact specifications of the generic configuration in $Var[IRR_1]$ with C generically chosen. This means that $Var[IRR_3]$ should be contained entirely in $Var[IRR_1]$. By Theorem 4.1 this may also be verified by computations via the Chow Form or Chow Basis as follows.

In the affine 5-space A_5 the generic points GZ_i of IRR_i and hence also of $Var[IRR_i]$ are readily seen to be:

$$GZ_1 = (X_4, X_5, X_2 * X_3/X_1, X_2 * X_4/X_1, X_2 * X_5/X_1),$$
$$GZ_2 = (X_4, -X_3 * (X_4 - X_1)/X_1, X_2 * X_3/X_1, X_7, X_3 * (X_2 - X_7)/X_1),$$
$$GZ_3 = (0, X_5, X_2 * X_3/X_1, 0, X_2 * X_5/X_1).$$

The condensed Chow Basis are found to be consisting of pols B_{ij} for IRR_i, $i = 1,2$, as follows.

$$B_{11} = X_1 * X_6 - X_2 * X_3,\ B_{12} = X_1 * X_7 - X_2 * X_4,\ B_{13} = X_1 * X_8 - X_2 * X_5;$$
$$B_{21} = -X_1{}^2 * X_6 + X_1 * X_2 * X_5 + X_2 * X_3 * X_4,$$
$$B_{22} = X_1{}^2 * X_8 - X_1 * X_3 * X_7 - X_1 * X_2 * X_5 - X_2 * X_3 * X_4,$$
$$B_{23} = -X_1 * X_8 - X_3 * X_7 + X_1 * X_6,$$
$$B_{24} = X_1 * X_8 + X_3 * X_7 - X_2 * X_3,$$
$$B_{25} = X_1 * X_5 + X_3 * X_4 - X_1 * X_3,$$
$$B_{26} = -X_1 * X_6 + X_2 * X_3.$$

It is readily verified that

$$Remdr(B_{1j}/IRR_3) = 0,\quad j = 1, 2, 3.$$

It follows therefore from Theorem 4.1 that $Var[IRR_3]$ is contained in $Var[IRR_1]$, as already signified. We get in particular the unique decomposition into irreducible components of the affine algebraic variety $Zero(HYP)$, viz.

$$Zero(HYP) = Var[IRR_1] + Var[IRR_2].$$

The Desargues Theorem is now proved to be true on the whole first component but not so on the second one.

We make now some comparisons between the two methods as exhibited above. The Method II gives a complete answer to the problem of theorem-proving at the cost of much more computations being needed. It seems that the Method I, incomplete as it is in appearence, is often to be preferred in general. The reason is this. Actually all theorems in elementary geometry are only true in the GENERIC sense and are stated in a form with implicit GENERICITY hypothesis utterly unmentioned. Our Method I furnishes, disregarding at all any possible degeneracies, just such a GENERICITY proof in case it is TRUE, which should mean GENERICALLY TRUE and is in agreement with the real character what one means by a geometrical theorem. It avoids thus the labyrinth of degeneracy conditions unless some geometrical interest or practical need urges one to do so. It is owing to this reason that in our mechanical geometry theorem proving the Method I alone has proved already a great success. See however the next section for further comments. Finally, we would like to point out that the knowledge of Chow Basis of an irreducible component, though of importance from a theoretical point of view, may serve as little use for the proving of the theorem in question on this component. This may be seen clearly from the Chow Basis of IRR_2 as explicitely given above. Of course one can deal with IRR_1 easily from its Chow Basis. But this is just because the Chow Basis happens to be already in the form of an asc-set which is a quite rare case.

§ 6. Mechanical Theorem-Proving in the Reducible Case.

In the last section we have pointed out that in applying the various decomposition theorems to MTP, the method based on Dec.Th. D1 is preferred in general. We have to make however the supplemen-tary remark that there may arise some difficulties in applying D1 in e.g. the so-called REDUCIBLE case. In fact, for a theorem $T = \{HYP, CONC\}$ let

$$Zero(HYP) = Zero(CS/J) + SUM_i Zero(HYP_i) \qquad (*)$$

as before. It may happen that

$$Remdr(CONC/CS) \neq 0.$$

If the char-set CS is irreducible then by the decomposition theorems the theorem T in question is not true generically and not true on the irreducible component

$Var[CS]$. If CS is however reducible then nothing can be concluded. The reducibility may however arise owing to an inadequate choice of the coordinate system and may be avoided by adopting other coordinate systems. Such a choice may be furnished by the method of separation of variables X_i into two parts to be denoted by U_i and Y_i resp. as in Sect 1. This is in fact the earliest method adopted by the author, as exhibited in [WU1,2]. It amounts to choose among the X's those as independent ones and the others as bounded or dependent ones. In many cases with a suitable choice the corresponding char-set will become irreducible and Dec. Th.D1 may well be applied. There are however no general guiding rules of such choice to follow which depends heavily on the mathematical understanding of the theorem in question. There are also cases for which no such choice is possible at all. The only way seems then to apply the other decomposition theorems other than D1. The following is a concrete example for such cases.

Example 2. Center of Similitude Theorem. The three centers of similitude of the three pairs of circles taken from three circles in the plane will lie on the same line.

Let us take the coordinates such that the centers of the three circles Ci are resp.

$$O_1 = (X_1, 0), \quad O_2 = (X_2, 0), \quad O_3 = (0, X_3).$$

Let the radius of the circles be X_4, X_5, and X_6 resp. We assume here the three circles to be in generic position so that in particular $X_1, ..., X_6$ are all $\neq 0$, $X_1 \neq X_2$, and X_4, X_5, X_6 mutually unequal. Denote the three centers of similitude for the pairs $(C_1, C_2), (C_1, C_3)$, and (C_2, C_3) to be resp.

$$S_1 = (X_7, 0), \quad S_2 = (X_8, X_9), \quad S_3 = (X_{10}, X_{11}).$$

Then the hypothesis set $HYP = \{H_1, ..., H_5\}$ with

$$\begin{aligned}
H_1 &= X_4{}^2 * (X_7 - X_2)^2 - X_5{}^2 * (X_7 - X_1)^2,\\
H_2 &= X_6{}^2 * ((X_8 - X_1)^2 + X_9{}^2) - X_4{}^2 * (X_8{}^2 + (X_9 - X_3)^2),\\
H_3 &= X_1 * X_9 + X_3 * X_8 - X_1 * X_3,\\
H_4 &= X_6{}^2 * ((X_{10} - X_2)^2 + X_{11}^2) - X_5{}^2 * (X_{10}{}^2 + (X_{11} - X_3)^2),\\
H_5 &= X_2 * X_{11} + X_3 * X_{10} - X_2 * X_3.
\end{aligned}$$

The conclusion pol is given by

$$CONC = X_{11} * (X_8 - X_7) - X_{10} * X_9 + X_9 * X_7.$$

Let us consider $X_1, \ldots, X_6$ as constants and $X_7, \ldots, X_{11}$ as variables. It is easy to see that no matter how we rename $X_7, \ldots, X_{11}$ by a permutation which amounts to a separation of these X into U and Y, the corresponding char-set will always be a reducible one, even if we replace H_1 by one of its factors. For the char-set CS of HYP as above it is consisting of 5 pols C_i of which

$$C_i = H_i \text{ for } i = 3, 5.$$

On the other hand

$$C_1 = C_1' * C_1", \quad C_2 = C_2' * C_2", \quad C_4 = C_4' * C_4"$$

with

$$\begin{aligned}
C_1' &= X_4 * (X_7 - X_2) - X_5 * (X_7 - X_1),\\
C_1" &= X_4 * (X_7 - X_2) + X_5 * (X_7 - X_1),\\
C_2' &= X_6 * (X_8 - X_1) - X_4 * X_8,\\
C_2" &= X_6 * (X_8 - X_1) + X_4 * X_8,\\
C_4' &= X_6 * (X_{10} - X_2) - X_5 * X_{10},\\
C_4" &= X_6 * (X_{10} - X_2) + X_5 * X_{10}.
\end{aligned}$$

Applying Dec.Ths 2-4 we get then

$$Zero(HYP) = SUM_k \; Zero(IRR_i) = SUM_k \; Var[IRR_i],$$

with k running from 1 to 8. The IRR_i are irreducible asc-sets each consisting of 5 pols C_{i1} to C_{i5} with $C_{i3} = C_3$, $C_{i5} = C_5$ as above while each C_{ij} is either C_{ij}' or $C_{ij}"$ for $j = 1$, 2, or 4. It is readily verified that $Remdr(CONC/IRR_i) = 0$ for four of these IRR_i and $\neq 0$ for the other four. The theorem in question can thus be more precisely expressed in the following form:

The 6 centers of similitude of pairs of cicles taken from 3 circles in generic position lie 3 by 3 on 4 lines.

There are many methods in dealing with the reducible case due to various authors like Chou, Gao, Wang, and the author himself. Cf. also an interesting paper [SH2] in this MM-Preprints. Besides the one in using Dec.Ths. D2-D4, the author has proposed, specially for theorems in elementary geometries, the method of oriented lines and oriented circles. For the Example 2 above we see that the 3 centers of similitude will be uniquely determined and lie on the same line once the 3 circles are each definitely oriented. There are in all 8 different ways of orienting the circles which may be divided into 4 pairs with same triple of centers of similitude for each pair. This accounts for the 4 lines of centers of similitude in the theorem.

For more examples we refer to [WU9]. Much harder theorems than the Example 2 above have been proved in this way with relative ease, including a difficult theorem of Thebault-Taylor-Chou.

REFERENCES.

[BU] B.Buchberger, Ein algorithmus Kriterium fuer die Loesbarkeit eines algebraischen Gleichungssystems, Aeq.Math. 4 (1970) 374-383.

[CH] S.C.Chou, Mechanical geometry theorem-proving, Reidel, (1988).

[C-W] Chow,W.L. & Van der Waerden, Ueber Zugeordnete Formen und algebraische Systeme von Mannigfaltigkeiten, Math. Ann. 113 (1937) 692-704.

[H-P] W.V.D.Hodge & D.Pedoe, Methods of algebraic geometry, vol.2, Cambridge, (1952).

[KP] D.Kapur, Using Groebner bases to reason about geometrical problems, J.Symb.Comp. 2 (1986) 399-408.

[K-S] B.Kutzler & S.Stifter, On the application of Buchberger's algorithm to automated geometry theorem proving, J.Symb.Comp. 2 (1986) 389-397.

[R1] Ritt,J.F., Differential equations from the algebraic standpoint, Amer. Math. Soc., (1932).

[R2] ----, Differential algebra, Amer.Math.Soc., (1950).

[SH1] Shi He, On Chern characters of algebraic hypersurfaces with arbitrary singularities, Acta Math.Sinica, New Ser. 4 (1988) 289-300.

[SH2] ----,On the resultant formula for mechanical theorem proving, in this MM-Preprints.

[VdW] Van der Waerden, Einfuerung in die algebraischen Geometrie, Dover Publ., (1945).

[W-G] D.M.Wang & X.S.Gao, Geometry theorems proved mechanically using Wu's method — Part on Euclidean geometry, in MM-Res.Preprints, No.2, (1987) 75-106.

[WU1] Wu Wen-tsun, Basic Principles of Mechanical Theorem Proving in Geometries (Part on Elementary Geometries), (in Chinese), Science Press, (1984).

[WU2] —, On the decision problem and the mechanization of theorem-proving in elementary geometry, Scientia Sinica, 21 (1978) 159-172. Re-published in W. W. Bledsoe & D. W. Loveland, Automated Theorem Proving: after 25 Years, Amer. Math. Soc. (1984) 213-234.

[WU3] —, Basic principles of mechanical theorem-proving in elementary geometrics, J.Sys.Sci. & Math.Scis.,4 (1984) 207-235. Re-published in J. of Automated Reasoning, 2 (1986) 221-252.

[WU4] —, On zeros of algebraic equations — an application of Ritt principle, Kexue Tongbao 31 (1986) 1-5.

[WU5] —, On reducibility problem in mechanical theorem proving of elementary geometries, Chinese Quart.J.Math., 2 (1987) 1-19. Also in MM-Res.Preprints, Institute of Systems Science, No.2, (1987) 18-36.

[WU6] —, On a projection theorem of quasi-varieties in elimination theory, in this MM-Preprints.

[WU7] —, A mechanization method of equations-solving and theorem-proving, to be published.

[WU8] —, On Chern characteristic systems of an algebraic variety, (in Chinese), Shuxue Jinzhan, 8 (1965) 395-401.

[WU9] —, On Chern numbers of algebraic varieties with arbitrary singularities, Acta Math.Sinica, New Ser., 3 (1987) 227-236.

MM Research Preprints, 1–20
KLMM, AMSS, Academia Sinica
Vol. 6, Feb 1991

Mechanical Theorem Proving of Differential Geometries and Some of its Applications in Mechanics [1)]

Wu Wen-tsun
Institute of Systems Science, Academia Sinica
Beijing 10080, China

Abstract. Based on a well-ordering principle for differential polynomial sets, principles of mechanical theorem proving (MTP) and mechanical theorem discovering (MTD) are formulated and discussed. Examples are given to show how these principles may be applied to problems in differential geometries and mechanics.

Keywords. Mechanical theorem proving, mechanical theorem discovering, d-characteristic set, Well-Ordering Principle.

1. Introduction

In 1977/8 the author has introduced a method of mechanical theorem proving of elementary geometries which has been proved to be a very efficient one. The method was later extended to mechanical theorem proving of differential geometries and the present paper is devoted to such an exposition. As the case of two or more independent variables is much more complicate than that of a single independent variable, we have separated the treatments into two parts. In Section 2 we explain with proofs omitted the basic principles underlying mechanical theorem proving of differential geometries involving only one independent variable. In Section 3 some examples in differential geometry of curves and in Sections 4 and 5 examples in mechanics are given for the sake of illustrating these general principles of mechanical theorem proving as well as mechanical theorem discovering. In particular, it is shown how Newton's Gravitational Laws can be mechanically proved and even automatically discovered from Kepler's Laws. In Section 6, the case of two or more independent variables is briefly described and some examples from theory of surfaces are also given to serve as illustrations.

2. Case of one Independent Variable

For a basis of the present section, we refer to [4, 5] and [6].

A DIFFERENTIAL FIELD (abbr.d-FIELD), say F, is a field of characteristic 0 which has, besides the usual arithmetic operations, a further operation of DIFFERENTIATION

[1)] The present paper to be published elsewhere is a summary of partial results of the author before 1988. It is partially supported by NSFC Grant JI85312

such that any element A of F has a DERIVATIVE D_1A verifying the usual rules. We write for simplicity $D_iA = D_1 \cdots D_1A$ with D_1 i times and call D_iA the i-th DERIVATIVE of A. The element A itself is also considered as 0-TH DERIVATIVE of $A : A = D_0A$.

An INDETERMINATE X is just a symbol having an infinity of DERIVATIVES D_iX none of which is zero. A DIFFERENTIAL POLYNOMIAL (abbr. d-POL) say P in INDETERMINATES $X_1, X_2, \cdots, X_n$ over F is a polynomial in D_iX_j with coefficients in F. For P we can then form successive DERIVATIVES D_iP as well as various PARTIAL DERIVATIVES $dP/d(D_iX_j)$ in the usual formal manner.

To any non-constant d-POL P will be associated a 4-tuple $[t\ c\ r\ d]$ of integers, to be called the INDEX-SET of P, in notation $ind(P)$, as follows.

t = number of actual terms in P,

c = the greatest subscript c for which X_c occurs actually in P, to be called the CLASS of P, and to be denoted as $cls(P)$,

r = the highest order r for which the r-th derivative D_rX_c of the above X_c occurs actually in P, to be called the ORDER of P and to be denoted as $ord(P)$,

d = the highest degree d of the above D_rX_c which occurs actually in P, to be called the DEGREE of P and to be denoted by $deg(P)$.

In case P is a non-zero constant, then we define $c = 0, r = 0$, and $d = 0$.

A d-pol Q will be said to be d-REDUCED w.r.t. a d-pol P of class $c > 0$, if either the highest order m of derivative D_mX_c appearing in Q is $< ord(P)$, or, if D_rX_c appears in Q where $r = ord(P)$,then the highest degree of D_rX_c in Q is $< deg(P)$. On the other hand Q is not d-reduced w.r.t. any d-pol P of class 0.

For a d-pol P with $cls(P) = c > 0, ord(P) = r$, and $deg(P) = d$, we shall call the derivative D_rX_c the LEAD of P, to be denoted by $lead(P)$. Let L be this lead. Then the coefficient of L^d, which is itself a d-pol, is called the INITIAL of P, to be denoted as $init(P)$. The formal partial derivative of P w.r.t L is then called the SEPARANT of P, to be denoted by $sep(P)$.

For non-zero d-pols in indeterminates $X_1, \cdots, X_n$ over the d-field F partial orderings $\ll$ may be defined in various ways. To fix the ideas only the following ordering will be considered in this paper. Let P_1, P_2 be non-zero d-pols with index sets $[t_1\ c_1\ r_1\ d_1]$ and $[t_2\ c_2\ r_2\ d_2]$ resp. We say then $P_1 \ll P_2$ or $P_2 \gg P_1$ if one of the following cases occurs:

(a) $c_1 < c_2$,

(b) $c_1 = c_2$, but $r_1 < r_2$,

(c) $c_1 = c_2, r_1 = r_2$, but $d_1 < d_2$.

If neither $P_1 \ll P_2$ nor $P_1 \gg P_2$ so that P_1 and P_2 are incomparable in this ordering, then we write $P_1 <> P_2$.

With respect to such a partial ordering of d-pols we can then introduce the notions of DIFFERENTIAL ASCENDING SET, DIFFERENTIAL BASIC SET, and DIFFERENTIAL CHARACTERISTIC SET just as in the case of ordinary polynomials as follows.

DEFINITION. A finite sequence of non-zero d-pols

$$(ASC) \quad P_1, P_2, \cdots, P_r$$

is called a d-ASCENDING SET (abbr. d-ASC-SET) if either

(a) $r = 1$ and $cls(P_1) = 0$, or

(b) $cls(P_1) > 0$, and for any $j > i, cls(P_j) > cls(P_i)$ and the initial of P_j is d-REDUCED w.r.t. P_i.

In case of (a) the d-asc-set is then said to be TRIVIAL.

For a non-trivial d-asc-set (ASC) as above let S_i and I_i be respectively the separant and initial of $P_i, i = 1, 2, \cdots, r$. A d-pol G will be said to be d-REDUCED w.r.t. (ASC) if it is d-reduced w.r.t each P_i in (ASC). In particular all separants S_i and initials I_i are d-reduced w.r.t. (ASC). For any d-pol G we have then the following REMAINDER FORMULA:

$$(\prod_i (I_i^{l_i} * S_i^{m_i})) * G = \sum_{jk} (Q_{jk} * D_j P_k) \ + \ R, \tag{I}$$

in which l_i and m_i are certain non-negative integers which will be taken to be as small as possible, and Q_{jk}, R d-pols with R d-reduced w.r.t. (ASC). The product $\prod$ is to be extended over $1 \leq i \leq r$, and the summation $\sum$ is extended over only a finite number of terms. The d-pol R in (I) is accordingly called the d-REMAINDER (abbr. d-REMDR) of G w.r.t. (ASC), to be denoted as *d-Remdr*(G/ASC).

The d-asc-sets will also be arranged in a partial ordering $\ll$ according to the following DEFINITION. Let a second d-asc-set

$$(ASC)' \quad Q_1, Q_2, \cdots, Q_s$$

be given. Then $(ASC) \ll (ASC)'$ or $(ASC)' \gg (ASC)$ if one of the following cases occurs:

(a) There is some $k \leq r$ and $\leq s$ such that for $j < k, P_j <> Q_j$, while $P_k \ll Q_k$.

(b) $r > s$ and $P_j <> Q_j$ for $j \leq s$.

If neither $(ASC) \ll (ASC)'$ nor $(ASC) \gg (ASC)'$ so that (ASC) and $(ASC)'$ are incomparable in this ordering, then we write $(ASC) <> (ASC)'$.

DEFINITION. For any system (DPS) of d-pols any d-asc-set ASC of lowest order for which each d-pol belongs to (DPS) is called a d-BASIC-SET (abbr. d-BAS-SET) of (DPS).

Let a d- FIELD F be given. A d-FIELD F' will be said to be a d-EXTENSION-FIELD (abbr. d-EXT-FIELD) of F if, as an algebraic field, it is an extension field of F in the ordinary sense, and moreover any element A of F' which is also in F will have the same p-th derivative for any $p > 0$ whether it is considered as an element of F or of F'.

Let the d-FIELD F and INDETERMINATEs $X_1, X_2, \cdots, X_n$ be now fixed in advance. Consider any finite set (DPS) of d-pols in $X_1, \cdots, X_n$ over F. The system of equations $P = 0$ for all P in (DPS) will be represented symbolically by $(DPS) = 0$.

A finite set of non-zero d-pols is called a DIFFERENTIAL POLSET (abbr. d-POLSET). Let such a d-polset DPS be given. A set (Z_i) of elements Z_i in an arbitrary d-ext-field F' of F will be called a d-ZERO of the set (DPS) if it makes $(DPS) = 0$ when Z_i are substituted for X_i. The totality of all such d-zeros will be denoted by *d-Zero*(DPS) and the totality of those which are not d-zero of a given d-pol G will be denoted by *d-Zero*(DPS/G).

Given a d-polset DPS we can deduce a certain d-asc-set of particular interest in a mechanical way according to the following scheme:

$$\begin{array}{ccccc} DPS = DPS_0 & DPS_1 & \cdots & DPS_m & \\ DBS_0 \gg & DBS_1 \gg & \cdots \gg & DBS_m & \\ DRS_0 & DRS_1 & \cdots & DRS_m = \text{empty}. & \end{array} \tag{II}$$

In (II) each DBS_i is a d-bas-set of DPS_i, each DRS_i is the set of non-empty d-remainders, if there are any, of d-pols in DPS_i w.r.t. DBS_i. Finally, each DPS_i is the preceding DPS_{i-1}, enlarged by adjoining to it all the d-pols in DRS_{i-1}, i.e.

$$DPS_i = DPS_{i-1} + DRS_{i-1}, \tag{III}$$

in which + means set union. It is easy to prove that the procedure will ultimately terminate with some $DRS_m = empty$.

DEFINITION. The final d-bas-set DBS_m in the scheme (II) is called a d-CHARACTERISTIC-SET (abbr. d-CHAR-SET) of the d-polset (DPS).

The importance of this d-char-set, say DCS, lies in the following

WELL-ORDERING PRINCIPLE. For the d-zeros of (DPS) we have

$$\textit{d-Zero}(DPS) = \textit{d-Zero}(DCS/K) + \sum_k \textit{d-Zero}(DQS_k), \tag{IV}$$

in which K is the product of all initials and separants of d-pols in the d-char-set DCS, and DQS_k are the enlarged DPS with one of the initials or the separants adjoined to it. We have besides

$$\textit{d-Zero}(DPS) \subset \textit{d-Zero}(DCS) \tag{V}$$

$$\textit{d-Zero}(DCS/K) = \textit{d-Zero}(DPS/K) \subset \textit{d-Zero}(DPS), \tag{VI}$$

More general than (V), we have also

$$\textit{d-Zero}(DPS) = \textit{d-Zero}(DPS_k) \subset \textit{d-Zero}(DBS_k) \tag{V$'$}$$

for any d-polset (DPS_k) and d-bas-set (DBS_k) appearing in scheme (II).

REMARK. During the procedure it is convenient to remove certain factors to make d-pols occurring in the reduction not too high in term numbers. In such case the $\sum_k$ in (IV) should run over also all factors removed besides initials and separants and the product K should include all such factors too.

Consider now a differential-geometrical statement (S) with hypothesis $(HYP) = 0$ and conclusion $CONC = 0$ both expressed in terms of d-pols. Any d-zero of (HYP) is then just a geometrical configuration, eventually in some imaginative extended space, which verifies the hypothesis of (S). Let us form now a d-char-set DCS of (HYP) and form the d-remainder

$$R = \textit{d-Remdr}(CONC/DCS).$$

Suppose that $R = 0$. From the remainder formula (I) and the relation (V) we see that $CONC = 0$ for any d-zero of (HYP) for which

$$ND_k \neq 0, \tag{VII}$$

with ND_k the totality of (non-trivial) initials, separants of d-pols in (DCS) and eventually also factors removed during the procedure. Accordingly we lay down now the following

MECHANICAL THEOREM PROVING (abbr. MTP) PRINCIPLE. A geometrical theorem with hypothesis d-polset (HYP) and conclusion d-pol $CONC$ is GENERICALLY TRUE under certain NON-DEGENERACY CONDITIONs (VII) if

$$d\text{-}Remdr(CONC/DCS) = 0. \tag{VIII}$$

The above MTP Principle furnishes us not only a mechanical method of proving KNOWN geometrical theorems, but also one of DISCOVERING in a certain sense yet UNKNOWN theorems. For example, we may make guess about certain probable conclusions and then just verify its truth by simply computing the corresponding d-remainders to see its generic truth. What is more important is the following: The formula (V′) shows that, if (DPS) is the hypothesis set (HYP), then any d-pol P in any d-polset (DPS_k) or any d-bas-set (DBS_k) appearing in the scheme (II) during the procedure, will furnish us a geometrical theorem $P = 0$ whenever it has a geometrical meaning. Moreover, by deliberately arranging the indeterminates in an order at the outset, we are able to get certain not previously known geometrical relations between the first few indeterminates. This will furnish us thus a general method of discovering unknown relations or new theorems. We may thus formulate the following

MECHANICAL THEOREM DISCOVERING (abbr. MTD) PRINCIPLE. For a hypothesis set $(DPS) = (HYP)$ any d-pol P occurring in (DPS_k) or (DBS_k) of scheme (II) will furnish us a geometrical theorem $P = 0$, whenever it can be endowed an intrinsic geometric meaning.

Illustrative examples for the applications of the above MTP and MTD Principles will be given in successive sections. We remark that there are various refinements of the Well-Ordering Principle in the form of Structure Theorems of a d-zero-set of a d-polset. Correspondingly there are also refined MTP Principles which would comprise the most general and also complete one about mechanical theorem proving of geometries. However, in view of the high complexity in applying such refined principles it seems that the partial MTP Principles would often be preferred which are already both efficient and fruitful in bringing about non-trivial concrete results, as may be seen from the examples to follow. For this reason we shall satisfy ourselves in this paper to the above special forms of Principles and leave the more general studies to the relevant papers of the author.

3. Some Examples

Let a differential-geometry theorem be expressed in the form of $T = \{HYP, CONC\}$ with HYP the hypothesis d-polset and $CONC$ the conclusion d-pol. Let DCS be a d-char-set of HYP in accordance with the scheme (II) in Section 2. Denote the initials, separants, and also eventually factors removed during the procedure by ND_k and K be the product of all of them. Denote also by HYP_k the enlarged d-polset of HYP in adjoining to it the d-pol ND_k. Then the Well-Ordering Principle (IV) of Section 2 applied on HYP will be of the form

$$d\text{-}Zero(HYP) = d\text{-}Zero(DCS/K) + \sum_k d\text{-}Zero(HYP_k).$$

Let the d-remainder of the conclusion d-pol $CONC$ w.r.t. DCS be

$$R = d\text{-}Remdr(CONC/DCS).$$

Suppose that $R \neq 0$. Then nothing can be concluded. It may be due to the fact that the theorem T in question is utterly untrue at all or that the d-char-set DCS is REDUCIBLE in certain sense which we shall not enter. Suppose however $R = 0$. Then the MTP-Principle based on the above Well-Ordering Principle says that the theorem T in question is generically true under the non-degeneracy conditions $ND_k \neq 0$, or T is true on the part $d\text{-}Zero(DCS/K) = d\text{-}Zero(HYP/K)$ of the totality $d\text{-}Zero(HYP)$ of geometrical configurations verifying hypothesis $(HYP) = 0$. If we are interested in knowing whether the theorem will remain true in one of the degeneracy case $ND_k = 0$, we may just add ND_k to HYP to form the enlarged hypothesis set HYP_k and proceed with HYP_k just as HYP before. We remark that though nothing can be concluded in case $R \neq 0$ and $R = 0$ gives only a sufficient condition of proving a differential-geometry theorem, the method is already of some consequence as may be seen from the illustrative examples below.

Ex. 1. Parallel Planar Curves.

Two planar curves C, C' are said to be PARALLEL if there is a (1-1)-correspondence $P \longleftrightarrow P'$ such that the joining lines PP' are common normals to the curves. Prove that for such pairs the distance r between the corresponding points is a constant.

To prove it let the coordinates of $P = (X_1, X_2)$ and $P' = (X_3, X_4)$ be expressed in terms of same parameter t for which the values are same for points P, P' in correspondence. Denote the distance of PP' by X_5. The hypothesis set is then consisting of 3 d-pols

$$\begin{aligned} H_1 &= (X_3 - X_1) * D_1X_1 + (X_4 - X_2) * D_1X_2, \\ H_2 &= (X_3 - X_1) * D_1X_3 + (X_4 - X_2) * D_1X_4, \\ H_3 &= X_5^2 - (X_3 - X_1)^2 - (X_4 - X_2)^2, \end{aligned}$$

while the conclusion d-pol is given by

$$CONC = D_1X_5,$$

in which D_1, D_2, etc. means derivatives w.r.t. t. The d-char-set DCS is easily found to be consisting of the 3 d-pols below:

$$\begin{aligned} C_1 &= ((D_1X_1)^2 + (D_1X_2)^2) * D_1X_2 * D_1X_3 + (X_3 - X_1) * D_1X_1 \\ &\quad * (D_2X_1 * D_1X_2 - D_1X_1 * D_2X_2) - D_1X_1 * D_1X_2 * ((D_1X_1)^2 + (D_1X_2)^2), \\ C_2 &= D_1X_2 * X_4 + (X_3 - X_1) * D_1X_1 - X_2 * D_1X_2, \\ C_3 &= X_5^2 - (X_3 - X_1)^2 - (X_4 - X_2)^2. \end{aligned}$$

The leads are D_1X_3, X_4, and X_5 while the initials, separants, and removed factors, which are all eventually split into factors, are 4 in number:

$$ND_1 = D_1X_2, \quad ND_2 = X_3 - X_1,$$
$$ND_3 = (D_1X_1)^2 + (D_1X_2)^2, \quad ND_4 = X_5.$$

The d-remainder of CONC is readily found to be 0 as shown by

$$2 * ND_1^3 * ND_4 * CONC = A_1 * D_1C_3 + A_2 * D_1C_2 + A_3 * C_2 + A_4 * C_1,$$

in which A_i are all d-pols. It follows that the theorem in question is true under the non-degeneracy conditions $ND_k \neq 0$.

The condition $ND_1 = 0$ means that the curve C degenerates into a line parallel to the X-axis or eventually a single point. If we want to know whether the theorem remains true in this degenerate case we just add ND_1 to (HYP) to form a new hypothesis set (HYP_1) consisting of the 4 d-pols H_i and ND_1. Let us consider the case $ND_2 \neq 0$ so that we may remove the factor $ND_2 = X_3 - X_1$ during the procedure in forming the d-char-set. Then the d-char-set DCS_1 will be found to be consisting of 4 d-pols

$$C_{11} = D_1X_1, \quad C_{12} = D_1X_2,$$
$$C_{13} = H_2, \quad C_{14} = H_3.$$

The new non-degeneracy conditions are now $ND_{1k} \neq 0$ with

$$ND_{11} = ND_2, ND_{12} = X_4 - X_2, ND_{13} = ND_4.$$

We find readily the d-remainder to be 0 again so that the theorem is still true under the conditions

$$ND_1 = 0, \quad ND_{11} * ND_{12} * ND_{13} \neq 0.$$

We remark that in the present case C degenerates into a single point P while C' becomes a circle with P as center and radius non-zero. The correspondence is no more (1-1) and the theorem is to be interpreted as to be true in some degenerate sense. By proceeding further in the same way, we may verify the truth of the theorem under all possible degeneracy cases, if we wish to do so.

Ex. 2. Curve pairs of Bertrand type.

For a space curve there is associated to any regular point X on it a triple of significant lines L, viz. the tangent T, the principal normal P, and the binormal B. Suppose to C there is associated some other curve C' in (1-1)-correspondence to it such that at corresponding regular points X, X' one of the significant lines L of C coincides with some other significant line L' of C'. There are in all 9 such possibilities for which $(L, L') = (P, P')$ is the case of classical Bertrand curve pairs.

For the study of such curve pairs of Bertrand type we shall use the Cartan method of moving trihedrals for which only entities of intrinsic geometrical interest will be involved. In fact, in contrast to the case of elementary geometries no coordinates having no geometrical significance will enter which saves thus the memory storage as well as computational labor.

Let us take arc lengths S, S' of the curves C, C' as parameters so that S' is a function of S under the correspondence. The derivatives w.r.t. S will be denoted by D_1, D_2, etc. and we put also $D_1S' = R$.

Attach now to points of C the trihedrals (X, E_1, E_2, E_3) with lines of $E_i = T, P, B$ respectively. The Frenet equations will be:

$$\begin{aligned} D_1X = E_1, \quad D_1E_1 = K \cdot E_2, \\ D_1E_2 = -K \cdot E_1 + T \cdot E_3, \quad D_1E_3 = -T \cdot E_2, \end{aligned} \tag{1}$$

in which K and T are respectively the curvature and torsion of C. For C' we have also similar trihedals (X', E_1', E_2', E_3') and equations.

Consider the general case in which the trihedrals at corresponding points are related by equations (with matrix (U_{ij}) an orthogonal one):

$$X' = X + A_1 \cdot E_1 + A_2 \cdot E_2 + A_3 \cdot E_3, \tag{2}$$

$$E_i' = \sum_j U_{ij}.E_j, \quad i = 1, 2, 3. \tag{3}$$

Differentiating now both sides of (2), utilizing the Frenet equations of both C, C', and comparing coefficients of E_i, we get readily equations $H_i = 0, i = 1, 2, 3$, with

$$\begin{aligned} H_1 &= R * U_{11} - 1 - D_1A_1 + A_2 * K, \\ H_2 &= R * U_{12} - D_1A_2 - A_1 * K + A_3 * T, \\ H_3 &= R * U_{13} - D_1A_3 - A_2 * T. \end{aligned}$$

Treating in the same manner the other equations (3) we get 9 further equations $H_i = 0, i = 4, \cdots 12$. Let K' and T' be the curvature and torsion of the curve C', then these d-pols H_i are given by:

$$\begin{aligned} H_4 &= R * K' * U_{21} - D_1U_{11} - K * U_{12}, \\ H_5 &= R * K' * U_{22} - D_1U_{12} - K * U_{11} + T * U_{13}, \\ H_6 &= R * K' * U_{23} - D_1U_{13} - T * U_{12}, \\ H_7 &= -R * K' * U_{11} + R * T' * U_{31} - D_1U_{21} + K * U_{22}, \\ H_8 &= -R * K' * U_{12} + R * T' * U_{32} - D_1U_{22} - K * U_{21} + T * U_{23}, \\ H_9 &= -R * K' * U_{13} + R * T' * U_{33} - D_1U_{23} - T * U_{22}, \\ H_{10} &= -R * T' * U_{21} - D_1U_{31} + K * U_{32}, \\ H_{11} &= -R * T' * U_{22} - D_1U_{32} - K * U_{31} + T * U_{33}, \\ H_{12} &= -R * T' * U_{23} - D_1U_{33} - T * U_{32}. \end{aligned}$$

Change now the notations as given by

$$(R, A_1, A_2, A_3, U_{11}, U_{12}, U_{13}, U_{21}, U_{22}, U_{23}, U_{31}, U_{32}U_{33}, K, T, K', T')$$

$$= (X_5, X_6, X_7, X_8, X_9, X_{10}, X_{11}, X_{12}, X_{13}, X_{14}, X_{15}, X_{16}, X_{17}, X_a, X_b, X_c, X_d)$$

Consider first the classical Bertrand case (P, P') so that $E_2' = +$ or $-E_2$. To fix the ideas let us take the + sign and similarly for the orthogonality relations between the U_{ij} 's so that we have:

$$
\begin{gathered}
A_1 = 0, \quad A_3 = 0, \\
U_{12} = U_{21} = U_{23} = U_{32} = 0, \quad U_{22} = 1, \\
U_{11} = U_{33}, \quad U_{13} = -U_{31}, \\
U_{11}^2 + U_{13}^2 = 1.
\end{gathered} \tag{4}
$$

Suppose we are interested in finding the possible relations between the torsions T, T' of C, C'. We may then choose $X_a, \cdots, X_d$ to be e.g.

$$
\begin{aligned}
K = X_a = X_{25}, &\quad T = X_b = X_2, \\
K' = X_c = X_{30}, &\quad T' = X_d = X_4.
\end{aligned}
$$

The hypothesis system (HYP) is then consisting of the 14 d-pols in the $X's$ below besides the trivial ones of (4), viz.

$$
\begin{gathered}
H_1' = X_5 * X_9 - 1 + X_{25} * X_7, \\
H_2' = -D_1 X_7, \\
H_3' = X_5 * X_{11} - X_2 * X_7, \\
H_4' = -D_1 X_9, \\
H_5' = X_{30} * X_5 - X_{25} * X_9 + X_2 * X_{11}, \\
H_6' = -D_1 X_{11}, \\
H_7' = X_{30} * X_5 * X_9 - X_4 * X_5 * X_{15} - X_{25}, \\
H_8' = X_{30} * X_5 * X_{11} - X_4 * X_5 * X_{17} + X_2, \\
H_9' = -D_1 X_{15}, \\
H_{10}' = -X_4 * X_5 - X_{25} * X_{15} + X_2 * X_{17}, \\
H_{11}' = -D_1 X_{17}, \\
H_{12}' = X_9 - X_{17}, \\
H_{13}' = X_{11} + X_{15}, \\
H_{14}' = X_9^2 + X_{11}^2 - 1.
\end{gathered}
$$

We remark that the derivation of these equations $H_i' = 0$ are entirely of a mechanical character. Now

$$
H_2' = 0, \quad H_4' = 0, \quad H_6' = 0, \quad H_9' = 0, \quad H_{11}' = 0,
$$

means that the distance $A_2 = X_7$ between corresponding points and $\cos A = X_9, \sin A = X_{11}$ and so the angle A between the corresponding tangents or binormals are all constants. Hence we have already discovered these classical theorems in an automatic manner.

The d-char-set of (HYP) is found to be a set of 10 d-pols C_1'-C_{10}' with

$$C_1' = X_4 * D_1X_2 + X_2 * D_1X_4, \text{etc.}$$

The equation $C_1' = 0$ means that $X_2 * X_4 = const.$ Thus we have discovered automatically the following

THEOREM of SCHELL. The product of torsions at corresponding points of a Bertrand pair of curves is a constant.

Suppose we are now interested in the probable restrictions in K and T about the curve C to have a Bertrand companion C'. For this purpose we may take e.g.

$$K = X_a = X_1, \quad T = X_b = X_2,$$
$$K' = X_c = X_{30}, \quad T' = X_d = X_{40}.$$

The d-char-set of the hypothesis system is again readily found with the first one given by

$$C_1' = F_1 * (D_1X_1 * D_2X_2 - D_1X_2 * D_2X_1),$$

where

$$F_1 = (X_1 * D_1X_2 - X_2 * D_1X_1) * (X_1 * D_1X_1 + X_2 * D_1X_2).$$

Under the non-degeneracy conditions $F_1 \neq 0, D_1X_1 \neq 0$, etc. we see that

$$D_1X_1 * D_2X_2 - D_1X_2 * D_2X_1 = 0,$$

or $X_1 = K, X_2 = T$ verify some linear relation. We have thus discovered automatically the following classical

THEOREM of BERTRAND. If a curve C has another curve C' to form a Bertrand pair, then generically the curvature and torsion of C are in some linear relation with constant coefficients.

As before we may study the degeneracy cases $F_1 = 0$, etc. if we are interested in doing so.

Let Z, Z' be now the centers of curvature of C, C' at corresponding points X, X', and let CR be the cross ratio of (X, Z, X', Z'). Then

$$1/CR = -(1 + A_2 * K') * (A_2 * K - 1) = G, \text{ say.}$$

W.r.t. any one of the above d-char-set DCS we find readily

$$d\text{-}Remdr(D_1G/DCS) = 0.$$

We have therefore proved the following

THEOREM of MANNHEIM. The cross ratio of any two corresponding points and the centers of curvature of a Bertrand pair of curves is a constant under the non-degeneracy conditions $K * K' \neq 0$ among others.

We have also treated in the same manner the other cases $(L, L') = (P, B)$, (B, P), (B, B) and re-discovered such theorems of Mannheim, etc. in an automatic manner. Moreover, for curves in affine space there are notions of affine principal normal, affine binormal, affine winding coefficients, and affine torsion. There are also Winternitz equations and Darboux equations connecting these affine invariants alike to the Frenet equations for curves in the ordinary space. However, it seems that there are no analogues of theorems so interesting as those of Schell or Mannheim as given above. Cf. [15].

4. Example 3. Kepler's Equation in Celestial Mechanics

Kepler's equation in celestial mechanics plays an important role in astronomical computations. As a further example of illustrating our general method we shall show how to solve Kepler's equation in a manner entirely different from the usual known ones.

For this purpose let us consider a planet P moving in an elliptic orbit with eccentricity e and period T. Let t be the time elapsed in starting from the perihelion. The angle to be determined E, is the angle spanned by the major axis toward the perihelion, and the radius vector from the center of the elliptical orbit to that point on the circumscribing circle which projects to the same point on the major axis as does the planet. The Kepler's equation which permits to determine E is then of the form

$$E - e * \sin E = M, \tag{1}$$

with

$$M = 2 * \pi * t/T.$$

The equation (1) is usually solved by numerical or graphical methods. A solution is also furnished by the Lagrange series of E in e with coefficients in terms of derivatives of powers of $\sin M$ w.r.t. M. The derivation of Lagrange series is somewhat intricate and the series converges only for $e(< 1)$ sufficiently small. We shall now give a simple device turning the equation into a form which permits to, again by our general method, determine E, as well as other simple functions of E like $\sin E$, etc. in the form of convergent series of M.

The underlying principle of this method is quite simple. Let us consider $\sin E$ as a separate indeterminate function by setting

$$F = \sin E. \tag{2}$$

The sine function, as also most of transcendental functions occurring in mathematics, will satisfy some differential equation. This differential equation, together with the original Kepler equation (1), will furnish us a d-polset consisting of two d-pols in E and F on the independent variable M. Applying our general method we get then equations in E or F alone which may then be solved in the form of series in M.

The computation is quite simple and runs as follows. Considering E and F as functions of M with primes denoting derivatives w.r.t. M, we get from (2)

$$F' = \cos E * E', \tag{2'}$$

$$F'' = -\sin E * E'^2 + \cos E * E''. \tag{2''}$$

Multiplying (2″) by E' and removing $\sin E, \cos E$ by means of (2) and (2′),we get

$$E' * F'' - F' * E'' + F * E'^3 = 0. \tag{3}$$

In order to solve for E we may now set

$$E = X_1, F = X_2. \tag{4}$$

Let us use again D_i to denote the i-th derivative w.r.t. M, then(1) and (3) will become

$$X_1 - e * X_2 - M = 0, \tag{1'}$$

$$D_1X_1 * D_2X_2 - D_1X_2 * D_2X_1 + X_2 * (D_1X_1)^3 = 0. \qquad (3')$$

The d-pols P_1' and P_2' in the left sides of (1′) and (3′) form now a d-polset DPS'. To form the d-char-set of DPS' we just form the d-remainder of P_2' w.r.t. P_1', getting thus a d-pol

$$R' = D_2X_1 + (X_1 - M) * (D_1X_1)^3,$$

which, together with (1′), will form a d-char-set of the d-polset DPS'. The equation $R' = 0$, or

$$E'' + (E - M) * E'^3 = 0 \qquad (5)$$

is then the equation of E in terms of M as required. The initial values of E and E' at $M = 0$ are obviously

$$E(0) = 0, \quad E'(0) = 1/(1 - e).$$

We get therefore a series of the form

$$E = (1/(1 - e)) * M - (e/(1 - e)^4) * M^3/3! + \cdots \qquad (6)$$

which is convergent in some neighborhood of $M = 0$ by the general theory of differential equations. It can in fact be proved that the series (6) is convergent for all values of M.

If we are interested in the determination of $F = \sin E$ we may set instead of (4)

$$E = X_2, \quad F = X_1.$$

The equations (1) and (3) become then

$$X_2 - e * X_1 - M = 0, \qquad (1'')$$

$$D_1X_2 * D_2X_1 + X_1 * (D_1X_2)^3 - D_1X_1 * D_2X_2 = 0. \qquad (3'')$$

We get then a d-polset DPS'' consisting of the left-side d-pols P_1'' and P_2'' of (1″) and (3″). Forming the d-remainder of P_2'' w.r.t. P_1'', we get a d-pol

$$R'' = D_2X_1 + X_1 * (e * D_1X_1 + 1)^3.$$

The equation $R'' = 0$, or

$$F'' + F * (e * F' + 1)^3 = 0$$

gives then $F = \sin E$ in terms of M. With obvious initial values

$$F(0) = 0, \quad F'(0) = 1/(1 - e),$$

we get then the series of $\sin E$ as required:

$$\sin E = (1/(1 - e)) * M - (1/(1 - e)^4) * M^3/3! + \cdots \qquad (7)$$

The series (7) can again be proved to be convergent for all values of M.

The above method is quite general and may be used to determine e.g. $CosE$, the sun-planet radius vector, the anomaly, etc. in terms of M. It can also be applied to the case of parabolic or hyperbolic orbits.

5. Example 4. Automatic Derivation of Newton's Gravitational Laws from Kepler's Observational Laws

It is an important historical event that Newton derived his gravitational laws from Kepler's laws. During a visit to Argonne National Laboratory in 1986 the author was told by Prof. Gabriel there about the significance of deducing Newton's Gravitational Laws from Kepler's Laws in a mechanical way for which Prof. Gabriel was already quite successful in applying his own automated reasoning method, cf. [3]. As a 4-th example of illustrating our general method we shall show below how our method can be applied to deal with such kind of problems.

For this purpose let us first formulate the Kepler's laws (K) and the Newton's laws (N) in the manner as given below:

(K_1) The planets move in elliptic orbits around the sun as focus.

(K_2) The vector from the sun to the planet sweeps equal areas in equal times.

(K_3) The squares of periods of planet motions are proportional to the cube of the major axis of the elliptic orbits.

(N_1) The acceleration of a planet is inversely proportional to the square of the distance from the sun to the planet.

(N_2) The acceleration vectors of planets are directed toward the sun.

(N_3) The proportinality factor of the inverse square law (N_1) is independent of the different planets.

In order to deduce mechanically, or even discover automatically the Newton's laws from Kepler's laws let us take first coordinates and transform the various laws into equation forms as follows.

Let us take e.g. rectangular coordinates (x, y) with the sun at the origin and the major axis of the elliptic orbit as the X-axis. Let r be the radius vector from the sun to the planet. Then the orbit will have an equation of the form

$$r = p + e * x. \tag{1}$$

The Kepler's law (K_1) will correspond then to the equation (1) and also (2)-(4) below taken together:

$$r^2 = x^2 + y^2, \tag{2}$$

$$p = \text{ const, or } p' = 0, \tag{3}$$

$$e = \text{ const, or } e' = 0, \tag{4}$$

in which the prime means derivation w.r.t. the time t. Similarly Kepler's law (K_2) will correspond to the equations (5), (6) below:

$$x * y' - y * x' = h, \tag{5}$$

$$h' = 0. \tag{6}$$

For (K_3) let T be the period for a planet to turn once around the Sun on its elliptical orbit and $2 * a$, $2 * b$ be the major and minor axis of this orbit. Then Kepler's Law (K_3) means that T^2/a^3 is a universal constant independent of the planets, though possibly dependent on the sun. Now according to the meaning of h we have

$$h * T = 2 * \pi * a * b,$$

Since $b^2/a = p$, we have

$$T^2/a^3 = 4 * \pi^2 * p/h^2.$$

It follows that

$$c = p/h^2 \tag{7}$$

is also a universal constant independent on the planets with $c' = 0$. On the other hand the Newton's laws $(N_1), (N_2)$, with A as the acceleration, will correspond to the following set of equations (8)-(10):

$$A^2 = x''^2 + y''^2, \tag{8}$$

$$r^2 * A = \text{ const, or since } r \neq 0, 2 * r' * A + r * A' = 0, \tag{9}$$

$$x * y'' = y * x''. \tag{10}$$

For the Newton's Law (N_3) the constant $r^2 * A$ as asserted by (N_1) is a universal one and there may exist thus some relations connecting this constant with c which we have to try to find out.

As Newton's Law (N_2) corresponding to (10) follows directly from Kepler's Law (K_2) corresponding to (5),(6) we shall consider below only the Laws (N_1) and (N_3).

For this purpose let us now introduce indeterminates in replacing the various functions by X's as given below:

$$(r,\ A,\ c,\ p,\ e,\ x,\ y,\ h)$$
$$= (X_{11}, X_{12}, X_{15}, X_{21}, X_{22}, X_{31}, X_{32}, X_{51}).$$

The various functions are so arranged that r, A, and c come as first few ones in order to discover possible relations between them which we suppose to be entirely ignorant. With this change of notations the equations (1)-(10) will turn to be the equations $P_i = 0$ with P_i given by (1')-(10') as shown below: (D_i means i-th derivative w.r.t. t):

$$P_1 = X_{11} - X_{31} * X_{22} - X_{21}, \tag{1'}$$

$$P_2 = X_{31}^2 + X_{32}^2 - X_{11}^2, \tag{2'}$$

$$P_3 = D_1 X_{21}, \tag{3'}$$

$$P_4 = D_1 X_{22}, \tag{4'}$$

$$P_5 = X_{31} * D_1 X_{32} - X_{32} * D_1 X_{31} - X_{51}, \tag{5'}$$

$$P_6 = D_1 X_{51}, \tag{6'}$$

$$P_7 = X_{15} * X_{51}^2 - X_{21}, \tag{7'}$$

$$P_8 = (D_2 X_{31})^2 + (D_2 X_{32})^2 - X_{12}^2, \tag{8'}$$

$$P_9 = 2 * D_1 X_{11} * X_{12} + X_{11} * D_1 X_{12}, \tag{9'}$$

$$P_{10} = X_{31} * D_2X_{32} - X_{32} * D_2X_{31}. \tag{10'}$$

Take now the d-polset DPS to be consisting of the d-pols (1′)-(8′) of the above set corresponding to the Kepler's Laws. Remark that the planets move in true non-degenerate elliptic and non-circular orbits so that we have

$$\begin{aligned} X_{21} = p \neq 0, \quad X_{22} = e \neq 0, \\ X_{11} = r \neq 0, \quad D_1X_{11} = r' \neq 0, \end{aligned} \tag{11}$$

and also

$$X_{15} = c \neq 0, X_{12} = A \neq 0. \tag{12}$$

In applying our algorithm for the finding of d-char-set DCS of DPS we can then remove any such factors during the procedure. The DCS is found to be the 7-th d-bas-set consisting of the 8 d-pols C_i with index sets given below:

[7 11 4 1], [4 12 0 2], [3 15 0 1], [3 21 0 1],

[7 22 0 2], [3 31 0 1], [3 32 0 2], [3 51 0 1].

Of the 8 d-pols C_i the first two are one in $X_{11} = r$ alone and the other in $X_{11} = r$ and $X_{12} = A$. The first one gives us thus a differential equation of derivative order 4 observed by the radius vector r. This equation and the second one between A and r are both too complicate and are of little interest. However, during the process the 4-th d-bas-set DBS_4 appears to be consisting of 6 d-pols B_{4i} of which the first one B_{41} is given by:

$$B_{41} = 2 * X_{12} * D_1X_{11} + D_1X_{12} * X_{11} = 2 * r' * A + r * A'.$$

By our general MTD-Principle, $B_{41} = 0$ should be a consequence of the original d-polset, i.e. a consequence of Kepler's Laws. The equation $B_{41} = 0$ is however nothing else but the Newton's inverse square law $r^2 * A = const.$ We have thus discovered in an automatic manner the Newton's Law (N_1) from the Kepler's Laws by means of our general principle. Furthermore, the 3-th d-bas-set during the process is consisting of 6 d-pols B_{3i} of which the first one is given by:

$$B_{31} = X_{15}^2 * X_{12}^2 * X_{11}^4 - 1 = c^2 * r^4 * A^2 - 1.$$

The equation $B_{31} = 0$ is a consequence of the Kepler's Laws too, which implies

$$c = + \text{ or } - 1/(r^2 * A). \tag{13}$$

It follows that the proportionality constant $r^2 * A$ as already asserted by Newton's Law (N_1) is equal to the reciprocal of the universal constant c up to a sign and is thus itself a universal constant. The Newton's Law (N_3) is thus also discovered in an automatic manner.

The ambiguity of the sign in (13) comes from our introduction of acceleration A by means of (8) which gives only its magnitude but not its direction. If we take the positive value of $Sqrt(A^2)$ as A with corresponding acceleration vector pointing to the sun, then in (13) the + sign is to be taken so that

$$r^2 * A = +1/c = +h^2/p$$

is the universal constant independent of the planets, though eventually dependent on the sun.

Let us assume now the attractive force between two masses M and m be proportional to both M and m. Let us assume also that the mass m under the action of a force F will undergo a motion with an acceleration A given by the general Law $F = m * A$ also due to Newton. Then we deduce immediately from the above Newton's Laws (N_1)-(N_3) the following general GRAVITATIONAL LAW of Newton:

The gravitational force between the sun of mass M and a planet of mass m is given by

$$F = G * M * m/r^2,$$

in which r is the distance between the sun and the planet while G is a universal constant independent of the planet. The universal constant G is in fact connected with the above c by the relation $G = +1/(M * c)$ which should be independent of the sun if the roles of sun and planet are to be considered as symmetric.

REMARK. In the above formulation we have supposed that Newton's Gravitational Laws are not known a priori and it turns out they will be discovered in an automatic manner by our general method. The computations are carried out on a SUN3/140 with running time = $15'58''$ and $maxt = 342$, where $maxt$ means the maximum number of terms of d-pols occurring during the procedure. In the previous drafts [16][17] different sets of coordinates and equations have been tried. Comparing the various trials it shows that the mere mechanical proving of Newton's Laws supposed known already would be somewhat simpler than the automatic discovering of these Laws supposed yet unknown.

6. Case of Two or More Independent Variables

Consider now the case of m independent variables with $m \geq 2$. Most of the notions in the case of one independent variable as described in Section 2, e.g. d-field, d-bas-set, etc. extend naturally to the present case and we shall keep the same terminologies and notations. However, the notion of d-CHAR-SET requires some modifications because of presence of integrability conditions.The general case is quite involved for which we refer to [18]. For the sake of simplicity of exposition let us restrict ourselves therefore to the case $m = 2$. The independent variables will be say t_1 and t_2 and the partial derivation i times w.r.t. t_2 and j times w.r.t. t_1 will be denoted by D_{ij}. The set of partial derivatives in indeterminates $X_1, \cdots, X_n$ will be ordered in some natural way. W.r.t. this order we shall define the LEAD of a d-pol, etc. in the usual way. Let a d-asc-set ASC be given. Suppose in ASC there are two d-pols F_1, F_2 with leads $D_{ij}X_c$ and $D_{hk}X_c$ in same X_c for which $i > h, j < k$. Differentiate $a = k - j$ times F_1 w.r.t. t_1 and $b = i - h$ times F_2 w.r.t. t_2 we get then equations of the form

$$S_1 * D_{0a}D_{ij}X_c = G_1, \quad S_2 * D_{b0}D_{hk}X_c = G_2,$$

in which S_1, S_2 are separants of F_1, F_2 respectively and G_1, G_2 are d-pols with all derivatives of lower order than $D_{ik}X_c = D_{0a}D_{ij}X_c = D_{b0}D_{hk}X_c$. The equality $S_2 * G_1 = S_1 * G_2$ follows as a consequence of the equations $(ASC) = 0$ and we shall lay down the following

DEFINITION. The d-pol

$$IC(F_1, F_2) = S_2 * G_1 - S_1 * G_2$$

is called the INTEGRABILITY-POL (abbr.INTEG-POL) of the d-pols F_1, F_2. We now modify the schemes (II) and (III) in Section 2 in the following way. For any i let DRS_{i-1} be the set of all non-zero d-remainders not only of d-pols in DPS_{i-1} but also of all integ-pols of pairs of d-pols in DBS_{i-1}, if there are any. With this DRS_{i-1} we form the schemes (II) and (III) as before. The ultimate d-bas-set DBS_m with corresponding $DRS_m = empty$ will then be called a d-CHAR-SET of the original d-polset DPS.

We have implemented the procedure of finding d-char-set of a d-polset on our computer. Experiments are not yet much done. It turns out that the formation of integ-pols requires often a large amount of memory storage. However, it seems that such formation may be avoided in some ways as shown by the example below.

Ex. 5. Surface pairs in 3-space.

For any pair of surfaces S, S' in ordinary 3-space in (1-1)- correspondence we may ask similar questions as in Ex.2. Thus, we may ask the geometrical conditions or interrelations of S, S' with corresponding points $X \leftrightarrow X'$ for the following cases:

(NN) XX' are normals to both S and S' (case of parallel surfaces).

(NT) XX' are normal to S at X and tangent to S' at X' (e.g. case of surfaces of centers).

(TT) XX' are both tangents to S and S' at X and X'.

Let us consider e.g. more in details the case (TT). For this purpose let us attach moving trihedrals $MT = (X, E_1, E_2, E_3)$ and $MT' = (X', E_1', E_2', E_3')$ to S and S' in such a way that E_3, E_3' are normals to S, S' and E_1, E_1' are along the common tangent line XX'. Denote the distance between X, X' by R and the angle between E_3, E_3' by A. Then we have

$$X' = X + R \cdot E_1, \tag{1}$$

$$E_1' = E_1, \tag{2}$$

$$E_2' = U \cdot E_2 + V \cdot E_3, \tag{3}$$

$$E_3' = -V \cdot E_2 + U \cdot E_3, \tag{4}$$

in which we have set $U = \cos A, V = \sin A$ so that

$$U^2 + V^2 = 1. \tag{5}$$

With W_i, W_{ij} $(i, j = 1, 2, 3)$ the Cartan exterior differential forms in parameters t_1, t_2 on S corresponding to the above moving trihedrals MT we have the following set of Cartan's structure equations ($\wedge$ means here exterior multiplication and d means exterior derivation):

$$dX = W_1.E_1 + W_2 \cdot E_2, \tag{6}$$

$$dE_i = \sum_j W_{ij} \cdot E_j, \tag{7}$$

$$W_{ij} = -W_{ji}, \tag{8}$$

$$W_3 = 0, (E_3 \text{ being normal to } S) \tag{9}$$

$$dW_1 = \sum_j W_j \wedge W_{j1} = -W_2 \wedge W_{12}, \tag{10}$$

$$dW_2 = \sum_j W_j \wedge W_{j2} = W_1 \wedge W_{12}, \tag{11}$$

$$dW_3 = 0 = \sum_j W_j \wedge W_{j3} = W_1 \wedge W_{13} + W_2 \wedge W_{23}, \tag{12}$$

$$dW_{12} = \sum_k W_{1k} \wedge W_{k2} = -W_{13} \wedge W_{23}, \tag{13}$$

$$dW_{13} = \sum_k W_{1k} \wedge W_{k3} = W_{12} \wedge W_{23}, \tag{14}$$

$$dW_{23} = \sum_k W_{2k} \wedge W_{k3} = -W_{12} \wedge W_{13}, \tag{15}$$

$$K \cdot W_1 \wedge W_2 = W_{13} \wedge W_{23}. \tag{16}$$

In the above equations all $\sum$'s are to be extended over 1,2,3 and K is the Gaussian curvature of S at X. Refer the corresponding points on S' to the same parameters t_1, t_2 we have also forms W_i', W_{ij}' for MT' and Cartan equations (6′)-(16′) similar to (6)-(16).

The first step toward the establishment of hypothesis d-pol-sets is consisting of deriving relations between various exterior forms W' and W. For this purpose let us consider e.g. the equation (1). Forming exterior derivatives of both sides, using equations (2)-(9), (6′)-(9′) and then comparing coefficients of E_i we find:

$$W_1' = W_1 + dR, \tag{17}$$

$$U \cdot W_2' = W_2 + R \cdot W_{12}, \tag{18a}$$

$$V \cdot W_2' = R \cdot W_{13}, \tag{18b}$$

$$V \cdot W_2 + R * V \cdot W_{12} = R * U \cdot W_{13}. \tag{19}$$

Treating in the same way equations (2)-(4) in turn we find further equations equivalent to:

$$W_{12}' = U \cdot W_{12} + V \cdot W_{13}, \tag{20}$$

$$W_{13}' = -V \cdot W_{12} + U \cdot W_{13}, \tag{21}$$

$$W_{23}' = W_{23} + U \cdot dV - V \cdot dU. \tag{22}$$

The relation (19) shows that there should be some geometrical conditions to be imposed on the surface S to have a companion S' in such a correspondence (TT) if A and R are given. In this respect let us consider the simplest case for which both A and R are constants. Then:

$$D_{01}U = 0, \quad D_{10}U = 0, \tag{23}$$

$$D_{01}V = 0, \quad D_{10}V = 0, \tag{24}$$

$$D_{01}R = 0, \quad D_{10}R = 0. \tag{25}$$

The correspondence $X \leftrightarrow X'$ under these assumptions is then called a BAECKLUND TRANSFORMATION. To study its geometrical properties let us express W_i and W_{ij} in terms of the parameters t_1, t_2 with $W_1 = 0, W_2 = 0$ as the parametric curves as follows (dt_1 and dt_2 are considered as exterior differential forms as in Cartan's formulation):

$$W_1 = X_{31} \cdot dt_1, \quad W_2 = X_{37} \cdot dt_2, \tag{26}$$

$$W_{12} = X_{52} \cdot dt_1 + X_{42} \cdot dt_2, \tag{27}$$

$$W_{13} = X_{53} \cdot dt_1 + X_{43} \cdot dt_2, \tag{28}$$

$$W_{23} = X_{56} \cdot dt_1 + X_{46} \cdot dt_2, \tag{29}$$

$$W_1' = X_{61} \cdot dt_1 + X_{66} \cdot dt_2, W_2' = X_{62} \cdot dt_1 + X_{67} \cdot dt_2, \tag{26'}$$

$$W_{12}' = X_{72} \cdot dt_1 + X_{82} \cdot dt_2, \tag{27'}$$

$$W_{13}' = X_{73} \cdot dt_1 + X_{83} \cdot dt_2, \tag{28'}$$

$$W_{23}' = X_{76} \cdot dt_1 + X_{86} \cdot dt_2, \tag{29'}$$

Set also

$$U = X_{21}, \quad V = X_{22}, \quad R = X_{23}, \quad K = X_{25}, \quad K' = X_{30}. \tag{30}$$

The equations (5), (10)-(30), and (10′)-(16′), (26′)-(29′) will give rise to a hypothesis system (HYP) consisting in all 34 d-pols in the X's. In determining the corresponding d-char-set it is rather complicate to carry out the computations up to the length owing to the complexity of the integ-pols involved. However, it is unnecessary to do so. In fact, there appears already in the 3-th d-bas-set (DBS_3) a d-pol of the form

$$B_6 = X_{23}^2 * X_{30} + X_{22}^2.$$

The vanishing of B_6 gives us a condition for $X_{30} = K'$ to be satisfied by above surface pair under some non-degeneracy conditions such as $R = X_{23} \neq 0$ among the others. As the relation between S and S' is a symmetric one the same is true for K. We have thus discovered automatically the following

Theorem of Baecklund. Two surfaces in correspondence of Baecklund have equal constant negative curvature given by:

$$K = K' = -\sin^2 A/R^2.$$

We remark that the method last described is also a mechanical as well as a general one. For example, we may ask the same questions for pairs of surfaces in similar relations as $(NN), (NT), (TT)$ in affine space since notions of affine normals, affine curvatures, etc. are also well-defined in the affine case. Experiments on such kind of problems are yet in progress. cf. [1]

References

[1] W.Blaschke, Vorlesungen ueber Differential Geometrie, I. Elementare Differentialgeometrie, Auflage 3, New York (1967); II. Affine Differentialgeometrie, Auflage 2, New York (1967).

[2] E.Cartan, Les systemes differentiels exterieurs et leurs applications geometriques, Paris, (1945).

[3] J.R.Gabriel, SARA - A small automated reasoning assistant, Preprint, Argonne National Laboratory (1986).

[4] J.F.Ritt, Differential equations from the algebraic standpoint, Amer.Math.Soc. Colloquium, 14,(1932).

[5] J.F.Ritt, Differential algebra, Amer.Math.Soc.,(1950).

[6] J.M.Thomas, Differential systems, Amer.Math.Soc.,(1937).

[7] Wu Wen-tsun, On the mechanization of theorem-proving in elementary and differential geometry, Scientia Sinica, Math. Supplement (I), (1979) 94-102. (in Chinese)

[8] Wu Wen-tsun, Mechanical theorem proving in elementary geometry and differential geometry, in Proc.1980 Beijing DD-Symposium, Beijing, v.2, (1982) 1073-1092.

[9] Wu Wen-tsun, Some remarks on mechanical theorem-proving in elementary geometry, Acta Math. Scientia 3 (1983) 357-360.

[10] Wu Wen-tsun, Basic principles of mechanical theorem-proving in elementary geometries, J.Sys.Sci. & Math.Scis.,4 (1984) 207-235. Re-published in J. Automated Reasoning, 2 (1986) 221-252.

[11] Wu Wen-tsun, A constructive theory of differential algebraic geometry, in Diff.Geom. & Diff. Eqs, Proc. Shanghai 1985, Lect. Notes in Math. , No.1255, Springer (1987) 173-189.

[12] Wu Wen-tsun, On zeros of algebraic equations — an application of Ritt principle, Kexue Tongbao 31 (1986) 1-5.

[13] Wu Wen-tsun, A mechanization method of geometry and its applications, I. Distances, areas, and volumes, J.Sys.Sci. & Math.Scis., 6 (1986) 204-216.

[14] Wu Wen-tsun, A mechanization method of geometry and its applications, I. Distances, areas, and volumes in euclidean and non-euclidean geometries, Kuxue Tongbao 32 (1986) 436-440.

[15] Wu Wen-tsun, A mechanization method of geometry and its applications, 2. Curve pairs of Bertrand type, Kuxue Tongbao 32 (1987) 585-588.

[16] Wu Wen-tsun, Mechanical derivation of Newton's Gravitational Laws from Kepler's Laws, MM-Reseach Preprints, No.2 (1987) 53-61.

[17] Wu Wen-tsun, Automated derivation of Newton's Gravitational Laws from Kepler's Laws, to appear in New Trends in Automated Mathematical Reasoning, Eds. A. Ferro et al.

[18] Wu Wen-tsun, On the foundation of algebraic differential geometry, MM-Res. Preprints, No.3 (1989) 1-26, also in Sys.Sci. & Math.Scis.,2 (1989) 289-312.

[19] Wu Wen-tsun, On the generic zero and Chow basis of an irreducible ascending-set, MM-Res. Preprints, No.4 (1989).

[20] Wu Wen-tsun, A mechanization method of equations solving and theorem proving, to appear in Issues in Robotics and Non-linear Geometry, Eds. Ch.Hoffmann et el.

[21] Wu Wen-tsun and Wu Tianjiao, A mechanization method of geometry and its applications, 5. Solving transcendental equations by algebraic methods, MM-Res. Preprints, No.3 (1989) 30-32.

MM Research Preprints, 1–18
MMRC, AMSS, Academia Sinica
No. 8, October 1992

On a Finiteness Theorem about Optimization Problems [1)]

Wu Wen-tsun
Institute of Systems Science, Academia Sinica

1. Introduction

The present paper is devoted to the study of the following optimization problem:

Optimization Problem P. Let $R^n(X)$ be the *real* euclidean space of dimension n in coordinates $X = (x_1, \cdots, x_n)$ and D a domain in R^n, R being the real field. Let $f, h_i, (i \in I = \{1, \cdots, r\}), g_j, j = 1, \cdots, s$ and g be all pols in $R[X]$. To determine the least or the greatest value, if it exists, of the pol f in the domain D under the equality constraints

$$h_i = 0, i \in I, \tag{1.1}$$

the inequality constraints

$$g_j \geq 0, \ (or \ \leq 0, > 0, < 0) \tag{1.2}$$

and the non-zero condition

$$g \neq 0. \tag{1.3}$$

Clearly, if D is closed and bounded and (1.2),(1.3) do not present, then such least and greatest values will necessarily exist. We shall restrict ourselves to such domains D which are closed and bounded with boundaries on a finite number of real algebraic surfaces. By introducing eventually new variables and new equations we may turn all the inequality constraints (1.2) into equality ones and turn the closed bounded domain D into a closed domain of rectangular form in the due euclidean space. So in what follows we shall suppose for the Problem P that the inequalities (1.2) are non-existant and that the domain D is of the rectangular form

$$D : a_i \leq x_i \leq b_i, i \in I. \tag{1.4}$$

Let us write as HS the polset of all pols $h_i, i \in I$. Let D' be an arbitrary domain in R^n, not necessarily closed or bounded. Denote now the set of all zeros of HS in D' for which $g \neq 0$ by $D'Zero(HS/g)$ and set by definition

$$D'Val_f(HS/g) = \{f(X)|X \in D'Zero(HS/g)\} \tag{1.5}$$

1)The present paper is partially supported by NSFC Grant JI85312 and TWAS Grant 86-93.

With D' the domain D in (1.4) our Problem P becomes the problem of determining *Least* $DVal_f(HS/g)$ and/or *Greatest* $DVal_f(HS/g)$, if exist, where *Least* resp. *Greatest* means the least resp. the greatest value of the set of values in question. We shall now give a solution of such a Problem P as expressed in the following theorem for which the proof will be given in the next section.

Finite Kernel Theorem. There is an algorithm which gives for any Problem P of above type a finite set

$$K \subset DVal_f(HS/g) \tag{1.6}$$

such that if the least or greatest value in question does exist, then K is non-empty and

$$\begin{aligned} Least\ K &= Least\ DVal_f(HS/g), \\ Greatest\ K &= Greatest\ DVal_f(HS/g). \end{aligned} \tag{1.7}$$

Definition. The finite set of real values K eventually empty whose existence is asserted by the Finiteness Theorem will be called a *kernel set* of the Problem P or one of the set $DVal_f(HS/g)$ to be denoted by

$$K = DKer_f(HS/g). \tag{1.8}$$

In Sect 2 we shall describe the proof of the Finiteness Theorem and in Sect 3 we shall give illusrative examples for the applications of our method to some concrete problems.

2. Proof of Finiteness Theorem

Let f, h_i, and g be as in Sect 1 but the domain D be an *open* one, say O. Let X^0 be a point in O verifying the conditions (1.1), (1.3) such that f attains its local minimum or maximum under conditions (1.1), (1.3). We shall say that X^0 is an *extremal zero* (abbr. *E-zero*) and $f(X^0)$ an *extremal value* (abbr. *E-value*) of the corresponding problem. Henceforth the set of all such E-zeros and E-values will be denoted respectively by

$$E_fOZero(HS/g) \text{ and } E_fOVal(HS/g). \tag{2.1}$$

Let us consider first the particular case for which the polset HS is an asc-set of either type 0 or type 1 with the pols h_i in the form below:

$$h_i \ = \ I_i * y_i^{d_i} + \ lower\ degree\ terms\ in\ y_i. \tag{2.2}$$

In (2.2) we have rearranged the variables so that

$$(x_1, \cdots, x_n) = Perm(u_1, \cdots, u_d,\ y_1, \cdots, y_r),\ (r + d = n), \tag{2.3}$$

for some permutation $Perm$ for which the ordering of $u_1, \cdots, u_d$ and the ordering of $y_1, \cdots, y_r$ are same as in that of $x_1, \cdots, x_n$. Besides the *initials* I_i of pols h_i in HS we are also interested in the so-called *separants* S_i of h_i defined by

$$\frac{\partial h_i}{\partial y_i} = S_i. \tag{2.4}$$

Remark that S_i coincides with the initial I_i of h_i in case $d_i = 1$. Set

$$IP = PROD_i\ I_i,\ SP = PROD_i\ S_i,\ and\ ISP = PROD_i\ (I_i * S_i), \tag{2.5}$$

to be called the *initial-product (abbr. I-product)*, the *separant-product (abbr. S-product)*, and the *initial-separant-product (abbr. IS-product)* respectively.

The solving of Problem P in the present case will now be done via the classical Lagrange method of multipliers as follows. Introduce *multipliers* $M = (m_1, \cdots, m_r)$ and form the *Lagrange pol*

$$L = f + SUM_i\ (m_i * h_i). \tag{2.6}$$

Set $N = (1, \cdots, n)$ and form now the following *Lagrange polset*

$$LAG = \{\frac{\partial L}{\partial x_k}, h_i \mid k \in N, i \in I\}. \tag{2.7}$$

Let $Proj$ be the natural projection of the euclidean space $R^{n+r}(X, M)$ to $R^n(X)$. Then we have the following

Lemma 1. Suppose that the polset HS is an asc-set as in (2.2) and the pol g is divisible by each of the initials I_i and also by each of the separants S_i. Then we have

$$E_f OZero(HS/g) \subset Proj\ OZero(LAG/g). \tag{2.8}$$

Proof. Let

$$X^0 = (x_1^0, \cdots, x_n^0) = Perm(u_1^0, \cdots, u_d^0, y_1^0, \cdots, y_r^0) \tag{2.9}$$

be an E-zero of f in $E_f OZero(HS/g)$. As each I_i and S_i is non-zero at X^0 we may apply the implicit function theorem to $h_1 = 0, \cdots, h_r = 0$ in succession at X^0. There will be thus some neighborhood V about X^0 contained in O and continuously differentiable functions $\phi_i(u_1, \cdots, u_d)$ in V such that for $i = 1, \cdots, r$ we have

$$y_i = \phi_i(u_1, \cdots, u_d)\ in\ V, \tag{2.10}$$

$$y_i^0 = \phi_i(u_1^0, \cdots, u_d^0),\ and \tag{2.11}$$

$$h_i(Perm(u_1, \cdots, u_d, \phi_1, \cdots, \phi_r)) = 0\ in\ V. \tag{2.12}$$

Let us set

$$f(Perm(u_1, \cdots, u_d, \phi_1, \cdots, \phi_r)) = F(u_1, \cdots, u_d). \tag{2.13}$$

We have then

$$\frac{\partial F}{\partial u_j} = SUM_i\ (\frac{\partial f}{\partial y_i} * \frac{\partial \phi_i}{\partial u_j}) + \frac{\partial f}{\partial u_j}\ in\ V, j = 1, \cdots, d. \tag{2.14}$$

Now f attains its extremal value at X^0 in V implies that $F(u_1, \cdots, u_d)$ attains its extremal value at $(u_1^0, \cdots, u_d^0)$. So we have also

$$\frac{\partial F}{\partial u_j} = 0 \ at \ X^0, j = 1, \cdots, d. \tag{2.15}$$

Differentiating now (2.12) we get for each pair of $i = 1, \cdots, r$ and $j = 1, \cdots, d$,

$$\frac{\partial h_i}{\partial u_j} + SUM_k \, (\frac{\partial h_i}{\partial y_k} * \frac{\partial \phi_k}{\partial u_j}) = 0 \ in \ V. \tag{2.16}$$

Differentiating (2.6) we get

$$\begin{aligned}
\frac{\partial L}{\partial y_1} &= \frac{\partial f}{\partial y_1} + m_1 * S_1 + m_2 * \frac{\partial h_2}{\partial y_1} + \cdots + m_{r-1} * \frac{\partial h_{r-1}}{\partial y_1} + m_r * \frac{\partial h_r}{\partial y_1}, \\
\frac{\partial L}{\partial y_2} &= \frac{\partial f}{\partial y_2} + 0 \quad + m_2 * S_2 + \cdots + m_{r-1} * \frac{\partial h_{r-1}}{\partial y_2} + m_r * \frac{\partial h_r}{\partial y_2}, \\
& \cdots\cdots\cdots\cdots\cdots \\
\frac{\partial L}{\partial y_r} &= \frac{\partial f}{\partial y_r} + 0 \quad + 0 \quad + \cdots + 0 \quad + m_r * S_r.
\end{aligned} \tag{2.17}$$

As $S_i \neq 0$ at X^0 we may solve the equations

$$\frac{\partial L}{\partial y_i} = 0, i = 1, \cdots, r \tag{2.18}$$

at X^0 to get $M = M^0 = (m_1^0, \cdots, m_r^0)$. For a fixed $j = 1, \cdots, d$ let us multiply the equations in (2.17) successively by $\frac{\partial \phi_i}{\partial u_j}, i = 1, \cdots, r$ and adding, then by (2.14)-(2.16) we get

$$\frac{\partial L}{\partial u_j} = 0 \ at \ (X^0, M^0). \tag{2.19}$$

From (2.18),(2.19) we see that (X^0, M^0) is in $OZero(LAG/g)$ and hence X^0 is in $Proj\ OZero(LAG/g)$. This proves (2.8).

Lemma 2. Let HS and g be as in Lemma 1. Suppose that

$$f(X) = x_1. \tag{2.20}$$

Let $Proj_1$ be the projection of $R^{n+r}(X)$ to R defined by

$$Proj_1 \ (X, M) = x_1. \tag{2.21}$$

Then there is a finite set of real values

$$OKer_{x_1}(HS/g) = Proj_1 \ OZero(LAG/g), \tag{2.22}$$

which may eventually be empty, such that

$$E_{x_1} OZero(HS/g) \subset OKer_{x_1}(HS/g) \subset OVal_{x_1}(HS/g). \tag{2.23}$$

Proof. Owing to the chosen ordering of variables x_1 can only be either u_1 or y_1. Consider first the case $x_1 = u_1$. Then by (2.20) we have

$$\frac{\partial f}{\partial y_1} = \frac{\partial f}{\partial y_2} = \cdots = \frac{\partial f}{\partial y_r} = 0, \; \frac{\partial f}{\partial u_1} = 1.$$

For any $(X^0, M^0) \in OZero(LAG/g)$ we have $\frac{\partial L}{\partial y_i} = 0$ at (X^0, M^0) and $S_i \neq 0$ at X^0 so that (2.17) gives $M^0 = 0$ or $m_i^0 = 0, i = 1, \cdots, r$. As $\frac{\partial L}{\partial u_1} = 0$ at (X^0, M^0) too this will be in contradiction to the equation

$$\frac{\partial L}{\partial u_1} = \frac{\partial f}{\partial u_1} + m_1 * \frac{\partial h_1}{\partial u_1} + \cdots + m_r * \frac{\partial h_r}{\partial u_1} = 1 \; at \; (X^0, M^0).$$

Hence $OZero(LAG/g) = empty$ in this case and (2.23) follows trivially from (2.8) and (2.22).

Consider next the case $x_1 = y_1$. Then by (2.20) we have

$$\frac{\partial f}{\partial y_1} = 1, \frac{\partial f}{\partial y_2} = \cdots = \frac{\partial f}{\partial y_r} = 0.$$

Let (X^0, M^0) be again any zero in $OZero(LAG/g)$. Then as before (2.17) will give us

$$m_2^0 = \cdots = m_r^0 = 0, \; m_1^0 = -\frac{1}{S_1^0} \neq 0,$$

where S_1^0 is the value of S_1 at X^0. Now $h_i = 0$ at X^0 for any $i = 1, \cdots, r$. Owing to our chosen ordering of variables h_1 will have no variables u_j involved and is a pol in $y_1 = x_1$ alone with constant coefficients. The possible values of x_1^0 for which $h_1 = 0$ is to be satisfied are thus finite in number. Then by Lemma 1 the set $OKer_{x_1}(HS/g)$ defined by (2.22) is a finite set of real values verifying (2.22). The Lemma 2 is now completely proved.

Consider now the case for which HS is a general polset in variables $x_1, \cdots, x_n$. Our char-set method with slight modifications (cf. e.g. [WU1-3]) will give rise to the following lemmas 3,4.

Lemma 3. For an arbitrary polset $PS \subset R[X]$ let CS be a char-set of PS. Let I_i and S_i be the initials and separants of pols in CS and ISP be the corresponding IS-product. Then we have

$$\begin{aligned} OZero(PS) = OZero(PS/ISP) + UNION_j \; OZero(PS + I_j) \\ + UNION_j \; OZero(PS + S_j). \end{aligned} \tag{2.24}$$

Lemma 4. There is an algorithm such that for an arbitrary real polset PS we shall arrive in a finite number of steps at a set of asc-sets AS_j such that

$$OZero(PS) = UNION_j \; OZero(AS_j/ISP_j), \tag{2.25}$$

in which for each j, ISP_j is the IS-product of AS_j.

From (2.25) for $PS - HS$ we have now for any $f \in R[X]$,

$$OVal_f(HS) = UNION_j \; OVal_f(AS_j/ISP_j). \tag{2.26}$$

It is clear that

$$E_f OVal(HS) \subset UNION_j \; E_f OVal(AS_j/ISP_j). \tag{2.27}$$

Lemma 5. Let HS be an arbitary polset and f be given by (2.20). Then there is a finite set of real values K such that

$$E_{x_1}OVal(HS) \subset K \subset OVal_{x_1}(HS). \tag{2.28}$$

Proof. Let us decompose $OZero(HS)$ as in (2.25). By Lemmas 1,2 we have then a finite set of real values K given by

$$K = UNION_j\ OKer_{x_1}(AS_j/ISP_j), \tag{2.29}$$

By (2.27) we have then (2.28) as to be proved.

We are now in a position to prove our main theorem as follows.

Proof of Finite Kernel Theorem.

Let us introduce a new variable x_0 and a new pol

$$h_0 = f(X) - x_0. \tag{2.30}$$

For any domain $D' \subset R^n(X)$ we shall write D'^+ for the domain in $R^{n+1}(x_0, X)$ defined by $X \in R^n$ while x_0 is arbitrary. Set also

$$HS^+ = HS + \{h_0\}. \tag{2.31}$$

Then it is clear that for D' open,

$$D'Val_f(HS/g) = D'^+Val_{x_1}(HS^+/g),\ and \tag{2.32}$$

$$E_fD'Val(HS/g) = E_{x_0}D'^+Val(HS^+/g). \tag{2.33}$$

Let s be now any set $(s_1, \cdots, s_n)$ with each s_i a sign $+, -$, or 0. For any such set s let O_s be the open domain in R^n defined by the set of equations

$$x_i = b_i,\ or\ x_i = a_i,\ or\ a_i < x_i < b_i,$$

according as $s_i = +,\ or\ -,\ or\ 0, i$ being from 1 to n. Consider any sign set s. By Lemma 5 we have a finite set of real values K_s, to be called the *kernel set* for the open domain O_s or O_s^+, such that

$$E_{x_0}O_s^+Val(HS^+/g) \subset K_s \subset O_s^+Val_{x_0}(HS^+/g). \tag{2.34}$$

Now it is clear that

$$\begin{aligned} E_{x_0}D^+Val(HS^+/g) &\subset UNION_s\ E_{x_0}O_s^+Val(HS^+/g),\ and \\ D^+Val_{x_0}(HS^+/g) &= UNION_s\ O_s^+Val_{x_0}(HS^+/g), \end{aligned} \tag{2.35}$$

in the $UNIONs$ is runing over all the possible 3^n sign sets. Set now

$$K = UNION_sK_s. \tag{2.36}$$

Then by (2.33)-(2.36) we get

$$E_f DVal(HS/g) \subset K \subset DVal_f(HS/g). \tag{2.37}$$

As the least or greatest value of f in question, supposed to exist, is necessarily one of the E-values in $E_f DVal(HS/g)$, we get readily (1.7) from (2.37). As the algorithm for arriving at the kernel set K is clear from the above context, the theorem is completely proved.

3. Some Examples

Problems in non-linear programming are typical optimization problems which have been dealt with by our methods in e.g. [Wu5] and [WTJ1,2]. Problems involving inequalities can also usually be reduced to optimization problems. A general method of inequalities-proving is furnished by the CAD method of Collins, cf. e.g. [Col] and [A]. On the other hand in the book [Wu1] has been described a method of proving geometrical inequalities of some special type which has been further exploited and extended by S.C.Chou and X.S.Gao, cf. e.g. [C-G]. The author has also exhibited a general method based on a classical theorem of elementary calculus (cf. [Cou], p.198) which has been applied to the proving of various algebraic and geometric inequalities, cf. [Wu4,6]. The method described in the preceding sections is a refinement and also a complement of the preceding method. The examples given below may give some idea about the efficiency of the present method. Compare the papers [Wu6] and [C-G].

Example 1. The Pasch Theorem. Given a triangle ABC and a line l passing none of the vertices and intersecting the 3 sides BC, AC, AB in the points D, E, F. Then either none or just two of D, E, F are inside the segments BC, AC, AB.

Proof. Let us suppose that D is inside the segment BC and E is outside the segment AC in the order of ACE. We have to prove that F is inside the segment AB. The other cases of the theorem may be deduced from this case by *reductio absurdo.*

For this purpose let us take oblique coordinates so that

$$A = (0,0), B = (a,0), C = (0,b), E = (0,rb), F = (x,0), \ and$$
$$D = (1-y).B + y.C = ((1-y)*a, y*b),$$

in which

$$a > 0, b > 0, r > 1, \ and$$
$$0 < y < 1. \tag{3.1}$$

Now instead of considering D as a fixed point in the open segment BC, let us take D to be a point varying on the closed segment BC including the two end points B, C. Instead of (3.1) with y fixed we shall have y a variable in the closed domain

$$D : 0 \leq y \leq 1. \tag{3.1'}$$

From collinearity of D, E, F we have $h = 0$ where

$$h = y * (r * a - x) - r * (a - x). \tag{3.2}$$

We shall take the ordering of variables to be

$$x \prec y.$$

The problem is now to determine the least and greatest values of x clearly exist for y varying in the closed domain D under the constraint equation $h = 0$. The variable x plays thus the role of x_0 in the proof of the proof of the Finite Kernel Theorem in Sect 2. For the above purpose let us split now the domain D into 3 open ones O_1, O_2, O_3 with O_1 defined by (3.1) and O_2, O_3 defined by $y = 0$ and $y = 1$ respectively. For O_2 we have $x = a$ and for O_3 we have $x = 0$, being the projections on x of the points $(x, y) = (a, 0)$ and $(0, 1)$ respectively. The kernel sets for O_2 and O_3 are thus

$$K_2 = \{a\}, K_3 = \{0\}. \tag{3.3}$$

Consider now the open domain $O_1 : 0 < y < 1$. We have here $HS = \{h\}$ and the initial or separant is given by

$$I = S = r * a - x.$$

As $r > 1$ we see that $O_1 Zero(HS + S_1) = \ empty$ so that

$$O_1 Zero(HS) = O_1 Zero(HS/S_1).$$

Now HS is its own asc-set with the leading variable y but not x, the variable to be optimized, By the general method it follows that the corresponding kernel set for O_1 is

$$K_1 = \ empty. \tag{3.4}$$

(3.3) and (3.4) give now the kernel set of our problem as

$$K = K_1 + K_2 + K_3 = \{0, a\} \subset R.$$

It follows that the least and greatest values of x are 0 and a corresponding to the positions $F = A, D = C$ and $F = B, D = B$ respectively. For other positions of D, i.e. for D inside the segment BC, we have then necessarily $0 < x < a$ or F inside the segment AB, which proves the theorem.

Example 2. Quadrilateral Convexity Theorem. Let $ABCD$ be a convex quadrilateral with points O, P, Q, R on the inside of the sides AD, AB, BC, and CD respectively. Then $OPQR$ is also a convex quadrilateral.

Proof. Owing to convexity of $ABCD$ oblique coordinates can be so taken that

$$O = (0, 0), A = (0, -a), B = (b_1, -b_2), C = (c_1, c_2), D = (0, d),$$

$$Q = (q, 0),\ q = \frac{b_1 * c_2 + b_2 * c_1}{b_2 + c_2},$$

$$P = r_0.B + (1 - r_0).A = (r_0 * b_1, -r_0 * b_2 - (1 - r_0) * a), 0 < r_0 < 1,$$
$$R = r.C + (1 - r).D = (r * c_1, r * c_2 + (1 - r) * d),\ \ 0 < r < 1.$$

The constants a, b_1, b_2, c_1, c_2, d will all be positive so that q is also positive. Let us set

$$k = \frac{Oriented\ Area(\triangle OQR)}{Oriented\ Area\ (\triangle OQP)}$$

so that $h = 0$ where

$$h = k * (r_0 * b_2 + (1 - r_0) * a) + (r * c_2 + (1 - r) * d).$$

We shall prove that $k < 0or\mathrm{R,P}areonoppositesidesoftheline\mathrm{OQ}$. To do this we shall, as in the case of Example 1, instead of considering R as a fixed point inside the open segment CD with r constant and $0 < r < 1$, take R as a point varying on the closed segment CD with r varying in the domain

$$D : 0 \leq r \leq 1.$$

Let us take ordering of variables to be

$$k \prec r.$$

We come then to the problem of optimizing k for the domain D under the constraint equation $h = 0$. Here k will play again the role of x_0 in Sect 2. Split now the domain D into open ones O_1, O_2, O_3 corresponding to the cases $0 < r < 1, r = 0$, and $r = 1$ respectively. Suppose first that $c_2 \neq d$ or CD is not parallel to OQ. It is then readily found that the kernel sets for the open domains O_i and for D are respectively ($Proj_k$ means projection on coordinate k)

$$\begin{aligned} K_1 &= \ empty, \\ K_2 &= \{k_2\} = Proj_k\ \{(k, r) = (k_2, 0)\}, \\ K_3 &= \{k_3\} = Proj_k\ \{(k, r) = (k_3, 1)\}, \\ K\ &= \ K_1 + K_2 + K_3 = \{k_2, k_3\}, \end{aligned}$$

in which

$$\begin{aligned} k_2 &= -\frac{d}{r_0 * b_2 + (1 - r_0) * a} < 0,\ and \\ k_3 &= -\frac{c_2}{r_0 * b_2 + (1 - r_0) * a} < 0. \end{aligned}$$

We see therefore both the least and the greatest value of k are negative. If $c_2 = d$ then the least and greatest value of k are both equal to the same negative value $k_2 = k_3$. In any way we have $k < 0$ so that R, P are on opposite sides of the line OQ. In the similar way we prove that O, Q are on the opposite sides of the line PR. It follows that the quadrilateral $OPQR$ is convex and the theorem is proved.

Example 3. The Median-Bisector Theorem. For a non-isosceles triangle ABC the median over the side AB is always greater than the interior bisector on the same side.

This theorem has been proved in the previous paper [Wu6] by means of the method given in that paper. We shall re-prove it by means of the method in the present paper to make a comparison of the two methods.

Let us take coordinates as in [Wu6] such that $A = (-a, 0), B = (a, 0), C = (x_0, y_0), (a > 0, y_0 > 0)$ and center of circumcircle at $(0, b)$. Let the radius of the circumcircle be $c > 0$ so that $c^2 = a^2 + b^2$. Let $U = (0, c + b), V = (0, b - c)$ be the two extremities of the diameter of the circumcircle through the mid-point $M = (0, 0)$ of the side AB. Let CV meets AB in $X = (x, 0)$. Then CM is the median and CX is the interior bisector both on the side AB. Introduce uniformizing parameters u, t such that

$$x_0 = 2 * c * \frac{u}{1 + u^2}, \; y_0 = c * \frac{1 - u^2}{1 + u^2} + b, \tag{3.5}$$

$$c = a * \frac{1 + t^2}{2 * t}, \; b = a * \frac{1 - t^2}{2 * t}. \tag{3.6}$$

Set $d = |CM|^2 - |CX|^2$. Then we have $HS = \{h_1, h_2\} = 0$, where

$$h_1 = x - a * t * u, \tag{3.7}$$

$$h_2 = x^2 * t * (u^2 + 1) - 2 * x * a * u * (t^2 + 1) + d * t * (u^2 + 1). \tag{3.8}$$

Now consider C not as a fixed point but as a point varying on the closed arc BU. Here we admit the degenerate triangle ABC for which C coincides with B, the side AC coincides with AB and the side BC degenerates into the tangent line at B of the circumcircle. We are then led to the consideration of the problem of optimizing d for the above varying positions of C. Here d will play the role of x_0 in Sect 2 and the ordering of variables will be

$$d \prec u \prec x. \tag{3.9}$$

The domain D in question is defined by

$$0 \leq u \leq \frac{b + c}{a}. \tag{3.10}$$

Split now the domain D into 3 open ones O_1, O_2, O_3 corresponding to the ranges $0 < u < \frac{b+c}{a}, u = 0$, and $u = \frac{b+c}{a}$ respectively. Geometrically these ranges correspond to the cases for which C varies on the open arc between B and U, C takes the single position U, and C takes the single position B. The kernel sets for O_2, O_3 are readily seen to be

$$K_2 = \{d = 0\}, K_3 = \{d = a^2\}. \tag{3.11}$$

Consider now the case of O_1. We find

$$O_1 Zero(HS) = O_1 Zero(AS_1/S_1) + O_1 Zero(AS_2). \tag{3.12}$$

Here $AS_1 = \{A_{11}, A_{12}\}, AS_2 = \{A_{21}, A_{22}, A_{23}\}$ are asc-sets with

$$\begin{aligned} A_{11} &= a^2 * t^2 * u^4 - (a^2 * t^2 + 2 * a^2 - d) * u^2 + d, \\ A_{12} &= h_1, \\ S_1 &= \frac{\partial A_{11}}{\partial u} = 2 * u * (2 * a^2 * t^2 * u^2 - (a^2 * t^2 + 2 * a^2 - d)), \\ A_{21} &= d^2 - 2 * d * a^2 * (3 * t^2 + 2) + a^4 * (t^2 + 2)^2, \\ A_{22} &= \frac{S_1}{2 * u}, \\ A_{23} &= h_1. \end{aligned} \tag{3.13}$$

For $O_1 Zero(AS_1/S_1)$ it gives no contribution to the kernel set. For $O_1 Zero(AS_2)$ we see that there are two zeros $d = d_1, d_2$ of $A_{21} = 0$ which are both positive. Consequently for the kernel set K_1 for O_1 we have

$$K_1 \subset \{d_1, d_2\}. \tag{3.14}$$

From (3.11) and (3.14) we have therefore

$$\{0, a^2\} \subset K = K_1 + K_2 + K_3 \subset \{0, a^2, d_1, d_2\}$$

It follows that the least possible value of d is 0 which occurs when $C = U$ or when the triangle ABC is an isosceles one. This proves the theorem.

Example 4. The Equi-bisector Theorem. A triangle with two equal interior bisectors is an isosceles one.

This theorem is not at all trivial and has intrigued geometers of last century. Clearly the theorem follows from the following a little stronger theorem:

A triangle of unequal sides will have greater interior bisector for smaller angle.

The theorem in this strengthened from was in the first time proved by our general method of mechanical geometry theorem proving as a joint work of S.C.Chou and the present author (unpublished). The proof is again quite non-trivial. Cf. in this respect a popular pamphlet [W-L] in Chinese. Below we shall give a different proof based on the method of the present paper.

Proof. Consider a triangle ABC with $|AC| > |BC|$. Let the bisectors of the angles A, B be respectively AE, BF with E on BC and F on AC. We have to prove that $|AE| > |BF|$.

For this purpose let us construct an ellipse passing through C and having A and B as its two foci. Let the lengths of the sides AC, BC be respectively s_b, s_a, the lengths of AE, BF be respectively d_a, d_b and $d_0 = d_a^2 - d_b^2$. Let us take coordinates such that the equation of the ellipse is

$$\frac{X^2}{a^2} + \frac{Y^2}{b^2} = 1, \ while$$

$$A = (-c, 0), B = (+c, 0), C = (x, y), E = (x_a, y_a), F = (x_b, y_b),$$

in which

$$a > 0, b > 0, c > 0, \ and \ a^2 = c^2 + b^2.$$

The points E, F will be so determined that $\frac{|EB|}{|EC|} = \frac{|AB|}{|AC|}$ and $\frac{|FA|}{|FC|} = \frac{|AB|}{|BC|}$. Then we have a set of equations $HS = \{h_1, \cdots, h_{10}\} = 0$ with h_i given below:

$$\begin{aligned}
h_1 &= y^2 * a^2 + x^2 * (a^2 - c^2) - a^2 * (a^2 - c^2),\\
h_2 &= a * s_a - (a^2 - c * x),\\
h_3 &= a * s_b - (a^2 + c * x),\\
h_4 &= -x_a * s_b - 2 * x_a * c + s_b * c + 2 * x * c,\\
h_5 &= y_a * (s_b + 2 * c) - 2 * c * y,\\
h_6 &= d_a^2 - (x_a + c)^2 - y_a^2,\\
h_7 &= -x_b * s_a - 2 * x_b * c - s_a * c + 2 * x * c,\\
h_8 &= y_b * (s_a + 2 * c) - 2 * c * y,\\
h_9 &= d_b^2 - (x_b - c)^2 - y_b^2,\\
h_{10} &= d_a^2 - d_b^2 - d_0.
\end{aligned}$$

Let U and V be the points $(a, 0)$ and $(0, b)$ on the ellipse. As before let us consider C not as a fixed point but a point varying on the closed arc UV on the ellipse. Then we come to the problem of optimizing d_0 in the domain D defined by

$$0 \le x \le a, 0 \le y \le b,$$

for which optimal values of d_0 should exist. Again as before let us split the domain D into 3 open ones O_1, O_2, O_3 corresponding to

$$0 < x < a, 0 < y < b; (x, y) = (0, b);\ and\ (x, y) = (a, 0),$$

respectively. For O_2 and O_3 the kernel sets are readily seen to be

$$K_2 = \{d_0 = 0\}, K_3 = \{d_0 = 16 * c^2 * \frac{(a + c)^2}{(a + 3 * c)^2}\}$$

For the kernel set corresponding to O_1 let us take the ordering of variables to be

$$d_0 \prec x \prec y \prec s_a \prec s_b \prec x_a \prec y_a \prec d_a \prec x_b \prec y_b \prec d_b.$$

Then the char-set of HS is found to be $CS = \{C1, \cdots, C_{10}\}$ with

$$\begin{aligned}
C_i &= h_{i-1}\ for\ i = 2, \cdots 10,\ while\\
C_1 &= x^4 * d_0 * c^4\\
&+ 48 * x^3 * c^6 * a + 64 * x^3 * c^5 * a^2 + 16 * x^3 * c^4 * a^3\\
&- 8 * x^2 * d_0 * c^4 * a^2 - 8 * x^2 * d_0 * c^3 * a^3 - 2 * x^2 * d_0 * c^2 * a^4\\
&- 64 * x * c^6 * a^3 - 128 * x * c^5 * a^4 - 112 * x * c^4 * a^5\\
&- 64 * x * c^3 * a^6 - 16 * x * c^2 * a^7\\
&+ 16 * d_0 * c^4 * a^4 + 32 * d_0 * c^3 * a^5 + 24 * d_0 * c^2 * a^6\\
&+ 8 * d_0 * c * a^7 + d_0 * a^8.
\end{aligned}$$

The only initials and separants worthy of consideration are

$$I_1 = d_0 * c^4, S_1 = \frac{\partial C_1}{\partial x}, S_2 = 2 * y * a^2, S_7 = 2 * d_a, S_{10} = 2 * db.$$

Denote the IS-product of CS by ISP, then we have

$$\begin{aligned} O_1Zero(HS) &= O_1Zero(CS/ISP) + O_1Zero(HS + \{d_0\}) \\ &+ O_1Zero(HS + \{S_1\}/d_0) + O_1Zero(HS + \{y\}) \\ &+ O_1Zero(HS + \{d_a\}) + O_1Zero(HS + \{d_b\}). \end{aligned}$$

The set $O_1Zero(CS/ISP)$ has no contribution to the kernel set K_1 for O_1. It is clear that $O_1Zero(HS+\{y\}) = empty$. From $d_a = 0$ it would follow from C_7, C_6, C_2 that $y_a = 0, y = 0$, and $x = a$ so that $O_1Zero(HS + \{d_a\}) = empty$. Similarly we have $O_1Zero(HS + \{d_b\}) = empty$ too. So it remains only to consider $O_1Zero(HS + \{d_0\})$ and $O_1Zero(HS + \{S_1\}/d_0)$.

Consider first the case $d_0 = 0$. In the char-set of $HS + \{d_0\}$ the first pol is given by

$$C_{11} = 3 * x^2 * c^3 + x^2 * c^2 * a - 4 * c^3 * a^2 - 4 * c^2 * a^3 - 3 * c * a^4 - a^5.$$

From $C_{11} = 0$ it would follow $x < 0$ or $x > a$. Hence we have again

$$O_1Zero(HS + \{d_0\}) = empty.$$

Consider now the case of $S_1 = 0$. In the char-set of $HS + \{S_1\}$ the first two pols are given by

$$\begin{aligned} C_{21} &= 32 * d_0^4 * c^5 + 80 * d_0^4 * c^4 * a + 80 * d_0^4 * c^3 * a^2 \\ &+ 40 * d_0^4 * c^2 * a^3 + 10 * d_0^4 * c * a^4 + d_0^4 * a^5 \\ &+ 2304 * d_0^2 * c^9 + 9600 * d_0^2 * c^8 * a + 16480 * d_0^2 * c^7 * a^2 \\ &+ 15376 * d_0^2 * c^6 * a^3 + 8462 * d_0^2 * c^5 * a^4 + 2695 * d_0^2 * c^4 * a^5 \\ &+ 408 * d_0^2 * c^3 * a^6 - 14 * d_0^2 * c^2 * a^7 - 14 * d_0^2 * c * a^8 - d_0^2 * a^9 \\ &- 13824 * c^1 3 - 62208 * c^1 2 * a - 139392 * c^1 1 * a^2 - 208448 * c^1 0 * a^3 \\ &- 227584 * c^9 * a^4 - 188672 * c^8 * a^5 - 120832 * c^7 * a^6 - 59264 * c^6 * a^7 \\ &- 21760 * c^5 * a^8 - 5632 * c^4 * a^9 - 896 * c^3 * a^1 0 - 64 * c^2 * a^1 1, \\ C_{22} &= 48 * x * d_0^2 * c^6 + 88 * x * d_0^2 * c^5 * a + 82 * x * d_0^2 * c^4 * a^2 \\ &+ 53 * x * d_0^2 * c^3 * a^3 + 16 * x * d_0^2 * c^2 * a^4 + x * d_0^2 * c * a^5 \\ &+ 5184 * x * c^{10} + 18144 * x * c^9 * a + 27648 * x * c^8 * a^2 + 25248 * x * c^7 * a^3 \\ &+ 15168 * x * c^6 * a^4 + 5664 * x * c^5 * a^5 + 1152 * x * c^4 * a^6 + 96 * x * c^3 * a^7 \\ &- 16 * d_0^3 * c^4 * a - 32 * d_0^3 * c^3 * a^2 - 24 * d_0^3 * c^2 * a^3 - 8 * d_0^3 * c * a^4 \\ &- d_0^3 * a^5 - 1728 * d_0 * c^8 * a - 6144 * d_0 * c^7 * a^2 - 8936 * d_0 * c^6 * a^3 \\ &- 6908 * d_0 * c^5 * a^4 - 3064 * d_0 * c^4 * a^5 - 768 * d_0 * c^3 * a^6 \\ &- 96 * d_0 * c^2 * a^7 - 4 * d_0 * c * a^8. \end{aligned}$$

From $C_{22} = 0$ we see that $d_0 < 0$ would imply $x < 0$. Hence in O_1 any zero of $O_1 Zero(HS + \{S_1\}/d_0)$ have its d_0 positive. In particular K_1, if not empty, is consisting of only positive values. It follows that the final kernel set $K = K_1 + K_2 + K_3$ is consisting of $d_0 = 0$ and utmost other positive values and consequently

$$Least\ K = 0$$

which corresponds to the isosceles triangle ABC with C at V. For all other positions of C in arc UV we should have $|AE| > |BF|$ and the theorem is thus proved.

References

[A] D.S.Arnon, A bibliography of quantifier elimination for real closed fields, JSC 5 (1988) 267-274.

[C-G] S.C.Chou & X.S.Gao, On the mechanical proof of geometry theorems involving inequalities, Preprints TR-89-31, UTA (1989).

[Col] G.E.Collins, Quantifier elimination for real closed fields by cylindrical algebraic decomposition, in 2nd GI Conf.Automata Theory & Formal Languages, Springer Lect.Notes Comp.Sci. 33 (1975) 134-183.

[Cou] R.Courant, Differential and Integral Calculus, vol.2 London & Glascow, (1936).

[WTJ1] Wu Tian-jiao, Some test problems on applications of Wu's method in non-linear programming problems, MM-Res.Preprints, No.6 (1991) 144-155.

[WTJ2] Wu Tian-jiao, On a collision problem, MM-Res.Preprints, No.7 (1992) 96-104.

[WU2] Wu Wen-tsun, On zeros of algebraic equations — an application of Ritt principle, Kexue Tongbao 31 (1986) 1-5.

[WU3] Wu Wen-tsun, A zero structure theorem for polynomial equations -solving and its applications, MM-Res.Preprints, No.1 (1987) 2-12.

[WU4] Wu Wen-tsun, A mechanization method of geometry and its applications, 3. Mechanical proving of polynomial inequalities and equations-solving, MM-Res. Preprints, No.2, (1987) 1-17. Also in Sys.Sci.& Math.Scis., 1 (1988) 1-17.

[WU5] Wu Wen-tsun, On the chemical equilibrium problem and equations-solving, MM-Res.Preprints, No.4 (1989) 23-39. Also in Acta Math.& Sci., 10 (1990) 361-374.

[WU6] Wu Wen-tsun, On problems involving inequalities, MM-Res.Preprints, No.7 (1992) 1-13.

[W-L] Wu Wen-tsun & X.L.Lue, Triangles with equal bisectors, (in Chinese) People's Education Press, Beijing (1985).

MM Research Preprints
No. 10, 1993, 1-10

On Surface-Fitting Problem in CAGD.

WU WEN-TSUN
吴文俊
(MMRC, Institute of Systems Science, Academia Sinica)

1. The Problem.

In the present paper we shall give a general method of solving the following surface-fitting problem in computer-aided geometry design (CAGD):

Problem SF. Given in real 3-space R^3 three sets of irreducible algebraic curves C_i, C_j, C_k with $i \in I, j \in J, k \in K$ respectively, I, J, K being all finite sets of indices. Given also two sets of irreducible algebraic surfaces $S_j, S_k, (j \in J, k \in K)$ containing C_j, C_k respectively. To determine an irreducible algebraic surface S of given degree m verifying the following conditions:

(a) S contains all the curves C_i, C_j, C_k, for $i \in I, j \in J, k \in K$.

(b) S touches smoothly each of S_j, S_k along the curves C_j, C_k respectively, for $j \in J, k \in K$. More precisely, for each point on C_j or C_k which is regular for C_j, S, S_j or for C_k, S, S_k, S and S_j or S and S_k have same tangent planes at that point.

(c) S possesses same curvature as S_k along the curves C_k, for each $k \in K$. More precisely, for each point on C_k which is regular for C_k, S and S_k, S and S_k will have the same (Gaussian) curvature at that point.

We may also replace (c) by other more stringent conditions, e.g. on conditions about normal curvatures, etc. which we shall not enter.

The problem for the requirements (a), (b) and some further smooth requirements have already been solved by Bajaj et al by the method of interpolation based on the theorem of Bezout and its extensions, cf. [B] and [B-I1,2]. We shall solve the above problem in all its generality based on some entirely different principle and method to be described in next sections.

We remark that in practice only real pieces of curves and surfaces C_i, S_j, etc. are actually given and only piece of real surface S is required. Our method bears however no influence on this restriction.

2. Basic Principles from Algebraic Geometry.

We recall first some fundamental concepts and facts of algebraic geometry. Cf. e.g. [H-P], [VdW], [WU1,2] as well as various papers of the author in MM-Res.Preprints, e.g. [WU3,4].

Let $\mathbf{K}$ be the basic field of characteristic 0, $\mathbf{K}^n$ the affine space of dimension n over $\mathbf{K}$, and $X = (x_1, \cdots, x_n)$ with x_i independent indeterminates.

Def. *Extended point* in $\mathbf{K}^n$:= point in $\mathbf{K}'^n$ with $\mathbf{K}'$ some extension field of $\mathbf{K}$.

Any extended point $\Xi = (\xi_1, \cdots, \xi_n)$ in $\mathbf{K}^n$ may be represented by a set of polynomials

$$P_i \equiv I_{c_i,0} * x_{c_i}^{d_i} + I_{c_i,1} * x_{c_i}^{d_i - 1} + \cdots + I_{c_i,d_i}, \quad i = 1, \cdots, r, \tag{2.1}$$

such that $(e_1), (e_2), (e_3)$ below hold true:

(e_1) $0 < c_1 < \cdots < c_r$.

(e_2) $I_{c_i,j} \in \mathbf{K}[x_1, \cdots, x_{c_i - 1}]$.

(e_3) Let P_i' be the polynomial get from P_i by substituting ξ_j for $x_j, j = 1, \cdots, c_i - 1$, then P_i' is an irreducible polynomial in the field $\mathbf{K}(\xi_1, \cdots, \xi_{c_i - 1})$, and ξ_{c_i} is a root of the equation $P_i' = 0$ so that each equation $P_i' = 0$ is the defining equation of ξ_{c_i}.

Def. The set of polynomials P_i in (2.1) forms an irreducible asc-set in $\mathbf{K}[X]$ and is called the *defining asc-set* of the extended point $\Xi = (\xi_1, \cdots, \xi_n)$.

Def. $n - r$:= *Dimension* over $\mathbf{K}$ of the extended point with defining asc-set (2.1).

Remark that different extended points may have same defining asc-set.

Def. An extended point Ξ^0 in $\mathbf{K}^n$ is a *specialization* over $\mathbf{K}$ of an extended point Ξ in $\mathbf{K}^n$ if for any polynomial $P \in \mathbf{K}[X]$ with $P(\Xi) = 0$, we have also $P(\Xi^0) = 0$. In notation:

$$\Xi \longrightarrow_{\mathbf{K}} \Xi^0. \tag{2.2}$$

Notation. Set of all specializations over $\mathbf{K}$ of an extended point Ξ in $\mathbf{K}^n$:= $Spec(\Xi)$.

Ex. Let x_1, x_2 be independent indeterminates Then

$$(x_1, x_2, \frac{x_2}{x_1}) \longrightarrow_{\mathbf{K}} (0, 0, 1).$$

Remark. There is analogous concept of specialization for projective spaces and products of affine and projective spaces.

Def. An *affine algebraic variety* or simply a *variety* in $\mathbf{K}^n$:= zero-set $Zero(PS)$ of some polset $PS \subset \mathbf{K}[X]$.

Theorem 2.1. An algebraic variety $Zero(PS)$ in $\mathbf{K}^n$ is irreducible over $\mathbf{K}$ if and only if it possesses an extended point Ξ in $\mathbf{K}^n$ such that all points of $Zero(PS)$ are specicializations over $\mathbf{K}$ of that point Ξ, in other words, if and only if for some extended point Ξ in $\mathbf{K}^n$,

$$Zero(PS) = Spec(\Xi). \tag{2.3}$$

Def. The extended point Ξ in $\mathbf{K}^n$ verifying (2.3) := a *generic point* of the irreducible variety $Zero(PS)$.

The above theorem can then be re-phrased as

Theorem 2.1'. An algebraic variety in $\mathbf{K}^n$ is irreducible over $\mathbf{K}$ if and only if it has a generic point over $\mathbf{K}$.

Def. For an irreducible asc-set IRR let Ξ be any extended point with IRR as defining asc-set. Then the algebraic variety $Spec(\Xi)$ depends only on IRR and will be called the *algebraic variety associated to IRR*. In notation: $Var[IRR]$:

$$Var[IRR] = Spec(\Xi). \tag{2.4}$$

Remark. We have

$$Var[IRR] \subset Zero(IRR),$$

but in general

$$Var[IRR] \neq Zero(IRR).$$

For this reason we use square bracket [ ] for the associated variety but not the parenthesis ( ) to avoid confusion.

Def. *Dimension* of an irreducible algebraic variety := dimension of its generic point.

From the above we see that an irreducible algebraic variety is completely determined by any one of its generic points which in turn is in correspondence to its defining irreducible asc-set Hence irreducible algebraic varieties, generic points, and irreducible asc-sets may be considered as equivalent concepts which are different representations of same geometry entity.

Theorem 2.2. For $\mathbf{K}$ the real field $\mathbf{R}$ a generic point of an irreducible algebraic variety is a *simple point* of that variety, i.e. a point at which the variety will have a well-defined tangent space of same dimension as that of the variety.

In accordance to the terminologies of differential geometry, such a simple point is a *regular* point of the real algebraic variety.

Theorem 2.3. Let Ξ in $\mathbf{K}^n$ be a generic point of an irreducible algebraic variety $Zero(PS)$ over $\mathbf{K}$ with defining irreducible asc-set IRR. Then for any pol P we have

$$P(\Xi) = 0 <===> Remdr(P/IRR) = 0. \tag{2.5}$$

Moreover, if a pol P in $\mathbf{K}[X]$ is reduced w.r.t. the irreducible asc-set IRR, then $P(\Xi) \neq 0$.

3. The Methods.

Let $\mathbf{K} = \mathbf{R}$ be the real field and let $n = 3$ so that we are considering real curves and real surfaces with real traces in the ordinary real space $\mathbf{R}^3$. Let us replace x_1, x_2, x_3 by the usual x, y, z with the ordering

$$x \prec y \prec z.$$

Then an irreducible algebraic surface in $\mathbf{R}^3$ will be defined by an equation $P = 0$ irreducible in $\mathbf{R}$ where P is a real polynomial with leading variable either z or y or x. If the leading variable of P is x, then P is necessarilly linear in x and the surface is a plane $x = const$. On the other hand an irreducible algebraic curve in $\mathbf{R}^3$ will be defined by an irreducible asc-set $IRR = \{ P_1, P_2\}$. The leading variables of P_1, P_2 will be either x, y or x, z or y, z. In case the leading variable of P_1 is x, then P_1 is necessarily linear in x so that the curve lies wholly in some plane $x = const$. Remark that the irreducible curve C determined by the irreducible asc-set IRR, is in general different from the curve $Zero(IRR)$, which is in general reducible.

In what follows let C be an irreducible algebraic curve with generic point defined by the irreducible asc-set $IRR = \{P_1, P_2\}$ and S, S' be irreducible algebraic surfaces defined by $P = 0$ and $P' = 0$ respectively all in $\mathbf{R}^3$.

From theorems in Sect 2 we have now the following

Theorem 3.1. The surface S will contain wholly of C if and only if

$$Remdr(P/IRR) = 0. \tag{3.1}$$

Proof. Let Ξ be a generic point of C. Then by Theorem 2.3 (3.1) is equivalent to $P(\Xi) = 0$. For any point Ξ^0 of C we would have then $P(\Xi^0) = 0$ so that $C \subset S$. This proves the theorem.

Theorem 3.2. Suppose that C is contained wholly in S and also in S' but is not contained wholly in the singularity part of S or S'. Set $P_x = \frac{\partial P}{\partial x}, P'_x = \frac{\partial P'}{\partial x}, etc.$ and form the pols

$$\begin{aligned} D_1 &= P_x * P'_y - P_y * P'_x, \\ D_2 &= P_x * P'_z - P_z * P'_x, \\ D_3 &= P_y * P'_z - P_z * P'_y. \end{aligned} \tag{3.2}$$

Then S and S' will touch smoothly along C.

$$Remdr(D_i/IRR) = 0, \;\; for\; i = 1, 2, 3. \tag{3.3}$$

Proof. Consider a generic point $\Xi = (\xi, \eta, \varsigma)$ of C. Then Ξ is a simple or regular point of C. As C is not wholly contained in the singularity part of S, Ξ is a simple or regular point of S, or P_x, P_y, P_z are not all equal to 0 at the point Ξ. Similarly P'_x, P'_y, P'_z are not all equal to 0 at the point Ξ. The normal of S at Ξ is then well-defined and has its direction cosines proportional to the values of P_x, P_y, P_z at Ξ not all 0. Similarly the normal of S' at Ξ is also well-defined and has its direction cosines proportional to the values of P'_x, P'_y, P'_z at Ξ not all 0. The surfaces S, S' will have same tangent plane at Ξ if and only if (P_x, P_y, P_z) at Ξ is proportional to (P'_x, P'_y, P'_z) at Ξ or if and only if $D_i = 0, i = 1, 2, 3$, at Ξ, or (3.3) by Theorem 2.3. This will imply then $D_i = 0$ at all points Ξ^0 of C which are regular for all of C, S, S'. This implies in turn that S, S' will have same tangent planes at all such points Ξ^0. Hence S will touch smoothly S' along C and the theorem is thus proved.

Lemma. For the irreducible algebraic surface S given by $P = 0$ the curvature κ at a regular point is given by

$$\kappa = \frac{V}{H^2}, \;\; where \tag{3.4}$$

$$H = P_x^2 + P_y^2 + P_z^2, \tag{3.5}$$

$$\begin{aligned} V =& P_{xx} * P_{yy} * P_z^2 + P_{xx} * P_{zz} * P_y^2 + P_{yy} * P_{zz} * P_x^2 \\ &- 2 * P_x * P_y * P_{xy} * P_{zz} - 2 * P_x * P_z * P_{xz} * P_{yy} - 2 * P_y * P_z * P_{yz} * P_{xx} \\ &+ 2 * P_x * P_y * P_{xz} * P_{yz} + 2 * P_x * P_z * P_{xy} * P_{yz} + 2 * P_y * P_z * P_{xy} * P_{xz} \\ &- P_x^2 * P_{yz}^2 - P_y^2 * P_{xz}^2 - P_z^2 * P_{xy}^2. \end{aligned} \tag{3.6}$$

In H and V all the partial derivatives $P_x, P_{xx} = \frac{\partial^2 P}{\partial x^2}, etc.$ take values at that regular point.

Theorem 3.3. Let C, S, S' be as before which satisfy the conditions in Theorem 3.2. Then S, S' will have same curvatures at points regular to C, S, S' if and only if

$$Remdr(H^2 * V' - H'^2 * V/IRR) = 0. \tag{3.7}$$

In (3.7) H' and V' are pols for P' similar to H and V for P.

Proof. From (3.7) we have by Theorem 2.3

$$H^2 * V' = H'^2 * V \tag{3.8}$$

at a generic point of C. (3.8) will then hold for any point regular to C, S, S'. Now at such a point $H \neq 0, H' \neq 0$. Hence from (3.4) we have $\kappa = \kappa'$ at all such points where κ and κ' are the corresponding curvatures of S and S'. The converse is clearly true and the theorem is thus proved.

Our method of solving the Problem SF may now be described as follows.

For the irreducible algebraic surface S of degree m defined by $P = 0$ let us write now P in the form

$$P = \Sigma_m \; u_{ijk} * x^i * y^j * z^k, \tag{3.9}$$

in which Σ_m is to be extended over triples (i, j, k) verifying

$$i \geq 0, \; j \geq 0, \; k \geq 0, \; i + j + k \leq m. \tag{3.10}$$

We shall denote by U_m the set of all u_{ijk} verifying (3.10).

Consider the conditions that S contains as a whole the irreducible algebraic curve C with generic point defined by the irreducible asc-set IRR. Let $X = (x, y, z)$ be such a generic point. By Theorem 3.1 the conditions become

$$R = Remdr(P/IRR) = 0. \tag{3.11}$$

Let the polset formed of the coefficients of various power-products of x, y, z in R be $US_a \subset \mathbf{R}[U_m]$. Then the condition (3.11) is identical to the conditions

$$US_a = 0. \tag{3.12}$$

For the requirement (a) in Problem SF let us form, in accordance to US_a of (3.12), the polsets $US_{ai}, US_{aj}, US_{ak}$ for each of the curves C_i, C_j, C_k. Similarly for the requirement (b) in Problem SF Theorem 3.2 will give us polsets US_{bj}, US_{bk} and for the requirement (c) Theorem 3.3 give us polsets US_{ck} whose vanishing are the corresponding conditions to be verified. Remark that all pols in $US_{ai}, US_{aj}, US_{a_k}, US_{bj}, US_{bk}$ are linear in the u's, while those in US_{ck} are in general non-linear in the u's. Combining now all the above polsets into a single one $US \subset \mathbf{R}[U_m]$, then all possible solutions S verifying the requirements of Problem

SF are furnished by the real zeros of $Zero(US)$. In next section we shall give some examples to illustrate the above method.

4. Examples.

A general cubic surface S in $\mathbf{R}^3$ is of the form

$$\begin{aligned} f(x,y,z) \equiv & u_{300} * z^3 + z^2 * (u_{210} * y + u_{201} * x + u_{200}) \\ & + z * (u_{120} * y^2 + u_{111} * y * x + u_{102} * x^2 + u_{110} * y + u_{101} * x + u_{100}) \\ & + u_{030} * y^3 + u_{021} * y^2 * x + u_{012} * y * x^2 + u_{003} * x^3 \\ & + u_{020} * y^2 + u_{011} * y * x + u_{002} * x^2 \\ & + u_{010} * y + u_{001} * x + u_{000} = 0, \end{aligned} \tag{4.1}$$

in which the u's are all in the real field $\mathbf{R}$. We give below some examples illustrating our method of determining such cubic surfaces S meeting some requirements as described in Sect 1.

Ex.1. In $\mathbf{R}^3$ consider two circular cylinders CYL_1, CYL_2 with x-axis and y-axis as their axis and two circular sections C_1, C_2 by planes orthogonal to these axis respectively. Then these circular sections will have generic points with defining asc-sets $AS_1 = \{C_{11}, C_{12}\}, AS_2 = \{C_{21}, C_{22}\}$ given by

$$C_{11} = x - d_1, \ C_{12} = z^2 + y^2 - r_1^2, \tag{4.2}$$

$$C_{21} = y - d_2, \ C_{22} = z^2 + x^2 - r_2^2. \tag{4.3}$$

The two circular cylinders are given by equations $C_{12} = 0, C_{22} = 0$ respectively. Naturally we shall assume that d_1, d_2, r_1, r_2 are all non-zero. We now ask for the determination of such cubic surfaces (4.1) which will contain the circles C_1, C_2 and touch the cylinders CYL_1, CYL_2 smoothly along these circles.

Our method gives now a set of 28 equations in U_3 for the solution. It is readily found by the package *wsolve* of D.K.Wang of MMRC implemented in the MAPLE system of some SPARC2 that such solution will exist only if

$$r_1^2 + d_1^2 = r_2^2 + d_2^2. \tag{4.4}$$

In that case the only possible cubic surface is then given by

$$\begin{aligned} & z^2 * (y * d_2 + x * d_1 - d_2^2 - d_1^2) + (y^3 * d_2 + x^3 * d_1) \\ & + y * x * (y * d_1 + x * d_2 - 2 * d_2 * d_1) - (y^2 + x^2) * (d_2^2 + d_1^2) \\ & + y * d_2 * (d_1^2 - r_1^2) + x * d_1 * (d_2^2 - r_2^2) \\ & + (r_1^2 + d_1^2) * (r_2^2 + d_2^2) - 2 * d_1^2 * d_2^2 = 0. \end{aligned} \tag{4.5}$$

Compare for this example [B-I1] Ex.5.2.

Ex.2. Consider besides circles C_1, C_2 as in Ex.1 a third circle C_3 with center on z-axis and orthogonal to that axis. Let the defining asc-set of a generic point of C_3 be $\{C_{31}, C_{32}\}$ with

$$C_{31} = y^2 + x^2 - r_3^2, \; C_{32} = z - d_3. \tag{4.6}$$

As before we assume that all d_i, r_i are non-zero. Suppose first that these d_i, r_i are otherwise arbitrary. Then our method shows that the only cubic surface containing all the 3 circles is the trivial one which degenerates into 3 planes containing the 3 circles respectively. If the d_i, r_i satisfy the relation (4.4) but otherwise arbitrary, then besides the trivial surface of 3 planes, there are the only surfaces which degenerate into the plane $z = d_3$ containing the circle C_3 and one in a family of quadrics through the circles C_1, C_2 given by the equation below, u being a parameter,

$$(d_1*d_2+u)*(x^2+y^2+z^2)+(r_1^2+d_1^2)*y*x-(r_1^2+d_1^2)*(d_1*y+d_2*x+u) = 0. \tag{4.7}$$

Suppose now there exists the relation

$$r_1^2 + d_1^2 = r_2^2 + d_2^2 = r_3^2 + d_3^2 = k^2, \; say. \tag{4.8}$$

between d_i, r_i but otherwise arbitrary. Then clearly the sphere of center the origin and radius $|k|$ will contain all the 3 circles C_i. Together with an arbitrary plane, there will be a family of ∞^3 degenerate cubic surfaces containing all the 3 circles. However, our method shows that there are in fact 4 families of $\infty^2, \infty^3, \infty^3, \infty^4$ cubic surfaces containing the 3 circles. The cubic surfaces consisting of the sphere and an arbitrary plane form only a subfamily of the family of ∞^4 surfaces. In fact, the last family is defined by 15 equations and depends on the parametric ratios $u_{100} : u_{002} : u_{010} : u_{001} : u_{000}$. Setting $u_{011} = 0$, then we get $u_{111} = u_{110} = u_{101} = 0$ too and the other equations show that the family degenerates into the subfamily of ∞^3 degenerate cubic surfaces consisting of the sphere $z^2+y^2+x^2 = k^2$ and the plane $u_{100} * z + u_{010} * y + u_{001} * x + u_{000} = 0$.

Our method shows that the above are the only possible cubic surfaces which contain the 3 circles. As each non-degenerate family of cubic surfaces given above depends on several parameters, we may determine, if we like, such ones among them which will meet further requirements as in Sect 1 by our method. However, if CYL_i are the cylinders with x, y, z axis as axis and bounded by $C_i, i = 1, 2, 3$ respectively, then in view of the form of (4.5), there cannot exist any cubic surface containing all the 3 circles and touching smoothly the 3 cylinders along these circles.

The above two examples concern curves like circles and surfaces like circular cylinders which are easily parametrized. With such parametrizations the problems in these examples can be dealt with by other known methods, cf. e.g. [B-I1]. However, curves and surfaces are generally non-parametrizable except in the very rare case of rational ones. This is already so for general cubic curves and cubic surfaces. The following is such an example which cannot be treated by means of parametrizations.

Ex.3. Let $f_1(x,y,z)$ and $f_2(x,y,z)$ be as $f(x,y,z)$ in (4.1) with coefficients u_{ijk} replaced by a_{ijk} and b_{ijk} respectively. Let S_1, S_2 be the irreducible cubic surfaces defined by $f_1 = 0$ and $f_2 = 0$ respectively and C_1, C_2 be the irreducible cubic curves having generic points with respective defining asc-sets $IRR_1 = \{y, g_1\}, IRR_2 = \{g_2, z\}$, where $g_1 = f_1(x,0,z), g_2 = f_2(x,y,0)$. Our problem is to determine cubic surfaces S as in (4.1) which will contain both C_1 and C_2 and meet eventually further requirements. Now C_1, C_2 will intersect the x-axis $y = z = 0$ in points with x-coordinates given by the respective equations

$$\begin{aligned} a_{003} * x^3 + a_{002} * x^2 + a_{001} * x + a_{000} = 0, \\ b_{003} * x^3 + b_{002} * x^2 + b_{001} * x + b_{000} = 0. \end{aligned} \tag{4.9}$$

For a cubic surface S containing C_1, C_2 to exist it is necessary that these two triples of intersection points should be the same. So we assume at the outset that

$$a_{003} = b_{003},\ a_{002} = b_{002},\ a_{001} = b_{001},\ a_{000} = b_{000}. \tag{4.10}$$

For the sake of simplifying the computations we shall assume that the coincident triple of intersection points will none of them be the origin or the point at infinity on the x-axis so that

$$a_{000} = b_{000} \neq 0,\ a_{003} = b_{003} \neq 0. \tag{4.11}$$

We assume further that neither the z-axis will be asymptotic to the curve C_1 nor the y-axis will be asymptotic to the curve C_2 so that

$$a_{300} \neq 0,\ b_{030} \neq 0. \tag{4.12}$$

Under the conditions (4.10)-(4.12) our method, in applying the package *wsolve* , shows that the only cubic surfaces S of (4.1) containing the the two cubic curves will form a family of ∞^4 cubic surfaces depending on the parametric ratios u_{110} : $u_{111} : u_{120} : u_{210} : u_{300}$ and are defined by, besides (4.10), the equations below:

$$\begin{aligned}
&u_{000} * a_{300} - u_{300} * a_{000} = 0,\ u_{001} * a_{300} - u_{300} * a_{001} = 0,\\
&u_{010} * a_{300} - u_{300} * b_{010} = 0,\ u_{002} * a_{300} - u_{300} * a_{002} = 0,\\
&u_{011} * a_{300} - u_{300} * b_{011} = 0,\ u_{020} * a_{300} - u_{300} * b_{020} = 0,\\
&u_{003} * a_{300} - u_{300} * a_{003} = 0,\ u_{012} * a_{300} - u_{300} * b_{012} = 0,\\
&u_{021} * a_{300} - u_{300} * b_{021} = 0,\ u_{030} * a_{300} - u_{300} * b_{030} = 0,\\
&u_{100} * a_{300} - u_{300} * a_{100} = 0,\ u_{101} * a_{300} - u_{300} * a_{101} = 0,\\
&u_{102} * a_{300} - u_{300} * a_{102} = 0,\ u_{200} * a_{300} - u_{300} * a_{200} = 0,\\
&u_{201} * a_{300} - u_{300} * a_{201} = 0.
\end{aligned} \tag{4.13}$$

We may determine subfamilies meeting further requirements by our method if required. Complete set of families of cubic surfaces containing the two cubic curves with some of the intersection points at infinity or (4.1) not observed have also been determined.

Further much more complicate examples have been treated by D.K.Wang.

Bibliography.

[B] C.L.Bajaj, Surface fitting using implicit algebraic surface patches, in Curve and Surface Modelling (ed. H.Hagen), SIAM Publ., (1991) 1-31.

[B-I1] C.L.Bajaj & I.Ihm, Hermite interpolation using real algebraic surfaces, Proc.5th Symp. on Computational Geometry, West Germany, (1989)

[B-I2] C.L.Bajaj & I.Ihm, Smoothing polyhedra using implicit algebraic splines, Proc. of ACM SIGGRAPH'92, Computer Graphics, 1992.

[E] L.P.Eisenhart, A treatise on the differential geometry of curves and surfaces, (1909).

[H-P] W.V.D.Hodge & D.Pedoe, Methods in algebraic geometry, vol.2, Cambridge (1952).

[VdW] Van der Waerden, Einleitung in die algebraischen Geometrie, Springer, (1939).

[WDK] D.K.Wang, Doctoral Thesis, MMRC, (1993).

[WU1] Wu Wen-tsun,Basic principles of mechanical theorem-proving in geometries, Part I. Elementary geometries, (in Chinese), Science Press, Beijing (1984).

[WU2] —, On zeros of algebraic equations — an application of Ritt principle, Kexue Tongbao 31 (1986) 1-5.

[WU3] —, A zero structure theorem for polynomial-equations-solving and its applications, MM-Res.Preprints, No.1 (1987) 2-12.

[WU4] —, On the generic zero and Chow basis of an irreducible ascending set, MM-Res. Preprints, No.4 (1989) 1-21.

MM Research Prprints
No. 13. 1995, 1-14

Central Configurations in Planet Motions and Vortex Motions

WU WEN-TSUN
吴文俊
(Institute of Systems Science, Academia Sinica)

Abstract *The determination of central configurations in planet motion may be reduced to a problem of polynomial equations-solving. We determine thus these configurations in the case of three planets by the char-set method. It shows that the only solutions are the classical ones due to Euler and Lagrange. The same method permits also to determine the rigid configurations formed by three parallel filaments in an incompressible nonviscous fluid extending to infinity moving under their own influences.*

1. Central Configurations in Planet Motions.

Notations 1.1.

$J_n := \{1, \cdots, n\}$;

$J_n^2 := \{(i,j) \mid i, j \in J_n, i \neq j\}$;

$m_1, \cdots, m_n :=$ masses of n particles moving under mutual Newtonian gravitational attractions.

$\mathbf{r}_1, \cdots, \mathbf{r}_n :=$ positions of these masses at a certain momemt, with $\mathbf{r}_i \neq \mathbf{r}_j$ for $i \neq j$.

$\begin{bmatrix} m_1 & m_2 & \cdots & m_n \\ \mathbf{r}_1 & \mathbf{r}_2 & \cdots & \mathbf{r}_n \end{bmatrix}$:= a *configuration* formed of masses m_i at positions $\mathbf{r}_i, i = 1, \cdots, n$.

Definition 1.1. The configuration $\begin{bmatrix} m_1 & m_2 & \cdots & m_n \\ \mathbf{r}_1 & \mathbf{r}_2 & \cdots & \mathbf{r}_n \end{bmatrix}$ is a *rigid configuration* (respectively a *central configuration* with respect to the masses $m_1, \cdots, m_n$:= There are initial velocities of the masses m_i such that under the Newtonian gravitational attractions the configurations formed by the masses during the motion will remain *congruent* (respectively *similar*) to the initial one.

As a direct consequence of Newtonian mechanics for a rigid or a central configuration $\begin{bmatrix} m_1 & m_2 & \cdots & m_n \\ \mathbf{r}_1 & \mathbf{r}_2 & \cdots & \mathbf{r}_n \end{bmatrix}$ the center of mass of the masses $m_1, \cdots, m_n$ may be considered to be fixed during the motion.

Definition 1.2. An *inertial coordinate system* associated to a rigid or a central configuration $\begin{bmatrix} m_1 & m_2 & \cdots & m_n \\ \mathbf{r}_1 & \mathbf{r}_2 & \cdots & \mathbf{r}_n \end{bmatrix}$:= A cartesian coordinate system for which the origin is at the fixed center of mass of the masses $m_1, \cdots, m_n$.

Again as a direct consequence of Newtonian mechanics, we have the following proposition:

Proposition 1.1. For a rigid or a central configuration $\begin{bmatrix} m_1 & m_2 & \cdots & m_n \\ \mathbf{r}_1 & \mathbf{r}_2 & \cdots & \mathbf{r}_n \end{bmatrix}$ we have in any associated inertial coordinate system

$$\Sigma_i \, m_i * \mathbf{r}_i = 0. \tag{1.1}$$

Definition 1.3. Two rigid configurations with respect to same masses $m_1, \cdots,$ m_n belong to the same *configuration class* := they have same center of mass and there exists an orientation-preserving similarity transformation which keeps the center of mass fixed and tranforms one configuration to the other.

Some Historical Account about Central Configurations.

Clearly a rigid configuration is also a central configuration with same set of masses. The converse is known to be true, see e.g. [Wint], Sect.355-382. It follows that a central configuration determines a class of rigid configurations and vice versa.

In what follows only inertial coordinate systems are considered.

Notations 1.2.

$q_3(n) = q_3(n; m_1, \cdots, m_n)$:= number of classes of rigid configurations with given masses $m_1, \cdots, m_n$.

$q_2(n) = q_2(n; m_1, \cdots, m_n)$:= number of classes of rigid configurations with given masses $m_1, \cdots, m_n$ for which the masses m_i are situated in the same plane.

$q_1(n) = q_1(n; m_1, \cdots, m_n)$:= number of classes of rigid configurations with given masses $m_1, \cdots, m_n$ for which the masses m_i are situated in the same line.

Remark 1.1. Our notations differ from those of Wintner in his book [Wint] in that q_2, q_3 of that book correspond to $q_2 - q_1, q_3 - q_2$ here.

Results known. For given masses $m_1, \cdots, m_n$ we have:

$q_1(2) = q_2(2) = q_3(2) = 1;$

$q_1(3) = 3.$ (Euler 1767);

$q_2(3) = q_3(3) = 4$. (Lagrange 1772);

$q_1(n) = \frac{n!}{2}$. (Moulton 1910);

$q_3(n) > q_2(n)$ for each $n > 4$ with some particular sets of $m_1, \cdots, m_n$. (Waldvogel 1972).

It is clear that

$$q_1(n; m_1, \cdots, m_n) \le q_2(n; m_1, \cdots, m_n) \le q_3(n; m_1, \cdots, m_n). \qquad (1.2)$$

It is known that $q_1(n; m_1, \cdots, m_n)$ is finite for all n and masses $m_1, \cdots, m_n$ owing to the results of Moulton.

Wintner Conjecture. $q_3(n; m_1, \cdots, m_n)$ and hence also $q_2(n; m_1, \cdots, m_n)$ are finite for all n and masses $m_1, \cdots, m_n$.

In recent years Smale and his followers have studied central configurations with topological methods via Morse critical point theory. See e.g. [Sm1,2]. It seems that Wintner's conjecture remains open. In this section we shall restrict ourselves to the actual determination of rigid or central configurations in treating it as a problem of polynomial equations solving. For this purpose we shall adopt an inertial coordinate system associated to the rigid or central configuration $\begin{bmatrix} m_1 & m_2 & \cdots & m_n \\ \mathbf{r}_1 & \mathbf{r}_2 & \cdots & \mathbf{r}_n \end{bmatrix}$ and the following notations will be used:

Notations 1.3.

$\mathbf{r}_i = (x_i, y_i), i \in J_n;$

$x_{ij} = x_i - x_j,\ y_{ij} = y_i - y_j,\ (i,j) \in J_n^2;$

$r_{ij} = \sqrt{x_{ij}^2 + y_{ij}^2},\ (i,j) \in J_n^2.$

Fundamental Equations for Rigid Configurations.

For a rigid configuration $\begin{bmatrix} m_1 & m_2 & \cdots & m_n \\ \mathbf{r}_1 & \mathbf{r}_2 & \cdots & \mathbf{r}_n \end{bmatrix}$ in a plane it is known that during the motion the plane will be fixed to keep the configuration congruent to the original one, cf. [Wint] again. Take now planar inertial coordinate system (x, y) with origin O at the center of mass. Then the configuration will move around O with an angular velocity ω such that

$$\begin{aligned} \omega^2 * x_i &= \Sigma_{j \ne i}\, m_j * \frac{x_{ij}}{r_{ij}^3}, \\ \omega^2 * y_i &= \Sigma_{j \ne i}\, m_j * \frac{y_{ij}}{r_{ij}^3}, \end{aligned} \qquad i \in J_n. \qquad (1.3)_n$$

Up to factors $m_i the right-hand sides$ represent the total Newtonian attractive forces exerted on the mass m_i by all the other masses, while the left-hand side represent the corresponding centrifugal forces. The equations $(1.3)_n$ show in particular that ω is a constant. Furthermore, the equations are invariant under orientation-preserving similarity transformations with origin keeping fixed. The constant ω may then be changed.

We have assumed $r_{ij} \neq 0, (i,j) \in J_n^2$, which means that collisions are out of consideration. The determination of rigid configurations and hence also central configurations amounts then to the solving of polynomial equations obtained from $(1.3)_n$ by clearing of fractions. To achieve this we shall introduce some preliminary transformations of $(1.3)_n$ to make easier the solving.

Let us take the complex number field $\mathbf{C}$ as the basic field and introduce complex variables u_i, v_i, z_{ij}, w_{ij} as follows.

$$\begin{aligned} &u_i = x_i + \mathbf{i} * y_i, \ v_i = x_i - \mathbf{i} * y_i, \ i \in J_n, \\ &z_{ij} = \frac{1}{(u_i - u_j) * r_{ij}}, \ w_{ij} = \frac{1}{(v_i - v_j) * r_{ij}}, \ (i,j) \in J_n^2, \\ &m_0 = -\omega^2. \end{aligned} \tag{1.4$_n$}$$

From (1.3) we have now the following set of polynomial equations:

$$\left.\begin{aligned} m_0 * u_i + \Sigma_{j \neq i} \, m_j * w_{ij} = 0, \\ m_0 * v_i + \Sigma_{j \neq i} \, m_j * z_{ij} = 0. \end{aligned}\right. \quad i \in J_n. \tag{1.5$_n$}$$

$$\begin{aligned} &(u_i - u_j) * z_{ij} * r_{ij} - r^3 = 0, \\ &(v_i - v_j) * w_{ij} * r_{ij} - r^3 = 0, \qquad (i,j) \in J_n^2. \\ &r_{ij}^2 - (u_i - u_j) * (v_i - v_j) = 0. \end{aligned} \tag{1.6$_n$}$$

$$r_{ij} - r_{ji} = 0, \ z_{ij} + z_{ji} = 0, \ w_{ij} + w_{ji} = 0. \ (i,j) \in J_n^2. \tag{1.7$_n$}$$

To these we may also add

$$\begin{aligned} \Sigma_i \, m_i * u_i = 0, \\ \Sigma_i \, m_i * v_i = 0, \end{aligned} \tag{1.8$_n$}$$

which follows from (1.1) or (1.3).

Remark 1.2. In the above equations Σ is to be extended either on $i \in J_n$ or $(i,j) \in J_n^2$ as is evident from the context and the variable r is introduced to render the equations homogeneous in the relevant variables.

Remark 1.3. In the above equation u_i, v_i, etc. should satisfy the following conditions to meet the reality of the situation:

—4—

$m_0, m_1, \cdots, m_n$ are all real numbers, with $m_0 < 0, m_i > 0, i \in J_n$.

Actually $r = 1$, and all $r_{ij}, (i,j) \in J_n^2$, are real and positive.

u_i, v_i for $i \in J_n$, are conjugate complex numbers, **i** being $\sqrt{-1}$.

$z_{ij} \neq 0, w_{ij} \neq 0$, for $(i,j) \in J_n^2$.

Definition 1.4. For a rigid or central configuration $\begin{bmatrix} m_1 & m_2 & \cdots & m_n \\ \mathbf{r}_1 & \mathbf{r}_2 & \cdots & \mathbf{r}_n \end{bmatrix}$ *reality conditions* := the conditions in Remark 1.3.

Determination of Rigid or Central Configurations for $n = 3$.

The determination of rigid or central configurations for $n = 2$ is trivial. Let us proceed to the case $n = 3$ which amounts to the problem in the example below.

Example 1. To solve the system of equations $(1.5)_3 - (1.8)_3$ for the variables u_i, v_i, m_0, r_{ij} with $i \in J_3, (ij) \in J_3^2$ in terms of the parameters $m_i, i \in J_3$.

For the solving let us first replace the variables u_2, u_3, v_2, v_3 by $u_{12}, u_{13}, v_{12}, v_{13}$ in setting

$$\begin{aligned} u_{12} &= u_1 - u_2, \ u_{13} = u_1 - u_{13}, \\ v_{12} &= v_1 - v_2, \ v_{13} = v_1 - v_{13}, \end{aligned} \tag{1.9}$$

so that

$$u_{23} = u_{13} - u_{12}, \quad v_{23} = v_{13} - v_{12}. \tag{1.10}$$

To solve $(1.5)_3 - (1.8)_3$ is then easily seen to be equivalent to the solving of the system $QS = 0$ below for variables $u_1, u_{12}, u_{13}, v_1, v_{12}, v_{13}, m_0, r_{12}, r_{13}, r_{23}$ in terms of known parameters m_1, m_2, m_3. Here $QS = \{q_1, \cdots, q_9\}$ with

$$\begin{aligned}
q_1 &= u_1 * (m_3 + m_2 + m_1) - u_{13} * m_3 - u_{12} * m_2, \\
q_2 &= v_{13}^2 * r_{23} * r_{13} * (r_{12}^3 * m_0 + m_2 + m_1) \\
&\quad - v_{13} * v_{12} * r_{23} * r_{13} * (r_{12}^3 * m_0 + m_2 + m_1) \\
&\quad + v_{13} * v_{12} * r_{12} * m_3 * (r_{23} - r_{13}) \\
&\quad - v_{12}^2 * r_{23} * r_{12} * m_3, \\
q_3 &= v_{13}^2 * r_{23} * r_{13} * m_2 \\
&\quad + v_{13} * v_{12} * r_{23} * r_{13} * (r_{13}^2 * r_{12} * m_0 - m_2) \\
&\quad + v_{13} * v_{12} * r_{12} * (r_{23} * m_3 + r_{23} * m_1 + r_{13} * m_2) \\
&\quad - v_{12}^2 * r_{23} * r_{12} * (r_{13}^3 * m_0 + m_3 + m_1), \\
q_4 &= v_1 * u_{13} * u_{12} * r_{13} * r_{12} * m_0 \\
&\quad + u_{13} * r_{13} * m_2 + u_{12} * r_{12} * m_3,
\end{aligned}$$

—5—

$$
\begin{aligned}
q_5 &= u_{13}^2 * r_{23} * r_{13} * (r_{12}^3 * m_0 + m_2 + m_1) \\
&- u_{13} * u_{12} * r_{23} * r_{13} * (r_{12}^3 * m_0 + m_2 + m_1) \\
&+ u_{13} * u_{12} * r_{12} * m_3 * (r_{23} - r_{13}) \\
&- u_{12}^2 * r_{23} * r_{12} * m_3, \\
q_6 &= u_{13}^2 * r_{23} * r_{13} * m_2 \\
&+ u_{13} * u_{12} * r_{23} * r_{13} * (r_{13}^2 * r_{12} * m_0 - m_2) \\
&+ u_{13} * u_{12} * r_{12} * (r_{23} * m_3 + r_{23} * m_1 + r_{13} * m_2) \\
&- u_{12}^2 * r_{23} * r_{12} * (r_{13}^3 * m_0 + m_3 + m_1), \\
q_7 &= v_{12} * u_{12} - r_{12}^2, \\
q_8 &= v_{13} * u_{13} - r_{13}^2, \\
q_9 &= (v_{13} - v_{12}) * (u_{13} - u_{12}) - r_{23}^2.
\end{aligned}
\tag{1.11}
$$

Remark that the reality conditions have to be observed. In particular $r_{12}, r_{13}, r_{23}, m_0, m_1, m_2, m_3$ are all non-zero. Let NZ be the product of all these non-zero variables and parameters. Then the problem is reduced to the determination of

$$Zero_{(rc)}(QS/NZ) \subset Zero_{C}(QS/NZ), \tag{1.12}$$

in which (rc) means that the zeros are restriceted to those for which the reality conditions are observed.

To determine (1.12) let us first arrange the variables and parameters involved in the following order:

$$m_1 \prec m_2 \prec m_3 \prec r_{12} \prec r_{13} \prec r_{23} \prec m_0 \prec u_{12} \prec u_{13} \prec v_{12} \prec v_{13} \prec u_1 \prec v_1. \tag{1.13}$$

Let us apply now the Replacement Rules in replacing q_2, q_3 by the remainder q_{10} of q_2 with respect to q_3 and the resultant q_{11} of q_2 and q_3. The resultant q_{11} has an index set $[57\ m_0\ 3]$ while q_{10} is of the form

$$q_{10} = v_{13} * g_0 + v_{12} * h_0, \tag{1.14}$$

where g_0, h_0 have index sets $[12\ m_0\ 2]$ and $[8\ m_0\ 2]$ respectively. Similarly, let us replace q_5, q_6 by the remainder q_{12} of q_5 with respect to q_6 and the resultant q_{13} of q_5 and q_6. It turns out that q_{13} is the same as q_{11} so that it may be removed while q_{12} is of the form

$$q_{12} = u_{13} * g_0 + u_{12} * h_0, \tag{1.15}$$

Forming remainders of q_8 with respect to q_{10}, q_{12}, q_7 in eliminating v_{13}, u_{13}, v_{12} successively we get a pol q_{14} which may be factored as

$$\begin{aligned} q_{14} &= f_a * f_b, \\ f_a &= -r_{13} * g_0 + r_{12} * h_0, \\ f_b &= +r_{13} * g_0 + r_{12} * h_0. \end{aligned} \tag{1.16}$$

It follows from $q_{14} = 0$ that either $f_a = 0$ or $f_b = 0$. Form next the remainders of q_9 with respect to q_8, q_7, q_{10}, q_{12} we get a pol q_{15} of the form

$$q_{15} = -r_{23}^2 * g_0 + r_{13}^2 * g_0 + r_{12}^2 * g_0 + 2 * r_{12}^2 * h_0. \tag{1.17}$$

Consider first the case $f_a = 0$. Then after reduction it turns out that q_{15} will be factored into three factors $f_{ai}, i = 1, 2, 3$ as shown below:

$$\begin{aligned} f_{a1} &= g_0, \\ f_{a2} &= r_{23} - r_{13} - r_{12}, \\ f_{a3} &= r_{23} + r_{13} + r_{12}. \end{aligned} \tag{1.18}$$

Consider next the case $f_b = 0$. Then q_{15} after reduction will be factored into factors $f_{bi}, i = 1, 2, 3$ given below:

$$\begin{aligned} f_{b1} &= g_0, \\ f_{b2} &= r_{23} - r_{13} + r_{12}, \\ f_{b3} &= r_{23} + r_{13} - r_{12}. \end{aligned} \tag{1.19}$$

It follows that $Zero_{(rc)}(QS/NZ)$ is included in the union of $Zero_{(rc)}(AS_k/NZ)$ and $Zero_{(rc)}(BS_k/NZ)$ for $k = 1, 2, 3$ in which

$$\begin{aligned} AS_k &= QS + \{q_{10}, q_{11}, q_{12}\} + \{f_a, f_{ak}\}, \\ BS_k &= QS + \{q_{10}, q_{11}, q_{12}\} + \{f_b, f_{bk}\}, \end{aligned} \quad k = 1, 2, 3. \tag{1.20}$$

It is easy to see that this inclusion is actually an identity so that we have

$$Zero_{(rc)}(QS/NZ) = \bigcup_k Zero_{(rc)}(AS_k/NZ) + \bigcup_k Zero_{(rc)}(BS_k/NZ). \tag{1.21}$$

In (1.21) the union is to be extended over $k = 1, 2, 3$ Let us denote for simplicity the six zero sets in (1.20) by ZA_k and ZB_k respectively. Then owing to the reality conditions $f_{a3} \neq 0$ so that we have

$$ZA_3 = Zero_{(rc)}(AS_3/NZ) = \emptyset. \tag{1.22}$$

On the other hand by our general method we have

$$\begin{aligned} ZA_1 &= Zero_{(rc)}(AS_1/NZ) = Zero_{(rc)}(CS_{a1}/NZ), \\ ZA_2 &= Zero_{(rc)}(AS_2/NZ) = Zero_{(rc)}(CS_{a2}/NZ), \end{aligned} \tag{1.23}$$

In (1.23) $CS_{ak} = \{c_{ak1}, \cdots, c_{ak8}\}, k = 1, 2$ are triangulated or asc-sets with pols $c_{aki}, i = 1, \cdots, 8$ given below:

$$\begin{aligned} c_{a11} &= r_{13} - r_{12}, \\ c_{a12} &= r_{23} - r_{12}, \\ c_{a13} &= m_0 * r_{12}^3 + (m_3 + m_2 + m_1), \\ c_{a14} &= u_{13}^2 - u_{13} * u_{12} + u_{12}^2, \\ c_{a15} &= v_{12} * u_{12} - r_{12}^2, \\ c_{a16} &= v_{13} * u_{13} - r_{12}^2, \\ c_{a17} &= u_1 * (m_3 + m_2 + m_1) - u_{13} * m_3 - u_{12} * m_2, \\ c_{a18} &= v_1 * u_{13} * u_{12} * r_{12} * m_0 + u_{13} * m_2 + u_{12} * m_3. \end{aligned} \tag{1.24}$$

$$\begin{aligned} c_{a21} &= r_{13}^5 * (m_2 + m_1) \\ &\quad + r_{13}^4 * r_{12} * (3 * m_2 + 2 * m_1) \\ &\quad + r_{13}^3 * r_{12}^2 * (3 * m_2 + m_1) \\ &\quad - r_{13}^2 * r_{12}^3 * (3 * m_3 + m_1) \\ &\quad - r_{13} * r_{12}^4 * (3 * m_3 + 2 * m_1) \\ &\quad - r_{12}^5 * (m_3 + m_1), \\ c_{a22} &= r_{23} - r_{13} - r_{12}, \\ c_{a23} &= m_0 * r_{13}^2 * r_{12}^3 * (r_{13} + r_{12})^2 \\ &\quad + r_{13}^2 * (r_{13} + r_{12})^2 * (m_2 + m_1) \\ &\quad - 2 * r_{13} * r_{12}^3 * m_3 - r_{12}^4 * m_3, \\ c_{a24} &= u_{13} * r_{12} + u_{12} * r_{13}, \end{aligned} \tag{1.25}$$

and c_{a2k} same as c_{a1k}, for $k = 5, 6, 7, 8$ From $c_{a12}, c_{a13}, c_{a22}$ it is readily seen that ZA_1 gives the central configuration of Lagrange while ZA_2 gives that of Euler for which the particles m_1, m_2, m_3 are collinear with m_1 lying between m_2 and m_3. In particular, c_{a21} gives a quintic equation in $\frac{r_{13}}{r_{12}}$ which has one and only one real solution positive owing to the reality conditions as originaly given by Euler.

The sets $ZB_1 = Zero_{(rc)}(BS_1/NZ)$ coincides with $ZA_1 = Zero_{(rc)}(AS_1/NZ)$ and the sets $ZB_k = Zero_{(rc)}(BS_k/NZ), k = 2,3$ are similar to $ZA_2 = Zero_{(rc)}$ (AS_2/NZ) which correspond to the other two collinear central configurations of Euler.

The set of all possible central configurations in case of $n = 3$ is thus completely determined which is consisting of the classical ones due to Euler and Lagrange but no others.

2. Rigid Configurations in Vortex Motions.

Consider now vortex movements in an incompressible and nonviscous fluid extending to infinity.

Notations 2.1.

$F_1, \cdots, F_n$:= n parallel rectilinear vortex filaments moving under their own influences.

$k_1, \cdots, k_n$:= strengths of the n vortex filaments F_i.

P := a fixed plane perpendicular to all the n filaments F_i with a coordinate system (x, y).

$\mathbf{r}_1, \cdots, \mathbf{r}_n$:= traces of vortex filaments F_i on the plane P, or vectors from the origin O to that trace.

$\mathbf{e}$:= unit vector orthogonal to the fixed plane P.

r_{ij} := distance between the parallel filaments F_i and F_j.

$\begin{bmatrix} k_1 & k_2 & \cdots & k_n \\ \mathbf{r}_1 & \mathbf{r}_2 & \cdots & \mathbf{r}_n \end{bmatrix}$:= vortex configuration of the n filaments F_i.

Hypothesis K.

$$\Sigma_i\ k_i \neq 0. \tag{2.1}$$

Under Hypothesis K it is known that there will exist a *center* in the finite part of the plane P such that, if the center is taken as the origin O, then we will have

$$\Sigma_i\ k_i * \mathbf{r}_i = 0,\ \ or$$

$$\begin{aligned} \Sigma_i\ k_i * x_i &= 0, \\ \Sigma_i\ k_i * y_i &= 0. \end{aligned} \tag{2.2}$$

Consider now a vortex configuration $\begin{bmatrix} k_1 & m_2 & \cdots & k_n \\ \mathbf{r}_1 & \mathbf{r}_2 & \cdots & \mathbf{r}_n \end{bmatrix}$.

Definition 2.1.

$\begin{bmatrix} k_1 & m_2 & \cdots & k_n \\ \mathbf{r}_1 & \mathbf{r}_2 & \cdots & \mathbf{r}_n \end{bmatrix}$ is a *fixed configuration* := the configuration remains fixed during the motion under its own influences.

Definition 2.2.

$\begin{bmatrix} k_1 & k_2 & \cdots & k_n \\ \mathbf{r}_1 & \mathbf{r}_2 & \cdots & \mathbf{r}_n \end{bmatrix}$ is a *rigid configuration* := the configuration remains congruent to the original one during the motion under its own influences.

Remark 2.1. There is no analogue of fixed configurations for particles moving under mutual Newtonian attractions.

Some Elementary Properties of Rigid Vortex Configurations.

Below $\begin{bmatrix} k_1 & k_2 & \cdots & k_n \\ \mathbf{r}_1 & \mathbf{r}_2 & \cdots & \mathbf{r}_n \end{bmatrix}$ will be a vortex configuration verifying Hypothesis K so that the center of the system exists which is taken to be the origin of the coordinate system.

Proposition 2.1. For a rigid vortex configuration which verifies Hypothesis K the whole configuration will rotate about the fixed center with a uniform angular velocity.

Proof. Let the origin O be at the fixed center. Let us denote by $\frac{c(t)}{2*\pi}$ the angular velocity about O of the whole rigid configuration at time t. Then the velocity $\mathbf{v}_i(t)$ of i-th filament at time t will be

$$\mathbf{v}_i(t) = \frac{c(t)}{2*\pi} * (\mathbf{e} \times \mathbf{r}_i(t)).$$

Now the velocity of filament F_i due to filament F_j at time t is given by

$$\mathbf{v}_{ji}(t) = \frac{1}{2*\pi} * k_j * \frac{\mathbf{e}\times(\mathbf{r}_i(t) - \mathbf{r}_j(t))}{r_{ij}^2}.$$

Hence the velocity of F_i at time t is

$$\mathbf{v}_i(t) = \frac{1}{2*\pi} * \Sigma_{j\neq i}\ k_j * \frac{\mathbf{e}\times(\mathbf{r}_i(t) - \mathbf{r}_j(t))}{r_{ij}^2}.$$

Comparing the two expressions of $\mathbf{v}_i(t)$, we get

$$\Sigma_{j\neq i}\ k_j * \frac{\mathbf{r}_i(t) - \mathbf{r}_j(t)}{r_{ij}^2} = c(t) * \mathbf{r}_i(t).$$

As $\mathbf{r}_i(t)$ differs from $\mathbf{r}_i(0) = \mathbf{r}_i$ by a rotation independent of i, we see that $c(t)$ is independent of t. Hence $\frac{c(t)}{2*\pi} = \frac{c}{2*\pi}$ is the uniform angular velocity of the whole rigid configuration as to be proved. Moreover we have

$$\Sigma_{j\neq i}\ k_j * \frac{\mathbf{r}_i - \mathbf{r}_j}{r_{ij}^2} \ = \ c * \mathbf{r}_i. \tag{2.3}$$

Remark 2.2. From the proof we see that $c = 0$ corresponds to the case of a fixed vortex configuration.

Proposition 2.2. For a fixed vortex configuration it is necessary that

$$\Sigma_{i\neq j}\ k_i * k_j \ = \ 0. \tag{2.4}$$

Proof. From (2.3) we get

$$c * \Sigma_i\ k_i * \mathbf{r}_i^2 = \Sigma_{i\neq j}\ k_i * k_j.$$

As $c = 0$ for a fixed vortex configuration, we get (2.4).

Fundamental Equations of Rigid Vortex Configurations.

From (2.3) we get

$$\begin{aligned} \Sigma_{j\neq i}\ k_j * \frac{x_{ij}}{r_{ij}^2} \ &= \ c * x_i, \\ \Sigma_{j\neq i}\ k_j * \frac{y_{ij}}{r_{ij}^2} \ &= \ c * y_i. \end{aligned} \qquad i \in J_n. \tag{2.5$_n$}$$

Let us introduce u_i, v_i, z_{ij}, w_{ij} as given below:

$$\begin{aligned} &u_i = x_i + \mathbf{i} * y_i,\ v_i = x_i - \mathbf{i} * y_i,\ i \in J_n, \\ &u_{ij} = u_i - u_j,\ v_{ij} = v_i - v_j,\ (i,j) \in J_n^2, \\ &z_{ij} = \frac{1}{u_{ij}},\ w_{ij} = \frac{1}{v_{ij}},\ (i,j) \in J_n^2, \\ &k_0 \ = \ -\ c. \end{aligned} \tag{2.6$_n$}$$

Then $(2.5)_n$ will be transformed into the equations below:

$$\begin{aligned} k_0 * u_i + \Sigma_{j\neq i}\ k_j * w_{ij} = 0, \\ k_0 * v_i + \Sigma_{j\neq i}\ k_j * z_{ij} = 0. \end{aligned} \qquad i \in J_n. \tag{2.7$_n$}$$

$$
\begin{aligned}
&u_{ij} * z_{ij} - r^2 = 0, \\
&v_{ij} * w_{ij} - r^2 = 0, \quad (i,j) \in J_n^2. \\
&r_{ij}^2 - u_{ij} * v_{ij} = 0.
\end{aligned}
\qquad (2.8)_n
$$

$$
r_{ij} - r_{ji} = 0,\ z_{ij} + z_{ji} = 0,\ w_{ij} + w_{ji} = 0.\ (i,j) \in J_n^2. \qquad (2.9)_n
$$

To these we may also add

$$
\Sigma_i\ k_i * u_i = 0,\ \Sigma_i\ k_i * v_i = 0. \qquad (2.10)_n
$$

The variables and parameters involved in $(2.7)_n - (2.10)_n$ are subject to some *reality conditions*, viz.

$k_1, \cdots, k_n$ are real and non-zero, and k_0 is real.

u_i, v_i, are complex conjugates for $i \in J_n$.

r_{ij} are real and positive, for $(ij) \in J_n^2$.

$u_{ij}, v_{ij}, z_{ij}, w_{ij}$ are all non-zero, for $(ij) \in J_n^2$.

r is actually equal to 1.

Determination of Rigid Vortex Configurations in Case of $n = 3$.

In comparing $(2.7)_n - (2.10)_n$ with $(1.3)_n - (1.8)_n$ we see that the determination of rigid vortex configurations will be analogous to that of rigid planet configurations. Consider in particular the case $n = 3$ or the problem below:

Example 2. To solve the system of equations $(2.7)_3 - (2.10)_3$ for variables u_i, v_i, k_0, r_{ij} with $i \in J_3, (ij) \in J_3^2$ in terms of the parameters $k_i, i \in J_3$.

For fixed configurations we have $k_0 = -c = 0$ and the solving is quite easy. So in what follows we shall consider only the case $k_0 \neq 0$.

For this purpose let us replace as before u_2, u_3, v_2, v_3 by $u_{12}, u_{13}, v_{12}, v_{13}$ by means of (1.9), (1.10). Then the solving of $(2.7)_3 - (2.10)_3$ is equivalent to the solving of $QS = 0$ under the relevant reality conditions where $QS = \{q_1, \cdots, q_9\}$ with q_i given below:

$$
\begin{aligned}
q_1 &= u_1 * (k_3 + k_2 + k_1) - u_{13} * k_3 - u_{12} * k_2,\\
q_2 &= v_{13}^2 * (k_0 * r_{12}^2 + k_2 + k_1)\\
&\quad - v_{13} * v_{12} * (k_0 * r_{12}^2 + k_2 + k_1) - v_{12}^2 * k_3,\\
q_3 &= v_{13}^2 * k_2\\
&\quad + v_{13} * v_{12} * (k_0 * r_{13}^2 + k_3 + k_1) - v_{12}^2 * (k_0 * r_{13}^2 + k_3 + k_1),\\
q_4 &= v_1 * u_{13} * u_{12} * k_0 + u_{13} * k_2 + u_{12} * k_3,\\
q_5 &= u_{13}^2 * (k_0 * r_{12}^2 + k_2 + k_1)\\
&\quad - u_{13} * u_{12} * (k_0 * r_{12}^2 + k_2 + k_1) - u_{12}^2 * k_3,\\
q_6 &= u_{13}^2 * k_2\\
&\quad + u_{13} * u_{12} * (k_0 * r_{13}^2 + k_3 + k_1) - u_{12}^2 * (k_0 * r_{13}^2 + k_3 + k_1),\\
q_7 &= v_{12} * u_{12} - r_{12}^2,\\
q_8 &= v_{13} * u_{13} - r_{13}^2,\\
q_9 &= (v_{13} - v_{12}) * (u_{13} - u_{12}) - r_{23}^2.
\end{aligned}
\tag{2.11}
$$

We have to determine

$$
Zero_{(rc)}(QS/NZ) \subset Zero_{\mathbf{C}}(QS/NZ), \tag{2.12}
$$

in which (rc) means that the zeros should be chosen among those verifying the relevant reality conditions and NZ is the product of the non-zero variables and parameters $k_0, k_1, k_2, k_3, r_{12}, r_{13}, r_{23}$.

Proceed now as in Example 1 we get finally the result below:

$$
Zero_{(rc)}(QS/NZ) = \bigcup_{k=1,2,3,4} Zero_{(rc)}(CS_k/NZ), \tag{2.13}
$$

In (2.13) we have $CS_k = \{c_{k1}, \cdots, c_{k8}\}$ for each k with c_{ki} given below:

$$
\begin{aligned}
c_{11} &= r_{13} - r_{12},\\
c_{12} &= r_{23} - r_{12},\\
c_{13} &= k_0 * r_{12}^2 + (k_3 + k_2 + k_1),\\
c_{14} &= u_{13}^2 - u_{13} * u_{12} + u_{12}^2,\\
c_{15} &= v_{12} * u_{12} - r_{12}^2,\\
c_{16} &= v_{13} * u_{13} - r_{12}^2,\\
c_{17} &= u_1 * (k_3 + k_2 + k_1) - u_{13} * k_3 - u_{12} * k_2,\\
c_{18} &= v_1 * u_{13} * u_{12} * r_{12} * k_0 + u_{13} * k_2 + u_{12} * k_3.
\end{aligned}
\tag{2.14}
$$

$$
\begin{aligned}
c_{21} &= r_{13}^3 * (k_2 + k_1) + r_{13}^2 * r_{12} * (2 * k_2 + k_1) \\
&\quad - r_{13} * r_{12}^2 * (2 * k_3 + k_1) - r_{12}^3 * (k_3 + k_1), \\
c_{22} &= r_{23} - r_{13} - r_{12}, \\
c_{23} &= k_0 * r_{13} * r_{12}^2 * (r_{13} + r_{12}) + r_{13}^2 * (k_2 + k_1) \\
&\quad + r_{13} * r_{12} * (k_2 + k_1) - r_{12}^2 * k_3, \\
c_{24} &= u_{13} * r_{12} + u_{12} * r_{13},
\end{aligned}
\tag{2.15}
$$

and c_{2i} same as c_{1i}, for $i = 5, 6, 7, 8$. The polsets CS_3, CS_4 are similar to CS_2. The zero-set of CS_1 is analogous to the Lagrange case while those of CS_2, CS_3, CS_4 are analogous to the Euler cases of planet motions.

Remark 2.3. As k_1, k_2, k_3, though non-zero, may be either positive or negative, so from $c_{21} = 0$ we will get either none, or 1, or 2, or 3 positive real roots for the corresponding rectilinear vortex motions, in comparing with the single one rectilinear planet motion in that case.

References.

[Mou] F.R.Moulton, The straight line solutions for the problem of n-bodies, Annals of Math., 12 (1910) 1-17.

[Sm1] S.Smale, Topology and mechanics, I,II. Invent.Math. 10 (1970) 305-331; 11 (1970) 45-64.

[Sm2] S.Smale, Problems on the nature of relative equilibria in celestial mechanics, in Manifolds-Amsterdam 1970, 194-198.

[Som] A.Sommerfeld, Mechanics of deformable bodies, 1950.

[Syn] J.L.Synge, On the motion of three vortices, Canadian J.Math., 1 (1949) 257-270.

[Wald] J.Waldvogel, Note concerning a conjecture by A.Wintner, Celestial Mechanics, 5 (1972) 37-40.

[Wint] A.Wintner, Analytic foundations of celestial mechanics, 1941.

Vol. 17 No. 2 Journal of Systems Science and Complexity Apr., 2004

ON ALGEBRICO-DIFFERENTIAL EQUATIONS-SOLVING*

WU Wenjun(Wu Wen-tsun)
(*Institute of Systems Science, Academy of Mathematics and Systems Science, Chinese of Academy Sciences, Beijing* 100080, *China*)

Abstract. The char-set method of polynomial equations-solving is naturally extended to the differential case which gives rise to an algorithmic method of solving arbitrary systems of algebrico-differential equations. As an illustration of the method, the Devil's Problem of Pommaret is solved in details.

Key words. Algebrico-differential equations, (Differential) Zero-decomposition theorem, Riquier-Janet theory and method, integrability d-polynomial, compatibility d-polynomial, Pommaret's devil problem.

1 Introduction

Let $y, u_j, j \in J$, be infinitely differentiable functions in independent variables $X = \{x_k, k = 1, 2, \cdots .n\}$. A polynomial in various derivatives of y and u_j with respect to x_k with coefficients in the differential field of rational functions of X will be called an *algebrico-differential polynomial.* Suppose given a finite system of such polynomials $DPS = \{DP_i \mid i \in I\}$. Let us consider the associated system of partial differential equations of y with u_j supposed known:

$$DPS = 0, \quad \text{or} \quad DP_i = 0, \ i \in I.$$

Our problem is to determine the integrability conditions for y to be solvable in terms of x_k, u_j and in affirmative case to determine the set of all possible formal solutions of y.

Criteria and even algorithmic methods of solving the above problem were known in quite remote times for which we may cite particularly Riquier[1,2], Janet[3,4], and E. Cartan[5,6]. In recent years J. F. Pommaret had given a systematic *formal intrinsic* way of treatment and had published several voluminous treatises, cf. e.g. [7, 8, 9]. On the other hand, the present author had given an alternative method in following essentially the steps of Riquier and Janet, cf. [10]. The present paper is actually a simplified version of the above paper. An example due to Pommaret, the so-called *Devil's Problem* will be treated in details to illustrate the procedure of our method.

For the illustration of our procedure let us first recall our *char-set* method of solving arbitrary purely algebraic polynomial equations. Thus, consider a set of variables $X = \{x_1, x_2, \cdots, x_n\}$ and polynomials in X with coefficients in the complex field $\boldsymbol{C}$. We shall introduce some partial ordering among all finite systems of such polynomials. For this purpose let us arrange first the variables x_k in natural ascending order. Any non-constant polynomial P in $\boldsymbol{C}[X]$ may then be written in a canonical form

$$P = I_d * x_c^d + I_{d-1} * x_c^{d-1} + \cdots + I_0,$$

Received October 29, 2003.

*The present paper is in honor of late Professor R.Thom as a great mathematician, a great scientist, and also a great thinker of modern times.

in which I_i are themselves polynomials in variables $x_1, x_2, \cdots, x_{c-1}$ with $I_d \neq 0$. We call then c the *class* of P, d the *degree* of P, x_c the *leading variable* of P, and I_d the *leading coefficient* or *initial* of P. We then introduce a partial ordering of non-zero polynomials first according to class and then to degree, the non-zero constants being considered as polynomials of lowest ordering.

For the partial ordering of polynomial systems let us consider first such polynomial sets well-arranged in the following sense. The polynomials in the set are non-constant ones and may be so arranged with classes c_i steadily increasing:

$$0 < c_1 < c_2 < \cdots < c_r.$$

The leading coefficient or the initial of the i-th polynomial in the set is either a non-zero constant or has a class less than c_i which, if it is of class $c_j, 1 \leq j < i$, should have a degree less than that of j-th polynomial in the set. Such a polynomial set is then called an *ascending set* (abbr. *asc-set*). Some partial ordering is then introduced among the system of all such asc-sets, with the set consisting of a single non-zero constant considered as a *trivial asc-set* to be arranged in the lowest ordering.

Consider now arbitrary finite systems of non-zero polynomials. For such a polynomial system, any asc-set of lowest ordering contained wholly in the given system is called a *basic set* (abbr. *bas-set*) of the system. A partial ordering is then unambiguously introduced among all non-empty polynomial systems according to the partial ordering of their basic sets. Any polynomial system containing a non-zero constant polynomial will be clearly one of lowest ordering.

After the introduction of partial ordering among all finite polynomial systems let us consider now such a given system PS and consider the scheme (S) shown below:

$$\begin{array}{rlllllll} PS = & PS^0 & PS^1 & \cdots & PS^i & \cdots & PS^m & \\ & BS^0 & BS^1 & \cdots & BS^i & \cdots & BS^m & = CS \\ & RS^0 & RS^1 & \cdots & RS^i & \cdots & RS^m & = \emptyset. \end{array} \qquad \text{(S)}$$

In the scheme (S) each BS^i is a basic set of PS^i, each RS^i is the set of non-zero remainders, if any, of polynomials in $PS^i \setminus BS^i$ with respect to BS^i, and $PS^{i+1} = PS \cup BS^i \cup RS^i$ if RS^i is non-empty. It is easily proved that the sequence of BS^i is a steadily decreasing sequence:

$$BS^0 \succ BS^1 \succ \cdots \succ BS^r \succ \cdots$$

Such a sequence cannot be an infinite one and should terminate at certain stage m with $RS^m = \emptyset$. The corresponding basic set $BS^m = CS$ is then called a *characteristic set* (abbr. *char-set*) of the given polynomial system PS. The *zero-set* of $PS, Zero(PS)$, consisting of all possible complex solutions or *zeros* of the system of polynomial equations $PS = 0$, is closely connected with that of CS by the Well-Ordering Principle in the form below:

$$Zero(PS) = Zero(CS/IP) \bigcup Zero(PS \bigcup \{IP\}),$$

in which IP is the product of all initials of polynomials in CS and $Zero(CS/IP) = Zero(CS) \setminus Zero(IP)$.

Now $PS \cup \{IP\}$ is easily seen to be a polynomial set of lower ordering than PS. If we apply the Well-Ordering Principle to $PS \cup \{IP\}$ and proceed further and further in the same way we should stopped in a finite number of steps and arrived at the following

Zero-Decomposition Theorem *For any finite polynomial system PS there is an algorithm which will give in a finite number of steps a finite set of asc-sets CS^s with initial-product*

IP^s such that

$$Zero(PS) = \bigcup_s Zero(CS^s \ / \ IP^s). \tag{Z}$$

Now CS^s are all asc-sets. Hence all zero-sets $Zero(CS^s)$ and all $Zero(CS^s/\ IP^s)$ may be considered as well-determined in some natural sense. The formula (Z) gives thus actually an explicit determination of $Zero(PS)$ for all finite polynomial systems PS which serves for the solving of arbitrary systems of polynomial equations.

Mathematics should be incessantly faced with the solving of various kinds of problems, both theoretical and practical ones. Such problems are abundant in nature, in sciences, in reconstructions, in engineering, in administrative works, etc., besides those in mathematics herself. As the data given and results to be found are usually connected by some form of equations, so equations-solving becomes naturally one of the main concern of mathematics. As the algebraic polynomial equations and differential equations, ordinary or partial, appear as the usual form of equations which arise most often, so the solving of such kinds of equations become naturally our most urgent task to deal with. In this section we have presented an algorithmic method of solving arbitrary systems of polynomial equations. It is conceivable that there will be various kinds of applications of this general method. This is really the case. Among the applications we may cite in particular the mechanical proving on computers of theorems in various kinds of geometries, those of euclidean geometry in particular. See e.g. the book [11] of S.C.Chou. For the methods of our theory as well as their applications we refer to the author's book [12] and the references there. For the extension to differential case we refer to the next sections.

Remark finally that, instead of theorem-proving, polynomial equations-solving occupies a central position throughout the long history of thousands of years of development of Chinese ancient mathematics. In fact, the above general method of polynomial equations-solving had its origin in some of our ancient classic due to the scholar Zhu Shijie in Yuan Dynasty (1271–1368 A.D.), see [13]. Of course, there are many defects in Zhu's work. However, the main lines of thought and treatment are sound and the above Well-Ordering Principle is just a modified reformulation of Zhu's work in applying the modern techniques even terminologies of the works of J. F. Ritt, see [14, 15]. For more details we refer to various writings of the present author, notably the book [12].

2 Partial Ordering of Algebrico-Differential Polynomials and Algebrico-Differential Polynomial Systems

Let us consider now the case of *algebrico-differential polynomials* (abbr. *ad-pol* or simply *d-pol*) and such polynomial sets (abbr. *ad-polset* or simply *d-polset*) with notations X, x_k, y, u_j, DPS, DP_i, etc. as in the beginning of Section 1. A d-pol with no y or its partial derivatives actually occuring in it will be called a *trivial d-pol*. For each tuple of n non-negative integers $\mu = (i_1, i_2, \cdots, i_n)$ let us write $||\mu||$ for $i_1 + i_2 + \cdots i_n$ and ∂_μ for the partial derivative $\frac{\partial^{||\mu||}}{\partial x_1^{i_1} \cdots \partial x_n^{i_n}}$. We shall arrange all the partial derivatives $\partial_\mu y$ of y in the usual lexicographical order of μ. For any non-trivial d-pol DP the highest derivative occuring in DP is then called the *leading derivative* or simply the *lead* of DP. If the lead is $\partial_\mu y$, then μ is called the *class* of DP. If the highest degree of $\partial_\mu y$ occuring in DP is d (> 0), then d is called the *degree* of DP, with class and degree undefined for trivial d-pols. For non-trivial d-pol DP with class μ and degree d (> 0), we can write DP in the form

$$DP = I * (\partial_\mu y)^d + \text{lower degree terms in } \partial_\mu y.$$

The coefficient I ($\neq 0$) of $(\partial_\mu y)^d$ in DP , which is itself either a trivial d-pol or non-trivial d-pol in partial derivatives of lower ordering than $\partial_\mu y$, is called the *initial* of DP. The formal partial derivative of DP with respect to $\partial_\mu y$ is called the *separant* of DP. Clearly the separant is the same as the initial when degree d of DP is 1, and of the same class but of lower degree than DP if $d > 1$.

We now introduce a partial ordering among all non-trivial d-pols by first according to their class and then to their degree, with trivial d-pols in the lowest ordering. The following proposition is now clear from the very definitions:

Proposition 1 *Any sequence of d-pols steadily decreasing in ordering*

$$DP_1 \succ DP_2 \succ \cdots \succ DP_r \succ \cdots$$

is necessarily finite.

For two non-trivial d-pols DP, DQ we say that DQ is *reduced* with respect to DP if no proper derivative of the lead of DP occurs in DQ and the lead itself is either not occuring in DQ, or occuring in DQ with a degree less than the degree of DP.

For the introduction of partial ordering among arbitrary d-polsets let us consider first that of particular d-polsets called *ad-ascending sets* (abbr. *ad-asc-sets* or simply *d-asc-sets*) defined as follows. A d-polset is called a *d-asc-set* if it is either consisting of a single trivial d-pol or a d-polset for which the d-pols are all non-trivial ones and may be arranged in a sequence of d-pols in increasing ordering such that each one in the sequence is reduced with respect to the preceding ones. In the case of a single trivial d-pol the corresponding d-asc-set is then called a *trivial* d-asc-set.

Consider now two non-trivial d-asc-sets DAS, DBS with d-pols arranged in increasing ordering as follows:

$$DAS: \quad DA_1 \prec \cdots \prec DA_r,$$

$$DBS: \quad DB_1 \prec \cdots \prec DB_s.$$

We shall say that DAS is of *higher ordering* than DBS or DBS is of *lower ordering* than DAS if either (a) or (b) below holds true:

(a) There is some k ($\leq \min(r, s)$) such that for each $i < k$, DA_i and DB_i are incomparable in ordering while $DA_k \succ DB_k$ as d-pols.

(b) $r < s$ and DA_i, DB_i are incomparable in ordering as d-pols for all $i \leq r$.

It is easy to see that the above definition introduces really a partial ordering among all d-asc-sets with trivial d-asc-sets considered to be in the lowest ordering. As Proposition 1 we have also the proposition below for d-asc-sets:

Proposition 2 *Any sequence of d-asc-sets steadily decreasing in ordering*

$$DAS_1 \succ DAS_2 \succ \cdots \succ DAS_r \succ \cdots$$

is necessarily finite.

Consider now an arbitrary d-polset DP. Any d-asc-set wholly contained in DP will be called an *ad-basic-set* (abbr. *ad-baset* or simply *d-baset*) of DP. We shall introduce now partial ordering among all d-polsets according to the partial ordering of their d-basets. It is easily seen that this is unambiguously well-defined independent of d-basets chosen from the d-polsets.

The above completes the introduction of partial ordering among d-pols and d-polsets. After this preparation we shall show how to solve arbitrary systems of algebrico-differential equations in next section. We remark however that the partial ordering in the present section is only one of many possible ways which will meet our purposes.

3 (Differential) Characteristic-Set Formation and Solving of Arbitrary Algebrico-Differential Polynomial Equations

To extend the notion of char-set and the method of solving polynomial equations in the ordinary case to differential case we need two fundamental procedures of remainder formation and integrability-condition formation to be described in what follows.

For this purpose let us consider a non-trivial d-asc-set DAS as given below:

$$DAS : DA_1 \prec DA_2 \prec \cdots \prec DA_r. \tag{dA}$$

With respect to DAS we have then the following theorem due to J. F. Ritt[14,15] which is fundamental for the whole theory:

Ritt's Remainder Theorem *For any non-trivial d-pol DP there are for each $a \in \{1, \cdots, r\}$ integers s_a, t_a and certain partial derivatives $\partial_{\tau_{ab}}$ and d-pols C_{ab} for $b \in \{1, 2, \cdots, b_a\}$, such that*

$$DR = S_1^{s_1} * \cdots * S_r^{s_r} * I_1^{t_1} * \cdots * I_r^{t_r} * DP - \sum_{a,b} C_{ab} * \partial_{\tau_{ab}} DA_a \tag{dR}$$

is reduced with respect to DAS, i.e. reduced with respect to each d-pol in DAS.

The above formula (dR) will be called the *d-Remainder Formula* of DP and the procedure of getting DR from DP the *reduction* of DP with respect to DAS.

Consider now a pair of nontrivial d-pols DP, DQ with classes $\mu = (i_1, i_2, \cdots, i_n), \nu = (j_1, j_2, \cdots, j_n)$ respectively. Suppose that neither $i_k \geq j_k$ nor $j_k \geq i_k$ for all $k \in \{1, 2, \cdots, n\}$. Such a pair will be called a *legal pair* with respect to DAS. Now for each k let $m_k = \max(i_k, j_k), p_k = m_k - i_k, q_k = m_k - j_k$. Set $\xi = (p_1, p_2, \cdots, p_n), \eta = (q_1, q_2, \cdots, q_n)$ and $\zeta = (m_1, m_2, \cdots, m_n)$. Then $\partial_\xi DP$ and $\partial_\eta DQ$ have the same lead $\partial_\zeta y$ with degree 1. Let the initials of $\partial_\xi DP$ and $\partial_\eta DQ$ be I_ξ and I_η respectively. Then the difference $I_\eta * \partial_\xi DP - I_\xi * \partial_\eta DQ$, after eventually reduction with respect to DAS, will be called the eventually *reduced integrability d-pol* of the legal pair DP and DQ with respect to the d-asc-set DAS.

With the above notions of d-remainder and integrability d-pol, eventually reduced or not, with respect to a non-trivial d-asc-set we can now extend our procedures in ordinary polynomial case as given in scheme (S) of Section 1 to the differential case as shown in the scheme (dS) below:

$$\begin{array}{llllll} DPS = DPS^0 & DPS^1 & \cdots . & DPS^i & \cdots & DPS^m \\ & DBS^0 & DBS^1 & \cdots & DBS^i & \cdots & DBS^m = DCS \\ & DRIS^0 & DRIS^1 & \cdots & DRIS^i & \cdots & DRIS^m = \emptyset \\ & DCPS^0 \cup DCPS^1 \cup \cdots \cup DCPS^i \cup \cdots \cup DCPS^m = DCPS. \end{array} \tag{dS}$$

In the scheme (dS) DPS is the given d-polset. For each i, DBS^i is a d-baset of DPS^i, and the set $DRIS^i$ is the union of two parts. One is the set of all possible non-zero d-remainders formed from d-pols in $DPS^i \setminus DBS^i$ with respect to DBS^i, while the other is the set of integrability d-pols formed from all possible legal-pairs of d-pols in DPS^i, eventually reduced with respect to DBS^i, so far they contain actually y or its derivatives. On the other hand those containing no y or its derivatives but containing possibly u_j or their derivatives will form a set of *compatibility d-pols* for which the vanishing will form *compatibility conditions* in order that the given set of equations $DPS = 0$ will have solutions. In case $DRIS^i$ is non-empty, then the union $DPS \cup DBS^i \cup DRIS^i$ will form the next d-polset DPS^{i+1}.

As is easily verified, the d-basets DBS^i will form a sequence of steadilly decreasing ordering:

$$DBS^0 \succ DBS^1 \succ \cdots \succ DBS^s \succ \cdots$$

By Proposition 2 this sequence can only be a finite one, to be stopped at a certain stage m with $DRIS^m = \emptyset$. The corresponding d-baset $DBS^m = DCS$ is then called an *ad-* or simply *d-characteristic set* (abbr. *d-charset*) of the given d-polset DPS. The union $DCPS$ of all sets $DCPS^i, i = 1, 2, \cdots, m$, will form the totality of all possible compatibility d-pols whose vanishing form the compatibility conditions to guarantee the existence of solutions of the partial differential equations $DPS = 0$.

For any d-polset DPS and d-pol DG let $dZero(DPS)$ be the set of all possible solutions of the partial differential equations $DPS = 0$ and $dZero(DPS/DG) = dZero(DPS) \setminus dZero(DG)$. Then we have from scheme (dS) the following *(differential) Well-Ordering Principle*:

$$dZero(DPS) = dZero(DCS \ / \ DISP) \bigcup dZero\Big(DPS \bigcup \{DISP\}\Big), \qquad \text{(dW)}$$

in which $DISP$ is the product of all initials and separants of d-pols in DCS, so far the compatibility conditions $DCPS = 0$ are assumed to be verified.

As in the case of ordinary polynomial equations-solving, we deduce by successive applications of the above (differential) Well-Ordering Principle the following theorem, which is at the basis of solving arbitrary systems of algebrico-differential polynomial equations:

(Differential) Zero-Decomposition Theorem *For any finite d-polset DPS there is an algorithm which will give in a finite number of steps a finite set of d-asc-sets DCS^s with initial-separant-products $DISP^s$ as well as sets of compatibility d-pols $DCPS^s$ such that*

$$dZero(DPS) = \bigcup_s dZero(DCS^s \ / \ DISP^s), \qquad \text{(dZ)}$$

so far some compatibility conditions are supposed to be verified.

In order to give formal explicit solutions of the partial differential equations $DPS = 0$ for d-polset DPS let us consider first the case of a non-trivial d-asc-set DAS as given by (dA). Let the classes of DA_a in (dA) be $\mu_a, a = 1, 2, \cdots, r$. Then all partial derivatives of the leads $\partial_{\mu_a} y$, proper or improper, will be called *principal derivatives*, and all the others *paramatric* ones.

Consider now any set of constants $c_k \in \boldsymbol{C}, k = 1, 2, \cdots, n$, and also constants $c_\tau \in \boldsymbol{C}$ for each parametric derivative $\partial_\tau y$. The values $x_k = c_k$ will give definite values to the known functions $u_j, j \in J$, as well as their derivatives. We suppose that the above values will not render zero the initial-separant product ISP of DAS. The set of these constant values c_k, c_τ will then be called an *admissible preliminary constant set* with respect to the d-asc-set DAS.

With values of such admissible preliminary constants substituted in the equations $DA_a = 0, a = 1, 2, \cdots, r$, we can solve for them to get values for the proper principal derivative c_{μ_a} for classes μ_a. Let us take any set of such values for each μ_a. By differentiating DA_a and substituting the preliminary values as well as the chosen values of c_{μ_a}, we get also definite constant values c_ν for arbitrary improper principle derivatives $\partial_\nu y$. With these constant values we form now a Formal Taylor Series FTS of the form below:

$$FTS = \sum_\tau \frac{c_\tau}{\|\tau\|} * \prod_k (x_k - c_k)^{t_k},$$

in which τ runs over all integer-tuples $\tau = (t_1, t_2, \cdots, t_n)$ with $\|\tau\| = t_1 + t_2 + \cdots + t_n$.

It is easy to verify the following

Formal Taylor Series Theorem *With a given admissible preliminary value set all possible solutions of the partial differential equations $DAS = 0$ for which the initial-separant product of DAS is non-zero are given by formal Taylor series of above form FTS.*

Consider now an arbitrary non-trivial d-polset DPS with a non-trivial d-charset DCS as given in the scheme (dS). Then for any admissible preliminary value set for the d-asc-set DCS, for which the corresponding initial-separant product ISP of DCS is not zero, we will get totality of solutions of the partial differential equations $DPS = 0$, so far all compatibility d-pols in $DCPS$ are rendered zero. We may treat in the same way the d-charsets DCS^s in the (differential) Zero-Decomposition Theorem (dZ) of DPS and get the totality of solutions of $DPS = 0$ in the form of Formal Taylor Series, so far the preliminary value set will not render zero the corresponding initial-separant product ISP^s, while render zero all the d-pols in the corresponding compatibility d-polset.

Remark that the above is only a special case of our general theory and method which we refer to the previously cited paper [10]. In fact, in the general theory there may be several unknown functions y_1, y_2, etc. instead of a single one y. Moreover, the coefficient field may be an arbitrary differential field with differential operators obeying the usual differential rules. Furthermore, instead of pairwise determination of integrability d-pols, we may apply the device of multiplicativity of variables originated by Riquier and Janet to make smaller the number of pairs of forming the integrability d-pols. However, as the essence of these methods are actually the same as in the above particular case so we shall not enter into them.

Besides, the relations between the notions of d-charset, d-asc-set, etc. in our procedure and those of passiveness, prolongations, involutiveness, etc. in the procedures of Riquier, Janet, and Cartan require some clarification which we shall do in later occasions. Instead we shall show as an illustration how to solve a particular example by our procedure which will be described in details in the next section.

It is clear that our general method of algebrico-differential equations-solving will have an immense variety of applications as in the case of ordinary polynomial equations-solving. In particular, as in the ordinary case, we have applied our general method of differential equations-solving to the mechanical proving of differential geometry theorems, and to the automated determination of explicit form of relations for which only the existence of the relations is known. We refer these to the author's papers [16–18] and the paper [19] of S. C. Chou and X. S. Gao as well as the references in these papers. On the other hand, we shall leave the studies of various other applications to later occasions.

4 An Example: Pommaret's Devil Problem

For the sake of illustration of his formal intrinsic method of treatment of algebraic partial differential equations, Pommaret had exhibited in details an example, what he called the *Devil's Problem*, in his paper [7], his treatise [9], as well as in various courses or lectures taken place in France, in Germany, in Beijing, and elsewhere. We shall treat this Devil's Problem also in details by our own method as exhibited in Section 3. For this purpose let us reproduce the original statement of the Devil's Problem as well as its final solution from Pommaret's writings as given below.

Devil's Problem Let u, v, y be 3 functions of the cartesian coordinates x^1, x^2, x^3 on euclidean spaces related by the following system of 2 PDE where $\partial_{33}y = \frac{\partial^2 y}{\partial_{x^3}\partial_{x^3}}, \cdots$:

$$\begin{aligned} &\partial_{33}y - x^2\partial_{11}y - u = 0, \\ &\partial_{22}y - v = 0. \end{aligned} \tag{D}$$

(1) If $u = v = 0$ the space of solutions of the resulting linear system of PDE for y is a vector space over the constants. What is its dimension?

(2) Otherwise, what kind of compatibility conditions must be satisfied by u and v in order to insure the existence of solutions for y?

(3) Does there exist a "general" way to solve such problems?

The final result of Pommaret on this problem may be described as follows:

The final general solution is furnished by a "Good" Set consisting of 4 "good" algebraic differential polynomials $G_1, \cdots, G_4$ below:

$$\begin{aligned} G_1 &= \partial_{1111}y - z, \\ G_2 &= \partial_{112}y - w, \\ G_3 &= \partial_{22}y - v, \\ G_4 &= \partial_{33}y - x^2\partial_{11}y - u. \end{aligned} \tag{G}$$

In (G) w and z are given by

$$w = \frac{1}{2}(\partial_{33}v - x^2\partial_{11}v - \partial_{22}u), \qquad z = \partial_{33}w - \partial_{112}u - x^2\partial_{11}w.$$

Pommaret solved the above question (2) in deriving two *Compatibility Conditions* $A = 0, B = 0$ with A, B given below:

$$\begin{aligned} \frac{1}{2}A &= \partial_2 w - \partial_{11}v, \\ B &= \partial_{3333}w - 2x^2\partial_{1133}w + (x^2)^2\partial_{1111}w - \partial_{11233}u + x^2\partial_{11112}u - \partial_{1111}u. \end{aligned} \tag{CC}$$

Pommaret showed further that the two compatibility conditions $A = 0, B = 0$ are not independent of each other. They are in fact connected by the following differential identity

$$\partial_{3333}A - 2x^2\partial_{1133}A + (x^2)^2\partial_{1111}A - 2\partial_2 B = 0.$$

In the case of $u = 0, v = 0$ so that $w = 0, z = 0$ too the "good" sets become simply $\partial_{1111}y = \partial_{112}y = \partial_{22}y = \partial_{33}y - x^2\partial_{11}y = 0$. It follows that there remain only 12 derivatives below which can take arbitrary values, viz.:

$$y, \partial_1 y, \partial_{11}y, \partial_{111}y, \partial_2 y, \partial_{12}y, \partial_3 y, \partial_{13}y, \partial_{113}y, \partial_{1113}y, \partial_{23}y, \partial_{123}y.$$

As a consequence Pommaret solves the question (1) in giving a vector space of dimension 12 spanned by the above vectors.

Let us now apply our own method to the solving of the Devil's Problem. For this purpose let us first change the notations in order to make mainly in accordance with those in Section 2. Thus, instead of x^1, x^2, x^3 we shall write x_1, x_2, x_3 and instead of $\partial_{33}y$ for $\frac{\partial^2 y}{\partial x^3 \partial x^3}$ we shall write $\partial_{200}y$, etc., as in Section 2. In this way the system of 2 PDE (D) of the Devil's Problem will be $DPS = 0$ with $DPS = \{DP_1, DP_2\}$ in which

$$\begin{aligned} DP_1 &= \partial_{200}y - x_2 * \partial_{002}y - u, \\ DP_2 &= \partial_{020}y - v. \end{aligned}$$

As in the scheme (dS) of Section 3, we have then $DPS^0 = DPS = \{DP_1^0, DP_2^0\}$ with

$$DP_1^0 = DP_1, \; DP_2^0 = DP_2.$$

The d-baset chosen from DPS^0 is then $DBS^0 = \{DB_1^0, DB_2^0\}$ with

$$DB_1^0 = DP_2^0, \; DB_2^0 = DP_1^0$$

in increasing ordering. There are clearly no d-remainders but there is one integrability d-pol eventually reduced to be determined from the legal pair (DP_1^0, DP_2^0). To determine it let us first form

$$\begin{aligned} \partial_{200}DP_2^0 - \partial_{020}DP_1^0 &= x_2 * \partial_{022}y + 2 * \partial_{012}y - \partial_{200}v + \partial_{020}u \\ &= \Delta \text{ , } say. \end{aligned}$$

We reduce now Δ with respect to the d-baset DBS^0 to get the reduced integrability d-pol

$$DI_{12}^0 = \Delta - x_2 * \partial_{002} DB_1^0 = 2 * \partial_{012} y - 2 * w,$$

in which

$$w = \frac{1}{2}(\partial_{200} v - x_2 * \partial_{002} v - \partial_{020} u).$$

We see that $\frac{1}{2} DI_{12}^0$ is actually the same as the "good" set G_2 and w is actually the same as the w given by Pommaret but in different way of notations.

As the integrability d-pol DI_{12}^0 contains actually derivatives of y it will form the set $DRIS^0$ with the corresponding compatibility d-polset $DCPS^0$ empty. Hence, in adjoining DI_{12}^0 or simpler $\frac{1}{2} * DI_{12}^0$ to DPS^0 we get the new d-polset $DPS^1 = \{DP_1^1, DP_2^1, DP_3^1\}$ with

$$\begin{aligned} DP_1^1 &= DP_1^0 &&= \partial_{200} y - x_2 * \partial_{002} y - u, \\ DP_2^1 &= DP_2^0 &&= \partial_{020} y - v, \\ DP_3^1 &= \frac{1}{2} * DI_{12}^0 &&= \partial_{012} y - w. \end{aligned}$$

The d-baset chosen from DPS^1 is then $DBS^1 = \{DB_1^1, DB_2^1, DB_3^1\}$ with

$$DB_1^1 = DP_3^1, \quad DB_2^1 = DP_2^1, \quad DB_3^1 = DP_1^1$$

in increasing ordering. Again there are no d-remainders but there are 3 legal pairs (DP_1^1, DP_2^1), $(DP_1^1, DP_3^1), (DP_2^1, DP_3^1)$. The first pair has already been considered so there remain only two new ones to be considered which give two integrability d-pols eventually reduced below:

$$\begin{aligned} DI_{13}^1 &= \partial_{200} DP_3^1 - \partial_{012} DP_1^1 - x_2 * \partial_{012} DB_3^1 \\ &= \partial_{004} y - z, \\ DCP_{23}^1 &= \partial_{010} DP_3^1 - \partial_{002} DP_2^1 \\ &= - \partial_{010} w + \partial_{002} v, \end{aligned}$$

in which z is given by

$$z = \partial_{200} w - x_2 * \partial_{002} w - \partial_{012} u.$$

It is readily seen that DI_{13}^1 is the same as the "good" set "G_1" and z is same as that given before, both by Pommaret in different notations. We see that DI_{13}^1, containing actually derivatives of y, is to be a d-pol in $DRIS^1$, while DCP_{23}^1, containing only derivatives in u, v, but not y, is to form the compatibility d-polset $DCPS^1$. Furthermore the compatibility d-pol DCP_{23}^1 is readily seen to be equal to $-\frac{1}{2}A$, where A is the compatibility d-pol also already given by Pommaret in different notations.

Adjoin now DI_{13}^1 to DPS^1 to get $DPS^2 = \{DP_1^2, DP_2^2, DP_3^2, DP_4^2\}$ with

$$DP_1^2 = DP_1^1, \; DP_2^2 = DP_2^1, \; DP_3^2 = DP_3^1, \; DP_4^2 = DI_{13}^1.$$

The d-baset is then $DBS^2 = \{DB_1^2, DB_2^2, DB_3^2, DB_4^2\}$ with

$$\begin{aligned} DB_1^2 &= DP_4^2 = DI_{13}^1, \\ DB_2^2 &= DP_3^2, \\ DB_3^2 &= DP_2^2, \\ DB_4^2 &= DP_1^2 \end{aligned}$$

in increasing ordering. Again there is no d-remainders but there are 3 new legal pairs to be considered: (DP_4^2, DP_1^2), (DP_4^2, DP_2^2), (DP_4^2, DP_3^2) which give 3 compatibility d-pols DCP_{14}^2, DCP_{24}^2, DCP_{34}^2 as given below:

$$\begin{aligned} DCP_{14}^2 &= \partial_{200}DP_4^2 - \partial_{004}DP_1^2 - x_2 * \partial_{002}DB_1^2 \\ &= -\ \partial_{200}z + x_2 * \partial_{002}z + \partial_{004}u, \\ DCP_{24}^2 &= \partial_{020}DP_4^2 - \partial_{004}DP_2^2 \\ &= -\ \partial_{020}z + \partial_{004}v, \\ DCP_{34}^2 &= \partial_{010}DP_4^2 - \partial_{002}DP_3^2 \\ &= -\ \partial_{010}z + \partial_{002}w. \end{aligned}$$

As $DRIS^2 = \emptyset$, the procedure ends at the stage $m = 2$ so that the scheme (dS) in the present case becomes (dS)′ below:

$$\begin{array}{llll} DPS = DPS^0 & DPS^1 & DPS^2 & \\ \quad DBS^0 & DBS^1 & DBS^2 & = DCS \\ \quad DRIS^0 & DRIS^1 & DRIS^2 & = \emptyset \\ \quad DCPS^0 \cup & DCPS^1 \cup & DCPS^2 & = DCPS. \end{array} \qquad \text{(dS)}'$$

Remark that DCS here is the same as the "good set" $\{G1, G2, G3, G4\}$ given by Pommaret, while the compatibility d-polset here is given by

$$\begin{aligned} DCPS &= DCPS^0 \cup DCPS^1 \cup DCPS^2, \text{ with} \\ DCPS^0 &= \emptyset, \\ DCPS^1 &= \{DCP_{23}^1\}, \\ DCPS^2 &= \{DCP_{14}^2, DCP_{24}^2, DCP_{34}^2\}. \end{aligned}$$

The compatibility d-pols in $DCPS$ are not independent of each other. To determine their interrelations let us first turn these d-pols to be in v and u by means of the expressions of w and z in v and u. Let us consider now these $DCP's$ as d-pols in v as unknown function while u as known function in $X = \{x_1, x_2, x_3\}$. Then the leads in v of the d-pols $DCP_{23}^1, DCP_{14}^2, DCP_{24}^2$, DCP_{34}^2 are seen to be respectively $\partial_{210}v, \partial_{600}v, \partial_{420}v, \partial_{410}v$. Let us treat now $DCPS$ as a d-polset in v as we have treated DPS as a d-polset in y by means of our general method. The d-baset is then seen to be consisting of the two d-pols DCP_{23}^1, DCP_{14}^2 with leads $\partial_{210}v, \partial_{600}v$. It is readily found that the d-remainders of DCP_{24}^2, DCP_{34}^2 with respect to the d-baset are both zero with corresponding d-remainder formula given below:

$$\begin{aligned} DCP_{24}^2 &= \partial_{210}DCP_{23}^1 - x_2 * \partial_{012}DCP_{23}^1, \\ DCP_{34}^2 &= \partial_{200}DCP_{23}^1 - x_2 * \partial_{002}DCP_{23}^1. \end{aligned}$$

On the other hand there is only one legal pair (DCP_{23}^1, DCP_{14}^2) for which the integrability d-pol after reduction with respect to the d-baset is found to be zero with the corresponding interrelation given below:

$$\partial_{010}DCP_{14}^2 - \partial_{400}DCP_{23}^1 + 2 * x_2 * \partial_{202}DCP_{23}^1 - x_2^2 * \partial_{004}DCP_{23}^1 = 0.$$

It follows that the procedure ends at the stage $m = 1$ with the d-charset consisting of the two d-pols DCP_{23}^1, DCP_{14}^2. It is also easily verified that these two d-pols are respectively equal to $-A/2$ and $-B$ of Pommaret, and the interrelations between DCP_{23}^1, DCP_{14}^2 are the same as that between the two compatibility d-pols A, B given by Pommaret, only in different notations.

We see that Pommaret's results are complete for the Devil Problem and our results too, though by different ways of treatments. Remark that Pommaret uses a method with quite involved logical reasonings and modern techniques in applying exact sequences, diagram-chasing, etc. On the other hand our method is highly computational with little mental efforts, and the computations are almost straightforward.

References

[1] Ch. Riquier, *La Méthode des Fonctions Majorantes et les Systèmes d'Equations aux Dérivées Partielles*, Gauthier Villars, 1910.
[2] Ch. Riquier, *Les Systèmes d'Equations aux dérivées Partielles*, Gauthier-Villars, 1910.
[3] M. Janet, Les systèmes d'equations aux dérivées partielles, *J. Math. Pures Appl.*, 1920, **3**: 65.
[4] M. Janet, *Leçons sur les Systèmes d'Equations aux Dérivées Partielles*, Gauthiers-Villars, 1929.
[5] E. Cartan, Sur la structure des groupes infinis de transformations, *Ann. Ec. Norm. Sup.*, 1904, **21**: 153–206.
[6] E. Cartan, *Les Systèmes Différentiels Extérieurs et Leurs Applications Géomé Triques*, Hermann, 1945.
[7] J. F. Pommaret & A. Haddack, Effective Methods for Systems of Algebraic Partial Differential Equations, in T. Mora & C. Traverso (Eds), *Effective Methods in Algebraic Geometry*, Birkehauser, 1991, 441–426.
[8] J. F. Pommaret, *Systems of Partial Differential Equations and Lie Pseudogroups*, New York, 1978.
[9] J. F. Pommaret, *Partial Differential Equations and Group Theory*, Kluwer Acad. Publishers, 1994.
[10] Wen-tsun Wu, On the foundation of algebraic differential geometry, *Math. Mech. Res. Preprints*, Inst. Sys. Scis., 1989, **3**: 1–26.
[11] S. C. Chou, *Mechanical Geometry Theorem-Proving*, Reidel, 1988.
[12] Wen-tsun Wu, *Mathematics Mechanization, Mechanical Geometry Theorem-Proving, Mechanical Geometry Problem-Solving and Polynomial Equations-Solving*, Science Press / Kluwer Academic Publishers, 2000.
[13] Shijie Zhu, *Jade Mirror of Four Elements*, 1303.
[14] J. F. Ritt, *Differential Equations from the Algebraic Standpoint*, Amer. Math. Soc., 1932.
[15] J. F. Ritt, *Differential Algebra*, Amer. Math. Soc., 1950.
[16] Wen-tsun Wu, A mechanization method of geometry and its applications, II. curve pairs of Bertrand type, *Kexue Tongbao*, 1989, **32**: 585–588.
[17] Wen-tsun Wu, Mechanical Derivation of Newton's Gravitational Laws from Kepler's Laws, *Math. Mech. Res. Preprints*, Inst. Sys. Scis., 1987, **1**: 53–61.
[18] Wen-tsun Wu, Mechanical theorem-proving of differential geometries and some of its applications, *J. Aut. Reasoning*, 1991, **7**: 171–191.
[19] S. C. Chou, X. S. Gao, Automated reasoning in differential geometry and mechanics using the characteristic set method, Parts I, II, in *J. of Automated Reasoning*, 1993, **10**, 161–172, 173–189; Part III, in *Automated Reasoning* (Editor Z. Shi), Elsevier Science Publishers, 1992, 1–12.

Reprinted with permission from *Differential Equations with Symbolic Computation*, (Birkhäuser Basel, 2005), pp. 351–368.

On the Construction of Groebner Basis of a Polynomial Ideal Based on Riquier–Janet Theory

Wenjun Wu (Wen-tsün Wu)

Abstract. As a consequence of a previous study of algebraic differential geometry ([see WU1]) there may be associated to certain special kinds of differential ideals some well-behaved basis enjoying some well-behaved properties. If the differential ideals are further specialized so that they correspond to ordinary polynomial ideals then such a well-behaved basis will become the usual Groebner basis of the polynomial ideals while the latter is not known for differential ideals.

Mathematics Subject Classification (2000). Primary 13P10; Secondary 13N10.

Keywords. Polynomial ideal, differential ideal, well-behaved basis, Groebner basis.

0. Introduction

Riquier and Janet have created a theory of PDE which has been further developed by Ritt and Thomas and is closely related to the corresponding theory of E. Cartan. Based on such a theory the author has shown in a previous paper [WU1] how to construct a d-char-set *DCS* of a d-polset *DPS* for which their d-zero-sets are closely connected according to the following decomposition formula:

$$\text{d-Zero}(DPS) = \text{d-Zero}(DCS/J) + \text{SUM}_i\ \text{d-Zero}(DPS_i).$$

In the formula J is the product of all initials and separants of the d-pols in *DCS*, and DPS_i are the enlarged d-polsets of *DPS* in adjoining to it one of such initials

The present paper is partially supported by NSFC Grant JI85312. It is reprinted, with permission and with minor editorial changes, from *Systems Science and Mathematical Sciences* 4/3 (1991): 193–207, originally received on September 4, 1990.

or separants. In the particular case for which all these initials and separants are non-zero constants in the basic d-field, the above formula becomes simply

$$\text{d-Zero}(DPS) = \text{d-Zero}(DCS).$$

Moreover, denoting the differential ideal with *DPS* as a basis by d-Ideal(*DPS*), we see from the construction of *DCS* and theorems proved in that paper that *DCS* is also a basis of this ideal, or

$$\text{d-Ideal}(DPS) = \text{d-Ideal}(DCS).$$

Furthermore, this basis *DCS* possesses the following well-behaved property:

A d-pol *DP* will belong to the differential ideal d-Ideal(*DPS*) if and only if the d-remainder of *DP* w.r.t. the basis *DCS* is 0:

$$\text{d-Remdr}(DP/DCS) = 0.$$

For this reason we shall call the corresponding d-char-set *DCS*, in the above particular case, a *well-behaved basis* of the differential ideal d-Ideal(*DPS*) with *DPS* as a given basis.

Let us consider now the further specialized d-polset *DPS* with the following restrictions:

1. The basic d-field is one with trivial differentiations so that it is just an ordinary field of characteristic 0.

2. The independent variables are still $X_1, \ldots, X_n$ while there is only one dependent Y.

3. Each d-pol *DP* in the d-polset *DPS* is of the form

$$DP = \text{SUM}_t C_t * \text{DER}_t Y,$$

in which t runs over a finite set of n-tuples of nonnegative integers and C_t are non-zero constants in the basic field.

We are thus in the situation of a system $DPS = 0$ of linear PDE with constant coefficients. Now to each partial derivative $\text{DER}_t Y$ we may make a corresponding monomial $X^t = X_1^{i_1} * \cdots * X_n^{i_n}$ in which $(i_1, \ldots, i_n)$ is the n-tuple t. Under the correspondence the d-pols will then be turned into ordinary pols in $X_1, \ldots, X_n$ with coefficients in an ordinary field of characteristic 0. The above theory will then give a well-behaved basis of an ordinary polset *PS*. It turns out that this well-behaved basis is, in the present non-differential case, just the usual Groebner basis of the corresponding polynomial ideal Ideal(*PS*). This offers thus an alternative method of constructing a Groebner basis of a polynomial ideal different from that of Buchberger.

In the present paper we shall consider the last case of ordinary polsets alone. Our exposition will be so given that it is independent of the Riquier-Janet theory and the previous paper [WU1], though it will follow closely the steps exhibited in that paper. In studying the properties of the well-behaved basis of a polynomial ideal introduced in this way it will follow that this basis is just the Groebner basis of that ideal. We prove now several well-known beautiful properties of the

Groebner basis in a way along the line of the thoughts of the previous paper based on Riquier-Janet theory. The proofs are thus somewhat different from the known ones scattered in the literature. These proofs may in fact be carried over to the differential case as stated above for the well-behaved basis, while the Groebner basis is undefined in that case. We remark in passing that our theory will give a unique expression for an arbitrary pol w.r.t. such a basis of a polynomial ideal, while for the usual theory of the Groebner basis such an expression is unique only modulo the basis in some way. Finally, we give a concrete example for which the Groebner basis is determined by the present method in using the REDUCE implemented in our machine SUN3/140. Further examples are yet to be studied and a complexity study of the present method is required.

1. Tuples of Integers

Let n be a positive integer fixed throughout the present paper.

DEF. An ordered sequence of n non-negative integers

$$t = (I_1, \ldots, I_n)$$

is called an n-tuple or simply a *tuple*. I_i is then called the i-th *coordinate* of t, to be denoted by

$$\mathrm{COOR}_i(t) = I_i.$$

DEF. The particular tuple with all coordinates $= 0$ will be called the 0-*tuple*, to be denoted as 0.

Notation. For any tuple u and any integer $i \geq 1$ and $\leq n$, the tuple u' with

$$\mathrm{COOR}_i(u') = \mathrm{COOR}_i(u) + 1,$$
$$\mathrm{COOR}_j(u') = \mathrm{COOR}_j(u), \quad j \neq i,$$

will be denoted by ui or iu.

DEF. For any two tuples u and v, we say u is a *multiple* of v or v is a *divisor* of u, if

$$\mathrm{COOR}_i(u) \geq \mathrm{COOR}_i(v), \quad i = 1, \ldots, n.$$

We write then

$$u \gg v \ \text{ or } \ v \ll u.$$

DEF. For any two tuples u and v, their *product* $uv = vu$ is the tuple with

$$\mathrm{COOR}_i(uv) = \mathrm{COOR}_i(u) + \mathrm{COOR}_i(v), \quad i = 1, \ldots, n.$$

We introduce now an ordering among all the n-tuples according to the following

DEF. For any two tuples u and v we say that u is *higher than* v or v is *lower than* u if there is some $k > 0$ and $\leq n$ such that

$$\mathrm{COOR}_i(u) = \mathrm{COOR}_i(v), \quad i > k,$$
$$\mathrm{COOR}_k(u) > \mathrm{COOR}_k(v).$$

We write then

$$u > v \text{ or } v < u.$$

DEF. A set of tuples T is said to be *autoreduced* if no t in T is a multiple of anther t' in T.

The following two lemmas are already known or easily deduced from known results.

Lemma 1. *Any sequence of tuples steadily decreasing in order is finite.*

Lemma 2. *Any autoreduced set of tuples is finite.*

DEF. For any finite set of tuples T, the *maximum* of T, to be denoted by $\text{Max}(T)$, is the tuple defined by

$$\text{Max}(T) = n\text{-tuple}(\text{MAX}_1(T), \ldots, \text{MAX}_n(T)), \quad \text{with}$$
$$\text{MAX}_i(T) = \text{Max}\{\text{COOR}_i(t)/\ t \in T\}.$$

DEF. For any finite set of tuples T, the *completion* of T, to be denoted by $\text{Comp}(T)$, is the set of tuples defined by

$$\text{Comp}(T) = \{u/\ u \ll \text{Max}(T) \text{ and } u \gg t \text{ for some } t \text{ in } T\}.$$

DEF. For any finite set of tuples T and any tuple $t \ll \text{Max}(T)$, the integer $i\,(\geq 1, \leq n)$ is called a *multiplier* of t w.r.t. T if

$$\text{COOR}_i(t) = \text{MAX}_i(T).$$

Otherwise i is called a *non-multiplier* of t w.r.t. T. In that case we have

$$\text{COOR}_i(t) < \text{MAX}_i(T).$$

Notation. For any finite set of tuples T and any tuple $t \ll \text{Max}(T)$, we shall set

$$\text{Mult}(t/T) = \text{set of all multipliers of } t \text{ w.r.t. } T,$$
$$\text{Nult}(t/T) = \text{set of all non-multipliers of } t \text{ w.r.t. } T.$$

DEF. For $t \ll \text{Max}(T)$, the set of all multiples tu of t with

$$\text{COOR}_i(u) = 0 \text{ for } i \text{ in } \text{Nult}(t/T)$$

is called the *total multiple set* of t w.r.t. T, to be denoted by

$$\text{TMU}(t/T) = \{tu/\ \text{COOR}_i(u) = 0 \text{ for } i \text{ in } \text{Nult}(t/T)\}.$$

Lemma 3. Let T be a finite set of tuples. For any tuple v there is a unique tuple $t \ll \text{Max}(T)$ such that v is in $\text{TMU}(t/T)$. Moreover, if v is a multiple of some tuple in T, then t is in $\text{Comp}(T)$.

Proof. t is determined as $\text{COOR}_i(t) = \text{Min}(\text{COOR}_i(v), \text{MAX}_i(T))$. □

Tuple-decomposition Theorem. *For any finite tuple set T the totality of tuples each of which is a multiple of some tuple in T is the disjoint union of sets* $\text{TMU}(t/T)$ *with t running over* $\text{Comp}(T)$.

Proof. This follows directly from Lemma 3. □

We shall now introduce an ordering in the totality of autoreduced sets as follows.

Let us consider any two autoreduced sets AS and AS' with tuples arranged in increasing order:

$$(AS): \quad t_1 < t_2 < \cdots < t_r,$$
$$(AS)': \quad t'_1 < t'_2 < \cdots < t'_s.$$

DEF. The autoreduced set (AS) is said to be *higher than* the autoreduced set $(AS)'$, or $(AS)'$ *lower than* (AS), if either of the two following cases holds true:

(a) There is some $k \le r$ and $\le s$ such that

$$t_i = t'_i \text{ for } i < k, \text{ while } t_k > t'_k.$$

(b) $r < s$ and $t_i = t'_i$ for $i \le r$.

In notation, we shall set then

$$(AS) > (AS)', \text{ or } (AS)' < (AS).$$

Lemma 4. *Any sequence of autoreduced sets steadily decreasing in order is finite.*

Proof. Let the sequence be

$$(S): \quad AS_1 > AS_2 > \cdots$$

and suppose the contrary that it is infinite. For each autoreduced set AS_i let its tuples be arranged in increasing order. By Lemma 1 the sequence $as_{11}, as_{21}, \ldots, as_{i1}, \ldots$ of which as_{i1} is the first tuple of AS_i should consist of the same tuple, say t_1, from a certain stage onwards. Denote the corresponding infinite sequence of autoreduced sets from that stage onwards by

$$(S1): \quad AS_{11} > AS_{12} > \cdots.$$

Again by Lemma 1 the sequence of second tuples in AS_{1i}, should consist of the same tuple, say t_2, from a certain stage onwards. Denote the corresponding infinite sequence of autoreduced sets from that stage onwards by

$$(S2): \quad AS_{21} > AS_{22} > \cdots.$$

The above reasoning can be repeated indefinitely so that we get an infinite sequence of tuples

$$(T): \quad t_1 < t_2 < \cdots,$$

which is clearly an autoreduced set. This contradicts however Lemma 2 and hence (S) is finite. □

From the very definition of the ordering we have also

Lemma 5. *Let T be an autoreduced set and u be a tuple which is not a multiple of any tuple in T. Let T' be the autoreduced set obtained by adjoining u to T and then removing all tuples in T which are multiples of u. Then T' is of lower order than T.*

2. Well-Arranged Basis of a Polynomial Ideal

Henceforth throughout the paper there will be fixed an integer n, a set of variables $X_1, \ldots, X_n$, and a field K of characteristic 0. By a *pol* will then be meant, unless otherwise stated, a polynomial in $K[X_1, \ldots, X_n]$.

By a *monom* is meant a power-product in X_i of the form

$$X^t = X_n^{i_n} * \cdots * X_j^{i_j} * \cdots * X_1^{i_1},$$

in which the tuple $t = (i_1, \ldots, i_n)$ will be called the *degree-tuple* of the monom X^t. Any non-zero pol P can then be written in the unique *normal form*

$$P = A_1 * X^{t_1} + A_2 * X^{t_2} + \cdots + A_r * X^{t_r},$$

with A_i non-zero in K, and the degree-tuples t_i in decreasing order, viz.

$$t_1 > t_2 > \cdots > t_r.$$

We call $A_1 * X^{t_1}, X^{t_1}, A_1$, and t_1 resp. the *leading term*, the *leading monom*, the *leading coefficient*, and the *leading degree-tuple*, to be denoted resp. by

$$\text{Lterm}(P),\ \text{Lmonom}(P),\ \text{Lcoef}(P),\ \text{and}\ \text{Ldeg}(P).$$

DEF. For two non-zero pols P_1 and P_2, P_1 is said to be *higher than, lower than, or incomparable to* P_2 according as whether the leading degree-tuple of P_1 is *higher than, lower than, or identical to* that of P_2. In notation, we shall write resp.

$$P_1 > P_2,\ P_1 < P_2,\ \text{and}\ P_1 <> P_2.$$

DEF. For a finite polset *PS* of non-zero pols the set of leading degree-tuples of pols in *PS* will be called the *degree-tuple-set* of *PS* to be denoted by DTS(*PS*).

DEF. A finite polset *PS* of non-zero pols is said to be *autoreduced* if its degree-tuple-set is autoreduced.

DEF. Let *AS* be an autoreduced polset of non-zero pols and T = DTS(*AS*) be its degree-tuple-set. A non-zero pol P is said to be *reduced* w.r.t. *AS* if for each term in P, the corresponding degree-tuple is not a multiple of any tuple in T. The autoreduced *AS* itself is said to be *reduced* if each pol of *AS* is reduced w.r.t. the autoreduced polset formed from *AS* by removing that pol.

For any autoreduced polset *AS* consisting of non-zero pols F_i there may be different ways of putting P into a form

$$P = \text{SUM}_i E_i * F_i + R, \tag{2.1}$$

in which E_i, R are pols and R, if not zero, is reduced w.r.t. *AS*. We shall now proceed in the following way to get a unique R from P as follows. Write P in the normal form. Let $c * X^t$, c in K, be the non-zero term in P, if it exists, such that t is of highest order with t a multiple of some tuple u in T, u being chosen to be the highest one in T. Write $t = uv$ and let the pol in *AS* having u as its leading-degree-tuple be

$$F_i = a * X^u + F_i',$$

with $a * X^u$ as its leading term, $a \neq 0$ being in K. Set

$$P_1 = P/c - X^v * F_i/a.$$

Then P_1 is such a pol that the term of highest order in P_1 having its degree-tuple t_1 as a multiple of some tuple in T, if it exists, is of lower order than t above. We can then apply the same procedure to P_1 as above to get a pol P_2. The procedure can be continued until we get a pol P_s which is reduced w.r.t. AS. This pol P_s will then be the R required.

DEF. The unique pol R reduced w.r.t. the given autoreduced set AS got from P in the above manner will be called the *rest* of P w.r.t. AS , to be denoted by

$$R = \text{Rest}(P/AS).$$

DEF. The autoreduced polset AS is said to be *higher than* the autoreduced polset AS', or AS' *lower than* AS if

$$T = \text{DTS}(AS) > T' = \text{DTS}(AS').$$

Given an arbitrary finite polset PS of non-zero pols let us form now a scheme (SA) below:

$$\begin{array}{llll} PS = PS_0 & PS_1 & \cdots & PS_r \\ AS_0 & AS_1 & \cdots & AS_r \\ RS_0 & RS_1 & \cdots & RS_r = \text{Empty}. \end{array} \qquad \text{(SA)}$$

The scheme is formed in the following manner:

For each i AS_i is an autoreduced polset with pols chosen from PS_i such that the degree-tuple of any remaining pol in PS_i is a multiple of the degree-tuple of some pol in AS_i. Each RS_i is then the polset of all non-zero rests, if it exists, of the pols in $PS_i - AS_i$ w.r.t. AS_i. The polset PS_{i+1} is just the union of the previous AS_i and RS_i:

$$PS_{i+1} = AS_i + RS_i.$$

From the construction we see by Lemma 5 of Sect. 1 that the autoreduced sets AS_i are steadily decreasing in order:

$$AS_0 > AS_1 > \cdots .$$

By Lemma 4 of Sect. 1 the sequence is finite so that the procedure has to stop at a certain stage with its corresponding rest-set RS_r = Empty as shown in the diagram (SA).

Theorem 1. *The final autoreduced polset AS_r in the scheme* (SA) *forms a basis for the ideal* Ideal(PS) *with PS as a basis. In other words,*

$$\text{Ideal}(PS) = \text{Ideal}(AS_r).$$

Proof. Let AS_0 consist of pols P_i and the other pols in PS_0 be Q_j so that Ideal(PS) has a basis consisting of pols P_i and Q_j. Let $R_j = \text{Rest}(Q_j/AS_0)$. Then by definition of rest, it is clear that the ideal Ideal(PS_0) has also a basis consisting of pols P_i and those R_j which are non-zero, or

$$\text{Ideal}(PS) = \text{Ideal}(PS_0) = \text{Ideal}(AS_0 + RS_0) = \text{Ideal}(PS_1).$$

In the same way we have

$$\text{Ideal}(PS_1) = \text{Ideal}(PS_2) = \cdots = \text{Ideal}(PS_r).$$

Hence Ideal(PS) = Ideal(PS_r) = Ideal(AS_r) as to be proved. □

DEF. The final autoreduced polset AS_r in the scheme (SA) will be called a *well-arranged basis* of the ideal Ideal(PS).

3. Well-Behaved Basis of a Polynomial Ideal

Let AS be an autoreduced polset with degree-tuple set T. For any tuple u in Comp(T) let $u = tv$ with t the highest tuple in T which is a divisor of u. Let F_t be the pol in AS with t as its degree-tuple and let us set $H_u = X^v * F_t$. In particular, if u is itself in T, then $u = t$ and v is the 0-tuple so that H_u is just the pol F_t of AS.

DEF. The pol H_u defined above will be called the *completed pol* of AS *relative to* u. The polset consisting of all such completed pols will be called the *completed polset* of AS.

DEF. A product of the form $M*H_u$ in which H_u is the completed pol of AS relative to u in Comp(T), and M a monom X^w for which each i with $\text{COOR}_i(w) \neq 0$ is a multiplier of u will be called an *M-product* of AS.

DEF. A finite linear combination of M-products of AS with coefficients in K will be called an *M-pol* of AS.

Theorem 2. *Any pol P can be written uniquely in the form*

$$P = MP + N, \tag{3.1}$$

in which MP is an M-pol of AS and N is reduced w.r.t. AS.

Proof. Suppose that P is not reduced w.r.t. AS. Then in P there will be a term $a * X^u$ of highest order with u a multiple of some tuple in T, $a \neq 0$ being in K. By Lemma 3 of Sect. 1 there is a unique t in Comp(T) with u=vt such that each i with $\text{COOR}_i(v) \neq 0$ is a multiplier of t. Let H_t be the completed pol of AS relative to t with leading term $\text{Lterm}(H_t) = b * X^t$. Set

$$P_1 = P/a - X^v * H_t/b,$$

or

$$P = c_1 * MP_1 + b_1 * P_1, \;\; (c_1 = a/b,\, b_1 = a)$$

with $MP_1 = X^v * H_t$ an M-product. If P_1 is not reduced w.r.t. AS, then there will be a term $a_1 * X^{u_1}$ of highest order in P_1 with u_1 a multiple of some tuple in T and u_1 is of lower order than u. Apply now the preceding procedure to P_1 and we get a pol P_2 so that $P_1 = c_2 * MP_2 + b_2 * P_2$, with b_2, c_2 in K and MP_2 an M-product of lower order than MP_1. The procedure can be continued to get pols P_3, etc. until we arrive at some pol P_r which is reduced w.r.t. AS. We may then write P in the form (3.1) with MP an M-pol and $N = b_r * P_r$ reduced w.r.t. AS as required. That the decomposition of form (3.1) is unique follows also easily from Lemma 3 of Sect. 1. □

DEF. The pols MP and N in (3.1) will be called resp. the *M-part* and the *N-part* of the pol P w.r.t. AS.

Consider now any u in $\text{Comp}(T)$ with corresponding completed pol H_u and any non-multiplier i of u. Then $ui = v$ is also in $\text{Comp}(T)$ and the decomposition of $X_i * H_u$ into the M- and N-parts can be put in the form

$$X_i * H_u = a * H_v + MP_{ui} + N_{ui}, \tag{3.2}$$

in which $a \neq 0$ is in K, MP_{ui} is an M-pol with each M-product in it of lower order than H_v or $X_i * H_u$, and N_{ui} is the N-part of $X_i * H_u$. Note that N_{ui} is reduced w.r.t. AS. Owing to its importance we shall lay down the following

DEF. The N-part N_{ui} of pol $X_i * H_u$ in (3.2) will be called the *N-pol* of AS relative to the tuple u in $\text{Comp}(T)$ and the non-multiplier i of u.

Consider now a finite polset PS and let us form the scheme (SB) below:

$$\begin{array}{llll} PS = PS_0 & PS_1 & \cdots & PS_s \\ WS_0 & WS_1 & \cdots & WS_s \\ NS_0 & NS_1 & \cdots & NS_s = \text{Empty}. \end{array} \tag{SB}$$

The scheme is formed in the following way:

For each i WS_i is a well-arranged basis of the ideal $\text{Ideal}(PS_i)$, determined from PS_i as in Sect. 2 with scheme (SA) applied to PS_i, and NS_i is the set of all non-zero N-pols of WS_i, if it exists. Finally, the polset PS_{i+1} is the union of the preceding sets WS_i and NS_i, or

$$PS_{i+1} = WS_i + NS_i.$$

As in the case of scheme (SA), the sequence of autoreduced sets WS_i is steadily decreasing in order so that the above procedure will end in a certain stage with corresponding NS_s = Empty as shown in the diagram (SB).

Theorem 3. *The final polset WS_s in the scheme* (SB) *is a basis of the ideal* $\text{Ideal}(PS)$, *or*

$$\text{Ideal}(PS) = \text{Ideal}(WS_s).$$

Proof. By Theorem 1 of Sect. 2 we have Ideal(PS_0) = Ideal(WS_0). Now each pol N in NS_0 is the N-part of some pol $X_i * H_u$ with H_u the completed pol of WS_0 relative to the tuple u in Comp(T_0) where T_0 is the degree-tuple-set of WS_0 and i a non-multiplier of u so that $X_i * H_u = MP + N$ with MP an M-pol of WS_0. As both H_u and MP are clearly pols in the ideal Ideal(WS_0), the same is for N. Hence

$$\text{Ideal}(PS_0) = \text{Ideal}(WS_0) = \text{Ideal}(WS_0 + NS_0) = \text{Ideal}(PS_1).$$

Proceeding further in the same way we get then successively

$$\text{Ideal}(PS) = \text{Ideal}(PS_1) = \cdots = \text{Ideal}(PS_s) = \text{Ideal}(WS_s),$$

as to be proved. □

DEF. The final autoreduced polset WS_s in the scheme (SB) will be called a *well-behaved basis* of the ideal Ideal(PS).

In the next section it will be shown that the notion of well-behaved set coincides with the usual notion of Groebner basis.

4. Identification of Well-Behaved Basis with Groebner Basis

Consider any ideal ID for which the well-behaved basis, say WB, has been determined as in Sect. 3 so that ID = Ideal(WB).

Theorem 4. *Any pol in the ideal ID is an M-pol of its well-behaved basis WB, or the N-part of any such pol is* 0.

Proof. Let T be the degree-tuple set of WB. For any u in Comp(T) let H_u be the corresponding completed pol. It is enough to prove that any product of the form $M * H_u$ with M a monom and u in Comp(T) is an M-pol. We shall prove this by induction on the order of $M * H_u$ as well as on the number of X's in the monom M as follows.

If each i for which X_i appears in the monom M is a multiplier of u, then $M * H_u$ is already an M-pol and nothing is to be proved. Suppose therefore $M = M' * X_i$ with i a non-multiplier of u. As WB is the well-behaved basis of the ideal, the N-pol relative to u and i is 0 so that (3.2) of Sect. 3 may be written as

$$X_i * H_u = a * H_v + MP, \tag{4.1}$$

in which $v = ui$, and MP is an M-pol of lower order than H_v or $X_i * H_u$. It follows that $M * H_u = a * M' * H_v + M' * MP$, of which $M' * MP$ is of lower order than $M * H_u$ and M' has a smaller number of X's than M. By induction $M' * H_v$ and each term in $M' * MP$ are M-pols and so is $M * H_u$. The theorem is thus proved. □

Theorem 5. *The rest of any pol P w.r.t. the well-behaved basis WB coincides with the N-part of P w.r.t. WB.*

Proof. The rest is determined as the pol R in $P = \text{SUM}_k C_k * W_k + R$, in which W_k are the pols in WB, C_k are pols too, and R is reduced w.r.t. WB. By Theorem 4 $\text{SUM}_k C_k * W_k$ is an M-pol so that R is the N-part of P, as to be proved. □

From Theorems 4 and 5 we get the following

Theorem 6. *A pol P belongs to an ideal ID if and only if its rest w.r.t. the well-behaved basis WB of ID is 0:*

$$P \in ID \iff \text{Rest}(P/WB) = 0.$$

The previous results may be further put into a strengthened form as follows.

Theorem 7. *A well-behaved basis WB with degree-tuple-set T of an ideal ID possesses the following well-behaved property:*

Any pol P in $K[X_1, \ldots X_n]$ has a unique expression

$$P = \text{SUM}_u a_u * M_u * H_u + N, \tag{4.2}$$

in which H_u are completed H-pols with u running over the completion Comp*(T) of T, M_u are monoms in these X_i with each i a multiplier of u, a_u are constants in K, and N is reduced w.r.t. WB. Moreover, P is in the ideal ID if and only if $N = 0$.*

From the unique expression (4.2) for any pol in ID w.r.t. WB we get immediately the following theorem due to Macaulay, cf. [M]:

Theorem 8. *The Hilbert function of an ideal is completely determined by the degree-tuple-set of a well-behaved basis of the ideal.*

Theorem 9. *Let the well-behaved basis WB of an ideal ID consist of the pols $W_1, \ldots, W_r$. For any completed pol H_u of WB and any non-multiplier i of WB w.r.t. u let us rewrite* (4.1) *in the form*

$$\text{SUM}_i S_{uij} * W_j = 0. \tag{4.3}$$

Then the sets $S_{ui} = (S_{ui1}, \ldots, S_{uir})$ form a basis of the linear space of possible solutions $(S_1, \ldots, S_r)$ in pols for the syzygy equation

$$\text{SUM}_j S_j * W_j = 0. \tag{4.4}$$

Proof. Consider any solution of equation (4.4) in pols S_j. Denote the left-hand side of (4.4) by S. Then S is a pol belonging to the ideal ID with a well-behaved basis WB. From the proof of Theorem 4 we see that S can be shown to be 0 by successive reductions in the form of (4.1) or (4.3). Hence $S = 0$ is a consequence of equations (4.3) or S is a linear combination of S_{ui} with pols as coefficients, as to be proved. We remark only that the solutions S_{ui} are not necessarily independent ones.

If the ideal ID is given a basis $F_1, \ldots F_m$, then each F_i is a linear combination with pol-coefficients of W_j in the well-behaved basis WB and vice versa, which can be explicitly determined by means of the constructions in schemes (SA) and

(SB). Hence the above will furnish a method of deriving a basis of the solutions $(S_1, \ldots S_m)$ of the syzygy equation

$$\mathrm{SUM}_k S_k * F_k = 0. \qquad \square$$

Theorem 10. *The reduced well-behaved basis WB of an ideal ID* = Ideal(*PS*) *with polset PS as a basis is uniquely determined up to constant multiples by the following two properties:*

(*a*) *WB is a reduced autoreduced basis of ID.*

(*b*) *Let T be the degree-tuple-set of WB. Then for any tuple u in* Comp(T) *with completed pol H_u and any non-multiplier i of WB w.r.t. u, the N-part of $X_i * H_u$ is* 0.

Proof. We have shown how to determine from *PS* by schemes (SA) and (SB), by a further reduction if necessary, a well-behaved basis *WB* of *ID* verifying properties (a) and (b). From the proofs of Theorems 5 and 6 we see that there will follow also the following property (c).

(c) The rest of any pol P in the ideal *ID* w.r.t. *WB* is 0.

Consider now any polset WB' verifying the analogous properties $(a)'$, $(b)'$ and hence also $(c)'$. There is no loss of generality in assuming that all the pols in *WB* and WB' have been normalized to have their leading coefficients = 1. We are to prove that WB' coincides with *WB*.

To see this let us arrange the pols in *WB* and WB' both in decreasing order, viz.

$$(WB): \quad W_1 > W_2 > \cdots > W_r,$$
$$(WB)': \quad W_1' > W_2' > \cdots > W_s'.$$

By (c) we have $\mathrm{Rest}(W_1'/WB) = 0$ and by the corresponding rest formula we see that the leading degree-tuple of W_1' should be a multiple of the leading-degree-tuple of some pol in *WB* , say W_i. In the same way, by $(c)'$ the leading-degree-tuple of W_i should be a multiple of the leading-degree-tuple of some W_j' of WB'. As WB' is autoreduced it will only be possible that W_j' coincides with W_1'. Then W_1' will have the same leading-monom as W_i. Applying the same reasoning to W_1 we see that W_1 should have the same leading monom as some W_k' of WB'. This is only possible when $W_i = W_1, W_k' = W_1'$ and W_1, W_1' have the same leading-monoms.

Applying now the same reasoning to W_2 and W_2' we see that they should have the same leading monoms. Continuing we see then *WB* and WB' should have the same number of pols or $r = s$ and each pair W_i and W_i' should have the same leading monoms.

Consider now the last two pols W_r and W_r' in *WB* and WB'. As W_r' has the same leading monom as W_r and W_r' has rest 0 w.r.t. *WB* we see that W_r' should be identical to W_r. Let us consider the pair W_{r-1} and W_{r-1}'. As the rest of W_{r-1}' w.r.t. *WB* is 0 we should have an identity of the form

$$W_{r-1}' = W_{r-1} + M_r,$$

in which M_r is an M-pol constructed from W_r. Now W'_{r-1} and W_{r-1} have the same leading monoms and no other monoms in W'_{r-1} and W_{r-1} can be multiples of the leading monom of W_r. It follows from the Tuple Decomposition Theorem that this will be possible only when $M_r = 0$ or W'_{r-1} is identical to W_{r-1}. Applying now the same reasoning to the other pairs of pols in *WB* and *WB*′ successively in the reverse order we see that all the pairs should be identical to each other. The theorem is thus proved. □

Consider now an ideal *ID* with a reduced well-behaved basis *WB*. For $m < n$ let *ID*′ be the ideal of all pols in *ID* in $X_1, \dots, X_m$ alone. Let *WB*′ be the autoreduced polset consisting of such pols in *WB* in $X_1, \dots, X_m$ alone too. Then we have the following.

Theorem 11. *Let WB be a reduced well-behaved basis of an ideal*

$$ID \subset K[X_1, \dots, X_n].$$

Then the autoreduced polset

$$WB' = WB \cap K[X_1, \dots, X_m]$$

is a reduced well-behaved basis of the ideal

$$ID' = ID \cap K[X_1, \dots, X_m].$$

Proof. Let T be the degree-tuple-set of *WB* and T' that of *WB*′. Consider now any pol P in *ID*′. Let us consider P as a pol in *ID* and write it in the form (4.2). By Theorem 6 we have $N = 0$. By the Tuple Decomposition Theorem we see that in (4.2) for each term in H_u we should have $\mathrm{COOR}_k(u) = 0$ for $k > m$. Let $\mathrm{Max}'(T)$ be the m-tuple got from $\mathrm{Max}(T)$ by deleting the last $n - m$ coordinates. It is clear that $\mathrm{Max}(T') \ll \mathrm{Max}'(T)$. It follows that for each H_u in (4.2) for which $\mathrm{COOR}_k(u) = 0$ for $k > m$, each i with X_i occurring in M_u which is a multiplier of u w.r.t. *WB* should also be a multiplier u w.r.t. *WB*′. Hence the N-part of P, considered as a pol in *ID*′, is 0 too w.r.t. *WB*′. This implies in particular property (b) in Theorem 10 corresponding to *WB*′ of *ID*′. By Theorem 10 again *WB*′ is thus a reduced well-behaved basis of the ideal *ID*′. This completes the proof of the theorem. □

Finally, in comparing with the usual definition of Groebner basis of a polynomial ideal we see readily from Theorem 6 the following.

Theorem 12. *Any well-behaved basis of a polynomial ideal ID is a Groebner basis of ID. If the well-behaved basis is reduced and the leading coefficient of each pol in the basis is normalized to* 1, *then the basis is coincident with the reduced Groebner basis of ID.*

The well-behaved basis of a polynomial ideal, being nothing else but the usual Groebner basis of *ID*, will enjoy the various already well-known properties of Groebner basis. Some of such properties have been restated and reproved in the form of well-behaved basis as given above. The treatments and proofs are however done along the line of the thoughts of previous sections, giving thus alternative proofs of these known theorems about Groebner basis different from the known ones. Moreover, the proofs are given in order that they may be readily transferred to the differential case as described in the Introduction for which the corresponding notion of Groebner basis is non-existent. Furthermore, the above treatment shows that any pol in $K[X_1, \ldots, X_m]$ will have a *unique* expression w.r.t. a well-behaved basis, i.e. a Groebner basis, of a polynomial ideal in the form of equation (4.2), which is a property more precise than the corresponding known one for a Groebner basis under the usual known treatment.

5. An Example

The schemes (SB) and (SA) in the previous sections give an algorithm for the determination of a well-behaved basis, i.e. a Groebner basis of an ideal Ideal(PS) with a given basis PS. As an illustrative example (Problem 9(b) in [CG]) let us consider the following polset $PS = \{P_1, P_2, P_3\}$ with

$$\begin{cases} P_1 = X^2 + Y * Z + D * X + 1, \\ P_2 = Y^2 + Z * X + E * Y + 1, \\ P_3 = Z^2 + X * Y + F * Z + 1. \end{cases}$$

Introduce now an ordering among the various indeterminates by

$$Z > Y > X > D > E > F.$$

This amounts to equating these indeterminates to X_i such that $X_i > X_j$ if and only if $i > j$. We shall retain however the usual notations of Z, etc. as it will not cause misunderstandings.

According to the scheme (SB) we form first the well-behaved set (in decreasing order) WS_0 consisting of W_1, W_2, W_3 with

$$W_1 = P_3, \quad W_2 = P_1, \quad W_3 = P_2.$$

The leading-degree-tuple set of WS_0 is

$$T = \{(0,0,2), (0,1,1), (1,0,1)\}$$

so that $\mathrm{Max}(T) = (1,1,2)$. The completed H-pols arranged in descending order are thus 7 in number, viz.

$$H_1 = Y * X * W_1, \quad H_2 = Y * W_1, \quad H_3 = X * W_1,$$
$$H_4 = W_1, \quad H_5 = X * W_2, \quad H_6 = W_2, \quad H_7 = W_3.$$

Let the N-part of an H-pol H w.r.t. a non-multiplier i be denoted by $N(H/X_i)$. Most of the N-parts may be directly seen to be 0 by definition and the only non-zero N-parts are readily found to be the following ones:

$$\begin{aligned}
N_1 &= -N(H_5/Z)\\
&= -Z * H_5 + H_1 - F * H_5 + X^2 * H_7 + D * X * H_7 + H_7\\
&= 2 * Y^2 * X^2 + Y^2 * X * D + Y^2 + Y * X^2 * E + Y * X * D * E\\
&\quad + Y * X + Y * E - X^3 * F - X^2 * D * F + X^2 + X * D - X * F + 1,\\
N_2 &= N(H_6/Z)\\
&= Z * H_6 - H_2 + F * H_6 - X * H_7 - D * H_7\\
&= Z - 2 * Y^2 * X - Y^2 * D - Y * X * E - Y * D * E - Y\\
&\quad + X^2 * F + X * D * F - X - D + F,\\
N_3 &= N(H_7/Y)\\
&= Y * H_7 - H_5\\
&= Y^3 + Y^2 * E + Y - X^3 - X^2 * D - X,\\
N_4 &= N(H_7/Z)\\
&= Z * H_7 - H_3 - Y * H_6 - E * H_6 + F * H_7\\
&= Z + Y^2 * F - 2 * Y * X^2 - Y * X * D + Y * E * F - Y\\
&\quad - X^2 * E - X * D * E - X - E + F.
\end{aligned}$$

The polset $PS_1 = WS_0 + NS_0$ thus consists of 7 pols, W_i and N_j. We proceed to form a well-arranged basis WS_1 of Ideal(PS_1) according to scheme (SA) in starting from $QS_0 = PS_1$, viz.

$$\begin{array}{llll}
QS_0 & QS_1 & \ldots & QS_r,\\
AS_0 & AS_1 & \ldots & AS_r,\\
RS_0 & RS_1 & \ldots & RS_r.
\end{array}$$

It is found that for $r = 6$ the polset QS_6 consists of 4 pols Q_i below:

$$\begin{aligned}
Q_1 &= Z + Y^2 * F - 2 * Y * X^2 - Y * X * D + Y * E * F - Y - X^2 * E\\
&\qquad - X * D * E - X - E + F,\\
Q_2 &= Y^2 * D * F + Y^2 * F^2 + 2 * Y^2 + \cdots,\\
Q_3 &= Y * X^2 * G_3 + \cdots,\\
Q_4 &= Y * X * G_4 + \cdots,
\end{aligned}$$

in which G_3, G_4 are pols in D, E, and F alone. The number of the terms of Q_3 and Q_4 are resp. 90 and 314 and Q_4 is non-factorizable. To make computations not too complicated we shall consider the special case of $F = E$ which will not influence the computations already done. It turns out that in this case of $F = E$

the pols Q_i will be simplified to the following ones:

$$\begin{aligned} Q_1 &= Z + Y^2 * E - 2 * Y * X^2 - Y * X * D + Y * E^2 - Y - X^2 * E \\ &\quad - X * D * E - X, \\ Q_2 &= Y^2 * D * E + Y^2 * E^2 + 2 * Y^2 + 4 * Y * X^3 + 2 * Y * X^2 * D \\ &\quad - 2 * Y * X^2 * E - Y * X * D * E - Y * X * E^2 + 2 * Y * X \\ &\quad + Y * D * E^2 + Y * E^3 + 2 * Y * E + 2 * X^3 * E + 2 * X^2 * D * E \\ &\quad - 2 * X^2 * E^2 + 2 * X^2 - 2 * X * D * E^2 + D * E - E^2 + 2, \\ Q_3 &= F_1 * F_2, \\ Q_4 &= F_1 * F_3. \end{aligned}$$

The pols Q_3 and Q_4 split into factors F_i with

$$\begin{aligned} F_1 &= 2 * X^2 + X * D - X * E + 2, \\ F_2 &= Y * D * E - Y * E^2 - 2 * X^3 - 3 * X^2 * D - X * D^2 - X - E, \\ F_3 &= 2 * X^4 + 3 * X^3 * D + 2 * X^3 * E + X^2 * D^2 + 2 * X^2 * D * E \\ &\quad + X^2 * E^2 + X^2 + X * D * E^2 + 2 * X * E + E^2. \end{aligned}$$

The polset QS_6 is now already an autoreduced one and may be taken as the corresponding well-arranged set AS_6. Let us denote the completed H-pols by

$$\begin{aligned} H_{1ij} &= X^i * Y^j * Q_1, \quad i \le 6, j \le 2, \\ H_{2i} &= X^i * Q_2, \quad i \le 6, \\ H_{3i} &= X^i * Q_3, \quad i \le 4, \\ H_4 &= Q_4. \end{aligned}$$

The variables corresponding to the non-multipliers are then resp. at most

$$\begin{aligned} X, Y &\text{ for } H_{1ij}, \\ X, Z &\text{ for } H_{2i}, \\ X, Y, Z &\text{ for } H_{3i}, \\ Y, Z &\text{ for } H_4. \end{aligned}$$

To determine the N-pols let us consider first $N(H_{30}/Y)$ where $H_{30} = Q_3$. By direct computation we find

$$(D*E+E^2+2)*E*(D-E)*Y*H_{30} = P_2*Q_4+P_3*Q_3+E^2*(D-E)^2*F_1*Q_2,$$

in which P_2, P_3 are pols in X of degree 2 and 3 resp. As all terms X^i*Q_j occurring in the right-hand side of the above equation are M-products, it follows that

$$N(H_{30}/Y) = 0.$$

Consider now $N(H_4/Y)$ where $H_4 = Q_4$. Write F_2 in the form $F_2 = Y * E * (D - E) - P$ with P a pol in X of degree 3. Then by simply rewriting we get

$$E * (D - E) * Y * H_4 = F_3 * Q_3 + P * Q_4,$$

in which all terms $X^i * Q_j$ on the right-hand side are M-products. We thus again have

$$N(H_4/Y) = 0.$$

That other N-pols are all 0 may be deduced from the above ones or directly by rewriting in a similar way almost without computation. In conclusion it follows that QS_6 is already a well-behaved basis or a Groebner basis GB of the ideal Ideal(PS).

The zero-set Zero(PS) = Zero(GB) may be determined as follows. As F_1, F_3 are easily seen to be prime to each other (D, E are independent indeterminates) so $Q_4 = 0$ has 6 zeros of X, 2 from $F_1 = 0$ and 4 from $F_3 = 0$. For each zero of $F_1 = 0, Q_3$ will be 0 too and $Q_2 = 0$, $Q_1 = 0$ will give 2 zeros of GB. On the other hand, for each zero of $F_3 = 0$ we have $F_1 \neq 0$ and the resultant of Q_2 and F_2 is found to be 0, so such a zero of F_3 will be extended to only one zero of GB determined by $F_2 = 0$ and $Q_1 = 0$. In all we have 8 zeros of GB or PS. We remark that in the present case each zero of $Q_4 = 0$ can be extended to at least one zero of GB. This is however not the case in general. Cf. e.g. [WU2] and [L].

References

[BU1] Buchberger, B., Ein algorithmisches Kriterium fuer die Loesbarkeit eines algebraishes Gleichungs-systems, *Aeq. Math.*, **4** (1970), 374–383.

[BU2] Buchberger, B., Some properties of Groebner bases for polynomial ideals, *ACM SIGSAM. Bull.*, **10** (1976), 19–24.

[BU3] Buchberger, B., An algorithmic method in polynomial ideal theory, in N. K. Bose (Ed.), Recent trends in multidimensional systems theory, Reidel, 184–232, 1985.

[CAR] Cartan, E., Les systèmes différentiels extérieurs et leurs applications géometriques, Paris, 1945.

[J] Janet, M., Leçons sur les systèmes d'équations aux dérivées partielles, Gauthier Villars, 1920.

[K] Kolchin, E. R., Differential algebra and algebraic groups, Academic Press, 1973.

[L] Li, Ziming, Automatic implicitization of parametric objects, MM-Res. Preprints, No. 4 (1989), 54–62.

[M] Macaulay, F. S., Some properties of enumeration in the theory of modular systems. *Proc. London Math. Soc.*, **26** (1927), 531–555.

[RQ] Riquier, C. H., La méthode des fonctions majorantes et les systèmes d'équations aux dérivées partielles, Gauthier Villars, 1928.

[RITT1] Ritt, J. F., Differential equations from the algebraic standpoint, Amer. Math. Soc., 1932.

[RITT2] Ritt, J. F., Differential algebra, Amer. Math. Soc., 1950.

[TH] Thomas, J. M., Differential systems, Amer. Math. Soc., 1937.

[TR] Trinks, W., On Buchberger's method for solving systems of algebraic equations, *J. Number Theory*, **10** (1978), 475–488.

[WA] Wall, B., On the computation of syzygies, *ACM SIGSAM Bull*, **23** (1989), 5–14.

[WU1] Wu, Wen-tsün, On the foundation of algebraic differential geometry, MM-Res. Preprints, No. 3 (1989), 1–26, also in *Sys. Sci. & Math Scis.*, **2** (1989), 289–312.

[WU2] Wu, Wen-tsün, On a projection theorem of quasi-varieties in elimination theory, MM-Res. Preprints, No. 4 (1989), 40–48, also in *Chin. Ann. of Math.*, **11** B:2 (1990), 220–226.

Wenjun Wu (Wen-tsün Wu)
Institute of Systems Science
Academia Sinica
Beijing 100080, China

Reprinted with permission from *Differential Equations with Symbolic Computation*, (Birkhäuser Basel, 2005), pp. 343–350.

On "Good" Bases of Algebraico-Differential Ideals

Wen-tsün Wu

Abstract. The characteristic set method of polynomial equations-solving is naturally extended to the differential case, which gives rise to an algorithmic method for solving arbitrary systems of algebrico-differential equations. The existence of "good bases" of the associated algebrico-differential ideals is also studied in this way. As an illustration of the method, the Devil problem of Pommaret is studied in detail.

Mathematics Subject Classification (2000). Primary 13P10; Secondary 13N10, 35G05.

Keywords. Algebrico-differential equations, (differential) zero-decomposition theorem, integrability differential polynomial, compatibility differential polynomial, Pommaret's Devil problem, "good basis".

1. Introduction

In the seminar DESC held in Beijing, April 14–16, 2004, the present author gave a talk bearing the title "On 'Good Bases' of Polynomial Ideals" [10]. The present paper is an extended form of that talk in extending the notion of "good bases" of polynomial ideals to that of algebrico-differential ideals.

To begin with, let us consider a finite polynomial set PS in the polynomial ring $R = K[x_1, \dots, x_n]$, K being a coefficient field of characteristic 0. Then there are two important problems to be studied, viz:

Problem P1. Determine the totality of solutions of $PS = 0$ in all conceivable extension fields of K, to be denoted by Zero(PS) in what follows.

Problem P2. For the ideal Ideal(PS) with basis PS, determine some kind of *good* basis which will enjoy some *good* properties to be made precise.

We shall show how to solve Problem 1 in the polynomial case in Section 2, explain how to extend the solution to the algebrico-differential case in Section 3, and solve Problem 2 in the polynomial case in Section 4 by using the method developed in Section 3. In Section 5 we shall study Problem 2 in the algebrico-differential case and introduce the notion of probably existing "good basis" for certain algebrico-differential ideals. Finally in Section 6 we shall provide a solution to the Devil problem of Pommaret as an illustrative example.

2. Problem 1 in the Polynomial Case

For Problem P1 the present author has given a method for determining Zero(PS) completely, which may be described briefly as follows.

Arrange the variables $x_1, \ldots, x_n$ in the natural order; then any non-constant polynomial $P \in R$ may be written in the canonical form

$$P = I_0 x_c^d + I_1 x_c^{d-1} + \cdots + I_d,$$

in which all the I_j are either constants or polynomials in $x_1, \ldots, x_{c-1}$ alone with *initial* $I_0 \neq 0$. With respect to *class* c and *degree* d, we may introduce a partial ordering $\prec$ for all non-zero polynomials in R, with non-zero constant polynomials in the lowest ordering. Consider now some polynomial set, which either consists of a single non-zero constant polynomial, or in which the polynomials may be arranged with classes all positive and steadily increasing. We call such polynomial sets *ascending sets*. Then we may introduce a partial ordering $\prec$ among all such ascending sets, with the trivial ones consisting of single non-zero constant polynomials in the lowest ordering. For a finite polynomial set consisting of non-zero polynomials, any ascending set wholly contained in it and of lowest ordering is called a *basic set* of the given polynomial set. A partial ordering among all finite polynomial sets may then be unambiguously introduced according to their basic sets.

For any finite polynomial set $PS \subset R$, consider now the scheme

$$\begin{array}{lllllll} PS = PS^0 & PS^1 & \cdots & PS^i & \cdots & PS^m & \\ \quad\;\; BS^0 & BS^1 & \cdots & BS^i & \cdots & BS^m = CS & \qquad (S) \\ \quad\;\; RS^0 & RS^1 & \cdots & RS^i & \cdots & RS^m = \emptyset. & \end{array}$$

In this scheme, each BS^i is a basic set of PS^i, each RS^i is the set of non-zero remainders, if any, of the polynomials in $PS^i \setminus BS^i$ with respect to BS^i, and $PS^{i+1} = PS \cup BS^i \cup RS^i$ if RS^i is non-empty. It may be easily proved that the sequences in the scheme should terminate at certain stage m with $RS^m = \emptyset$. The corresponding basic set $BS^m = CS$ is then called a *characteristic set* (abbreviated *char-set*) of the given polynomial set PS. The *zero set* of PS, Zero(PS), which is the collection of common zeros of all the polynomials in PS, is closely connected with that of CS by the *well-ordering principle* in the form

$$\text{Zero}(PS) = \text{Zero}(CS/IP) \cup \text{Zero}(PS \ \cup \ \{IP\}),$$

in which IP is the product of all initials of the polynomials in CS and

$$\text{Zero}(CS/IP) = \text{Zero}(CS) \setminus \text{Zero}(IP).$$

Now $PS \cup \{IP\}$ is easily seen to be a polynomial set of lower ordering than PS. If we apply the well-ordering principle to $PS \cup \{IP\}$ and proceed further and further in the same way, we should stop in a finite number of steps and arrive at the following

Zero-Decomposition Theorem. There is an algorithm which may compute, from any finite polynomial set PS and in a finite number of steps, a finite set of ascending sets CS^s with initial-product IP^s such that

$$\text{Zero}(PS) = \bigcup_s \text{Zero}(CS^s/IP^s). \tag{Z}$$

Now all CS^s are ascending sets. Hence all the zero sets $\text{Zero}(CS^s)$ and all $\text{Zero}(CS^s/IP^s)$ may be considered as well-determined in some natural sense. The formula (Z) gives thus actually an explicit determination of $\text{Zero}(PS)$ for all finite polynomial sets PS, which serves for the solving of arbitrary systems of polynomial equations. This solves Problem 1 in the polynomial case.

3. Extension to Algebrico-Differential Systems

The above method of solving arbitrary systems of polynomial equations has been extended to arbitrary systems of *algebrico-differential* equations, either ordinary or partial ones, which will be explained below.

Let $y, u_j, j \in J$, be infinitely differentiable functions in independent variables $X = \{x_1, \ldots, x_n\}$. A polynomial in various derivatives of y and u_j with respect to x_k with coefficients in the differential field of rational functions of X will be called an *algebrico-differential polynomial.* Suppose that we are given a finite set of such polynomials $DPS = \{DP_i \mid i \in I\}$. Let us consider the associated system of partial differential equations of y with u_j supposed known:

$$DPS = 0, \quad \text{or} \quad DP_i = 0, \ i \in I.$$

Our problem is to determine the integrability conditions in terms of x_k, u_j for y to be solvable and in the affirmative case to determine the set of all possible *formal* solutions of y.

Criteria and even algorithmic methods for solving the above problem in some sense were known in quite remote times, for which we may cite in particular the work of C. H. Riquier, M. Janet, and E. Cartan. The method of Riquier and Janet was reformulated by J. F. Ritt in his books [5, 6]. In recent years, J. F. Pommaret has given a systematic *formal intrinsic* way of treatment and published several voluminous treatises. On the other hand, the present author has given an alternative method in following essentially the steps of Riquier and Janet as reformulated by Ritt [7]. The method consists in first extending naturally the notions of ascending sets, basic sets, remainders, etc. in the ordinary case to the present algebrico-differential case. Orderings among all derivatives and then partial orderings may then be successively introduced among all algebrico-differential polynomials, all differential-ascending sets, and finally all systems of algebrico-differential polynomial sets, somewhat analogous to the ordinary case.

For any system DPS of algebrico-differential polynomials, we may then form a scheme (dS) analogous to the scheme (S) in the ordinary case as shown below:

$$\begin{array}{llllllll} DPS = DPS^0 & DPS^1 & \cdots & DPS^i & \cdots & DPS^m & & \\ DBS^0 & DBS^1 & \cdots & DBS^i & \cdots & DBS^m & = & DCS \\ DRIS^0 & DRIS^1 & \cdots & DRIS^i & \cdots & DRIS^m & = & \emptyset \\ DCPS^0 \cup & DCPS^1 \cup & \cdots \cup & DCPS^i \cup & \cdots \cup & DCPS^m & = & DCPS. \end{array} \tag{dS}$$

In the scheme (dS), DPS is the given algebrico-differential polynomial set. For each i, DBS^i is a differential basic set of DPS^i. The set $DRIS^i$ is the union of two parts: one is the set of all possible non-zero differential remainders in the sense of Ritt formed from differential polynomials in $DPS^i \setminus DBS^i$ with respect to DBS^i, while the other is the set of integrability differential polynomials formed from certain pairs of differential polynomials in DPS^i, so far they contain actually y or its derivatives. Such pairs may be determined by the notions of *multiplicativity* and *non-multiplicativity* due to Riquier and Janet. On the other hand, those containing no y or its derivatives but containing possibly u_j or their derivatives form a set of *compatibility differential polynomials* whose vanishing gives the *compatibility conditions* under which the given system of equations $DPS = 0$ has solutions. In case $DRIS^i$ is non-empty, the union $DPS \cup DBS^i \cup DRIS^i$ forms the next differential polynomial set DPS^{i+1}.

As in the ordinary case the sequences will terminate at a certain stage m with $DRIS^m = \emptyset$. The corresponding differential basic set $DBS^m = DCS$ is then called a *differential characteristic set* (abbreviated *d-char-set*) of the given differential polynomial set DPS. The union $DCPS$ of all sets $DCPS^i, i = 1, \ldots, m$, will form the totality of all possible compatibility differential polynomials whose vanishing forms the compatibility conditions to guarantee the existence of solutions of the system of partial differential equations $DPS = 0$.

As in the ordinary case the above will lead finally to the formation of the totality of formal solutions of the given system of algebrico-differential equations under suitable initial data for which we refer to the paper [9].

4. Problem 2 in the Polynomial Case

Let us now consider the particular case for which the differential polynomials in DPS are all *linear* with *constant coefficients*. For each tuple of non-negative integers $\mu = (i_1, \ldots, i_n)$, let us write $\|\mu\|$ for $i_1 + \cdots + i_n$ and make the correspondence

$$\text{Partial derivative } \frac{\partial^{\|\mu\|}}{\partial x_1^{i_1} \cdots \partial x_n^{i_n}} \longleftrightarrow \text{Monomial } x_1^{i_1} \cdots x_n^{i_n}.$$

Then the partial differentiation of a derivative with respect to some x_j will correspond to the multiplication of the corresponding monomial with the variable x_j.

In this way a differential polynomial set *DPS* consisting of only linear differential polynomials with constant coefficients will become, under the above correspondence, a polynomial set *PS* in the ordinary sense. The scheme (dS) will then be turned into some scheme (W) for *PS* somewhat of the following form:

$$\begin{array}{llllllll} PS = & PS^0 & PS^1 & \cdots & PS^i & \cdots & PS^m & \\ & WS^0 & WS^1 & \cdots & WS^i & \cdots & WS^m = WS & \\ & IS^0 & IS^1 & \cdots & IS^i & \cdots & IS^m = \emptyset. & \end{array} \tag{W}$$

In the above scheme the WS^i are certain subsets of PS^i enjoying some well-arranged properties and each IS^i consists of remainders of the polynomials in $PS^i \setminus WS^i$ with respect to WS^i as well as those determined from certain pairs of polynomials in WS^i determined by the notions of multiplicativity and non-multiplicativity of Riquier and Janet. The union of WS^i, IS^i and eventually PS^0 will then be PS^{i+1} so far $IS^i \neq \emptyset$. It turns out that the final set *WS* is a basis of the given ideal Ideal(*PS*) and possesses many nice properties. It turns out too that this basis *WS* is just the well-known *Gröbner basis* of the given ideal Ideal(*PS*), which may now be found in some way different from the original one of B. Buchberger. Moreover, many known properties connected with the Gröbner basis which are dispersed in the literature have been proved in some simple and unanimous manner. We refer to the paper [8] for details. In particular, we have the following nice property of Gröbner bases which solves the important *membership problem*.

MP. A polynomial P in R belongs to the ideal Ideal(*PS*) if and only if the remainder of P with respect to the Gröbner basis of *PS* is 0.

It turns out that the Russian mathematician V. P. Gerdt has also found the Gröbner basis of a polynomial ideal essentially in the same way as above. He has used an alternative name of *involutive basis* and has given also a detailed analysis of various possible notions of *multiplicativity* and *non-multiplicativity* due to Riquier, Janet, Thomas, and Gerdt himself. For more details we refer to the paper [2] by Gerdt. At this point the author would like to express his hearty thanks to D. Wang who pointed out to the author the above-mentioned work of Gerdt.

5. Problem 2 in the Algebrico-Differential Case

Let us consider now Problem 2 of algebrico-differential systems in the general case. Let *DPS* be an arbitrary finite algebrico-differential polynomial set as before. The problem is to find some *finite* differential basis of the differential ideal dIdeal(*DPS*) that enjoys some nice properties as the Gröbner basis in the polynomial case and solves in particular the corresponding membership problem. It is natural to extend the method of Buchberger in the polynomial case to the present algebrico-differential case. Unfortunately, in 1986 G. Carrá Ferro showed in a well-known remarkable paper [1] that such a *finite* differential Gröbner basis does not exist in general. In later years the possibility of existence of such *finite* differential Gröbner bases was widely studied, notably by F. Ollivier (see [3]).

Now let us try to deal with this problem by our method explained in Section 3. Consider again the diagram (dS). We suppose naturally that all the compatibility conditions are verified. It is clear from the constructions that

$$\mathrm{dIdeal}(DPS) = \mathrm{dIdeal}(DPS^0) = \mathrm{dIdeal}(DPS^1) = \cdots = \mathrm{dIdeal}(DPS^m).$$

Suppose that for the final d-char-set *DCS* the following condition GC is verified.

GC. The initials and separants of the algebrico-differential polynomials in *DCS* are all constants.

By the differential remainder theorem of Ritt, it is readily seen that *DCS* is a differential basis of dIdeal(*DPS*) and a differential polynomial *DP* belongs to dIdeal(*DPS*) if and only if the differential remainder of *DP* with respect to *DCS* is 0.

It is thus seen that under the condition GC the final d-char-set *DCS* will serve as a *finite* differential basis of dIdeal(*DPS*), which solves the membership problem in a simple way. The condition GC is clearly less stringent than the condition of linearity and coefficients-constancy, which leads to the usual Gröbner basis in the polynomial case. On the other hand, the verification of the condition GC can be seen only after lengthy computations of d-char-set. In any way we may lay down the following definition.

Definition. An algebrico-differential polynomial set *DPS* verifying condition GC is called a *good set* and the corresponding differential basis formed by the final d-char-set is called a *good basis* of dIdeal(*DPS*).

In view of the significance and also the weakness of the above notion of *good basis*, we suggest now some problems for further study.

Problem 3. Try to find some *intrinsic* conditions for an algebrico-differential polynomial set to be "good" directly from the given set without passing to the final d-char-set.

Problem 4. Try to weaken the condition GC such that the differential ideal generated by the given algebrico-differential polynomials still has a *finite* differential basis that verifies some simple membership condition.

Problem 5. Compare our condition GC with other known conditions introduced by Ollivier and other authors.

6. Example: Pommaret's Devil Problem

To illustrate our treatment of algebrico-differential polynomial sets, let us consider the *Devil problem* of Pommaret, given for example in his paper [4]. We shall treat this Devil problem in detail by our method as exhibited in Section 3. For this purpose let us reproduce the statement of the Devil problem below.

Devil Problem. Let u, v, y be three functions of the Cartesian coordinates x_1, x_2, x_3 in Euclidean spaces related by the following two partial differential equations

$$\begin{aligned} DP_1 &= \partial_{200}y - x_2\partial_{002}y - u = 0, \\ DP_2 &= \partial_{020}y - v = 0, \end{aligned} \tag{D}$$

with the corresponding algebrico-differential polynomial set $DPS = \{DP_1, DP_2\}$. Note that here and below we use the notation $\partial_{i_3 i_2 i_1}$ for the partial derivative $\dfrac{\partial^{i_3+i_2+i_1}}{\partial x_3^{i_3} x_2^{i_2} x_1^{i_1}}$.

The functions u, v are supposed to be known. The problem consists in finding the compatibility conditions to be satisfied by u and v in order to insure the existence of solutions for y and to see whether the given algebrico-differential polynomial set DPS is a *good* one or not.

It turns out that our procedure ends at the stage $m = 2$ so that the scheme (dS) in the present case becomes

$$\begin{array}{lllllll} DPS = DPS^0 & & DPS^1 & & DPS^2 & & \\ DBS^0 & & DBS^1 & & DBS^2 & = DCS & \\ DRIS^0 & & DRIS^1 & & DRIS^2 & = \emptyset & \\ DCPS^0 & \cup & DCPS^1 & \cup & DCPS^2 & = DCPS. & \end{array} \tag{dS$'$}$$

The final d-char-set DCS is found to consist of 4 algebrico-differential polynomials

$$\begin{aligned} G_1 &= \partial_{004}y - z, \\ G_2 &= \partial_{012}y - w, \\ G_3 &= \partial_{020}y - v, \\ G_4 &= \partial_{200}y - x_2\partial_{002}y - u. \end{aligned} \tag{G}$$

In (G), w and z are given by

$$\begin{aligned} w &= \tfrac{1}{2}(\partial_{200}v - x_2\partial_{002}v - \partial_{020}u), \\ z &= \partial_{200}w - \partial_{012}u - x_2\partial_{002}w. \end{aligned}$$

The *compatibility conditions* are found to be $A = 0$ and $B = 0$ with

$$\begin{aligned} \tfrac{1}{2}A &= \partial_{010}w - \partial_{002}v, \\ B &= \partial_{400}w - 2x_2\partial_{202}w + x_2^2\partial_{004}w - \partial_{212}u + x_2\partial_{014}u - \partial_{004}u. \end{aligned} \tag{CC}$$

It may also be shown further that the two compatibility conditions $A = 0$ and $B = 0$ are not independent of each other. They are in fact connected by the differential identity

$$\partial_{400}A - 2x_2\partial_{202}A + x_2^2\partial_{004}A - 2\partial_{010}B = 0.$$

Naturally all the above were found by Pommaret by his method and in his notations, which are different from ours.

Now we see that the d-char-set *DCS* consisting of the 4 algebrico-differential polynomials $G_1, \ldots, G_4$ verifies the condition GC so that the given algebrico-differential polynomial set *DPS* is a *good* one with a *good* basis for the corresponding dIdeal(*DPS*).

References

[1] G. Carrá Ferro: Gröbner Bases and Differential Ideals. In: *Proc. AAECC-5*, Menorca, Spain, pp. 129–140. Springer-Verlag, Berlin Heidelberg (1987).

[2] V. P. Gerdt, Y. A. Blinkov: Minimal Involutive Bases. *Math. Comput. Simul.* **45**: 543–560 (1998).

[3] F. Ollivier: Standard Bases of Differential Ideals. In: *Proc. AAECC-8*, Tokyo, Japan, LNCS **508**, pp. 304–321. Springer-Verlag, Berlin Heidelberg (1990).

[4] J. F. Pommaret, A. Haddack: Effective Methods for Systems of Algebraic Partial Differential Equations. In: *Effective Methods in Algebraic Geometry* (T. Mora, C. Traverso, eds.), pp. 441–426. Birkhäuser, Basel Boston (1991).

[5] J. F. Ritt: *Differential Equations from the Algebraic Standpoint.* Amer. Math. Soc., New York (1932).

[6] J. F. Ritt: *Differential Algebra.* Amer. Math. Soc., New York (1950).

[7] W.-t. Wu: On the Foundation of Algebraic Differential Geometry. *Syst. Sci. Math. Sci.* **2**: 289–312 (1989).

[8] W.-t. Wu: On the Construction of Gröbner Basis of a Polynomial Ideal Based on Riquier–Janet Theory. *Syst. Sci. Math. Sci.* **4**: 193–207 (1991) [reprinted in this volume].

[9] W.-t. Wu: On Algebrico-Differential Equations-Solving. *J. Syst. Sci. Complex.* **17**(2): 153–163 (2004).

[10] W.-t. Wu: On "Good Bases" of Polynomial Ideals. Presented at the DESC seminar, Beijing, April 14–16, 2004.

Wen-tsün Wu
Key Laboratory of Mathematics Mechanization
Academy of Mathematics and System Sciences
Academia Sinica
Beijing 100080, China
e-mail: wtwu@mmrc.iss.ac.cn

Wu studying in France, 1948

Wu and his students, 1974. From right: Yu Yan-Lin, Jiang Jia-He, Li Pei-Xin, Wang Qi-Ming

北京海淀红艺

Wu's family. From left: Wu Yue-Ming, Wu Wen-Tsun, Wu Yun-Qi, Wu Tian-Jiao, Wu Xing-Xi, Chen Pi-He, 1978

Wu and Professor Chern Shiing-Shen, 1980

Wu in a welcome ceremony for Professor Henri Cartan (fourth from left) in Beijing, 1985

Wu received the National Supreme Award of Science and Technology from former Chinese President Jiang Ze-Min, 2001

Wu addressed at the opening ceremony of the International Congress of Mathematicians, 2002

Wu and members of the Mathematics Mechanization Research Center, 2006

Wu gave the Shaw Prize acceptance speech in Hong Kong, 2006

Chinese President Hu Jin-Tao chatted with Wu at Wu's home before the Chinese New Year, 2008